J. M. G.

POLYMERS: CHEMISTRY & PHYSICS OF MODERN MATERIALS

BLACKIE ACADEMIC & PROFESSIONAL
An Imprint of Chapman & Hall

Polymers: Chemistry and Physics of Modern Materials

Second Edition

J. M. G. COWIE
Professor of Chemistry
Heriot-Watt University
Edinburgh

BLACKIE ACADEMIC & PROFESSIONAL
An Imprint of Chapman & Hall
London · Glasgow · Weinheim · New York · Tokyo · Melbourne · Madras

Published by
Blackie Academic & Professional, an imprint of Chapman & Hall,
Wester Cleddens Road, Bishopbriggs, Glasgow G64 2NZ, UK

Chapman & Hall, 2-6 Boundary Row, London SE1 8HN, UK

Blackie Academic & Professional, Wester Cleddens Road, Bishopbriggs, Glasgow G64 2NZ, UK

Chapman & Hall GmbH, Pappelallee 3, 69469 Weinheim, Germany

Chapman & Hall USA, One Penn Plaza, 41st Floor, New York, NY10119, USA

Chapman & Hall Japan, ITP-Japan, Kyowa Building, 3F, 2-2-1 Hirakawacho, Chiyoda-ku, Tokyo 102, Japan

DA Book (Aust.) Pty Ltd, 648 Whitehorse Road, Mitcham 3132, Victoria, Australia

Chapman & Hall India, R. Seshadri, 32 Second Main Road, CIT East, Madras 600 035, India

---

First edition 1973
Second edition 1991
Reprinted 1993, 1994 (twice)

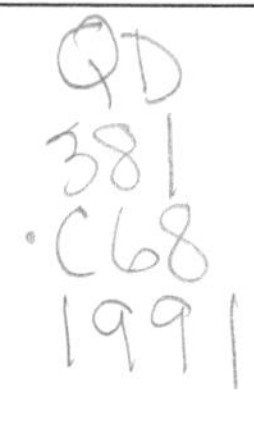

Typeset by Thomson Press (India) Limited, New Delhi
Printed in Great Britain by St Edmundsbury Press Ltd, Bury St Edmunds

ISBN 0 7514 0134 X

A Catalogue record for this book is available from the British Library

Library of Congress Cataloging-in-Publication Data available

Contents

Preface

When the first edition of this book appeared in 1973 it was meant to serve two major functions; the first was to provide a broadly based text on polymer science at an introductory level which would illustrate the interdisciplinary nature of the subject, and the second was to create a high information, inexpensive text that students would be able to afford. The response to the book over the intervening years has been both surprising and gratifying, and seems to indicate that the stated aims have been achieved. However, polymer science has moved on in dramatic fashion. Significant advances have been made in the synthesis and application of new high performance materials; the area of speciality polymers has blossomed and begun to flourish; the electronics industry has "discovered" how useful polymers can be, and synthetic polymer "metals" have become a reality. These exciting developments have made it necessary to update and expand the text to reflect the progress made. I have also responded to constructive comments made by colleagues over the years and thus have altered or incorporated various sections, and added several new chapters.

The interdisciplinary nature of polymer science is obvious. Polymers are materials with characteristic mechanical and physical properties which are controlled by the structure and the methods of synthesis. Consequently a scientist or engineer gains most from the subject if the interdisciplinary approach is emphasized from the beginning, but of course there must be a starting point. Bearing that in mind, this book is developed in the sequence: preparation, characterization, physical and mechanical properties, and culminates in a coverage of structure–property relations. Concluding chapters discuss growing areas of interest for applications of polymeric materials.

Of course there will always be aspects of the subject which are omitted from this type of book, nevertheless I trust that the revisions and additions that have been made will meet with general approval and that the text continues to serve the educational needs it was designed to meet.

My thanks are due to both Dr Keith Stead and Dr W. V. Steele for their constructive comments and criticisms. Finally I would like to dedicate this book to my long-suffering, but patient family—Ann, Graeme and Christian.

J.M.G.C.

CHAPTER 1

Introduction

1.1 Birth of a concept

What is a polymer? If you had asked that question during the latter half of the 19th century and the first quarter of the 20th, it would have been met with either a blank uncomprehending stare or, worse, by derision from sections of the scientific community. This question which is so pertinent today, asks for information about substances that are all pervasive in our everyday lives, that indeed we would have difficulty in avoiding, that may be handled, used, ignored, commented on, and normally taken for granted. Some of these substances are new and are recent products resulting from the ingenuity of the chemist, some are naturally occurring and have been used by man for several thousand years, some form parts of our bodies. All the substances referred to as polymers, or macromolecules, are giant molecules with molar masses ranging from several thousands to several millions.

Today the concept of a giant molecule is universally accepted by scientists but this was not always so, and the initial antagonism towards the idea that very large covalently bonded molecules could exist was deep-seated and difficult to dispel. It appears to have stemmed from the different approaches to the interpretation of colloidal behaviour. In 1861 the Scotsman Thomas Graham distinguished between *crystalloid* substances, which could diffuse easily when in solution, and *colloids* or glue-like substances, that refused to crystallize, that exhibited high viscosities in solution, and diffused slowly when dissolved in liquids. He explained this difference in behaviour by assuming that crystalloids were small particles whereas colloids were composed of large particles. This was acceptable to most scientists but disagreement became apparent when there were attempts at further analysis on the molecular level. This divergence of opinion is embodied in the physical approach as opposed to the chemical approach.

The chemical approach assumed that colloidal substances were in fact large molecules and that their behaviour could be explained in terms of the size of the individual molecules. The physical approach favoured the concept that the molecular sizes were no different in magnitude from those of the crystalloid materials, but that colloidal behaviour was a consequence of the formation of aggregates of these smaller molecules in solutions that were held together by physical forces rather than chemical bonds.

The physical approach prevailed because it suited the chemical methodology of the period. Classical organic chemistry demanded the careful preparation and investigation of pure substances with well-defined melting points, boiling points and molar masses. Even when experimental measurements pointed to the existence of large molecules the data were rationalized to fit the physical approach. Thus, while rubber latex, which showed colloidal behaviour, was assigned the correct structural formula I for the individual units, it was postulated to have the ring formation II.

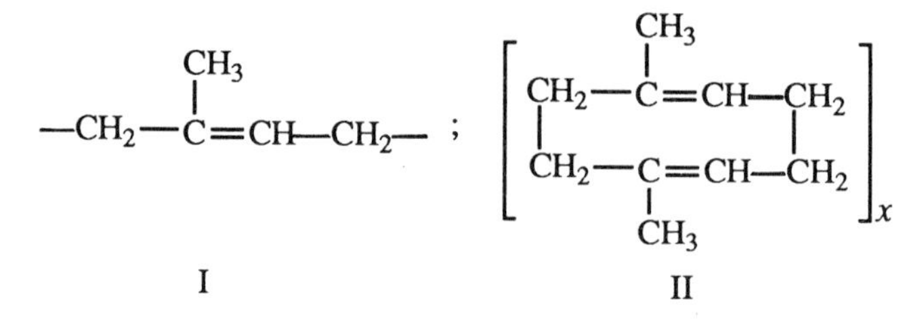

These rings were thought to form large aggregates in the latex particle. This idea was essential if particle masses of between 6500 and 10^5, which had been calculated from ebullioscopic and cryoscopic measurements of rubber particles in solution, were to be explained in accord with the physical approach to the problem.

So locked into the mind-warp of believing that only small molecules could exist as chemical entities, were the majority of scientists, that the possibility of the structure I forming long chains, rather than rings, as an alternative way of explaining the high molar masses does not seem to have been seriously considered. Similar work on starch, cellulose and in the protein field showed the existence of high molar mass species, but here too interpretation favoured the aggregate hypothesis.

One should not, however, be overly critical of this failure to accept a concept that to us may seem obvious. Received wisdom is a deceptively comfortable framework to work within and it takes a strong-minded, and perhaps equally dogmatic, person to break out of its strictures. The German organic chemist Hermann Staudinger proved to be that person. Building on observations by the English chemist Pickles (who was a fellow sceptic), which cast doubts on the presence of physical forces of aggregation in colloidal systems, and on his own work on the viscosity of materials exhibiting colloidal behaviour, he began a long battle of conversion. From 1927 onwards he started to convince other chemists, albeit slowly, that colloidal substances like rubber, starch and cellulose were in fact long, linear, threadlike molecules of variable length, composed of small definable molecular units, covalently bonded to each other to form macromolecules or polymers.

This was no easy task. He was asked by colleagues why he wished to abandon the "beautiful area of low molecular chemistry" and turn to work in Schmierenchemie (greasy chemistry). Even at the end of the 1920s he was given the following advice "Dear colleague, let me advise you to dismiss the idea of large molecules, there are no organic molecules with a molecular mass over 5000. Purify your products, like for instance rubber, and they will crystallize and reveal themselves as low molecular weight substances".

While this "greasy" chemistry image of polymer science was a difficult one to erase from some areas of chemical academia, the grease has turned out to be a rich vein of scientific gold. How rich can only be judged by digging more deeply into what has developed into one of the most exciting and diverse areas of science where the possibilities for innovation seem endless.

1.2 Classification

Because of the diversity of function and structure found in the field of macromolecules, it is advantageous to draw up some scheme which groups the materials under convenient headings, and one way of doing this is shown below.

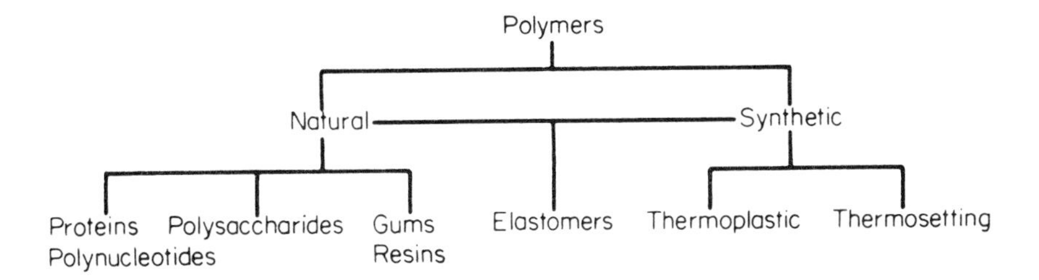

Natural polymers usually have more complex structures than synthetic polymers and we shall deal almost exclusively with the latter group. Elastomers can be either natural or man-made and are classified here as a common sub-group. The more general term elastomer is used to describe rubberlike materials, because there now exists a wide variety of synthetic products, whose structures differ markedly from the naturally occurring rubber, but whose elastic properties are comparable to, and sometimes better than, the original.

1.3 Some basic definitions

In order to place polymer science in the proper perspective we must examine the subject on as broad a basis as possible. It is useful to consider polymers first on the molecular level then as materials. These considerations can be interrelated by examining the various aspects in the sequence: synthesis, characterization, mechanical behaviour, and application, but before discussing the detailed chemistry and physics some of the fundamental concepts must be introduced to provide essential background to such a development. We need to know what a polymer is and how it is named and prepared. It is also useful to identify which physical properties are important and so it is necessary to define the molar mass, the molar mass distribution, obtain an appreciation of the molecular size and shape, and recognize the important transition temperatures.

A *polymer* is a large molecule constructed from many smaller structural units called *monomers*, covalently bonded together in any conceivable pattern. In certain cases it is more accurate to call the structural or repeat unit a *monomer residue* because atoms are eliminated from the simple monomeric unit during some polymerization processes.

The essential requirement for a small molecule to qualify as a monomer or "building block" is the possession of two or more bonding sites, through which each can be linked to other monomers to form the polymer chain. The number of bonding sites is referred to as the *functionality*. Monomers such as a hydroxyacid (HO—R—COOH) or vinyl chloride ($CH_2{=}CHCl$) are bifunctional. The hydroxyacid will condense with the other hydroxyacid molecules through the —OH and —COOH groups to form a linear polymer, and the polymerization reaction in this case consists of a series of simple organic reactions similar to

$$ROH + R'COOH \rightleftharpoons R'COOR + H_2O$$

The double bond of the vinyl compound is also bifunctional as activation by a free radical or an ion leads to polymer formation

$$CH_2{=}CHCl + R^{\bullet} \rightarrow RCH_2{-}CHCl{-}CH_2{-}CHCl \sim^{\bullet}$$

Bifunctional monomers form linear macromolecules but if the monomers are polyfunctional, *i.e.* have three or more bonding sites as in glycerol CH_2OH. $CHOH$. CH_2OH, branched macromolecules can be produced. These may even develop into large three-dimensional networks containing both branches and crosslinks.

When only one species of monomer is used to build a macromolecule the product is called a *homopolymer*, normally referred to simply as a polymer. If the chains are composed of two types of monomer unit, the material is known as a *copolymer*, and if three different monomers are incorporated in one chain a *terpolymer* results.

Copolymers prepared from bifunctional monomers can be subdivided further into four main categories:

(i) Statistical copolymers where the distribution of the two monomers in the chain is essentially random, but influenced by the individual monomer reactivities.

$$\sim \text{AAABABBABABBBBABAAB} \sim$$

(ii) Alternating copolymers with a regular placement along the chain.

$$\sim \text{ABABABABAB} \sim$$

(iii) Block copolymers comprised of substantial sequences or blocks of each.

$$\sim \text{AAAAAABBBBBAAAA} \sim$$

(iv) Graft copolymers in which blocks of one monomer are grafted on to a backbone of the other as branches.

```
B            B
B            B
B            B
B            B
AAAAAAAAAAAAAAAAA
      B
      B
      B
      B
```

1.4 Synthesis of polymers

A process used to convert monomer molecules into a polymer is called a *polymerization* and the two most important groups are step-growth and addition. A step-growth polymerization is used for monomers with functional groups such as —OH, —COOH, —COCl, *etc.* and is normally, but not always, a succession of condensation reactions. Consequently the majority of polymers formed in this way differ slightly from the original monomers because a small molecule is eliminated in the reaction, *e.g.* the reaction between ethylene glycol and terephthalic acid produces a polyester better known as terylene

$$n\mathrm{HO(CH_2)_2OH} + n\mathrm{HOOC{-}C_6H_4{-}COOH} \rightarrow \left(\mathrm{O(CH_2)_2O\,.\,\overset{\displaystyle O}{\overset{\|}{C}}{-}C_6H_4{-}\overset{\displaystyle O}{\overset{\|}{C}}} \right)_n + (2n-1)\,\mathrm{H_2O}$$

The addition polymerizations, for olefinic monomers, are chain reactions which convert the monomers into polymers by stimulating the opening of the double bond with a free radical or ionic initiator. The product then has the same chemical composition as the starting material, *e.g.* acrylonitrile produces polyacrylonitrile without the elimination of a small molecule.

$$n\mathrm{CH_2{=}CHCN} \rightarrow \sim\!\!(\mathrm{CH_2CHCN})_n\!\sim$$

The length of the molecular chains, which will depend on the reaction conditions, can be obtained from measurements of molar masses.

1.5 Nomenclature

The least ambiguous method of naming a polymer is based on its source. However, a wide variety of trade names are commonly used. The prefix poly is attached to the name of the monomer in addition polymers, and so polyethylene, polyacrylonitrile, polystyrene denote polymers prepared from these single monomers. When the monomer has a multi-worded name or has a substituted parent name then this is enclosed in parentheses and prefixed with poly, *e.g.* poly(methylmethacrylate), poly(vinyl chloride), poly(ethylene oxide), *etc.*

Polymers prepared by self-condensation of a single monomer such as ω-amino lauric acid, are named in a similar manner, but this polymer, poly(ω-amino lauric acid), (sometimes known as nylon-12) can also be prepared by a ring-opening reaction using lauryl lactam and could then be called poly(lauryl lactam). Both names are correct.

IUPAC has attempted to formalize the nomenclature of regular, single-stranded organic polymers and has proposed a set of procedures, some of which are described briefly as follows.

The first step is to select a *constitutional repeat unit*, CRU, which may contain one or more subunits. The name of the polymer is then the name of the CRU, in parentheses prefixed by poly. Before naming the CRU it must be orientated correctly. This involves placing the constituent parts in order of seniority with the highest to the left. In descending order this would be heterocyclic rings, chains with hetero atoms, carbocyclic rings and chains with only carbon atoms, if such an order is possible chemically.

Thus $\mathrm{(O{-}CH_2{-}CH_2)}$ would be poly(oxy ethylene) rather than $\mathrm{(CH_2CH_2{-}O)}$ poly(ethylene oxy). If there is a substituent on part of the CRU, then orientation will place the substituent closest to the left of the substituted portion,

TABLE 1.1. Nomenclature of some common polymers

Name	*Structure*	*Trivial name*
poly(methylene)	$-(CH_2CH_2)_n-$	polyethylene
poly(propylene)	$-(CH(CH_3)-CH_2)_n-$	polypropylene
poly(1,1-dimethylethylene)	$-(C(CH_3)_2-CH_2)_n-$	polyisobutylene
poly(1-methyl-1-butenylene)	$-(C(CH_3)=CHCH_2CH_2)_n-$	polyisoprene
poly(1-butenylene)	$-(CH=CHCH_2CH_2)_n-$	polybutadiene
poly(1-phenylethylene)	$-(CH(C_6H_5)-CH_2)_n-$	polystyrene
poly(1-cyanoethylene)	$-(CH(CN)-CH_2)_n-$	polyacrylonitrile
poly(1-hydroxyethylene)	$-(CH(OH)-CH_2)_n-$	poly(vinylalcohol)
poly(1-chloroethylene)	$-(CH(Cl)-CH_2)_n-$	poly(vinylchloride)
poly(1-acetoxyethylene)	$-(CH(OOCCH_3)-CH_2)_n-$	poly(vinylacetate)
poly(1,1-difluoroethylene)	$-(CF_2-CH_2)_n-$	poly(vinylidenefluoride)

TABLE 1.1. Nomenclature of some common polymers (*continued*)

Name	*Structure*	*Trivial name*
poly(1-(methoxycarbonyl) ethylene)	$-(\underset{\mid}{C}H-CH_2)_n-$ with $COOCH_3$ on CH	poly(methylacrylate)
poly(1-(methoxycarbonyl)-1-methyl-ethylene)	$-(C(CH_3)(COOCH_3)-CH_2)_n-$	poly(methylmethacrylate)
poly(oxymethylene)	$-(OCH_2)_n-$	polyformaldehyde
poly(oxyethylene)	$-(OCH_2CH_2)_n-$	poly(ethylene oxide) (sometimes called polyethylene glycol)
poly(oxyphenylene)	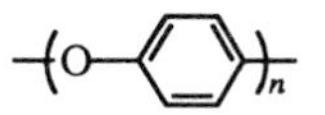 $-(O-C_6H_4)_n-$	poly(phenyleneoxide)
poly(oxyethylene-oxyterephthaloyl)	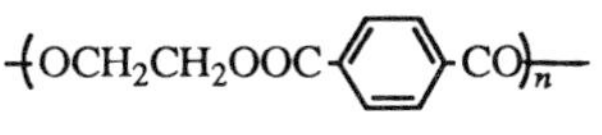 $-(OCH_2CH_2OOC-C_6H_4-CO)_n-$	poly(ethylene terephthalate)
poly(iminohexamethyl-eneiminoadipoyl)-	$-(NH(CH_2)_6NHCO(CH_2)_4CO)_n-$	poly(hexamethylene adipamide)
poly(difluoromethylene)	$-(CF_2-CF_2)_n-$	poly(tetrafluoroethylene)
poly((2-propyl-1,3-dioxane-4,6-diyl)-methylene)	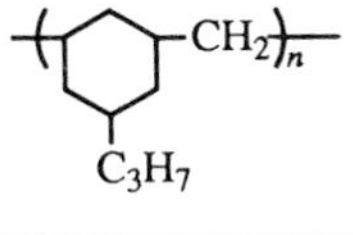 $-(C_4H_6O_2(C_3H_7)-CH_2)_n-$ (ring with C_3H_7 substituent)	poly(vinylbutyryl)

thus $+O-\underset{\substack{| \\ CH_3}}{CH}-CH_2+$, poly(oxy 1-methyl ethylene) is preferred, rather than $+O-CH_2-\underset{\substack{| \\ CH_3}}{CH}+$

Similarly, a more complex CRU might be orientated as

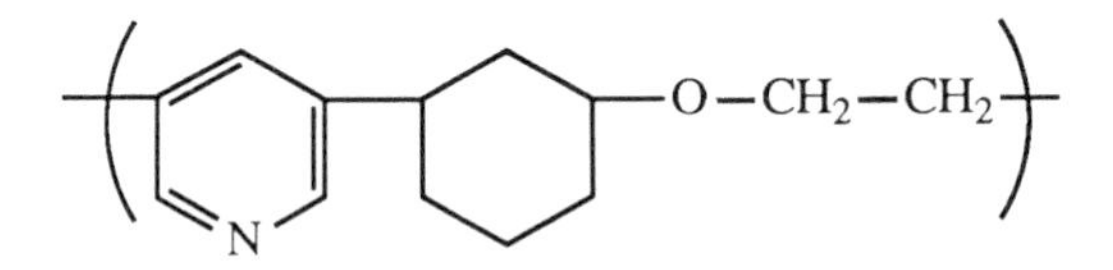

and named poly(3,5 pyridine diyl-1,3-cyclohexylene oxydimethylene). Other examples are shown in table 1.1 and the reader is referred to the references listed at the end of the chapter for a more comprehensive coverage.

1.6 Average molar masses and distributions*

One of the most important features which distinguishes a synthetic high polymer from a simple molecule is the inability to assign an exact molar mass to a polymer. This is a consequence of the fact that in a polymerization reaction the length of the chain formed is determined entirely by random events. In a condensation reaction, it depends on the availability of a suitable reactive group and in an addition reaction, on the lifetime of the chain carrier. Inevitably, because of the random nature of the growth process, the product is a mixture of chains of differing length – a *distribution* of chain lengths – which in many cases can be calculated statistically.

The polymer is characterized best by a molar mass distribution and the associated molar mass averages, rather than by a single molar mass. The typical distributions, shown in figure 1.1, can be described by a variety of averages. As the methods used for estimating the molar mass of polymers employ different averaging procedures, it is safer to use more than one technique to obtain two or more averages and by doing so characterize the sample more fully.

A colligative method, such as osmotic pressure, effectively counts the number of molecules present and provides a *number average* molar mass $\langle M \rangle_n$ defined by

$$\langle M \rangle_n = \frac{\Sigma N_i M_i}{\Sigma N_i} = \frac{\Sigma w_i}{\Sigma (w_i/M_i)} \tag{1.1}$$

where N_i is the number of molecules of species i of molar mass M_i. The brackets $\langle\ \rangle$ indicate that it is an average value, but by convention these are normally omitted.

The alternative expression is in terms of the mass $w_i = N_i M_i / N_A$ if required, where N_A is Avogadro's constant.

From light scattering measurements, a method depending on the size rather than the

*The quantity molar mass is used throughout this text instead of the dimensionless quantity molecular weight which is usual in polymer chemistry. All the equations in later sections evaluate molar mass rather than the dimensionless quantity molecular weight.

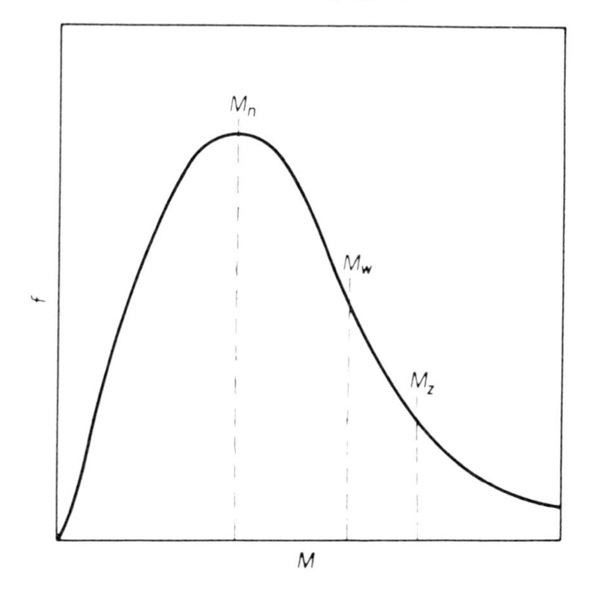

FIGURE 1.1. Typical distribution of molar masses for a synthetic polymer sample, where f is the fraction of polymer in each interval of M considered.

number of molecules, a weight average molar mass $\langle M \rangle_w$ is obtained. This is defined as

$$\langle M \rangle_w = \frac{\Sigma N_i M_i^2}{\Sigma N_i M_i} = \frac{\Sigma w_i M_i}{\Sigma w_i} \tag{1.2}$$

Statistically $\langle M \rangle_n$ is simply the first moment, and $\langle M \rangle_w$ is the ratio of the second to the first moment, of the number distribution.

A higher average, the z-average given by

$$\langle M \rangle_z = \frac{\Sigma N_i M_i^3}{\Sigma N_i M_i^2} = \frac{\Sigma w_i M_i^2}{\Sigma w_i M_i}, \tag{1.3}$$

can be measured in the ultracentrifuge which also yields another useful average, the $(z + 1)$-average,

$$\langle M \rangle_{z+1} = \frac{\Sigma N_i M_i^4}{\Sigma N_i M_i^3}, \tag{1.4}$$

often required when describing mechanical properties.

A numerical example serves to highlight the differences in the various averages. Consider a hypothetical polymer sample composed of chains of four distinct molar masses, 100 000, 200 000, 500 000, and 1 000 000 g mol^{-1} in the ratio 1:5:3:1, then

$$M_n/\mathrm{g\,mol^{-1}} = \frac{(1 \times 10^5) + (5 \times 2 \times 10^5) + (3 \times 5 \times 10^5) + (1 \times 10^6)}{1 + 5 + 3 + 1} = 3.6 \times 10^5$$

$$M_w/\mathrm{g\,mol^{-1}} = \frac{\{1 \times (10^5)^2\} + \{5 \times (2 \times 10^5)^2\} + \{3 \times (5 \times 10^5)^2\} + \{1 \times (10^6)^2\}}{(1 \times 10^5) + (5 \times 2 \times 10^5) + (3 \times 5 \times 10^5) + (1 \times 10^6)}$$

$$= 5.45 \times 10^5$$

and $M_z = 7.22 \times 10^5\ \mathrm{g\,mol^{-1}}$

The breadth of the distribution can often be gauged by establishing the *heterogeneity index* (M_w/M_n). For many polymerizations the most probable value is about 2.0, but both larger and smaller values can be obtained and it is at best only a rough guide.

An alternative method of describing the chain length of a polymer is to measure the *average degree of polymerization* x. This represents the number of monomer units or residues in the chain and is given by

$$x = M/M_0, \tag{1.5}$$

where M_0 is the molar mass of monomer or residue and M is the appropriate average molar mass. Hence the x average depends on which average is used for M. (To avoid confusion between the mole fraction x and the average degree of polymerization x, the latter will always be subscripted as x_n or x_w to indicate the particular M used in equation (1.5).)

1.7 Size and shape

Some measure of the polymer size is obtained from the molar mass, but what is the actual length of a chain and what shape does it adopt? We can begin to answer these questions by first considering a simple molecule such as butane and examining the behaviour when the molecule is rotated about the bond joining carbon 2 to carbon 3.

The Newman and "saw horse" projections show the *trans* position in figure 1.2a with the "dihedral angle" $\phi = 180°$. This is the most stable conformation with the greatest

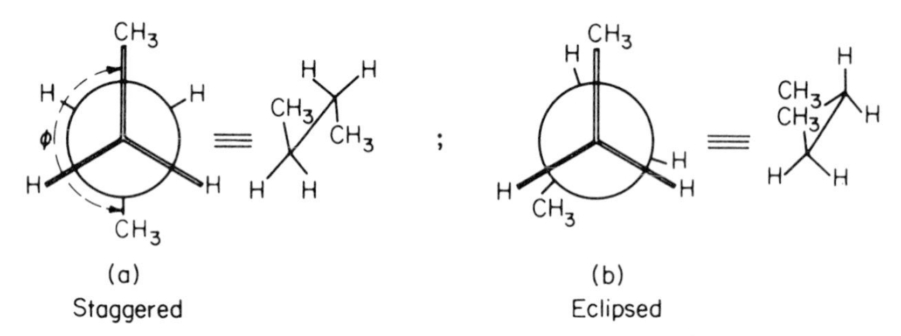

FIGURE 1.2. Newman and "saw horse" projections for n-butane, (a) a staggered state with $\phi = \pi$ and (b) an eclipsed position.

separation between the two methyl groups. Rotation about the C_2—C_3 bond alters ϕ and moves the methyl groups past the opposing hydrogen atoms so that an extra repulsive force is experienced when an eclipsed position (figure 1.2b) is reached.

The progress of rotation can be followed by plotting the change in potential energy $V(\phi)$ as a function of the dihedral angle, as shown in figure 1.3. The resulting diagram for butane exhibits three minima at $\phi = \pi$, $\pi/3$, and $5\pi/3$ called the *trans* and $\pm$ *gauche* states respectively, and the greater depth of the *trans* position indicates that this is the position of maximum stability. Although the *gauche* states are slightly less stable, all three minima can be regarded as discrete rotational states. The maxima correspond to the eclipsed positions and are angles of maximum instability. These diagrams will vary with the type of molecule and need not be symmetrical, but the butane diagram is very similar to that for the simple polymer polyethylene $\text{+}CH_2\text{—}CH_2\text{+}_n$, if the —$CH_3$ groups are replaced by the two sections of the chain adjoining the bond of rotation. The backbone of this polymer is composed of a chain of tetrahedral carbon atoms covalently bonded to each other so that the molecule can be represented as an extended all *trans*

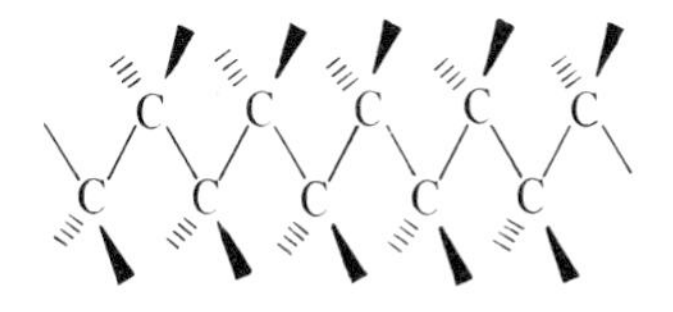

zig-zag chain. For a typical value of $M = 1.6 \times 10^5\,\mathrm{g\,mol^{-1}}$, the chain contains 10 000 carbon atoms; thus in the extended zig-zag state, assuming a tetrahedral angle of 109° and a bond length of 0.154 nm, the chain would be about 1260 nm long and 0.3 nm

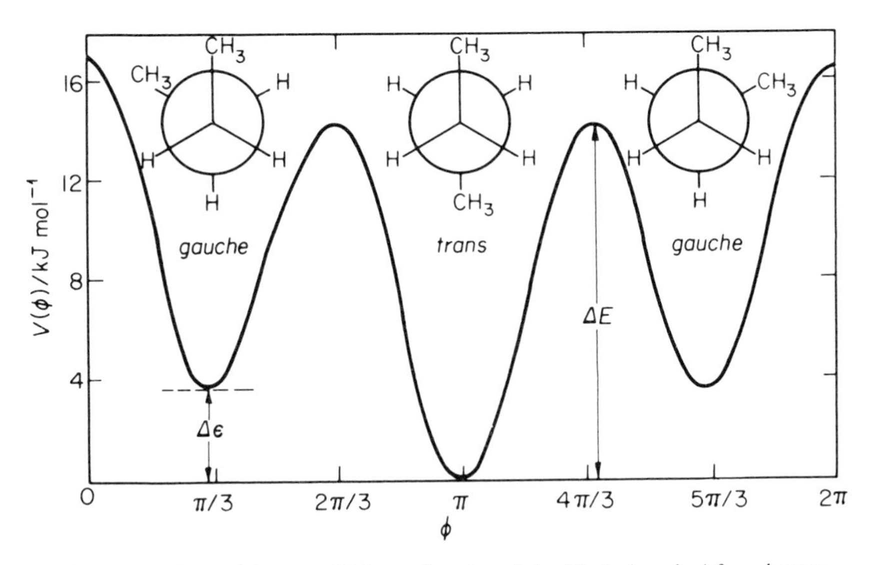

FIGURE 1.3. Potential energy $V(\phi)$ as a function of the dihedral angle ϕ for *n*-butane.

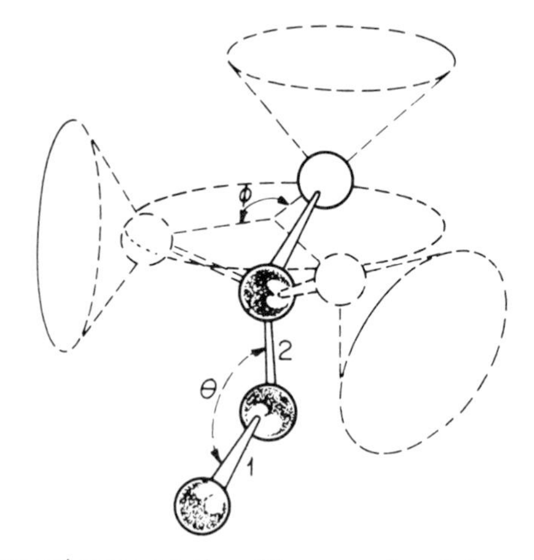

FIGURE 1.4. Diagrammatic representation of the cones of revolution available to the third and fourth bonds of a simple carbon chain with a fixed bond angle θ.

diameter. Magnified a million times, the chain could be represented by a piece of wire 126×0.03 cm. This means that polyethylene is a long threadlike molecule, but how realistic is the extended all *trans* conformation? As every group of four atoms in the chain has a choice of three possible stable rotational states, a total of 3^{10000} shapes are available to this particular chain, only one of which is the all *trans* state. So, in spite of the fact that the all *trans* extended conformation has the lowest energy, the most probable conformation will be some kind of randomly coiled state – assuming that no external ordering forces are present and that the rotation about the carbon bonds is in no way impeded. The many possible coiled forms are generated simply by allowing the chain to rotate into a *gauche* position which moves the atom out of the plane of the adjacent bonds. This is shown more clearly (see figure 1.4) by considering the various cones of revolution available to a chain over only two bonds. The distribution of *trans* (t) and *gauche* (g) states along a chain will be a function of the temperature and the relative stability of these states. Consequently there is an unequal distribution of each. The ratio of the number of *trans* n_t to *gauche* n_g states is then governed by a Boltzmann factor and

$$n_g/n_t = 2\exp(-\Delta\epsilon/kT), \tag{1.6}$$

where k is the Boltzmann constant, $\Delta\epsilon$ is the energy difference between the two minima, and the 2 arises because of the $\pm$ *gauche* states available. For polyethylene $\Delta\epsilon$ is about 3.34 kJ mol^{-1}, and values of (n_g/n_t) for 100, 200, and 300 K are 0.036, 0.264, and 0.524 respectively, showing that the chain becomes less extended and more coiled as the temperature increases. Because of the possibility of rotation about the carbon bonds, the chain is in a state of perpetual motion, constantly changing shape from one coiled conformation to another form, equally probable at the given temperature. The speed of this wriggling varies with temperature (and from one polymer to another) and dictates many of the physical characteristics of the polymer, as we shall see later.

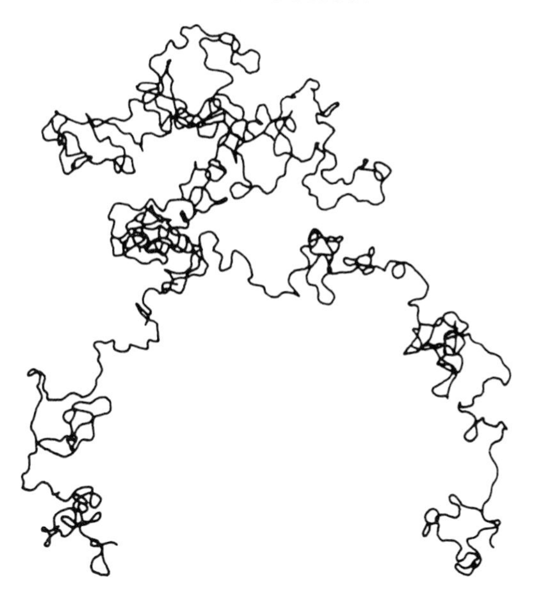

FIGURE 1.5. Random arrangement of a polyethylene chain containing 1000 freely rotating C–C bonds, in which each successive bond has been given a random choice of six equally spaced angular positions. (Treloar (1958), *Physics of Rubber Elasticity*.)

The height of the potential energy barrier ΔE determines the rate of bond interchange between the t and the g states and for polyethylene is about $16.7\,\mathrm{kJ\,mol^{-1}}$. When ΔE is very high (about $80\,\mathrm{kJ\,mol^{-1}}$), rotation becomes very difficult, but as the temperature is raised the fraction of molecules which possess energy in excess of ΔE increases and rotation from one state to another becomes easier.

Realistically then, a polymer chain is better represented by a loosely coiled ball (figure 1.5) than an extended rod. For the magnified polyethylene chain considered earlier a ball of about 4 cm diameter is a likely size.

The term *conformation* has been used here when referring to a three-dimensional geometric arrangement of the polymer, which changes easily when the bonds are rotated.

There is a tendency to use the term *configuration* in a synonymous sense, but as far as possible this will be reserved for the description of chains where the geometric variations can only be interchanged by breaking a bond.

1.8 The glass transition temperature T_g and the melting temperature T_m

At sufficiently low temperatures all polymers are hard rigid solids. As the temperature rises, each polymer eventually obtains sufficient thermal energy to enable its chains to move freely enough for it to behave like a viscous liquid (assuming no degradation has occurred).

There are two ways in which a polymer can pass from the solid to the liquid phase, depending on the internal organization of the chains in the sample. The different types

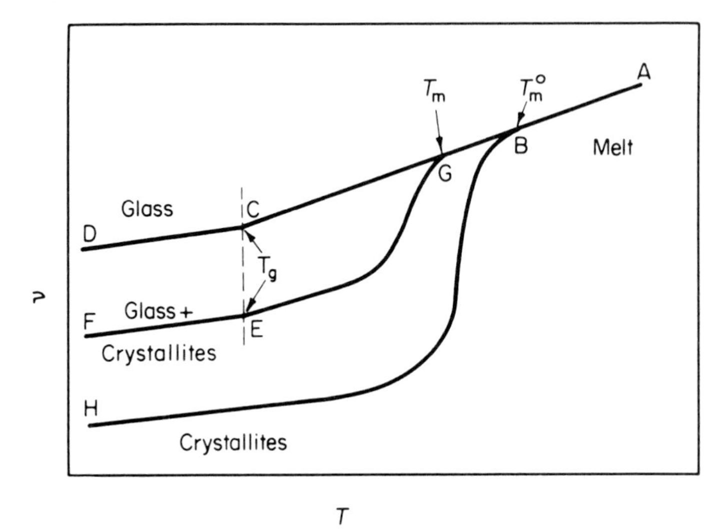

FIGURE 1.6. Schematic representation of the change of specific volume v of a polymer with temperature T for (i) a completely amorphous sample (A–C–D), (ii) a semi-crystalline sample (A–G–F), and (iii) a perfectly crystalline material (A–B–H).

of thermal response, illustrated by following the change in specific volume, are shown schematically in figure 1.6.

A polymer may be completely amorphous in the solid state, which means that the chains in the specimen are arranged in a totally random fashion. The volume change in amorphous polymers follows the curve A–D. In the region C–D the polymer is a glass, but as the sample is heated it passes through a temperature T_g, called the *glass transition temperature*, beyond which it softens and becomes rubberlike. This is an important temperature because it marks the point where important property changes take place, *i.e.* the material may be more easily deformed or become ductile above T_g. A continuing increase in temperature along C–B–A leads to a change of the rubbery polymer to a viscous liquid.

In a perfectly crystalline polymer, all the chains would be incorporated in regions of three-dimensional order, called crystallites, and no glass transition would be observed, because of the absence of disordered chains in the sample. The crystalline polymer, on heating, would follow curve H–B–A; at T_m°, melting would be observed and the polymer would become a viscous liquid.

Perfectly crystalline polymers are not encountered in practice and instead polymers may contain varying proportions of ordered and disordered regions in the sample. These semi-crystalline polymers usually exhibit both T_g and T_m, corresponding to the ordered and disordered portions and follow curves similar to F–E–G–A. As T_m° is the melting temperature of a perfectly crystalline polymer of high molar mass, T_m is lower and more often represents a melting range, because the semi-crystalline polymer contains a spectrum of chain lengths and crystallites of various sizes with many defects. These imperfections act to depress the melting temperature and experimental values of T_m can depend on the previous thermal history of the sample.

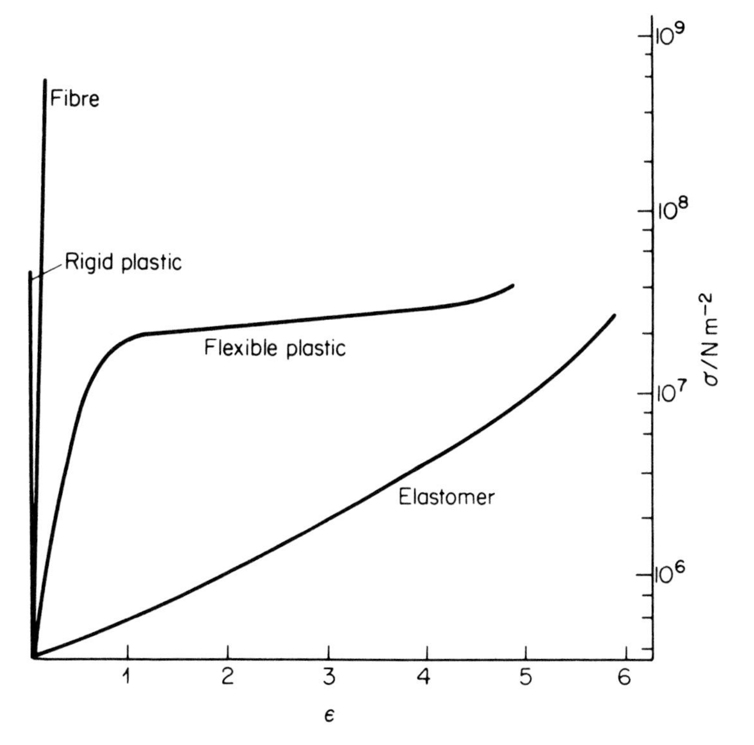

FIGURE 1.7. Typical stress-strain ($\sigma-\varepsilon$) plots for a rigid plastic, a fibre, a flexible plastic, and an elastomer.

Nevertheless, both T_g and T_m are important parameters, which serve to characterize a given polymer.

1.9 Elastomers, fibres, and plastics

A large number of synthetic polymers now exist covering a wide range of properties. These can be grouped into the three major classes, plastics, fibres, and elastomers, but there is no firm dividing line between the groups. However, some classification is useful from a technological viewpoint and one method of defining a member of these categories is to examine a typical stress-strain plot. Rigid plastics and fibres are resistant to deformation and are characterized by a high modulus and low percentage elongations. Elastomers readily undergo deformation and exhibit large reversible elongations under small applied stresses, *i.e.* they exhibit elasticity. The flexible plastics are intermediate in behaviour. An outline of the structure-property relations will be presented later, but before proceeding further with the more detailed science, we can profitably familiarize ourselves with some of the more common polymers and their uses. Some of these are presented in table 1.2 where an attempt is made to show that the lines of demarkation which are used to divide polymers into the three major groups are not clear cut.

TABLE 1.2. Some common plastics, elastomers, and fibres

Elastomers	*Plastics*	*Fibres*
polyisoprene	polyethylene	
polyisobutylene	polytetrafluoroethylene	
polybutadiene	polystyrene	
	poly(methylmethacrylate)	
	phenol-formaldehyde	
	urea-formaldehyde	
	melamine-formaldehyde	
← poly(vinyl chloride)	→	
← polyurethanes	→	
← polysiloxanes	→	
	← polyamide	→
	← polyester	→
	← polypropylene	→

A polymer normally used as a fibre may make a perfectly good plastic if no attempt is made to draw it into a filament. Similarly, a plastic, if used at a temperature above its glass transition and suitably crosslinked, may make a perfectly acceptable elastomer. In the following brief account of some of the more common plastics, fibres, and elastomers, the classification is based essentially on their major technological application under standard working conditions.

1.10 Fibre-forming polymers

While there are many fibre-forming polymers only a limited number have achieved great technological and commercial success. It is significant that these are polymers of long standing, and it has been suggested that further fibre research may involve the somewhat prosaic task of attempting to improve, modify, or reduce the cost of existing fibres, rather than to look for new and better alternatives. The commercially important fibres are listed in table 1.3; all are thermoplastic polymers.

The polyamides are an important group of polymers which include the naturally occurring proteins in addition to the synthetic nylons. The term nylon, originally a trade name, has now become a generic term for the synthetic polyamides, and the numerals which follow, *e.g.* nylon-6,6, distinguish each polymer by designating the number of carbon atoms lying between successive amide groups in the chain. Thus nylon-6,10 is prepared from two monomers and has the structure

$$\left[NH(CH_2)_6NHCO(CH_2)_8CO \right]_n$$

with alternative sequences of six and ten carbon atoms between the nitrogen atoms, while nylon-6 is prepared from one monomer and has the repeat formula $\left[NH(CH_2)_5CO \right]_n$ with regular sequences of six carbon atoms between the nitrogen atoms. A nylon with two numbers is termed *dyadic* indicating that it contains both dibasic acid (or acid chloride) and diamine moieties, where the first number represents the diamine and the second the diacid used in the synthesis. The *monadic* nylons have one number, indicating that synthesis involved only one type of monomer. This terminology means that a poly (α-amino acid) would be nylon-2.

Terylene is an important polyester. It exhibits high resilience, durability, and low moisture absorption, properties which contribute to its desirable "wash and wear" characteristics. The harsh feel of the fibre, caused by the stiffness of the chain, is overcome by blending it with wool and cotton.

The acrylics and modacrylics are among the most important of the amorphous fibres. They are based on the acrylonitrile unit —$CH_2CH(CN)$— and are usually manufactured as copolymers. When the acrylonitrile content is 85 per cent or higher, the polymer is an *acrylic* fibre, but if this drops to between 35 and 85 per cent it is known as a *modacrylic* fibre. Vinyl chloride and vinylidene chloride are the most important comonomers and the copolymers produce high bulk yarns which can be subjected to a controlled shrinking process after fabrication. Once shrunk the fibres are dimensionally stable.

Silk-like qualities have always been sought after by the synthetic fibre chemist. The new cycloaliphatic polyamide with the probable structure

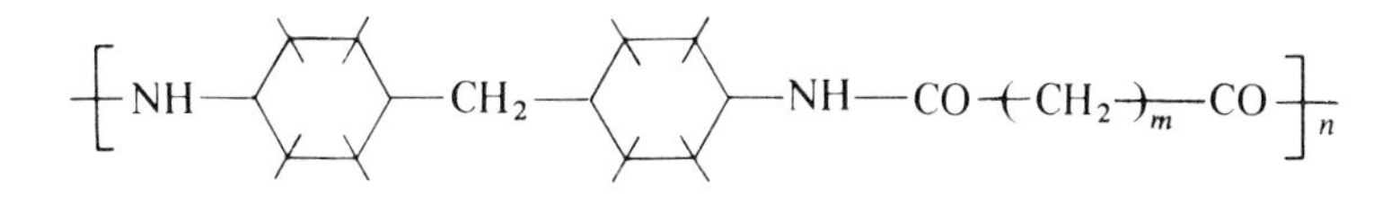

is said to have the aesthetic appeal of natural silk, and a silk-like fibre called "Chinon" has been prepared from a polyacrylonitrile-protein graft copolymer. This has been prepared by grafting acrylonitrile on to caseine and has many of the properties of natural silk.

1.11 Plastics

A plastic is rather inadequately defined as an organic high polymer capable of changing its shape on the application of a force and retaining this shape on removal of this force, *i.e.* a material in which a stress produces a non-reversible strain.

The main criterion is that plastic materials can be formed into complex shapes, often by the application of heat or pressure and a further sub-division into those which are *thermosetting* and those which are *thermoplastics* is useful. The thermosetting materials become permanently hard when heated above a critical temperature and will not soften again on reheating. They are usually crosslinked in this state. A thermoplastic polymer will soften when heated above T_g. It can then be shaped and on cooling will harden in this form. However, on reheating it will soften again and can be reshaped if required before hardening when the temperature drops. This cycle can be carried out repeatedly.

A number of the important thermoplastics are shown in table 1.4 together with a few examples of their more important uses, determined by the outstanding properties of each. Thus polypropylene, poly(phenylene oxide) and TPX have good thermal stability and can be used for items requiring sterilization. The optical qualities of polystyrene and poly(methyl methacrylate) are used in situations where transparency is a premium, while the low frictional coefficient and superb chemical resistance of poly(tetrafluoroethylene) make it useful in non-stick cookware and protective clothing. Low density polyethylene, while mechanically inferior to the high density polymer, has better impact resistance and can be used when greater flexibility is required, whereas the

TABLE 1.3. Chemical structure of synthetic fibres

Polymer	*Repeat unit*	*Trade names*
Step-growth		
POLYAMIDES (Nylons) (Uses: drip-dry fabrics, cordage, braiding, bristles, and surgical sutures.)		
polycaprolactam	$\left[NH(CH_2)_5CO \right]_n$	Nylon-6, Perlon
poly(decamethylene carboxamide)	$\left[NH(CH_2)_{10}CO \right]_n$	Nylon-11, Rilsan
poly(hexamethylene adipamide)	$\left[NH(CH_2)_6NHCO(CH_2)_4CO \right]_n$	Nylon-6,6. Bri-nylon
poly(*m*-phenylene isophthalamide)	$\left[NH-C_6H_4-NHCO-C_6H_4-CO \right]_n$	Nomex
POLYESTERS (Uses: fabrics, tyre-cord yarns, and yacht sails.)		
poly(ethylene terephthalate)	$\left[OC-C_6H_4-COO(CH_2)_2O \right]_n$	Terylene, Dacron
poly(cyclohexane 1,4-dimethylene terephthalate)	$\left[OCH_2-C_6H_{10}-CH_2OOC-C_6H_4-CO \right]_n$	Kodel
POLYUREAS		
poly(nonamethylene urea)	$\left[NHCONH(CH_2)_9 \right]_n$	Urylon

Addition

ACRYLICS (Uses: fabrics and carpeting.)		
polyacrylonitrile	$-(CH_2CHCN)_n-$ (often as copolymer with > 85 per cent acrylonitrile)	Orlon, Courtelle, Acrilan, Creslan
acrylonitrile copolymers	35 per cent < acrylonitrile < 85 per cent + vinyl chloride + vinylidene chloride	Dynel Verel
HYDROCARBONS (Uses: carpets and upholstery.)		
polyethylene	$-(CH_2CH_2)_n-$	Courlene. Vestolen
polypropylene (isotactic)	$-(CH_2-CH(CH_3))_n-$	Ulstron, Herculon, Meraklon
HALOGEN SUBSTITUTED OLEFINS (Uses: knitwear and protective clothing.)		
poly(vinyl chloride)	$-(CH_2CHCl)_n-$	Rhovyl, Valren
poly(vinylidene chloride)	$-(CH_2CCl_2)_n-$	Saran, Tygan
poly(tetrafluoroethylene)	$-(CF_2CF_2)_n-$	Teflon, Polifen
VINYL (Uses: fibres, adhesives, paint, sponges, films, and plasma extender.)		
poly(vinyl alcohol)	$-(CH_2CHOH)_n-$ (normally crosslinked)	Vinylon, Kuralon, Mewlon

TABLE 1.4. Thermoplastics

Polymer	*Repeat unit*	*Density* ($g\,cm^{-3}$)	*Uses*
Polyethylene (High Density)	$-(CH_2CH_2)-$	0.94 to 0.96	Household products, insulators, pipes, toys, bottles
(Low Density)		0.92	
Polypropylene	$-(CH_2CH(CH_3))-$	0.90	Waterpipes, integral hinges, sterilizable hospital equipment
Poly(4-methylpentene-1) (TPX)	$-(CH_2CH)-$ with side chain $-CH_2-CH(CH_3)_2$ (CH bearing two CH_3)	0.83	Hospital and laboratory ware
Poly(tetrafluoroethylene) (PTFE)	$-(CF_2CF_2)-$	2.20	Non-stick surfaces, insulation, gaskets
Poly(vinyl chloride) (PVC)	$-(CH_2CHCl)-$	1.35 to 1.45	Records, bottles, house siding and eaves
Polystyrene	$-(CH_2CH(C_6H_5))-$	1.04 to 1.06	Lighting panels, lenses, wall tiles, flower pots
Poly(methylmethacrylate) (PMMA)	$-(CH_2-C(CH_3)(COOCH_3))-$	1.17 to 1.20	Bathroom fixtures, knobs, combs, illuminated signs
Polycarbonates	$-(R.O.COO)-$	1.20	Cooling fans, marine propellors, safety helmets
Poly(2,6-dimethylphenylene oxide)	$-(C_6H_2(CH_3)_2-O)-$ (benzene ring with two CH_3 groups)	1.06	Hot water fittings, sterilizable, medical, and surgical equipment

popularity of poly(vinyl chloride) lies in its unmatched ability to form a stable, dry, flexible material when plasticized. The polyamides and terylene are also important thermoplastics.

1.12 Thermosetting polymers

The thermoset plastics generally have superior abrasion and dimensional stability characteristics compared with the thermoplastics which have better flexural and impact properties. In contrast to the thermoplastics, thermosetting polymers, as the name implies, are changed irreversibly from fusible, soluble products into highly intractable crosslinked resins which cannot be moulded by flow and so must be fabricated during the crosslinking process. Typical examples are:

Phenolic resins, prepared by reacting phenols with aldehydes. They are used for electrical fitments, radio and television cabinets, heat resistant knobs for cooking utensils, game parts, buckles, handles, and a wide variety of similar items.

Amino resins, which are related polymers formed from formaldehyde and either urea or melamine. In addition to many of the uses listed above, they can be used to manufacture lightweight tableware and counter and table surfaces. Being transparent they can be filled and coloured using light pastel shades, whereas the phenolics are already rather dark and consequently have a more restricted colour range.

Thermosetting *polyester resins* are used in paints and surface coatings where oxidation during drying forms a crosslinked film which provides a tough resistant finish.

Epoxy resins are polyethers prepared from glycols and dihalides and find extensive use as surface coatings, adhesives, and flexible enamel-like finishes because of their combined properties of toughness, chemical resistance, and flexibility.

1.13 Elastomers

The modern elastomer industry was founded on the naturally occurring product isolated from the latex of the tree *Hevea brasiliensis*. It was first used by South American Indians and was called caoutchouc from the Indian name, but later, simply rubber, when it was discovered, by Príestley, that the material rubbed out pencil marks.

From the early 20th century, chemists have been attempting to synthesize materials whose properties duplicate or at least simulate those of natural rubber, and this has led to the production of a wide variety of synthetic elastomers. Some of these have become technologically important and are listed in table 1.5 together with their general uses.

Although a large number of synthetic elastomers are now available, natural rubber must still be regarded as the standard elastomer because of the excellently balanced combination of desirable qualities. Presently it accounts for almost 36 per cent of the total world consumption of elastomers and its gradual replacement by synthetic varieties is partly a result of demand outstripping natural supply.

The most important synthetic elastomer is styrene-butadiene (SBR) which accounts for 41 per cent of the world market in elastomers. It is used predominantly for vehicle tyres when reinforced with carbon black. Nitrile rubber (NBR) is a random copolymer of acrylonitrile (mass fraction 0.2 to 0.4) and butadiene and it is used when an elastomer is required which is resistant to swelling in organic solvents. The range of properties can

TABLE 1.5. Some common elastomers and their uses

Polymer	*Formula*	*Uses*
Natural rubber (polyisoprene-*cis*)	$\text{-(}CH_2\text{—}C(CH_3)\text{=}CH\text{—}CH_2\text{)}_n\text{-}$	General purposes
Polybutadiene	$\text{-(}CH_2\text{—}CH\text{=}CH\text{—}CH_2\text{)}_n\text{-}$	Tyre treads
Butyl	$\text{-(}CH_2\text{—}C(CH_3)_2\text{)}_n\text{-}$	Inner tubes, cable sheathing, roofing, tank liners
SBR	$\text{-(}CH_2\text{—}CH\text{=}CH\text{—}CH_2\text{—}CH_2\text{—}CH(C_6H_5)\text{)}_n\text{-}$	Tyres, general purposes
ABS	$\text{-(}CH_2\text{—}CH(CN)\text{—}CH_2\text{—}CH\text{—}C_6H_5)_n$, with $\text{-(}CH\text{—}CH\text{=}CH\text{—}CH_2\text{)}_m\text{-}$ on the second CH	Oil hoses, gaskets, flexible fuel tanks
Polychloroprene	$\text{-(}CH_2\text{—}C(Cl)\text{=}CH\text{—}CH_2\text{)}_n\text{-}$	Used when oil resistance, good weathering, and inflammability characteristics are needed
Silicones	$\text{-(}O\text{—}SiR_2\text{)}_n\text{-}$	Gaskets, door seals, medical application flexible moulds
Polyurethanes	$\text{-(}R_1\text{—}NHCOOR_2OOCHN\text{)}_n\text{-}$	Printing rollers, sealing and jointing
EPR	$\sim\text{(-(}CH_2\text{—}CH_2\text{)}_m\text{(}CH_2\text{—}CH(CH_3)\text{)}_p\text{—)}_n\sim$	Window strips and channelling

be extended when styrene is also incorporated in the chain, forming ABS rubber. Butyl rubber (IIR) is prepared by copolymerizing small quantities of isoprene (3 parts) with isobutylene (97 parts). The elastic properties are poor but it is resistant to corrosive fluids and has a low permeability to gases. Polychloroprene possesses the desirable qualities of being a fire retardant and resistance to weathering, chemicals, and oils. More recently, ABA triblock copolymers of styrene-(ethene-*stat*-butene)-styrene have become commercially available. These are thermoplastic elastomers which are unaffected by polar solvents or non-oxidizing acids or alkalis.

Elastomers which fail to crystallize on stretching must be strengthened by the addition of filters such as carbon black. SBR, poly(ethylene-*stat*-propylene) and the

TABLE 1.6. Room temperature curing adhesives and sealants

Type	*General description*	*Particular advantages*	*Limitations (general operating temp. range)*
Moisture curing polyurethanes (PUs)	Sealants (rather than adhesives) which cure by an isocyanate reaction to atmospheric H_2O	Adhesive/sealing effect to a wide range of substrates	Slow curing Usually low modulus (– 80°C to + 120°C)
RTV (room temp. vulcanising) silicones	Sealants (rather than adhesives) which cure on exposure to atmospheric moisture by a condensation mechanism that results in release of side products such as acetic acid, alcohols or amines	Excellent thermal, oxidative and hydrolytic stability	Unpleasant side products Limited adhesion (– 80°C to + 200°C)
Anaerobic adhesives and sealants	Fluids which cure in the absence air and the presence of metals, heat of UV light by the free radical mechanism	Very good adhesion to metals and ceramics Resistant to organic solvents	Relatively brittle when cured Curing is sensitive to substrate and to joint geometry (– 50°C to + 150°C)
Cyanoacrylate adhesives	Relatively low viscosity which cure anionically in response to substrate-borne atmospheric moisture	Excellent adhesion to a wide range of substrates Very effective on rubber and on most plastics	Brittle when cured Limited thermal and hydrolytic stability (– 50°C to + 80°C)
Acrylics	Methacrylic adhesives which cure free radically in response to substrates treated with a primer/hardener. The adhesives usually contain rubber toughener	Forms durable adhesive joints to metallic substrates High peel strenght	Inhibited by atmospheric oxygen Limit cure-through-gap (– 50°C to + 100°C)

silicone elastomers fall into this category. While polyethylene is normally highly crystalline, copolymerization with propylene destroys the ordered structure and if carried out in the presence of a small quantity of non-conjugated diene (*e.g.* dicyclopentadiene) a crosslinking site is introduced. The material is an amorphous

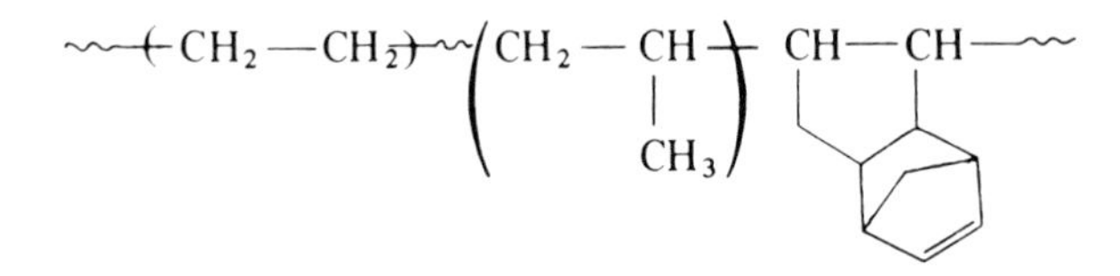

random terpolymer which when crosslinked forms an elastomer with a high resistance to oxidation. Unfortunately it is incompatible with other elastomers and is unsuitable for blending.

The silicone elastomers have a low cohesive energy between the chains which results in poor thermoplastic properties and an unimpressive mechanical response. Consequently they are used predominantly in situations requiring temperature stability over a range of 190 to 570 K when conditions are unsuitable for other elastomers.

Extensive use has been made of room temperature vulcanizing silicone rubbers. These are based on linear poly(dimethyl siloxane) chains, with M ranging from 10^4 to $10^5\ \mathrm{g\,mol^{-1}}$, and hydroxyl terminal groups. Curing can be achieved in a number of ways, either by adding a crosslinking agent and a metallic salt catalyst, such as tri- or tetra-alkoxysilane with stannous octoate, by exposure to light or by incorporating in the mixture a crosslinking agent sensitive to atmospheric water which initiates vulcanization. The products are good sealing, encapsulating, and caulking materials; they make good flexible moulds and are excellent insulators. They have found a wide application it the building, aviation, and electronics industries. Other room temperature curing adhesives and sealants are listed in table 1.6.

Having briefly introduced the diversity of structure and property encountered in the synthetic polymers, we can now examine more closely the fundamental chemistry and physics of these materials.

General Reading

T. Alfrey and E. F. Gurnee, *Organic Polymers*. Prentice-Hall (1967).

G. Allen and J. C. Bevington, Eds, *Comprehensive Polymer Science*, Vols 1–7. Pergamon Press (1989).

F. W. Billmeyer, *Textbook of Polymer Science*. John Wiley and Sons. 3rd Ed. 1984.

L. W. Chubb, *Plastics, Rubbers and Fibres*. Pan (1967).

E. W. Duck, *Plastics and Rubbers*. Butterworths (1971).

H. G. Elias, *Macromolecules*, Vols 1 and 2. Plenum Press (1984).

D. A. Hounshell and J. K. Smith, "The Nylon Drama," *American Heritage of Invention and Technology*, **4**, 40 (1988).

E. M. McCafferey, *Laboratory Preparation for Macromolecular Chemistry*. McGraw-Hill (1970).

F. M. McMillan, *The Chain Straighteners*, Macmillan Press (1979).

H. Morawetz, *Polymers: The Origins and Growth of a Science*. John Wiley (1985).

L. R. G. Treloar, *Introduction to Polymer Science*. Wykeham Publications (1970).

References

1. H. Zandvoort, *Studies in the History and Philosophy of Science*, **19 (4)**, 489 (1988).
2. "Nomenclature of Regular Single-Strand Organic Polymers," *Pure and Applied Chemistry*, **48 (3)**, 373 (1976).
3. R. A. Pethrick, Ed., *Polymer Yearbook 4*. Harwood Academic Publishers (1987). (Nomenclature rules).

CHAPTER 2

Step-growth Polymerization

The classical subdivision of polymers into two main groups was made around 1929 by W. H. Carothers, who proposed that a distinction be made between polymers prepared by the stepwise reaction of monomers and those formed by chain reactions. These he called:

(1) *Condensation polymers*, characteristically formed by reactions involving the elimination of a small molecule, such as water, at each step; and
(2) *Addition polymers*, where no such loss occurred.

While these definitions were perfectly adequate at the time, it soon became obvious that notable exceptions existed and that a fundamentally sounder classification should be based on a description of the chain growth mechanism. It is preferable to replace the term condensation with step-growth or step-reaction. Reclassification as *step-growth* polymerization, now logically includes polymers such as polyurethanes, which grow by a step reaction mechanism without elimination of a small molecule.

In this chapter we shall examine the main features of step-growth polymerization, beginning with the simpler reactions which produce linear chains exclusively. This type of polymerization is used to produce some of the industrially important fibres such as the nylons and terylene. A brief discussion of the more complex branching reactions follows to illustrate how the thermosetting plastics are formed.

2.1 General reactions

In any reaction resulting in the formation of a chain or network of high molar mass, the functionality (see section 1.3) of the monomer is of prime importance. In step-growth polymerization, a linear chain of monomer residues is obtained by the stepwise intermolecular condensation or addition of the reactive groups in bifunctional monomers. These reactions are analogous to simple reactions involving monofunctional units as typified by a polyesterification reaction involving a diol and a diacid.

$$\mathrm{HO{-}R{-}OH + HOOC{-}R'{-}COOH \rightleftharpoons HO{-}R{-}OCO{-}R'{-}COOH + H_2O}$$

If the water is removed as it is formed, no equilibrium is established and the first stage in the reaction is the formation of a dimer which is also bifunctional. As the reaction proceeds, longer chains, trimers, tetramers, and so on, will form through other esterification reactions, all essentially identical in rate and mechanism, until ultimately the reaction contains a mixture of polymer chains of large molar masses M. However, the formation of samples with significantly large values of M is subject to a number of rather stringent conditions which will be examined in greater detail later in this chapter.

Two major groups, both distinguished by the type of monomer involved, can be identified in step-growth polymerization. In the first group two polyfunctional monomers take part in the reaction, and each possesses only one distinct type of functional group as in the previous esterification reaction, or more generally:

$$\mathrm{A-A+B-B \rightarrow \!\!+\!\!\left(A-AB-B \right)\!\!+}$$

The second group is encountered when the monomer contains more than one type of functional group such as a hydroxyacid (HO—R—COOH), represented generally as A – B where the reaction is

$$n\mathrm{A-B} \rightarrow \left(\mathrm{AB} \right)_n$$

or

$$n(\mathrm{HO{-}R{-}COOH}) \rightarrow \mathrm{H}\left(\mathrm{ORCO} \right)_n \mathrm{OH}.$$

A large number of step growth polymers have the basic structure

—□—R—□—R—□—R—

where R can be $\left(\mathrm{CH_2} \right)_x$ or —⟨benzene ring⟩— and the link —□— is one of three important groups:

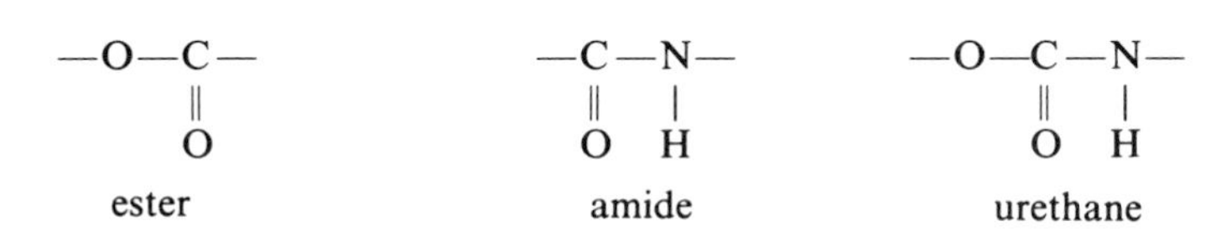

Other links and groups are involved in these reactions and some typical step-reaction polymers are shown in table 2.1.

2.2 Reactivity of functional groups

One basic simplifying assumption proposed by Flory, when analysing the kinetics of step-growth systems, was that all functional groups can be considered as being equally reactive. This implies that a monomer will react with both monomer or polymer species with equal ease.

The progress of the reaction can be illustrated in figure 2.1 where after 25 per cent reaction the number average chain length x_n is still less than two because monomers, being the most predominant species, will tend to react most often, and the reaction is mainly the formation of dimers and trimers. Even after 87.5 per cent of the reaction x_n will only be about eight and it becomes increasingly obvious that if long chains are required, the reaction must be pushed towards completion.

TABLE 2.1. Typical step-growth polymerization reactions

Polymer	*Reaction*
Polyester	$n\mathrm{HO(CH_2)_xCOOH} \rightarrow \mathrm{HO}\!\left[(\mathrm{CH_2})_x\!-\!\underset{\underset{\mathrm{O}}{\Vert}}{\mathrm{C}}\!-\!\mathrm{O}\right]_n\mathrm{H} + (n-1)\mathrm{H_2O}$
Polyamide	$n\mathrm{NH_2{-}R{-}COOH} \rightarrow \mathrm{H}[\mathrm{NH{-}R{-}CO}]_n\mathrm{OH} + (n-1)\mathrm{H_2O}$ $n\mathrm{NH_2{-}R{-}NH_2} + n\mathrm{HOOC{-}R'{-}COOH} \rightarrow \mathrm{H}[\mathrm{NH{-}R{-}NHCO{-}R'{-}CO}]\mathrm{OH} + (2n-1)\mathrm{H_2O}$
Polyurethanes	$n\mathrm{HO{-}R{-}OH} + \mathrm{OCN{-}R'{-}NCO} \rightarrow [\mathrm{OROCONH{-}R'{-}NHCO}]_n$
Polyanhydride	$n\mathrm{HOOC{-}R{-}COOH} \rightarrow \mathrm{HO}[\mathrm{OC{-}R{-}CO{\cdot}O}]_n\mathrm{H} + (n-1)\mathrm{H_2O}$
Polysiloxane	$n\mathrm{HO}\!-\!\underset{\mathrm{CH_3}}{\overset{\mathrm{CH_3}}{\mathrm{Si}}}\!-\!\mathrm{OH} \rightarrow \mathrm{HO}\!\left[\underset{\mathrm{CH_3}}{\overset{\mathrm{CH_3}}{\mathrm{Si}}}\!-\!\mathrm{O}\right]_n\!\mathrm{H} + (n-1)\mathrm{H_2O}$
Phenol-formaldehyde	n (phenol, OH on benzene ring) $+\ n\mathrm{CH_2O} \rightarrow$ (OH-substituted benzene ring)$\mathrm{-CH_2-}$[(OH-substituted benzene ring)$\mathrm{-CH_2}$]$_{n-1}$ $+\ (n-1)\mathrm{H_2O}$

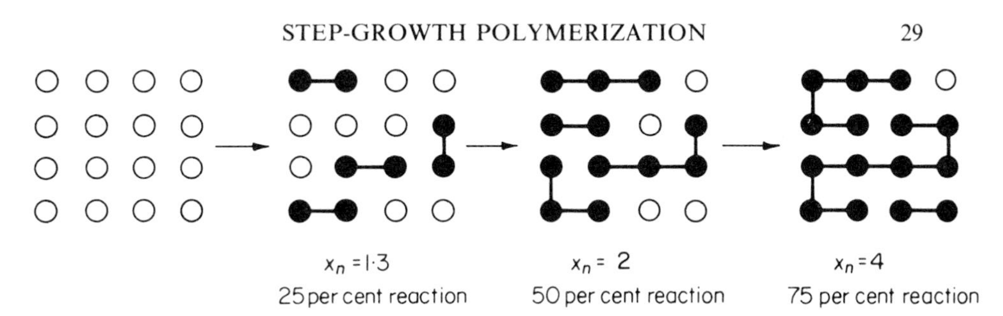

FIGURE 2.1. Diagrammatic representation of a step-growth polymerization.

2.3 Carothers equation

W. H. Carothers, the pioneer of step-growth reactions, proposed a simple equation relating x_n to a quantity p describing the extent of the reaction for linear polycondensations or polyadditions.

If N_0 is the original number of molecules present in an A–B monomer system and N the number of all molecules remaining after time t, then the total number of functional groups of either A or B which have reacted is $(N_0 - N)$. At that time t the extent of reaction p is given by

$$p = (N_0 - N)/N_0 \quad \text{or} \quad N = N_0(1 - p). \tag{2.1}$$

If we remember that $x_n = N_0/N$, a combination of expressions gives the *Carothers equation*,

$$x_n = 1/(1 - p). \tag{2.2}$$

This equation is also valid for an A–A + B–B reaction when one considers that in this case there are initially $2N_0$ molecules.

The Carothers equation is particularly enlightening when we examine the numerical relation between x_n and p; thus for $p = 0.95$ (*i.e.* 95 per cent conversion), $x_n = 50$ and when $p = 0.99$, then $x_n = 100$. In practical terms, it has been found that for a fibre forming polymer such as nylon-6.6 $\text{+NH(CH}_2)_6\text{NHCO(CH}_2)_4\text{CO+}_n$ the value of M_n has to be about 12 000 to 13 000 g mol^{-1} if a high tenacity fibre is to be spun, and as this corresponds to $x_n = 106$ to 116, the polymerization has to proceed beyond 99 per cent completion. Similarly for polyesters derived from ω-hydroxydecanoic acid, $\text{H+O(CH}_2)_9\text{CO+}_n\text{OH}$, x_n of about 150 is optimum for good fibres and so p must exceed 0.99.

Note that when using the Carothers equation for an A–A, B–B system, half the average molar mass of the [A–AB–B] repeat unit is used to calculate the degree of polymerization. This is because equation (2.2) gives a measure of the average number of monomers in a polymer and, as two different types are being used in systems such as the nylon 6,6 example, the average molar mass of these is $(114 + 112)/2 = 113$.

2.4 Control of the molar mass

Quite obviously the control of the molar mass of the product of these reactions is very important. Very high molar mass material may be too difficult to process, while low molar mass polymer may not exhibit the properties desired in the end product, and one must be able to stop the reaction at the required value of p. Consequently the reactions are particularly demanding with respect to the purity of the reagents and accurate control of the amount of each species in the mixture is cardinal. It is symptomatic of these critical requirements that only four types of reaction usefully produce linear polymers with $M_n > 25\,000\,\mathrm{g\,mol^{-1}}$.

(1) *Schotten–Baumann reaction.* This involves the use of an acid chloride in an esterfication or amidation; for example the so-called "nylon rope trick" reaction is an interfacial condensation between sebacoyl chloride and hexamethylenediamine, producing a polyamide known as nylon-6,10.

$$n\mathrm{ClCO(CH_2)_8COCl} + n\mathrm{H_2N(CH_2)_6NH_2} \rightarrow$$
$$\mathrm{-\!\!\left[CO(CH_2)_8CONH(CH_2)_6NH\right]\!\!-_{\mathit{n}}} + (2n-1)\mathrm{HCl}$$

The bifunctional acyl chloride is dissolved in CCl_4 and placed in a beaker. An aqueous alkaline solution of the bifunctional amine is layered on top and the nylon-6,10 which forms immediately at the interface can be drawn off as a continuous filament until the reagents are exhausted. The reaction has an S_N2 mechanism with the release of a proton to the base in the aqueous phase; the condensation is a successive stepwise reaction by this S_N2 route.

(2) *Salt dehydration.* Direct esterification requires high purity materials in equimolar amounts because esterfications rarely go beyond 98 per cent completion in practice. To overcome this, hexamethylene diamine and a dibasic acid such as adipic acid can be reacted to produce a nylon salt, hexamethylene diammonium adipate. A solution of 0.5 mol diamine in a mixture of 95 per cent ethanol (160 cm^3) and distilled water (60 cm^3) is added to 0.5 mol diacid dissolved in 600 cm^3 of 95 per cent ethanol over a period of 15 min. The mixture is stirred for 30 min during which time the nylon salt precipitates as a white crystalline solid. This can be recrystallized and should melt at 456 K. The pure salt can be converted into a polyamide by heating it under vacuum in a sealed tube, protected by wire gauze, at about 540 K in the presence of a small quantity of the diacid, *e.g.* 10 g salt to 0.55 g adipic acid is a suitable mixture. If a lower molar mass is desired, a monofunctional acid can replace the adipic acid and act as a chain terminator.

(3) *Ester interchange.* An alternative reaction is a trans-esterification in the presence of a proton donating or weak base catalyst such as sodium methoxide, *e.g.*

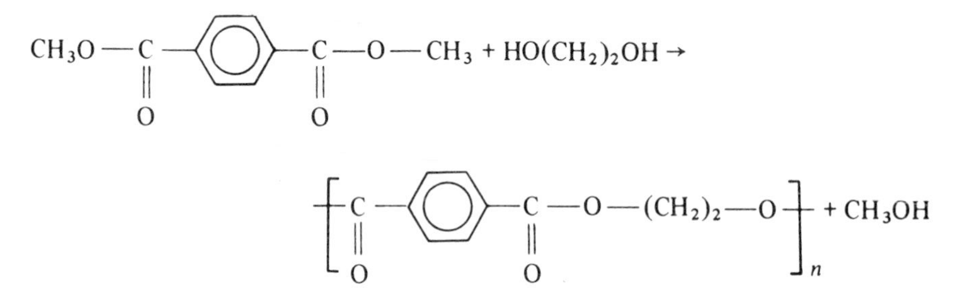

In this way ethylene glycol and dimethyl terephthalate produce terylene. Ester interchange is the most practical approach to polyester formation because of the faster

reaction rate and use of more easily purified products. The formation of poly(ethylene terephthalate) is essentially a two-stage process. The first stage, at 380 to 470 K, is the formation of dimers and trimers, each with two hydroxyl end-groups, and during this formation the methanol is being distilled off. To complete the reaction the temperature is raised to 530 K and condensation of these oligomers produces a polymer with large M_n. The major advantage is that the stoichiometry is self-adjusting in the second stage.
(4) *Step polyaddition* (*Urethane formation*). Polyurethanes with large M_n can be prepared using a method based on the Wurtz alcohol test. In the presence of a basic catalyst, such as a diamine, ionic addition takes place, for example between 1,4-butanediol and 1,6-hexanediisocyanate:

$$HO(CH_2)_4OH + OCN(CH_2)_6NCO \rightarrow \left[O(CH_2)_4{-}O{-}\underset{\underset{O}{\|}}{C}{-}\underset{\underset{H}{|}}{N}{-}(CH_2)_6NHCO \right]_n$$

This reaction produces a highly crystalline polymer, but a vast number of urethanes can be formed by varying the reactants and a series of polymers covering a wide spectrum of properties can be prepared.

2.5 Stoichiometric control of M_n

Chain growth, to produce high molar mass material, becomes increasingly difficult as the reaction proceeds. This is due to a number of reasons: (i) the difficulty in ensuring the precise equivalence of the reactive groups in the starting materials when two or more types of monomer are used, (ii) the decreasing frequency of functional groups meeting and reacting as their concentration diminishes, and (iii) the increasing likelihood of interference from side reactions.

Often it is preferable to avoid production of high molar mass polymer and control can be effected by rapidly cooling the reaction at the appropriate stage or by adding calculated quantities of monofunctional materials as in the preparation of nylon-6,6 from the salt.

More usefully, a precisely controlled stoichiometric imbalance of the reactants in the mixture can provide the desired result. For example, an excess of diamine over an acid chloride would eventually produce a polyamide with two amine end groups incapable of further growth when the acid chloride was totally consumed. This can be expressed in an extension of the Carothers equation as,

$$x_n = (1 + r)/(1 + r - 2rp), \tag{2.3}$$

where r is the ratio of the number of molecules of the reactants. Thus for a quantitative reaction ($p = 0.999$) between N molecules of phenolphthalein and $1.05N$ molecules of terephthaloyl chloride to form poly(terephthaloyl phenolphthalein) (see scheme on next page).

The value of $r = N_{AA}/N_{BB} = 1/1.05 = 0.952$

$$\text{and } x_n = (1 + 0.952)/(1 + 0.952 - 2 \times 0.999 \times 0.952) \approx 39,$$

rather than 1000 for $r = 1$. This reaction is an interfacial polycondensation whose progress may be followed by noting the colour change from the red phenolphthalein

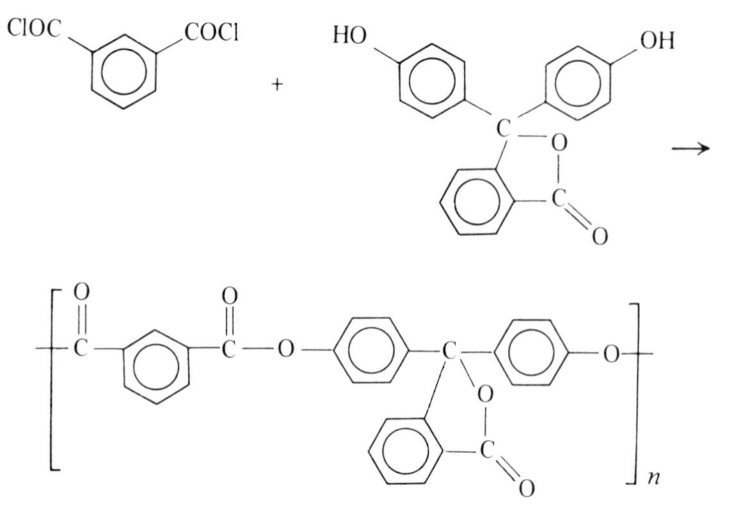

solution to the colourless polyester. The interface can remain stationary in the experiment but the uniformity of the polymer is improved by increasing the reaction surface using high-speed stirring.

However, the purity of the starting monomers is also crucial and if one of them contains only 95 per cent of the expected difunctional material, then $r = 0.95$ and the attainable x_n will again be ≈ 40.

In practice $p = 1$ is rarely achieved, nor is it a perfect stoichiometric balance. The consequences of this are shown in figure 2.2 where x_n is plotted as (a) a function of p, calculated from equation (2.2), and (b) as a function of the stoichiometric ratio for $p = 1.00$ and 0.998, calculated from equation (2.3).

The corresponding equation for a monofunctional additive is similar to equation (2.3) only now r is defined as the ratio $N_{AA}/(N_{BB} + 2N_B)$ where N_B is the number of monofunctional molecules added.

2.6 Kinetics

The assumption that functional group reactivity is independent of chain length can be verified kinetically by following a polyesterification. The simple esterification is an acid catalysed process where protonation of the acid is followed by interaction with the alcohol to produce an ester and water. If significant polymer formation is to be achieved, the water must be removed continuously from the reaction to displace the equilibrium and the water eliminated can be used to estimate the extent of the reaction. Alternatively the rate of disappearance of carboxylic groups can be measured by titrating aliquots of the mixture.

A typical apparatus is shown in figure 2.3 consisting of a reaction kettle, a Dean and Stark trap to collect the water, a stirrer, N_2 inlet, and thermometer. The reaction can be illustrated using ethylene glycol and adipic acid. A mixture of decalin (35 cm^3) and adipic acid (1 mol) is placed in the kettle and the trap is filled with decalin. The mixture is heated to 420 K and glycol (1 mol), preheated to this temperature, is added, followed by an acid catalyst such as p-toluene sulphonic acid (1 mmol). Nitrogen is bubbled

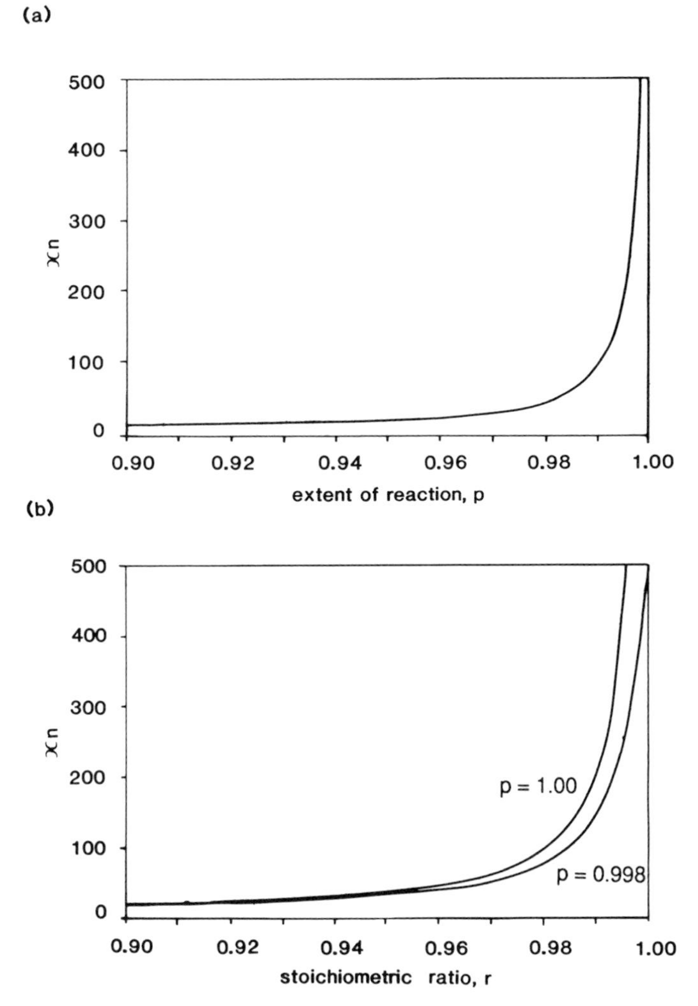

FIGURE 2.2. Variation of the degree of polymerization with (a) extent of reaction p, calculated from equation (2.2), and (b) the stoichiometric ratio r for $p = 1.00$, $x_n = \frac{1+r}{1-r}$, and 0.998, $x_n = \frac{1+r}{1+r-2rp}$, calculated from equation (2.3).

rapidly through the mixture which is quickly raised to reflux temperature. The water level in the trap is noted at regular intervals and small aliquots of the reaction mixture can be withdrawn, weighed, diluted with acetone, and titrated with methanolic KOH. If an activation energy is required, the temperature can be raised several times and the reaction followed for a time at each new temperature.

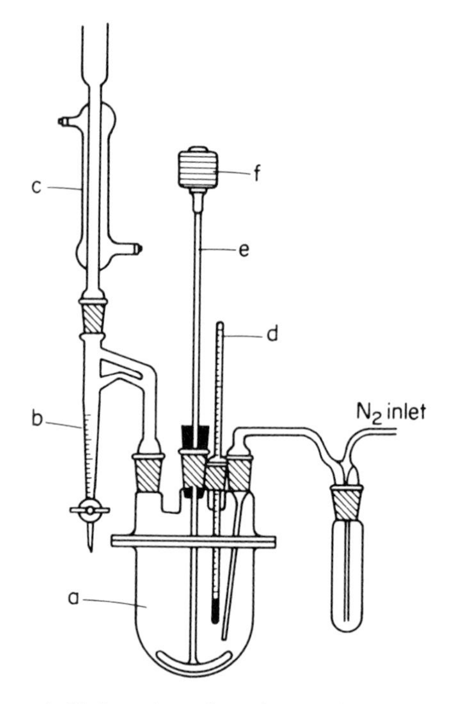

FIGURE 2.3. Apparatus suitable for polycondensation reactions: a, the reaction kettle; b, Dean and Stark trap; c, condenser; d, thermometer; e, stirrer; f, stirring motor.

Self-catalysed reaction. If no acid catalyst is added, the reaction will still proceed because the acid can act as its own catalyst. The rate of condensation at any time t can then be derived from the rate of disappearance of —COOH groups and

$$-\mathrm{d}[\mathrm{COOH}]/\mathrm{d}t = k[\mathrm{COOH}]^2[\mathrm{OH}]. \tag{2.4}$$

The second order [COOH] term arises from its use as a catalyst and k is the rate constant. For a system with equivalent quantities of acid and glycol the functional group concentration can be written simply as c and

$$-\mathrm{d}c/\mathrm{d}t = kc^3 \tag{2.5}$$

This expression can be integrated under the conditions that $c = c_0$ at time $t = 0$ and

$$2kt = 1/c^2 - 1/c_0^2. \tag{2.6}$$

The water formed is removed and can be neglected. From the Carothers equation it follows that $c = c_0(1 - p)$, leading to the final form

$$2c_0^2kt = 1/(1-p)^2 - 1. \tag{2.7}$$

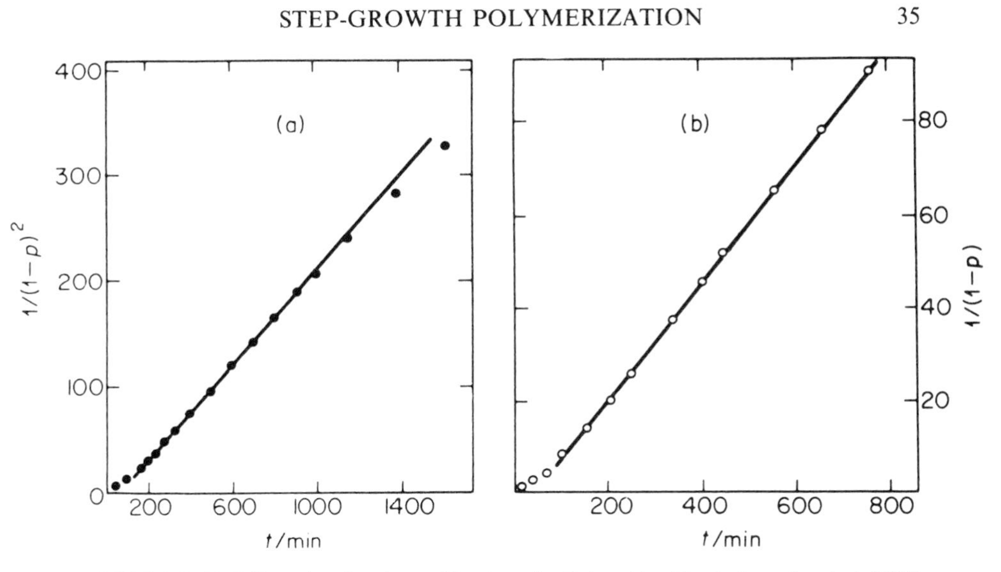

FIGURE 2.4. (a) Self-catalysed polyesterification of adipic acid with ethylene glycol at 439 K; (b) polyesterification of adipic acid with ethylene glycol at 382 K, catalysed by 0.4 mole per cent of *p*-toluene sulphonic acid. (From data by Flory.)

Acid-catalysed reaction. The uncatalysed reaction is rather slow and a high x_n is not readily attained. In the presence of a catalyst there is an acceleration of the rate and the kinetic expression is altered to

$$-\mathrm{d}[\mathrm{COOH}]/\mathrm{d}t = k'[\mathrm{COOH}][\mathrm{OH}], \tag{2.8}$$

which is kinetically first order in each functional group. The new rate constant k' is then a composite of the rate constant k and the catalyst concentration which also remains constant. Hence

$$-\mathrm{d}c/\mathrm{d}t = k'c^2, \tag{2.9}$$

and integration gives finally

$$c_0 k' t = 1/(1-p) - 1. \tag{2.10}$$

Both equations have been verified experimentally by Flory as shown in figure 2.4.

2.7 Molar mass distribution in linear systems

The creation of long chain polymers by the covalent linking of small molecules is a random process, leading to chains of widely varying lengths. Because of the random nature of the process, the distribution of chain lengths in a sample can be arrived at by simple statistical arguments.

The problem is to calculate the probability of finding a chain composed of x basic structural units in the reaction mixture at time t, for either of the two reactions

$$xA{-}A + xB{-}B \rightarrow A{-}A[B{-}BA{-}A]_{x-1}B{-}B,$$

$$\text{or} \quad xA{-}B \rightarrow A\{BA\}_{x-1}B,$$

i.e. to calculate the probability that a functional group A or B has reacted. For the sake of clarity we can consider one of the functional groups to be carboxyl and determine the probability that $(x-1)$ carboxyl groups have reacted to form a chain. This is p^{x-1}, where p is the extent of the reaction, defined in equation (2.1).

It follows that if a carboxyl group remains unreacted the probability of finding this uncondensed group is $(1-p)$ and so the probability P_x of finding one chain x units long (*i.e.* an x-mer) is

$$P_x = (1-p)p^{x-1}. \tag{2.11}$$

As the fraction of x-mers in any system equals the probability of finding one, the total

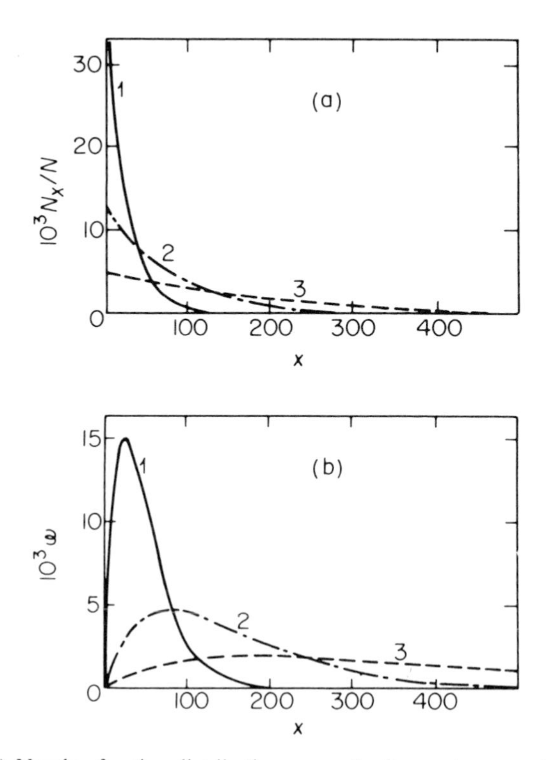

FIGURE 2.5. (a) Number fraction distribution curves for linear step growth polymerizations. Curve 1, $p = 0.9600$; Curve 2, $p = 0.9875$; Curve 3, $p = 0.9950$. (b) Corresponding weight fraction distribution for the same system.

number N_x present is given by

$$N_x = N(1 - p)p^{x-1}, \tag{2.12}$$

where N is the total number of polymer molecules present in the reaction. Substitution of the Carothers equation (2.2) gives

$$N_x = N_0(1 - p)^2 p^{x-1}, \tag{2.13}$$

where N_0 is the total number of monomer units present initially. The variation of N_x for various values of p and x is shown in figure 2.5. A slightly different set of curves is obtained if the composition is expressed in terms of mass fraction w, in this case $w_x = xN_x/N_0$ to give

$$w_x = x(1 - p)^2 p^{x-1}, \tag{2.14}$$

Both reveal that very high conversions are necessary if chains of significant size are to be obtained and that while monomer is normally the most numerous species, the proportion of low molar mass material decreases as p exceeds 0.95.

2.8 Average molar masses

Number and weight average molar masses can be calculated from the equations if M_0 is taken as the molar mass of the repeat unit. Thus

$$M_n = N_x\Sigma(M_0 N_x)/N = M_0/(1 - p), \tag{2.15}$$

and

$$M_w = M_0(1 + p)/(1 - p). \tag{2.16}$$

It can be seen that a heterogeneity index (M_w/M_n) for the most probable distribution when $p = 1$ is

$$(M_w/M_n) = \{M_0(1 + p)/(1 - p)\}\{(1 - p)/M_0\} = 2. \tag{2.17}$$

2.9 Characteristics of step-growth polymerization

It might be appropriate at this stage to summarize the main features of the step-reactions.

(1) Any two molecular species in the mixture can react.

(2) The monomer is almost all incorporated in a chain molecule in the early stages of the reaction, *i.e.* about 1 per cent of monomer remains unreacted when $x_n = 10$. Hence polymer yield is independent of the reaction time in the later stages.

(3) Initiation, propagation, and termination reactions are essentially idential in rate and mechanism.

(4) The chain length increases steadily as the reaction proceeds.

(5) Long reaction times and high conversions are necessary for the production of a polymer with large x_n.

(6) Reaction rates are slow at ambient temperatures, but increase with a rise in temperature although this has little effect on the chain length of the final product.

(7) Activation energies are moderately high and reactions are not excessively exothermic.

2.10 Typical step-growth reactions

Reactions are normally carried out in bulk in the temperature range 420 to 520 K to encourage fast reactions and promote the removal of the low molar mass condensation product. Activation energies are about 80 kJ mol^{-1}.

Low temperature polycondensations. The advantages of high temperature reactions are partly counteracted by the increasing danger of side-reactions, and room temperature reactions using highly energetic reactants provide routes to a variety of polymers. The use of the Schotten–Baumann reaction for polyamides has already been outlined. This is an example of an unstirred interfacial reaction in which the diamine is soluble in both phases and diffuses across the interface into the organic layer where polymerization takes place. Continuous polymer production is achieved by withdrawing the film formed at the interface to allow continued diffusion of the reactants. Alternatively, the continuity of the polymerization reaction can be maintained by stirring the system vigorously; this ensures a constantly changing interface and increases the surface area available for the reaction. As both methods are diffusion controlled the need for stringent stoichiometric control is obviated.

When the diamine used is aromatic only low molar mass polymer is formed because of the lower reaction rates. To produce longer chains, conditions must be readjusted so that both phases are polar and miscible and vigorous high-speed stirring of the system is necessary. This is used in the reaction between isophthaloyl chloride and *m*-phenylene diamine. These aromatic polyamides are particularly versatile materials and of considerable interest.

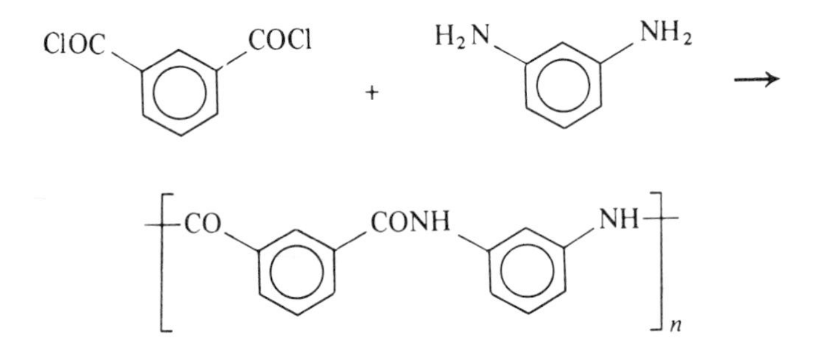

The ultimate extension to a homogeneous system with inert polar solvents is used in the synthesis of polyimides. Poly(methylene 4,4′-diphenylene pyromellitamide acid) is prepared by mixing equal amounts of pyromellitic dianhydride and bis(4-aminophenyl methane) in N,N′-dimethylformamide. The reaction is maintained at 288 K, with stirring, for an hour and the polymer is isolated by precipitation in vigorously stirred water.

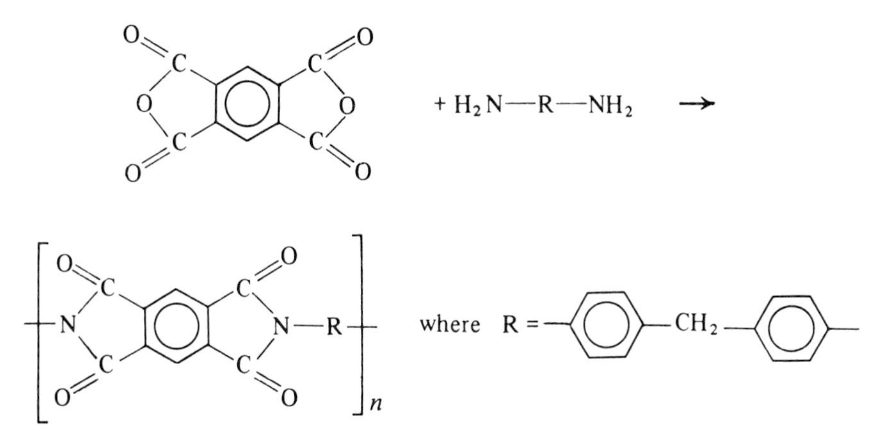

Polysulphonamides, polyanhydrides, and polyurethanes can also be formed in homogeneous low temperature reactions.

2.11 Ring formation

The assumption so far has been that all bifunctional monomers in step-growth reactions form linear polymers. This is not always true and competitive side reactions such as cyclization may occur, as with certain hydroxyacids which may form lactones, or lactams if an amino acid is used.

$$HO—R—COOH \rightarrow R\langle\begin{matrix} CO \\ | \\ O \end{matrix} + H_2O$$

lactone

To gauge the importance of such reactions consideration must be given to the thermodynamic and kinetic aspects of ring formation. A study of ring strain in cycloalkanes has shown that 3 and 4 membered rings are severely strained but that this decreases dramatically for 5, 6, or 7 membered rings, then increases again up to 11 before decreasing for very large rings. In addition to the thermodynamic stability, the kinetic feasibility of two suitable functional groups being in juxtaposition to react must also be considered. This probability decreases with increasing ring size and again 5, 6, or 7 membered rings are favoured and will form in preference to a linear chain when possible.

2.12 Non-linear step-growth reactions

In systems containing bifunctional monomers a high degree of polymerization is attained only when the reaction is forced almost to completion. The introduction of a trifunctional monomer into the reaction produces a rather startling change which is best illustrated using a modified form of the Carothers equation. A more general functionality factor f_{av} is introduced, defined as the average number of functional groups present per monomer unit. For a system containing N_0 molecules initially and equivalent numbers of two function groups A and B, the total number of functional groups is $N_0 f_{av}$. The number of groups that have reacted in time t to produce N molecules is then $2(N_0 - N)$ and

$$p = 2(N_0 - N)/N_0 f_{av}. \tag{2.18}$$

The expression for x_n then becomes

$$x_n = 2/(2 - p f_{av}), \tag{2.19}$$

but this is only valid when equal numbers of both functional groups are present in the system.

For a completely bifunctional system such as an equimolar mixture of phthalic acid and ethylene glycol, $f_{av} = 2$, and $x_n = 20$ for $p = 0.95$. If, however, a trifunctional alcohol, glycerol, is added so that the mixture is composed of 2 mol diacid, 1.4 mol diol, and 0.4 mol of glycerol f_{av} increases to

$$f_{av} = (2 \times 2 + 1.4 \times 2 + 0.4 \times 3)/3.8 = 2.1.$$

The value of x_n is now 200 after 95 per cent conversion but only a small increase to 95.23 per cent is required for x_n to approach infinity – a most dramatic increase. This is a direct result of incorporating a trifunctional unit in a linear chain where the unreacted hydroxyl provides an additional site for chain propagation. This leads to the formation of a highly branched structure and the greater the number of multi-functional units the faster the growth into ar. insoluble three-dimensional network. When this happens, the system is said to have reached its *gel point*.

If the stoichiometry of the system is unbalanced the definition of the average functionality must be modified and becomes

$$f' = 2r f_A f_B f_C / \{ f_A f_C + r\rho f_A f_B + r(1 - \rho) f_B f_C \}, \tag{2.20}$$

when monomers A and C have the same functional group but different functionalities f_A and f_C, and f_B is the functionality of monomer B. Also

$$r = (n_A f_A + n_C f_C)/n_B f_B \leqslant 1, \tag{2.21}$$

$$\rho = n_C f_C/(n_A f_A + n_C f_C), \tag{2.22}$$

and n_A, n_B and n_C are the amounts of each component. A commonly encountered system has monomers with $f_A = f_B = 2$ and $f_C > 2$. In this more general system the onset of gelation can be predicted by establishing the critical conversion limit beyond which a gel is sure to form. This is derived on the basis of the observation that x_n approaches infinity at or beyond the gel point. In the stoichiometric case the equation

$$p = (2/f_{av}) - (2/x_n f_{av}),$$

becomes at the gel point, when x_n tends to infinity,

$$p_G = 2/f_{av}, \tag{2.23}$$

whereas the critical extent of reaction p_G for the general case is

$$p_G = (1 - \rho)/2 + 1/2r + \rho/f_C. \tag{2.24}$$

2.13 Statistical derivation

An expression for p_G can be derived from statistical arguments. First a branching coefficient ζ is introduced, defined as the probability that a multifunctional monomer ($f > 2$) is connected either by a linear chain segment or directly to a second multifunctional monomer (or branch point) rather than to a chain segment terminating in a single functional group.

The critical value ζ_G necessary for incipient gelation can be calculated from the probability that at least one of the $(f-1)$ chain segments extending from a branch unit will be connected to another branch unit, and as this is simply $(f-1)^{-1}$ we have

$$\zeta_G = 1/(f-1), \tag{2.25}$$

where f is now the functionality of the branching unit and not an average as defined previously. If more than one monomer in the system is multifunctional, then an average of all monomers with $f \geqslant 3$ is used.

It follows that if a chain ends in a branch point, $\zeta(f-1)$ chains on the average will emanate from that point, and as ζ of these will also end in a branching unit a further $[\zeta(f-1)]^2$ are generated and so on. This occurs when $\zeta(f-1)$ is greater than 1 and the growth of the network is limited only by the boundaries of the system. For $\zeta(f-1) < 1$ no gelation is observed.

The arguments can be extended and for a system A—A + B—B + A(A)A (a trifunctional A unit), the critical extent of reaction of the A group at the gel point is

$$p_G = [r + r\rho(f-2)]^{-1/2}. \tag{2.26}$$

2.14 Comparison with experiment

The gel point in a branching system is usually detected by a rapid increase in viscosity η, indicated by the inability of bubbles to rise in the medium. It is also characterized by a rapid increase in x. The values of these quantities for the reaction of diethylene glycol + succinic acid + 1,2,3,-propanetricarboxylic acid are shown in figure 2.6. The increase in x_n is not as dramatic as x_w which can be identified with the η curve. The divergence of x_w and x_n is illustrated in table 2.2 for the above system when the reaction mixture has been adjusted hypothetically to provide a ratio of carboxyl to amine groups of 1. This situation can be achieved by assuming that the mixture contains 98.5 mol of diacid A, 100 mol of diol B, and 1 mol of triacid C. From equation (2.21) it follows that $r = 98.5 \times 2 + (1 \times 3)/(2 \times 100) = 1$. These reaction conditions lead to the appearance of a gel

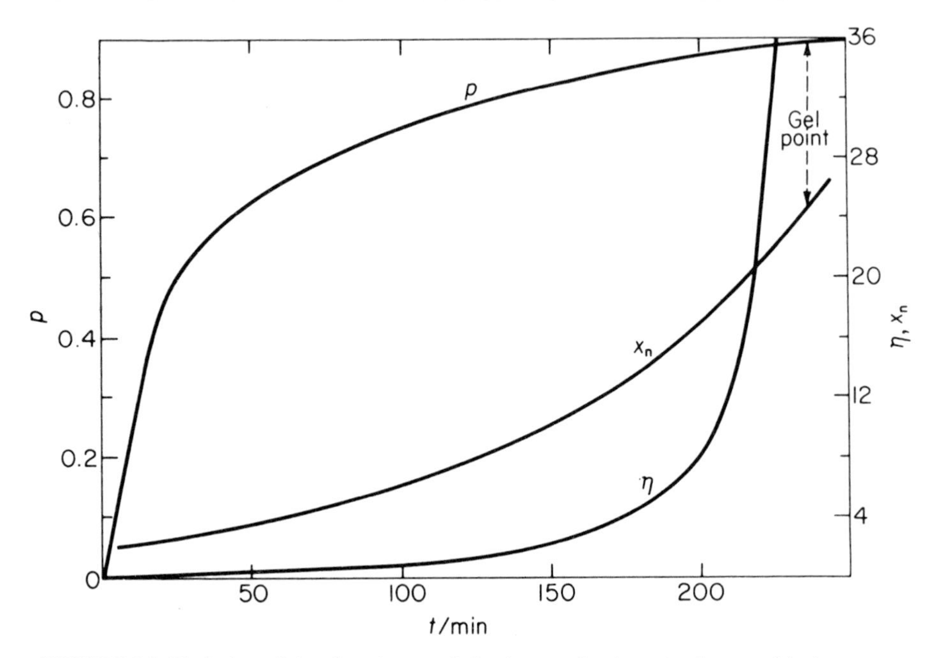

FIGURE 2.6. Variation of the viscosity η and the degree of polymerization x_n with the extent of network polymer formation in the system diethylene glycol + succinic acid + 1,2,3-propanetricarboxylic acid. (From data reported by Flory.)

TABLE 2.2. Branching system with $p_G = 0.9925$

p	x_w	x_w/x_n
0.100	1.2	1.1
0.500	3.0	1.5
0.700	5.8	1.7
0.900	20.4	2.0
0.950	45.6	2.2
0.980	153.8	2.7
0.990	747.2	5.6
0.992	3306.8	18.2

TABLE 2.3. 1,2,3-propanetricarboxylic acid, diethylene glycol, and adipic or succinic acid system

		p_G		
r	ρ	expt.	eqn (2.24)	eqn (2.26)
0.800	0.375	0.991	1.063	0.955
1.000	0.293	0.911	0.951	0.879
1.002	0.404	0.894	0.933	0.843

point at $p_G = 0.9925$ and the ratio (x_w/x_n) increases sharply as the reaction approaches the critical point. The distribution is readily compared with that for a totally bifunctional system for which $(x_w/x_n) = 1 + p$, and the broadening of the molar mass distribution, characteristic of these branching polymerizations is well illustrated.

A comparison of experimental and theoretical values of p_G has been reported by Flory for a mixture of a tricarboxylic acid with a diol and a diacid; the results are given in table 2.3.

The statistical equation under estimates p_G whereas equation (2.24) over estimates the experimental value. The Carothers equation leads to a high p_G because molecules larger than the observed x_n exist in the mixture and these undergo gelation before the predicted value is attained. This difficulty is overcome in the statistical treatment but now differences are attributable to intra-molecular cyclization in the system and the loops which are formed are non-productive in a branching sense. This means that the reaction must proceed further to overcome waste.

2.15 Polyurethanes

An important and versatile family of polymers, whose diverse uses include foams, fibres, elastomers, adhesives and coatings, is formed by the interaction of diisocyanates with diols.

$$\mathrm{OCN{-}R{-}NCO + HO{-}R^1{-}OH \longrightarrow \!\!\left(\!O{-}R^1{-}O{-}\underset{\underset{O}{\|}}{C}{-}\underset{\underset{H}{|}}{N}{-}R{-}\underset{\underset{H}{|}}{N}{-}\underset{\underset{O}{\|}}{C}\!\right)_{\!n}}$$

This leads to the production of linear polymers although branched or crosslinked structures can also be prepared by using multifunctional starting materials or, inadvertently, through side reactions. The nature and stiffness of the groups R and R^1 will control the type of material formed and its rigidity or flexibility. This, in turn, will control the properties, and ultimately the use of the polyurethanes formed.

The basic chain growth reaction is addition without elimination of a small molecule and is a consequence of the highly electrophilic nature of the carbon atom in the isocyanate group.

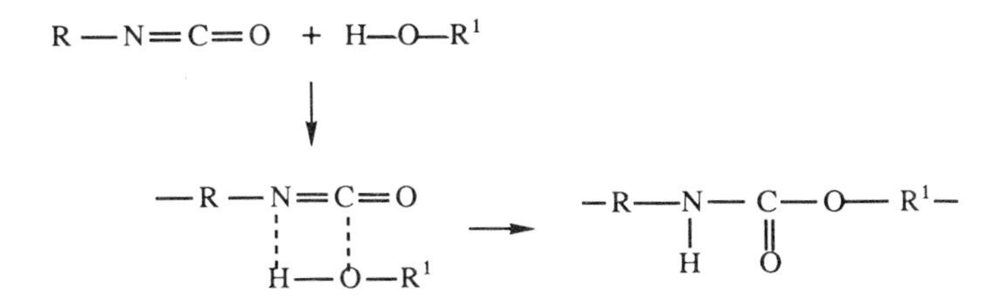

This makes it susceptible to attack by nucleophilic reagents such as alcohols, acids, water, amines and mercaptans. The electrophilicity of the carbon atom can be increased if R is an aromatic ring which can conjugate with the isocyanate group. Thus, aromatic diisocyanates are more reactive than aliphatic ones.

Isocyanate groups may not all have equal reactivity, for example in 1-methyl-2,4-diisocyanatobenzene, the isocyanate group nearest the methyl group has a different reactivity from the other; also when one isocyanate group is reacted, the susceptibility of the second to a reaction may be altered.

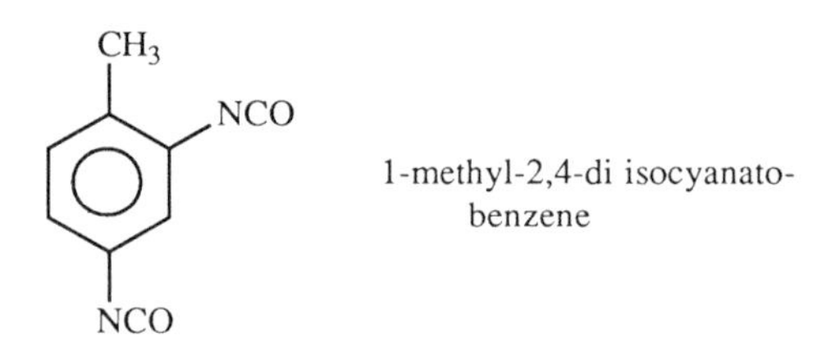

Although it is expected that difunctional monomers will give a linear polyurethane, the polymerization reaction is subject to possible side reactions. The formation of allophanate groups can occur, particularly if reaction temperatures exceed 400 K. Here, an isocyanate group adds onto the secondary amine in the urethane link and a branched or crosslinked structure is formed.

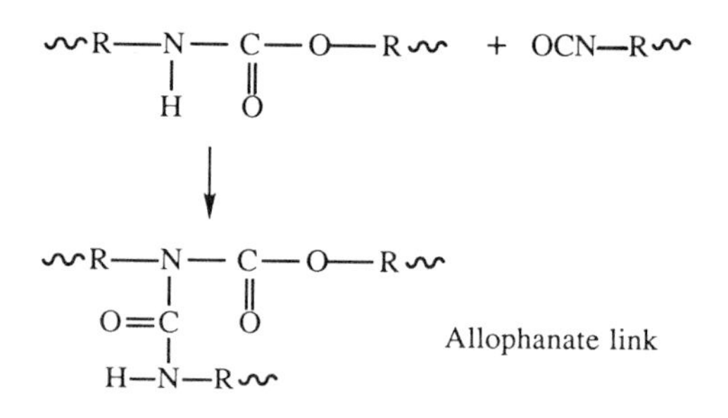

Polyurethanes are often prepared in two stages: firstly the production of a prepolymer and secondly a chain extension reaction. During the first step diisocyanate is reacted with a dihydroxy-terminated short chain polyether or polyester, such as poly(ethylene adipate), poly(ε-caprolactone), or poly(tetramethylene glycol) of approximate molar mass 1000–3000. Throughout this stage the diisocyanate is used in excess to give isocyanate end-capped blocks. These are then chain-extended by subsequent reaction with a short chain diol (*e.g.* ethylene glycol or 1,4-butane diol) or a diamine (see table 2.4). The reaction with the diamine gives a urea linkage and a poly(urethane-*co*-urea) structure which can also react with other (NCO) groups leading to biuret formation.

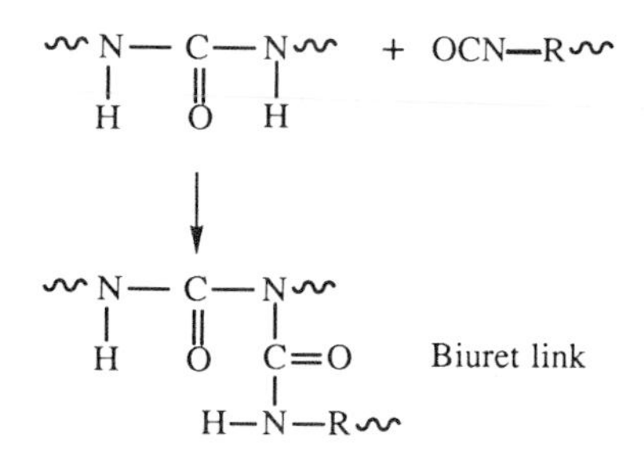

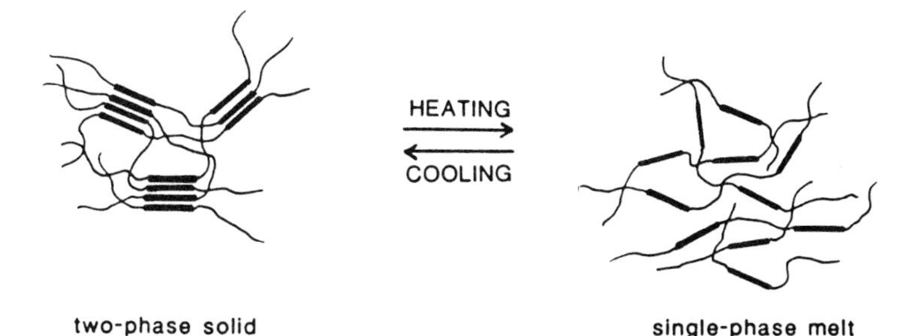

FIGURE 2.7. Schematic representation of a polyurethane, with hard blocks shown as the bold lines and soft blocks as the thin lines, phase separating in the solid state and undergoing disordering on heating. (Adapted from Pearson (1987).)

Crosslinked systems can also be obtained by the use of multifunctional monomers. In the polyurethanes such as those illustrated in table 2.4 and figure 2.7, the isocyanate monomer unit is regarded as the "hard" segment and the polyol forms the "soft" segment. The structures shown are particularly useful in biomedical research and have been demonstrated to be suitable for cardiovascular applications.

A major use of polyurethanes is in the manufacture of foams, both rigid and flexible. The chain extension reactions described above are normally used when elastomeric-type products are required, but for foams, the chain extender molecule can be omitted and polyols with an average functionality in excess of three are used. The reaction is base-catalysed by tertiary amines or organo tin compounds (*e.g.* stannous octanoate) and must also include a "blowing" agent in the reaction mixture. This can be achieved by adding controlled quantities of water to the system and making use of the reaction:

$$\sim RNCO + H_2O \longrightarrow \sim R{-}\underset{\displaystyle H}{\underset{|}{N}}{-}\underset{\displaystyle O}{\underset{\|}{C}}{-}OH \longrightarrow \sim RNH_2 + CO_2$$

where water reacts with an isocyanate group to form the unstable carbamic acid which decomposes to produce an amine with evolution of CO_2. The released gas forms spherical bubbles which increase in size, and eventually impinge on one another to form a polyhedral cell structure in the polymer matrix. A volatile liquid can be used as an alternative "blowing" agent, and Freon ($CFCl_3$) with a boiling point of 294 K can be included in the reaction mixture. As the polymerization is exothermic, liberating about 80 kJ mol^{-1}, the heat is sufficient to vapourize the $CFCl_3$ and create the foam. Use of these chlorofluoro carbon compounds is now discouraged on environmental grounds as they are believed to attack and destroy the ozone layer.

If a flexible foam is required, then longer, more flexible polyols and trifunctional crosslinking monomers are used, whereas higher crosslink densities and short chain polyols tend to form more rigid foams.

In the absence of a blowing agent and with conditions favouring the formation of linear chains, thermoplastic polyurethane elastomers can be formed. Materials with different properties can be obtained by altering the ratio of hard to soft blocks and it is also observed that, in many cases, the hard segments can crystallize. This can give rise

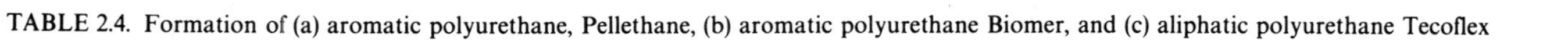

TABLE 2.4. Formation of (a) aromatic polyurethane, Pellethane, (b) aromatic polyurethane Biomer, and (c) aliphatic polyurethane Tecoflex

(a)

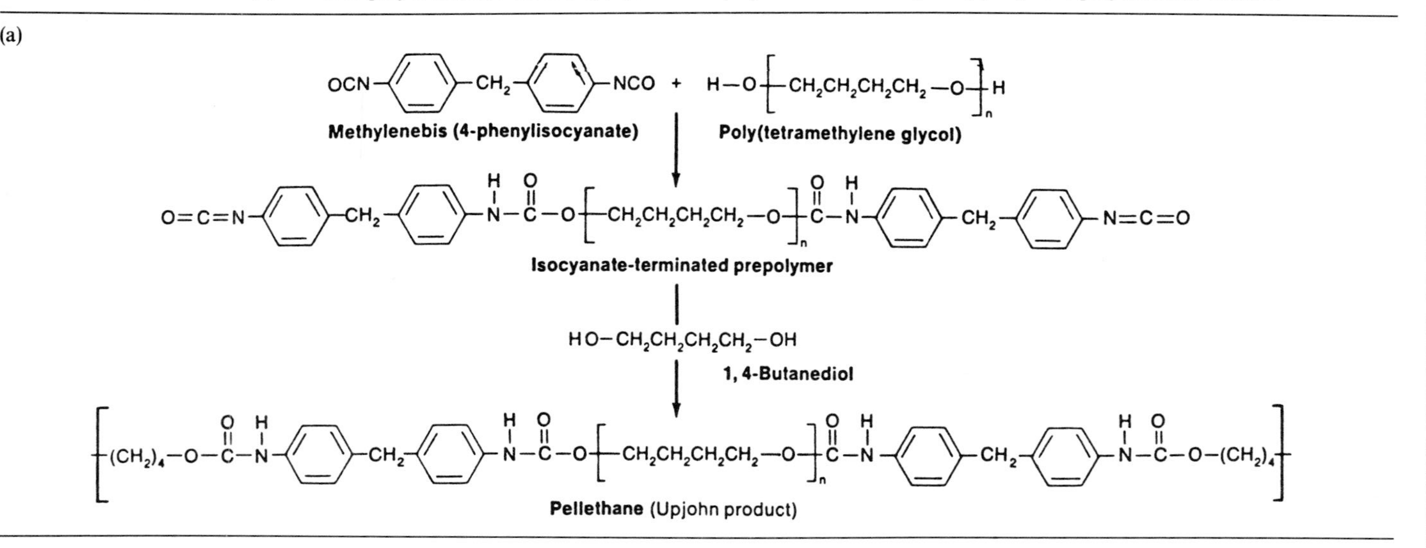

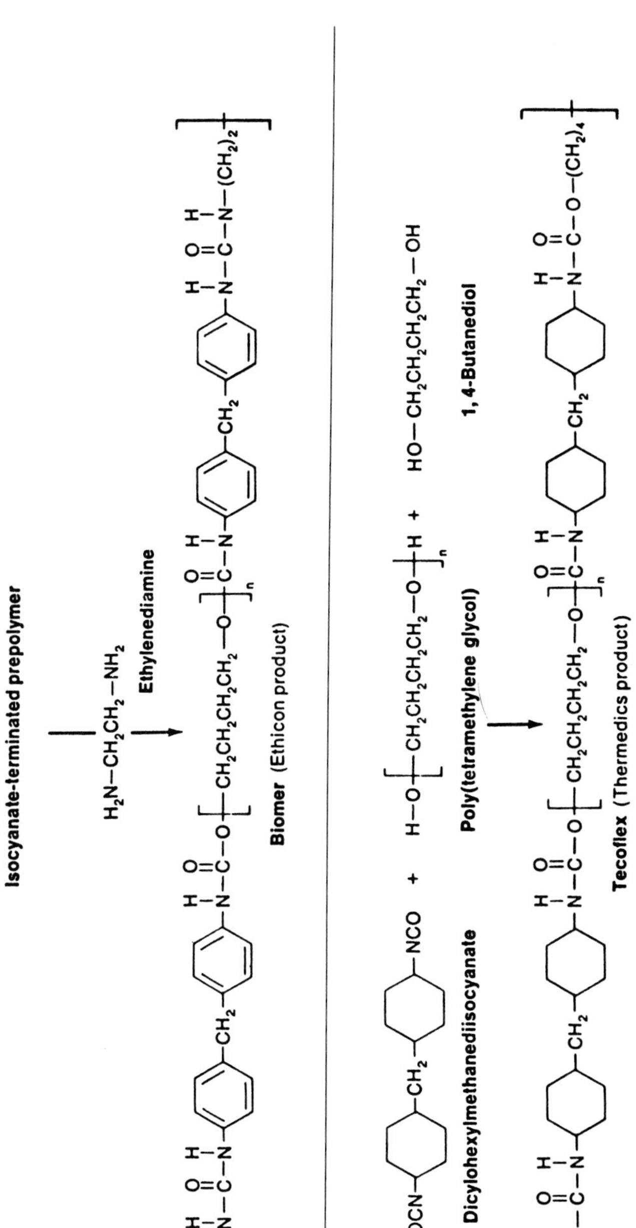
(b)
Isocyanate-terminated prepolymer
$H_2N-CH_2CH_2-NH_2$
Ethylenediamine
Biomer (Ethicon product)
(c)
Dicylohexylmethanediisocyanate
Poly(tetramethylene glycol)
$HO-CH_2CH_2CH_2CH_2-OH$
1, 4-Butanediol
Tecoflex (Thermedics product)

to phase-separated structures, shown schematically in figure 2.7, which behave like thermoplastic elastomers (see also section 15.6).

2.16 Thermosetting polymers

The production of highly branched network polymers is commercially important, but as crosslinking results in a tough and highly intractable material, the fabrication process is usually carried out in two stages.

The first stage is the production of an incompletely reacted prepolymer; this is either solid or liquid and of moderately low molar mass. The second stage involves conversion of this into the final crosslinked product *in situ* (*i.e.* a mould or form of some description). The prepolymers are either random or structoset and are discussed separately.

Phenol-formaldehyde. Random prepolymers are prepared by reacting phenol ($f = 3$ for the *ortho* and *para* positions in the ring) with bifunctional formaldehyde. The base catalysed reaction produces a mixture of methylol phenols.

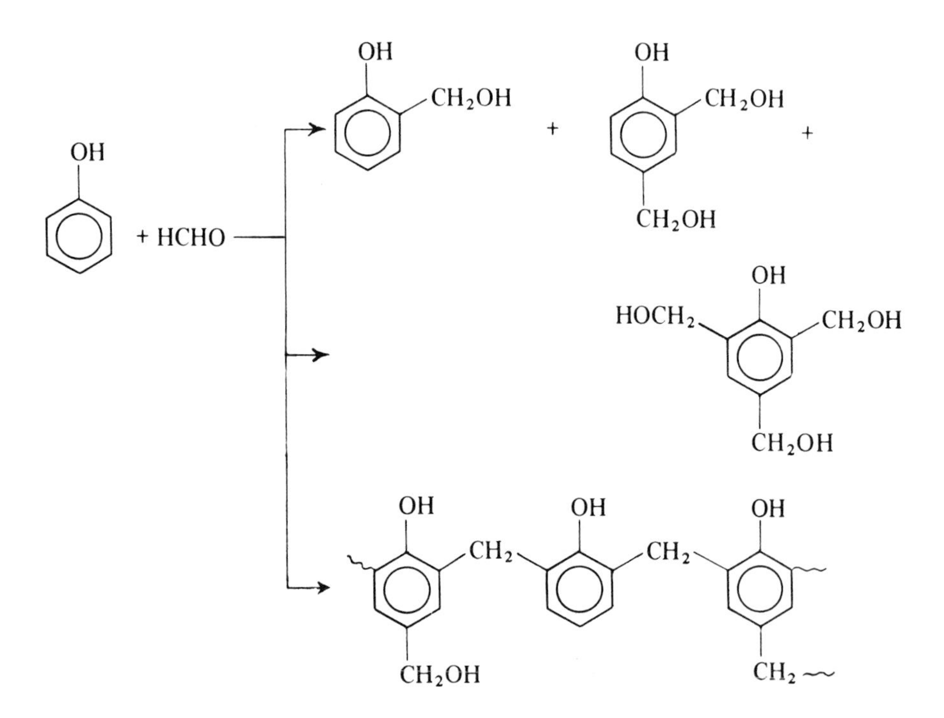

The composition of the mixture can be varied by altering the phenol to formaldehyde ratio. At this stage the methylol intermediates are dried and ground and in some cases a filler such as mica, glass fibre, or sawdust may be added. A crosslinking agent such as hexamine is mixed with the prepolymer together with CaO as a catalyst. When further heating takes place during moulding, the hexamine decomposes to form HCHO and ammonia which acts as a catalyst in the final crosslinking process by the HCHO.

Amino resins. A related family of polymers is made from random prepolymers prepared by reacting either urea or melamine (shown below) with HCHO. The products are known as aminoplasts.

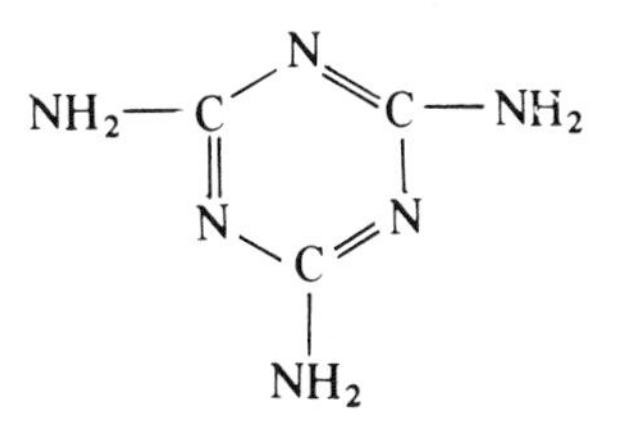

Epoxides. Structoset prepolymers are designed to have controlled and defined structures with functional groups located either at the chain ends – *structoterminal* – or located along the chain – *structopendant*. Epoxy resins make use of the epoxy group for the former, and hydroxyl groups for the latter. One of the major resin prepolymers formed by the reaction of bisphenol A and epichlorohydrin can be treated as either.

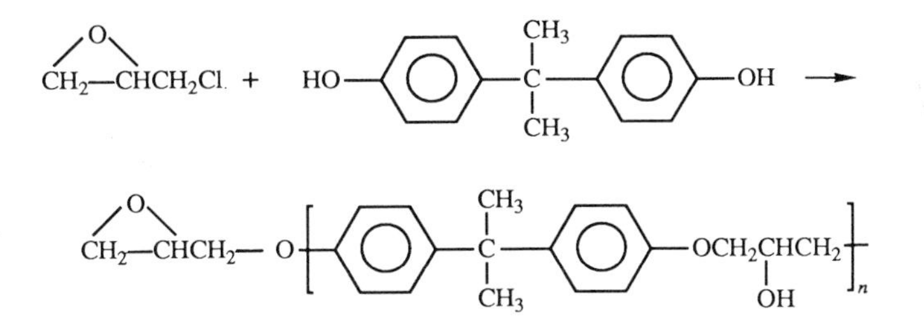

These prepolymers can subsequently react with a crosslinking (or curing) agent to turn them into strong intractable network structures. The most commonly used group of low cost, room temperature, crosslinking agents is the aliphatic amines, *e.g.* diethylene triamine ($H_2NCH_2CH_2NHCH_2CH_2NH_2$) and triethylene tetramine ($H_2N[CH_2CH_2NH]_2CH_2CH_2NH_2$). These react with the epoxy groups.

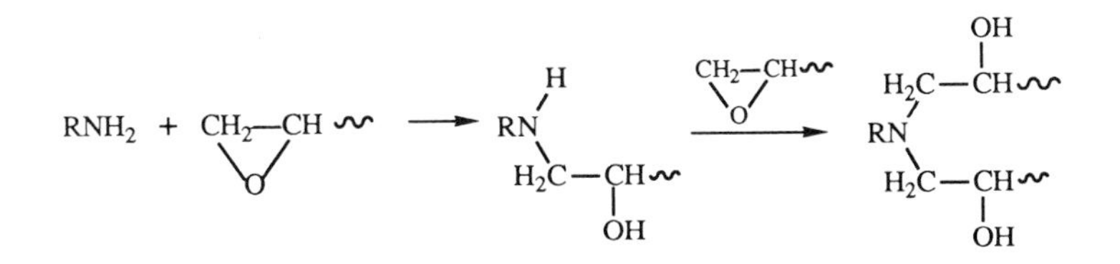

Anhydrides are also used as curing agents, and have the advantage that they do not irritate the skin as do some of the amines. The reaction usually proceeds best in the presence of a catalyst such as a Lewis base (or acid).

One proposed mechanism is that of a two step reaction with ring opening at stage one and addition of the epoxy at stage two. After this the anion reacts with another molecule of anhydride.

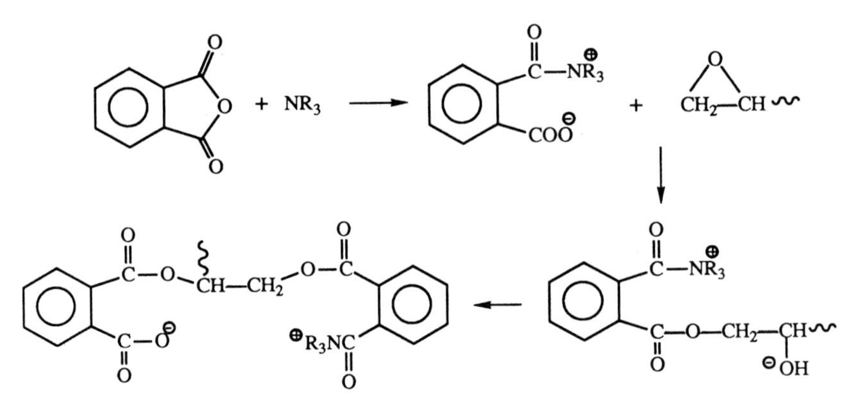

In the uncatalysed reaction, the anhydride ring is opened by reaction with a hydroxyl group but the reaction is much slower. The hydroxyl group also reacts with the epoxy group.

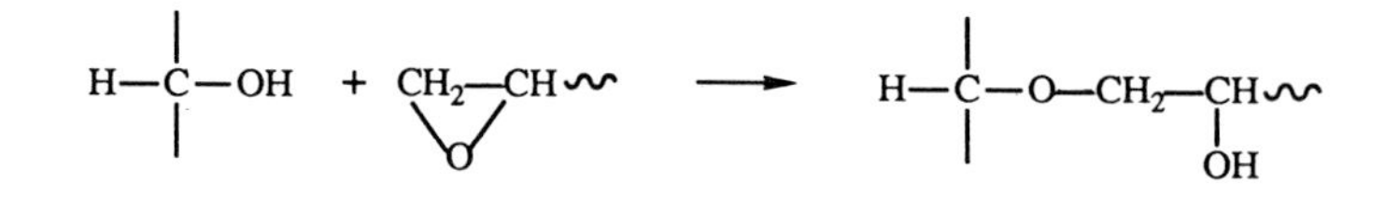

Other structures can be generated and an important group are the epoxy novolac resins which find use as moulding compounds.

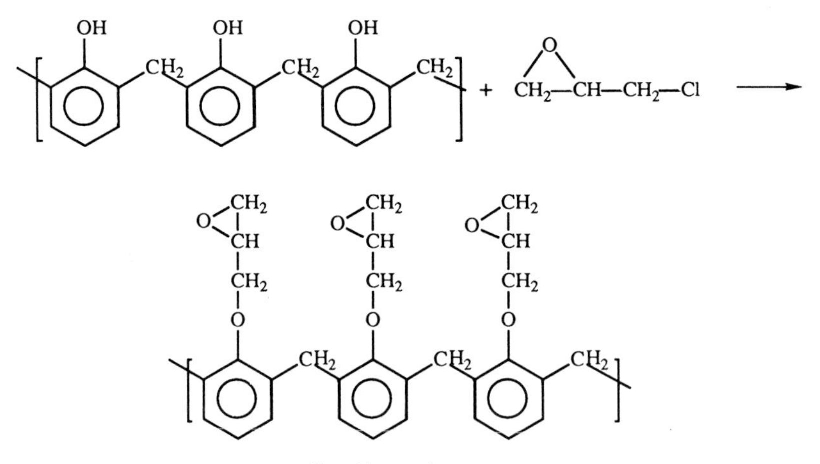

Epoxide novolac

The high crosslink densities that can be achieved in these systems make the resulting materials mechanically stable and resistant to high temperatures.

Flame retardant resins can be made using tetrabromobisphenol A as part of the prepolymer. When degraded at high temperatures, this compound releases halogen compounds which trap the free radicals formed, thereby helping to quench the flame.

Epoxy resins are used in tool making, as adhesives, insulators, tough surface coatings and, importantly, in composites in combination with reinforcing fibres. They have also found extensive use in the encapsulation of electronic devices.

General Reading

G. Allen and J. C. Bevington, Eds, *Comprehensive Polymer Science*, Vol. 5. Pergamon Press (1989).

P. J. Flory, *Principles of Polymer Chemistry*, Chapter 3. Cornell University Press, Ithaca, N.Y. (1953).

A. H. Frazer, *High Temperature Resistant Polymers.* Interscience (1968).

R. W. Lenz, *Organic Chemistry of Synthetic High Polymers*, Chapter 3. Interscience Publishers Inc. (1967).

H. F. Mark and G. S. Whitby, *The Collected Papers of Wallace Hume Carothers.* Interscience Publishers Inc. (1940).

P. W. Morgan, *Condensation Polymers by Interface and Solution Methods.* Interscience Publishers Inc. (1965).

P. O. Nielson, "Properties of epoxy resins, hardeners and modifiers." *Adhesives Age*, Vol. 42 (1982).

G. Odian, *Principles of Polymerization*, Chapter 2. John Wiley and Sons (1981).

G. Oertel, Ed., *Polyurethane Handbook.* Hanser Publishers, Munich (1985).

R. G. Pearson, in *Speciality Polymers.* Ed. R. W. Dyson. Blackie (1987).

W. G. Potter *Epoxide Resins.* Butterworth and Co. Ltd (1970).

P. Rempp and E. W. Merrill, *Polymer Synthesis.* Hüthig and Wepf, Basel (1986).

D. H. Solomon, *The Chemistry of Organic Film Formers.* John Wiley and Sons, Inc. (1967).

References

1. P. J. Flory, (a) *J. Amer. Chem. Soc.*, **61**, 3334 (1939);
   (b) **62**, 2261 (1940);
   (c) **63**, 3083 (1941).

CHAPTER 3

Free Radical Addition Polymerization

3.1 Addition polymerization

In step-growth polymerization reactions it is often necessary to use multifunctional monomers if polymers with high molar masses are to be formed; this is not the case when addition reactions are employed. Long chains are readily obtained from monomers such as vinylidene compounds with the general structure $CH_2{=}CR_1R_2$. These are bi-functional units, where the special reactivity of π-bonds in the carbon to carbon double bond makes them susceptible to rearrangement if activated by free radical or ionic initiators. The active centre created by this reaction then propagates a kinetic chain which leads to the formation of a single macromolecule whose growth is stopped when the active centre is neutralized by a termination reaction. The complete polymerization proceeds in three distinct stages: (i) *Initiation*, when the active centre which acts as a chain carrier is created; (ii) *Propagation*, involving growth of the macromolecular chain by a kinetic chain mechanism and characterized by a long sequence of identical events, namely the repeated addition of a monomer to the growing chain; (iii) *Termination*, whereby the kinetic chain is brought to a halt by the neutralization or transfer of the active centre. Typically the polymer formed has the same chemical composition as the monomer, *i.e.* each unit in the chain is a complete monomer and not a residue as in most step-growth reactions.

3.2 Choice of initiators

A variety of chain initiators is available to the polymer chemist. These fall into three general categories: free radical; cationic; and anionic. The choice of the most appropriate one depends largely on the groups R_1 and R_2 in the monomer and their effect on the double bond. This arises from the ability of the alkene π-bond to react in a different way with each initiator species to produce either heterolytic (I) or homolytic

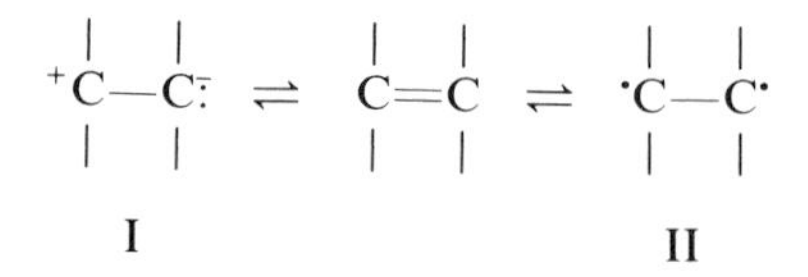

(II) fission. In most olefinic monomers of interest the group R_1 is either H or CH_3 and for simplicity we can consider it to be H. The group R_2 is then classifiable as an electron

TABLE 3.1. Effect of substituent on choice of initiator

Monomer	*Initiator*		
	Free radical	*Anionic*	*Cationic*
Ethylene, $CH_2{=}CH_2$	+	−	+
1,1′-Dialkylolefin, $CH_2{=}CR_1R_2$	−	−	+
Vinyl ethers, $CH_2{=}CHOR$	−	−	+
Vinyl halides, $CH_2{=}CH(Hal)$	+	−	−
Vinyl esters, $CH_2{=}CHOCOR$	+	−	−
Methacrylic esters, $CH_2{=}C(CH_3)COOR$	+	+	−
Acrylonitrile, $CH_2{=}CHCN$	+	+	−
Styrene, $CH_2{=}CHPh$	+	+	+
1,3-Butadiene $CH_2{=}CH{-}CH{=}CH_2$	+	+	+

withdrawing group

$$CH_2{=}\overset{\delta+}{CH}\rightarrow\overset{\delta-}{R_2}$$

or an electron donating group

$$\overset{\delta-}{CH_2}{=}CH\leftarrow\overset{\delta+}{R_2}$$

Both alter the negativity of the π-bond electron cloud and thereby determine whether or not a radical, an anion, or a cation will be stabilized preferentially.

In general, electron withdrawing substituents, $-CN$, $-COOR$, $-CONH_2$, reduce the electron density at the double bond and favour propagation by an anionic species. Groups which tend to increase the double-bond nucleophilicity by donating electrons, such as alkenyl, alkoxyl, and phenyl, encourage attack by cationic initiators and in addition the active centres formed are resonance stabilized. Alkyl groups do not stimulate cationic initiation unless in the form of 1,1′-dialkyl monomers or alkyl dienes and then heterogeneous catalysts are necessary. As resonance stabilization of the active centre is an important factor, monomers like styrene and 1,3-butadiene can undergo polymerization by both ionic methods because the anionic species can also be stabilized.

Because of its electrical neutrality, the free radical is a less selective and more generally useful initiator because most substituents can provide some resonance stabilization for this propagating species. Some examples are shown in table 3.1.

3.3 Free radical polymerization

A free radical is an atomic or molecular species whose normal bonding system has been modified such that an unpaired electron remains associated with the new structure. The radical is capable of reacting with an olefinic monomer to generate a chain carrier which can retain its activity long enough to propagate a macromolecular chain under the appropriate conditions.

$$R^{\cdot} + CH_2{=}CHR_1 \rightarrow RCH_2CHR_1^{\cdot}$$

3.4 Initiators

An effective initiator is a molecule which, when subjected to heat, electromagnetic radiation, or chemical reaction, will readily undergo homolytic fission into radicals of greater reactivity than the monomer radical. These radicals must also be stable long enough to react with a monomer and create an active centre. Particularly useful for kinetic studies are compounds containing an azonitrile group, as the decomposition kinetics are normally first order and the rates are unaffected by the solvent environment.

Typical radical producing reactions are:

(1) *Thermal decomposition* can be usefully applied to organic peroxides or azo compounds, *e.g.* benzoyl peroxide when heated eventually forms two phenyl radicals with loss of CO_2. A simpler one-stage decomposition is obtained when dicumyl

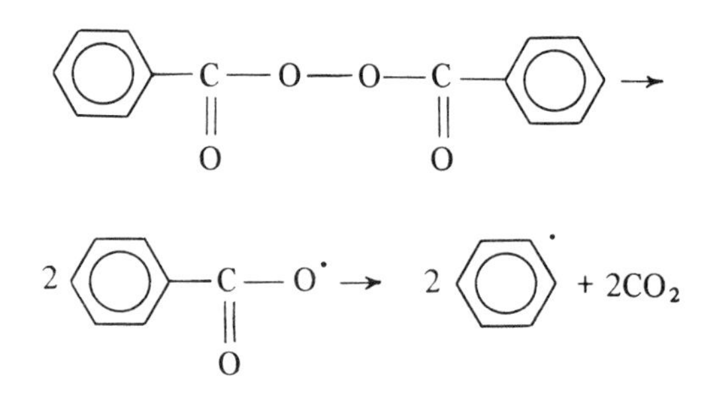

peroxide is used.

$$C_6H_5{-}C(CH_3)_2{-}O{-}O{-}(CH_3)_2C{-}C_6H_5 \rightarrow 2C_6H_5{-}C(CH_3)_2O^{\cdot}$$

(2) *Photolysis* is applicable to metal iodides, metal alkyls, and azo compounds, *e.g.* α,α'-azobisisobutyronitrile (AIBN) is decomposed by radiation with a wavelength of 360 nm.

$$(CH_3)_2{-}\underset{\displaystyle CN}{\underset{|}{C}}{-}N{=}N{-}\underset{\displaystyle CN}{\underset{|}{C}}{-}(CH_3)_2 \rightarrow 2(CH_3)_2{-}\underset{\displaystyle CN}{\underset{|}{C^{\cdot}}} + N_2$$

Note that in each reaction there are two radicals $R^{\cdot}$ produced from one initiator molecule I; in general:

$$I \rightarrow 2R^{\cdot} \tag{3.1}$$

(3) *Redox reactions*, *e.g.* the reaction between the ferrous ion and hydrogen peroxide in

solution produces hydroxyl radicals,

$$H_2O_2 + Fe^{2+} \rightarrow Fe^{3+} + OH^- + OH^{\bullet}.$$

Alkyl hydroperoxides may be used in place of H_2O_2. A similar reaction is observed when cerium(IV) sulphate oxidizes an alcohol:

$$RCH_2OH + Ce^{4+} \rightarrow Ce^{3+} + H^+ + RC(OH)H^{\bullet}$$

(4) *Persulphates* are useful in emulsion polymerizations where decomposition occurs in the aqueous phase and the radical diffuses into a hydrophobic, monomer containing, droplet.

$$S_2O_8^{2-} \rightarrow 2SO_4^{\bullet -}$$

(5) *Ionizing radiation* such as α, -β-, γ-, or X-rays may be used to initiate a polymerization, by causing the ejection of an electron followed by dissociation and electron capture to produce a radical.

Ejection:	$C \rightsquigarrow C^{\bullet} + e^-$
Dissociation:	$C^+ \longrightarrow A^{\bullet} + Q^{\bullet}$
e^--Capture:	$Q^+ + e^- \longrightarrow Q^{\bullet}$

Initiators that undergo thermal decomposition must be selected to ensure that at the polymerization temperature they provide a suitable source of free radicals which will sustain the reaction. For an initiator concentration of 0.1 M, a useful guide is that the rate of radical production should be 10^{-6} to 10^{-7} mol dm^{-3} s^{-1}, which is equivalent to a $k_d \approx 10^{-5}$ to 10^{-6} s^{-1} in the temperature range 320–420 K, normally regarded as the most suitable for radical polymerizations. The rate equations for several commonly used initiators are shown in table 3.2 and the operating temperatures are calculated on this basis. The half-lives are also shown in figure 3.1, plotted as a function of temperature.

TABLE 3.2. Some radical initiator decomposition rate equations and the corresponding suggested temperature range for use

Initiator	*Rate equation* (s^{-1})	*Temperature range* (K)
RC(O)OO(O)CR		
R = Et	$k = 10^{14} \exp(-146\,\text{kJ}/RT)$	382–402
R = Bu^t	$k = 6.3 \times 10^{15} \exp(-157\,\text{kJ}/RT)$	377–395
RN=NR		
R = $Me_2C(CN)$	$k = 1.8 \times 10^{15} \exp(-128.7\,\text{kJ}/RT)$	310–340
R = PhCHMe	$k = 1.3 \times 10^{15} \exp(-152.6\,\text{kJ}/RT)$	378–398
R = Me_2CH	$k = 5 \times 10^{13} \exp(-170.5\,\text{kJ}/RT)$	453–473

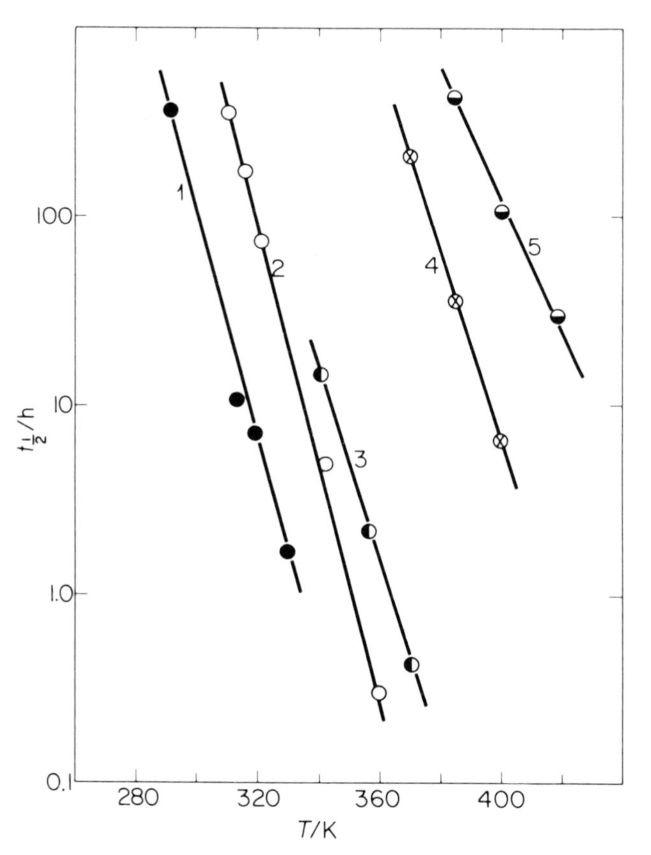

FIGURE 3.1. Half-lives $t_{1/2}$ of selected peroxide initiators. Curve **1**, isopropyl percarbonate; curve **2**, 2,2′-azo-bis-isobutyronitrile; curve **3**, benzoyl peroxide; curve **4**, di-tertiary butyl peroxide; curve **5**, cumene hydroperoxide.

Reference to these data provides the chemist with the information needed to make the correct selection for the chosen polymerization conditions. It should be noted, however, that the rate equations for the decomposition of the initiator molecules can depend on the solvent and the values shown in table 3.2 should be regarded as good approximations.

INITIATOR EFFICIENCY

Although the decomposition of an initiator molecule can be quantitative, chain initiation may be less than 100 per cent efficient. In a kinetic analysis the effective radical concentration is represented by an *efficiency factor* f which is less than unity when only a proportion of the radicals generated are effective in the creation of a kinetic chain. Inefficient chain propagation may be due to several side reactions.

Primary recombination can occur if the diffusion of radical fragments is impeded in the solution and a cage effect leads to

$$2(CH_3)_2\underset{\displaystyle CN}{\underset{|}{C^{\cdot}}} \rightarrow (CH_3)_2\underset{\displaystyle CN}{\underset{|}{C}}-\underset{\displaystyle CN}{\underset{|}{C}}(CH_3)_2.$$

The solvent usually plays an important part and the extent of the decomposition of benzoyl peroxide is limited to 35 per cent in tetrachloroethylene, 50 per cent in benzene, and 85 per cent in ethyl acetate, after refluxing for 4 h.

Further wastage occurs when induced decomposition is effected by the attack of an active centre

$$R^{\cdot} + R'-O-O-R' \rightarrow ROR' + R'O^{\cdot}$$

This effectively produces one radical instead of the three potential radical species.

For a 100 per cent efficient initiator $f = 1$, but most initiators have efficiencies in the range 0.3 to 0.8.

3.5 Chain growth

A chain carrier is formed from the reaction of the free radical and a monomer unit; chain propagation then proceeds rapidly by addition to produce a linear polymer.

$$RM_1^{\cdot} + M_1 \rightarrow RM_2^{\cdot} \tag{3.2}$$

$$RM_n^{\cdot} + M_1 \rightarrow RM_{n+1}^{\cdot} \tag{3.3}$$

The average life time of the growing chain is short, but a chain of over 1000 units can be produced in 10^{-2} to 10^{-3} s. Bamford and Dewar have estimated that the thermal polymerization of styrene at 373 K leads to chains of $x = 1650$ in approximately 1.24 s, *i.e.* a monomer adds on once every 0.75 ms.

3.6 Termination

In theory the chain could continue to propagate until all the monomer in the system had been consumed but for the fact that free radicals are particularly reactive species and interact as quickly as possible to form inactive covalent bonds. This means that short chains are produced if the radical concentration is high, because the probability of radical interaction is correspondingly high, and the radical concentration should be kept small if long chains are required. Termination of chains can take place in several ways: (1) the interaction of two active chain ends; (2) the reaction of an active chain end with an initiator radical; (3) termination by transfer of the active centre to another molecule which may be solvent, initiator, or monomer; (4) interaction with impurities (*e.g.* oxygen) or inhibitors.

The most important termination reaction is the first, a bimolecular interaction between two chain ends. Two routes are possible.

TABLE 3.3. Termination mechanisms for polymer radicals

Monomer	*Temperature*	*Mechanism*
Styrene	330–370	Combination
Acrylonitrile	330	Combination
Methyl methacrylate	273	Mainly combination
	>330	Mainly disproportionation
Methyl acrylate	360	Mainly disproportionation
Vinyl acetate	360	Mainly disproportionation

(a) *Combination* where two chain ends couple together to form one long chain.

$$\sim CH_2{-}\underset{\displaystyle Cl}{\underset{|}{CH^{\bullet}}} + \underset{\displaystyle Cl}{\underset{|}{HC^{\bullet}}}{-}CH_2 \sim \;\rightarrow\; \sim CH_2{-}\underset{\displaystyle Cl}{\underset{|}{CH}}{-}\underset{\displaystyle Cl}{\underset{|}{CH}}{-}CH_2 \sim$$

(b) *Disproportionation* with hydrogen abstraction from one end to give an unsaturated group and two dead polymer chains.

$$\sim CH_2{-}\underset{\displaystyle COOCH_3}{\underset{|}{\overset{\displaystyle CH_3}{\overset{|}{C^{\bullet}}}}} + {}^{\bullet}\underset{\displaystyle COOCH_3}{\underset{|}{\overset{\displaystyle CH_3}{\overset{|}{C}}}}{-}CH_2 \sim \;\rightarrow\; \sim CH{=}\underset{\displaystyle COOCH_3}{\underset{|}{\overset{\displaystyle CH_3}{\overset{|}{C}}}} + \underset{\displaystyle COOCH_3}{\underset{|}{\overset{\displaystyle CH_3}{\overset{|}{CH}}}}{-}CH_2 \sim$$

One or both processes may be active in any system depending on the monomer and polymerizing conditions. Experimental evidence suggests that polystyrene terminates predominantly by combination whereas poly(methyl methacrylate) terminates mainly by disproportionation when the reaction is above 333 K but by both mechanisms below this temperature (see table 3.3). The mechanism can be determined by measuring the number of initiator fragments per chain using a radioactive initiator. One fragment per chain is counted when disproportionation is operative and two when combination occurs. Alternatively the number average molar mass of the product can be measured.

3.7 Steady-state kinetics

The three basic steps in the polymerization process can be expressed in general terms as follows:

Initiation is a two stage reaction. Initiator decomposition

$$I \xrightarrow{k_d} 2R^{\bullet}, \tag{3.4}$$

is followed by radical attack on a monomer unit to form a chain carrier,

$$R^{\bullet} + M \xrightarrow{k_i} RM^{\bullet}. \tag{3.5}$$

Since the initial decomposition is slow compared with both the rate of addition of a primary radical to a monomer and the termination reaction, it is the rate determining step. The rate of initiation v_i is then the rate of production of chain radicals

$$v_i = d[RM^\bullet]/dt = 2k_d f[I], \tag{3.6}$$

where the factor 2 is introduced because two potentially effective radicals are produced in the decomposition: f also measures the ability of these to propagate chains. This expression is valid for a thermo-initiation, but many reactions can be photo-initiated when the monomer absorbs radiation and acts as its own initiator. The rate v_{ip} is then dependent on the intensity of light absorbed

$$v_{ip} = 2\phi I_a. \tag{3.7}$$

The quantum yield ϕ replaces f and defines the initiator efficiency; I_a is related to the incident light intensity I_o, the monomer concentration, and the extinction coefficient ϵ so that

$$v_{ip} = 2\phi\epsilon I_o[M]. \tag{3.8}$$

When the monomer is a poor absorber of radiation small quantities of a photosensitizer may be added to absorb the energy and then transfer this to the monomer to create an active centre. In this case [M] is replaced by the concentration of photosensitizer.

Propagation is the addition of monomer to the growing radical

$$RM_n^\bullet + M_1 \xrightarrow{k_p} RM_{n+1}^\bullet \tag{3.9}$$

The rate of bimolecular propagation is assumed to be the same for each step so that

$$v_p = k_p[M][M^\bullet], \tag{3.10}$$

where $[M^\bullet]$ represents the concentration of growing ends; and $[M^\bullet]$ is usually low at any particular time. The reaction is essentially the conversion of monomer to polymer and can be followed from the rate of disappearance of monomer.

Termination is also a bimolecular process depending only on $[M^\bullet]$; v_t for both mechanisms is

$$v_t = 2k_t[M^\bullet][M^\bullet]. \tag{3.11}$$

The rate constant k_t is actually $(k_{tc} + k_{td})$ where the two mechanisms, combination and disproportionation are possible but will be written as k_t for convenience. If the chain reaction does not lead to an explosion, a steady state is reached where the rate of radical formation is exactly counterbalanced by the rate of destruction, *i.e.* $v_i = v_t$, and for a thermal reaction

$$2k_t[M^\bullet]^2 = 2k_d f[I]. \tag{3.12}$$

From this an expression for $[M^{\cdot}]$ is obtained

$$[M^{\cdot}] = \{f k_d[I]/k_t\}^{1/2}, \tag{3.13}$$

and because the concentration of radicals is usually too low to be determined accurately it is replaced in the kinetic expression. The overall rate of polymerization is then

$$v_p = k_p\{f k_d[I]/k_t\}^{1/2}[M], \tag{3.14}$$

which shows that the rate is proportional to the monomer concentration and to $[I]^{1/2}$ if f is high, but for a low efficiency initiator f becomes a function of [M] and the rate is then proportional to $[M]^{3/2}$.

By analogy the rate of a photo-polymerization becomes

$$v_{pp} = k_{pp}\{\phi\epsilon I_o/k_t\}^{1/2}[M]^{3/2}. \tag{3.15}$$

Two parameters of interest can be derived from this analysis, the kinetic chain length $\bar{v}$ and the average degree of polymerization x.

The kinetic chain length $\bar{v}$ is a measure of the average number of monomer units reacting with an active centre during its lifetime and is related to x_n through the mechanism of the termination. Thus combination means $x_n = 2\bar{v}$ but $x_n = \bar{v}$ if disproportionation is the only termination reaction. Under steady-state conditions

$$\bar{v} = v_p/v_i = v_p/v_t = k_p^2[M]^2/2k_t v_p. \tag{3.16}$$

This means that as $\bar{v}$ is inversely proportional to the rate of polymerization, an increase in temperature produces an increase in v_p and a corresponding decrease in chain length. The equation shows that $\bar{v}$ is inversely proportional to the radical concentration, hence the chain length is short for high radical concentrations and vice-versa. This means that a certain degree of control over the polymer chain length can be exercised by altering the initiator concentration.

3.8 High conversion bulk polymerizations

In many polymerizations a marked increase in rate is observed towards the end of the reaction instead of the expected gradual decrease caused by the depletion of the monomer and initiator. This *auto-acceleration* is a direct result of the increased viscosity of the medium and the effect is most dramatic when polymerizations are carried out in the bulk phase or in concentrated solutions. The phenomenon, sometimes known as the *Trommsdorff-Norrish* or *gel effect*, is caused by the loss of the steady state in the polymerization kinetics.

When the viscosity of the reaction medium increases, the different steps in the polymerization reaction become diffusion controlled at stages which depend on the activation energy for that step. The initiation and propagation reactions have higher activation energies relative to that of the termination step, and so only become diffusion controlled in the latter stages of the reaction. The termination step has the lowest activation energy and becomes diffusion controlled at lower viscosities (earlier in the reaction). It is also a bimolecular process requiring fruitful collision between two radical ends attached to long, highly entangled polymer chains, and with increase in the viscosity of the medium this chain movement becomes progressively more hindered, with the result that the active chain ends have greater difficulty finding one another. This leads to a significant reduction in the rate of termination and, as this is the main process responsible for removing radicals in the polymerization, the overall radical concentration gradually increases. A consequence of this is an increase in the number of propagation steps with the release of more heat. This speeds up the rate of decomposition of the initiator, thereby producing even more radicals and resulting in auto-acceleration. A major hazard of this sequence of events is that if sufficient energy is released, which is not dissipated rapidly and efficiently, then an explosion may occur.

The simple classical kinetic analysis outlined in section 3.7 is an oversimplified model based on the assumptions that (i) steady state conditions prevail throughout the reaction at all conversions, (ii) radical reactivity is independent of chain length, (iii) no chain transfer occurs (iv) there is no auto-acceleration or gel effect. A schematic diagram of the rate of a typical bulk polymerization is presented in figure 3.2(a) as a function of percentage conversion. The rate rises initially to the steady state condition at about 0.1 per cent conversion and obeys the steady state approximation till about 10 per cent conversion. It then dips through a minimum before the gel effect begins to manifest itself by a rapid acceleration in rate. If the polymer has a glass transition temperature which is higher than the polymerization temperature, then as the remaining monomer no longer depresses the polymer glass transition below the reaction temperature, the mixture becomes glassy and the rate falls. This is illustrated in figure 3.2(b) where the polymerization of poly(methyl methacrylate) in the bulk ceases when the glass transition temperature is no longer depressed by the plasticizing

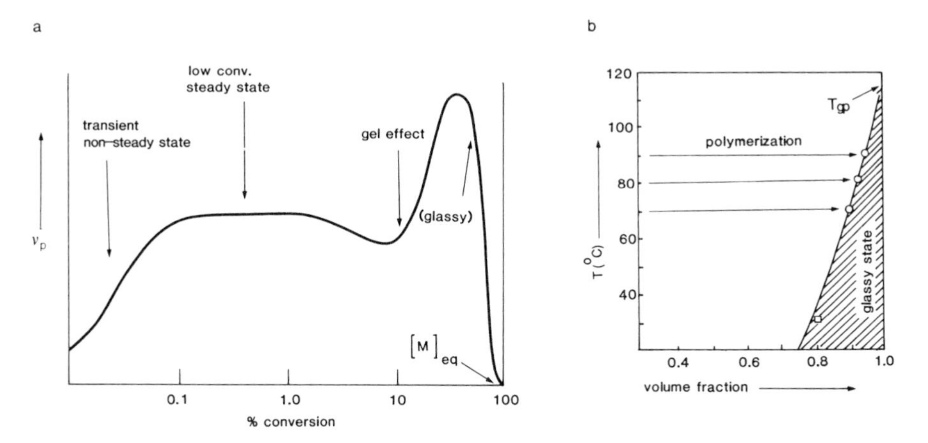

FIGURE 3.2. (a) Schematic representation of the polymerization rate versus conversion. (b) Schematic representation of the vitrification points for a bulk polymerization of poly(methyl methacrylate) at various polymerization temperatures. The solid line represents the volume fraction of polymer at which polymerization stops, as a function of polymerization temperature.

effect of the monomer below the polymerization reaction temperature, and vitrification occurs.

Careful experimental work has shown that the rate constant for termination is dependent on the chain length when $x_n \leqslant 100$, but is essentially independent of chain length for $x_n > 1000$. Close analysis of this effect suggests that k_t is then a function of coil size and will be large when the polymer coil is small or tightly bound. The consequence of this is that polymerization will be slower in a thermodynamically poor solvent than in a good one, and as the reaction progresses it ought to slow down not only because of monomer depletion, but also because the reaction medium will become progressively less favourable for the polymer. However, this effect may not be noticed.

O'Driscoll has proposed that the auto-acceleration can be modelled by recognizing that the termination reaction is diffusion controlled but will also depend on the size of the chain involved. The critical chain length for entanglement n_c then becomes an important parameter, and two termination rate constants can also be defined, one for chains smaller than n_c and one for large entangled chains. These are respectively k_t and k_{te}. If $\bar{v}$ is the kinetic chain length and v_p the conventional steady state polymerization, then the observed rate v_p^* is given by

$$\frac{v_p^*}{v_p} - 1 = \left[\left(\frac{k_t}{k_{te}}\right)^{1/2} - 1\right]\exp(-n_c\bar{v}) \tag{3.17}$$

There is good agreement between the predicted rate from equation (3.17) and the experimental values obtained for the bulk polymerization of methyl methacrylate at 343 K and two initiator concentrations (figure 3.3).

The auto-acceleration can be avoided by performing the polymerization in more dilute solutions or by stopping the reaction before the diffusion effect grows to

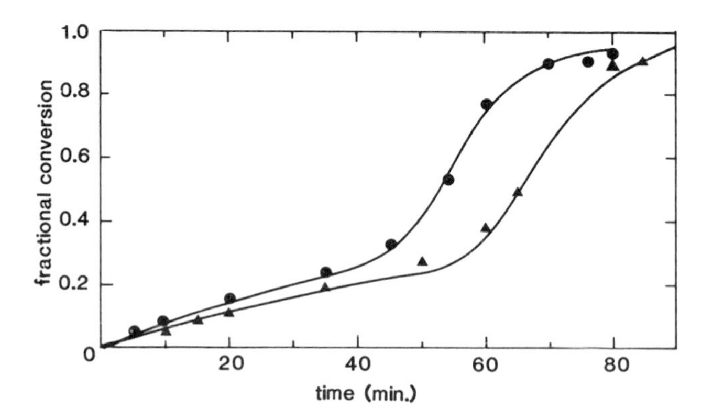

FIGURE 3.3. Fractional conversion curves plotted as a function of time for the bulk polymerization of poly(methyl methacrylate) at 343 K. Initial initiator concentrations are 0.5 per cent and 0.3 per cent. The solid lines are predicted from equation (3.17). (Adapted from J. N. Cardenas and K. F. O'Driscoll, *J. Polym. Sci. Polym. Chem. Ed.* **14**, 883 (1976). © John Wiley and Sons Inc., N.Y.)

noticeable proportions. However, polymerizations in solvents can be influenced by the choice of medium and this feature is treated in the chain transfer reactions.

3.9 Chain transfer

Termination in a free radical polymerization normally occurs by collision between two active centres attached to polymer chains, but the chain length of the product in many systems is lower than one would expect if this was the mechanism solely responsible for limiting the kinetic chain length $\bar{\nu}$. Usually x_n will lie within the expected limits of $\bar{\nu}$ (disproportionation) and $2\bar{\nu}$ (combination), but not always, and Flory found that attenuation of chain growth takes place if there is premature termination of the propagating chain by a transfer of activity to another species through a collision. This is a competitive process involving the abstraction of an atom by a chain carrier from an inactive molecule XY with replaceable atoms, and is dependent on the strength of the X—Y bond.

$$\sim M_m^{\bullet} + XY \rightarrow \sim M_mX + Y^{\bullet}$$

It is important to note that the free radical is not destroyed in the reaction, merely transferred, and if the new species is sufficiently active another chain will emanate from the new centre. This is known as *chain transfer* and is a reaction resulting in the exchange of an active centre between molecules during a bimolecular collision. Several types of chain transfer have been identified.

Transfer to monomer. The two important reactions in this group both involve hydrogen abstraction. Two competitive alternatives exist in the first group

$$R^{\bullet} + CH_2{=}CHX - \begin{cases} \nearrow RH + CH_2{=}C^{\bullet}X & \text{(I)} \\ \searrow RCH_2CHX^{\bullet} & \text{(II)} \end{cases}$$

If the radical formed in reaction (II) is virtually unstabilized by resonance, then the reaction with the parent unreactive monomer may produce little chain propagation due to the tendency for stabilization to occur by removal of hydrogen from the monomer. This leads to rapid chain termination and is known as *degradative transfer*. Allylic monomers are particularly prone to this type of reaction,

$$\sim R^{\bullet} + CH_2{=}CHCH_2OCOCH_3 \rightarrow \sim RH + \dot{C}H_2{\cdots}CH{\cdots}\dot{C}H{-}OCOCH_3,$$

where abstraction of the α-hydrogen leads to a resonance stabilized allylic radical capable only of bimolecular combination with another allyl radical. This is effectively an auto-inhibition by the monomer. Propylene also reacts in this manner and both monomers are reluctant to polymerize by a free radical mechanism.

A second group of transfer reactions can occur by hydrogen abstraction from the pendant group

The relevant kinetic expression is

$$\nu_{tr} = k_{tr}^{M}[M][M^{\bullet}]. \tag{3.18}$$

Transfer to initiator. Organic peroxides, when used as initiators, are particularly susceptible to chain transfer. Azo initiators are not vulnerable in this respect and are more useful when a kinetic analysis is required. For peroxides

$$v_{tr} = k_{tr}^{I}[I][M^{\bullet}]. \qquad (3.19)$$

Transfer to polymer. The transfer reaction with a polymer chain leads to branching rather than initiation of a new chain so that the average molar mass is relatively unaffected. The long and short chain branching detected in polyethylene is believed to arise from this mode of transfer.

Transfer to modifier. Molar masses can be controlled by addition of a known and efficient chain transfer agent such as an alkyl mercaptan.

$$\sim CH_2CHX^{\bullet} + RSH \rightarrow \sim CH_2CH_2X + RS^{\bullet}$$

$$RS^{\bullet} + CH_2{=}CHX \rightarrow RSCH_2CHX^{\bullet}$$

Mercaptans are commonly used because the S—H bond is weaker and more susceptible to chain transfer than a C—H bond.

Transfer to solvent. A significant decrease in polymer chain length is often found when polymerizations are carried out in solution rather than in the undiluted state and this variation is a function of both the extent of dilution and the type of solvent used. The effectiveness of a solvent in a transfer reaction depends largely on the amount present, the strength of the bond involved in the abstraction step, and the stability of the solvent radical formed. With the exception of fluorine, halogen atoms are easily transferred and the reaction of styrene in CCl_4 is a good example of this chain transfer.

$$\sim CH_2{-}\underset{\displaystyle C_6H_5}{\underset{|}{CH^{\bullet}}} + CCl_4 \rightarrow \sim CH_2\underset{\displaystyle C_6H_5}{\underset{|}{CH}}Cl + CCl_3^{\bullet} \qquad (I)$$

$$CCl_3^{\bullet} + CH_2{=}CHC_6H_5 \rightarrow Cl_3CCH_2{-}\underset{\displaystyle C_6H_5}{\underset{|}{CH^{\bullet}}} \qquad (II)$$

$$Cl_3C \sim \underset{\displaystyle C_6H_5}{\underset{|}{CH^{\bullet}}} + CCl_4 \rightarrow Cl_3C \sim \underset{\displaystyle C_6H_5}{\underset{|}{CH}}Cl + CCl_3^{\bullet} \qquad (III)$$

When the solvent is present in significant quantity, step (I) is of minor importance and the resulting polymer contains four chlorine atoms which can be detected by analysis.

Hydrogen is normally the atom abstracted and as radical stability enhances the transfer reaction, we find that toluene, which forms a primary radical, is less efficient than ethyl benzene, which forms a secondary radical, while both are inferior to isopropyl benzene which forms a tertiary radical. All are much better than *t*-butyl benzene, whose radical is unstable, so that virtually no chain transfer takes place in this solvent. It is interesting to note that even benzene acts as a chain transfer agent on a modest scale.

The kinetic expression is

$$v_{tr} = k^{S}_{tr}[M^{\bullet}][S]. \tag{3.20}$$

CONSEQUENCES OF CHAIN TRANSFER

The primary effect is a decrease in the polymer chain length, but other less obvious occurrences can be detected. If k_{tr} is much larger than k_p then a very small polymer is formed with x_n between 2 and 5. This is known as *telomerization*. The chain re-initiation process can also be slower than the propagation reaction and a decrease in v_p is observed. However, the influence on x_n is most important and it can be estimated by considering all the transfer processes in a form known as the Mayo equation.

$$1/x_n = (1/x_n)_0 + C_s[S]/[M] \tag{3.21}$$

This is a simplified form in which the main assumption is that solvent transfer

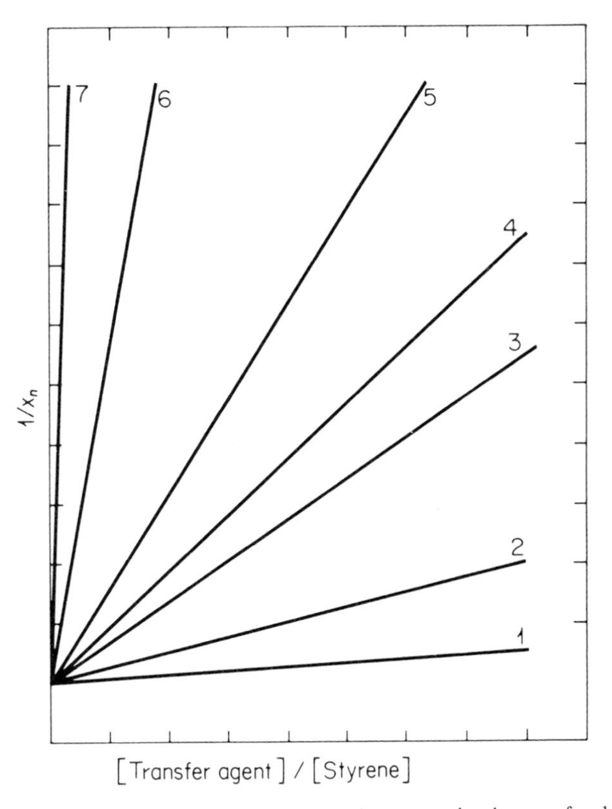

FIGURE 3.4. Effect of chain transfer to various solvents on the degree of polymerization of polystyrene at 333 K. 1, Benzene; 2, *n*-Heptane; 3, *sec*-Butyl benzene; 4, *m*-Cresol; 5, CCl_4; 6, CBr_4; 7, *n*-Butyl mercaptan.

TABLE 3.4. Chain transfer constants of various agents to styrene at 333 K

Agent	$10^4 C_s$
Benzene	0.023
n-Heptane	0.42
sec-Butyl benzene	6.22
m-Cresol	11.0
CCl_4	90
CBr_4	22 000
n-Butylmercaptan	210 000

predominates and all other terms are included in $(1/x_n)_0$. The chain transfer constant C_s is then (k_{tr}^s/k_p).

A plot of $1/x_n$ against $\{[S]/[M]\}$ for a variety of agents is shown in figure 3.4. The slope is a measure of C_s and the intercept is $(1/x_n)_0$. If the activation energy of the process is required, $\log C_s$ can be plotted against $1/T$.

3.10 Inhibitors and retarders

Chain transfer agents can lower the average chain length and in extreme cases, when used in large proportions, may lead to the formation of telomers.

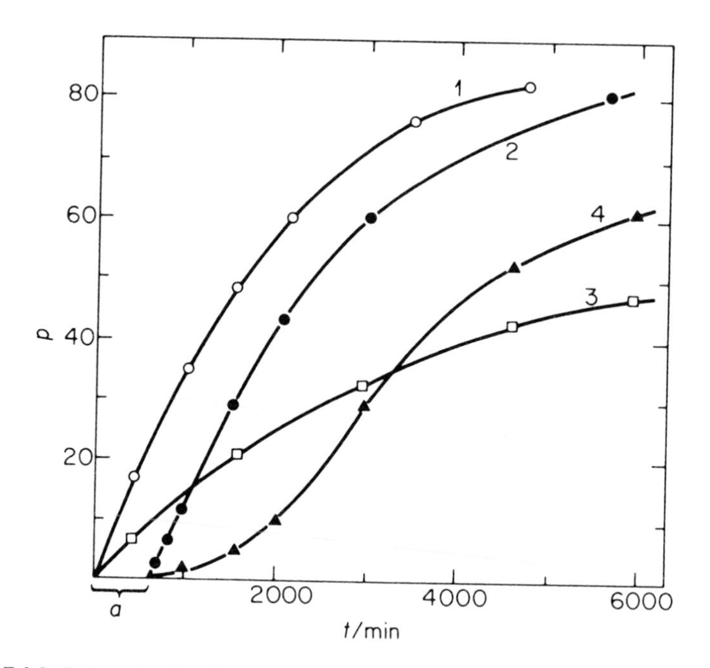

FIGURE 3.5. Polymerization of styrene at 373 K in the presence of: curve 1, no inhibitor; curve 2, 0.1 per cent benzoquinone; curve 3, 0.5 per cent nitrobenzene; curve 4, 0.2 per cent nitrosobenzene. The time t and percentage conversion p are plotted.

Some chain transfer agents yield radicals with low activity and if the re-initiation reaction is slow, the polymerization rate decreases because there is a build-up of radicals leading to increased termination by coupling. When this happens the substance responsible is said to be a *retarder, e.g.* nitrobenzene acts in this way with styrene.

In extreme cases an added reagent may suppress polymerization completely by reacting with the initiating radical species and converting them all efficiently into unreactive substances. This is known as inhibition, but the difference between an inhibitor and a retarder is merely in the degree of efficiency.

The phenomena are typified by the reaction of styrene with benzoquinone, nitrobenzene, and nitrosobenzene, studied by Schulz, and shown in figure 3.5. Curve 1 represents the polymerization of styrene in the absence of any agents. When benzoquinone is added the polymerization is completely inhibited until all the benzoquinone has been consumed, then the reaction proceeds normally (curve 2). The time interval a is the *induction period* and reflects the time taken for the benzoquinone to react with all the radicals formed until no more benzoquinone is left unreacted. In the presence of nitrobenzene, curve 3, the polymerization continues but at a much reduced rate. The action of nitrosobenzene is much more complex (curve 4). It acts first as an inhibitor but probably produces a substance during this period which then acts as a retarder and both effects are observed.

Monomers are usually transported and stored in the presence of inhibitors to prevent premature polymerization, and so must be redistilled and purified prior to use.

An excellent inhibitor is the resonance stabilized radical diphenyl picryl hydrazyl (DPPH), used extensively as a radical scavenger because the stoichiometry of the reaction is 1:1.

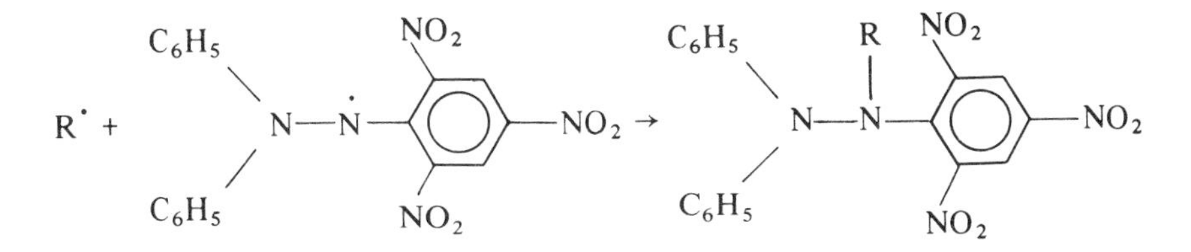

3.11 Experimental determination of individual rate constants

The parameters of interest are f, k_d, k_p, and k_t. Both k_d and f can be measured without resorting to polymerizations, but the steady-state kinetic scheme does not allow direct determination of the individual constants k_p and k_t, only the ratio (k_p^2/k_t). To separate these, non-steady-state conditions must be used, but first k_d and f can be measured.

Initiator decomposition and efficiency. An initiator is usually a compound with a labile bond whose dissociation energy lies in the range 105 to 170 kJ mol^{-1}. As the rate of decomposition of the initiator is normally the rate determining step this must be measured if v_i is required. The thermal dissociation of α,α'-azobisisobutyronitrile can be followed, in the absence of monomer, by following the rate of evolution of nitrogen during radical formation; a typical value for k_d is 1.2×10^{-5} s^{-1} at 333 K.

Thermo-initiated polymerizations suffer from the disadvantage that because of the large heat capacity of the system radical generation is difficult to control; also k_d measured in the absence of monomer may be invalid. Photo-initiation, where radical

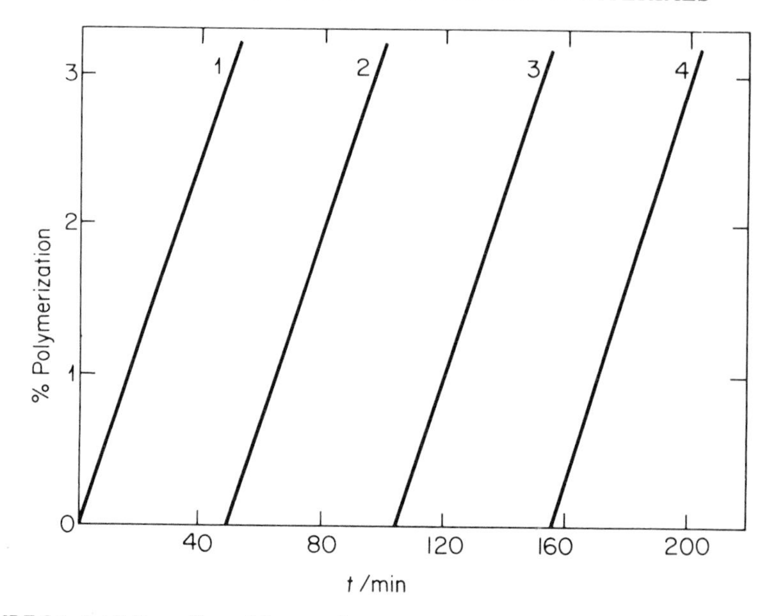

FIGURE 3.6. Inhibiting effect of benzoquinone on the photopolymerization of vinyl acetate (1) no quinone (2) 2.39 mg (3) 5.00 mg (4) 7.50 mg. (Adapted from Melville.)

generation is instantaneous, is preferred for kinetic work. The number of quanta absorbed by the system is estimated first by irradiating a uranyl oxalate solution in the reaction vessel. This reacts quantitatively and provides a measure of the number of radicals produced when the fraction of light absorbed is also known.

The initiator efficiency can be estimated by counting the number of chains formed, but this relies on knowledge of the termination mechanism and the absence of chain transfer reactions. A better technique is to make use of a radical scavenger such as DPPH, ferric chloride, or benzoquinone. As long as one is confident that a single molecule of inhibitor reacts with only one radical, a quantitative estimate of the number of radicals produced is possible. The addition of varying quantities of inhibitor to a system results in a corresponding number of induction periods whose durations are proportional to the number of radicals produced. This is illustrated in the photopolymerization of vinyl acetate in the presence of benzoquinone (figure 3.6).

Determination of k_p and k_t. To obtain individual values of k_p and k_t, a combination of two or more of the following measurements is required: (1) a steady-state measurement of v_p to give $(k_p^2 k_d/k_t)$; (2) an average radical life time at the steady state yielding $(k_d k_t)^{-1/2}$; (3) an estimate of x_n which gives $(k_p^2/k_t k_d)$.

MEASUREMENT OF v_p BY DILATOMETRY

The overall polymerization rate can be monitored through the change in a physical or chemical property of the system. Gravimetric analysis of the products or titration with bromine to measure the rate of disappearance of double bonds would both afford a means of following the reaction but present difficulties, especially if oxygen acts as an inhibitor. The choice of a physical property is more desirable. A change in the molar refractivity is related to the change in the number of double bonds, so that an increase in the refractive index could be used to follow v_p. Alternatively, one can make use of the fact that the density of the polymer is normally greater than that of the monomer and volume contractions as large as 27 per cent of the total have been obtained during the course of a reaction. This makes dilatometry a particularly useful technique. If only low degrees of conversion are measured the initiator concentration may be assumed to be constant and if L_0 is the height of the initial level in the dilatometer, L_t that after time t, and L_∞ the reading on completion of the reaction, a plot of $\log\{(L_0 - L_\infty)/(L_t - L_\infty)\}$ against t is linear if the reaction is first order in monomer, *i.e.* if $v_p = k_p(fk_d[\mathrm{I}]/k_t)^{1/2}[\mathrm{M}]$. The slope is then equal to the collection of constants and when k_d, f, and [I] are known, $(k_p/k_t^{1/2})$ is obtained.

RADICAL LIFETIMES BY THE ROTATING SECTOR TECHNIQUE

Under steady-state conditions the average life-time of a propagating chain τ_s is determined by the ratio of the concentration of radicals at any time to their rate of disappearance, given by equation (3.11),

$$\tau_s = [\mathrm{M}^\cdot]/2k_t[\mathrm{M}^\cdot]^2 = 1/2k_t[\mathrm{M}^\cdot]. \tag{3.22}$$

Substituting for the radical concentration (using equation (3.10)) gives

$$\tau_s = k_p[\mathrm{M}]/2k_t v_p \tag{3.23}$$

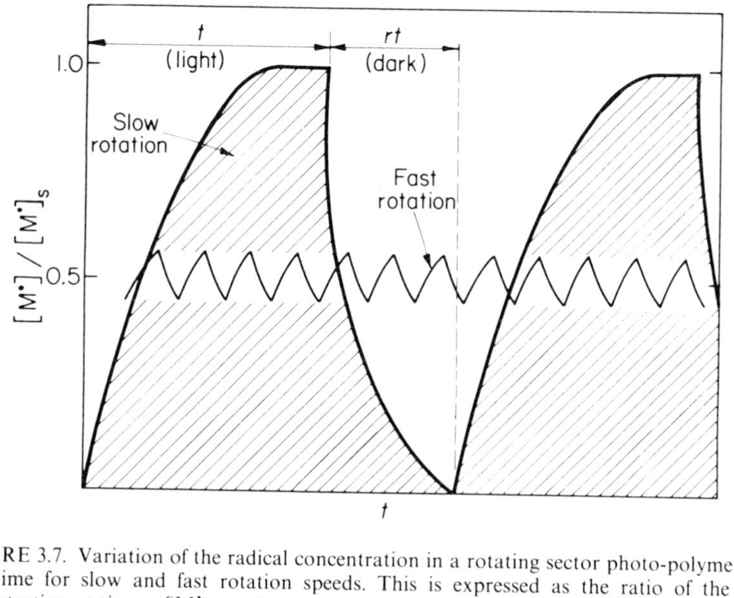

FIGURE 3.7. Variation of the radical concentration in a rotating sector photo-polymerization with time for slow and fast rotation speeds. This is expressed as the ratio of the radical concentration at time t, $[\mathrm{M}^\cdot]$, to the concentration under steady-state conditions $[\mathrm{M}^\cdot]_\mathrm{S}$.

and as $(k_p/k_t^{1/2})$ is obtained from v_p, knowledge of τ_s provides a means of separating k_p and k_t.

The radical life-time τ_s can be evaluated from non-steady-state conditions and this is conveniently carried out using a photo-initiated system where controlled instantaneous generation of radicals is possible. The reaction mixture is placed in a quartz cell dilatometer and thermostatted. The cell is then placed in a beam of UV radiation which can be interrupted to produce alternating periods of illumination and darkness, by rotating a disc with a cut-out sector in the light path. The ratio of the durations of the dark and light periods is r and a typical value is $r = 3$. The time of each period of illumination in a cycle can be varied by altering the rate of rotation of the disc and in this way both steady and non-steady-state conditions can be attained.

For slow speeds of rotation the illumination time t is large compared with τ_s, thereby allowing the radical concentration to build up to a steady state during each cycle (shaded areas in figure 3.7). If the cycle time is now increased until t is small compared with τ_s, then the radical concentration never reaches the steady state but remains low and almost constant. This in effect reduces the light intensity by a factor of $(1 + r)^{-1}$, and if v_p is measured for various rotation times, the ratio (v_p/v_{ps}) will change from a lower limit of $(1 + r)^{-1}$ for large values of t to an upper limit of $(1 + r)^{-1/2}$ for small t, as the illumination time passes from $t > \tau_s$ to $t < \tau_s$. Here v_p and v_{ps} are the average and steady-state rates respectively. To locate the intermediate speed for which $t = \tau_s$ a plot of (v_p/v_{ps}) against $\log t$ is constructed and compared with a theoretical curve of (v_p/v_{ps}) against $(\log t - \log \tau_s)$. The two curves are superimposed by displacing the theoretical curve along the abscissa and the magnitude of the horizontal shift is then equal to $\log \tau_s$.

MEASUREMENT OF k_p AND k_t USING A PULSED LASER TECHNIQUE

Recently, Olaj and his coworkers have described a method for measuring the individual rate constants, which replaces the rotating sector with a pulsed laser light source. The experimental set up is shown schematically in figure 3.8. An Nd:YAG laser is used to generate pulses of electromagnetic radiation with a wavelength of 355 nm. These pass into a solution of monomer and initiator contained in a pyrex reaction vessel. Within this, each pulse creates a finite radical concentration, which subsequently decays as

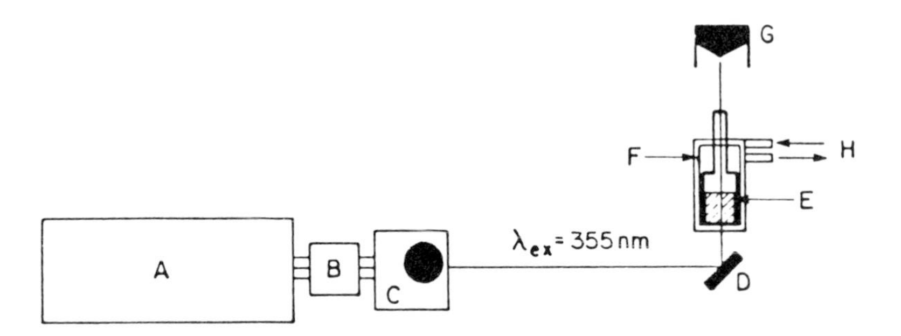

FIGURE 3.8. Schematic diagram of the experimental apparatus for the pulsed laser technique. A = Nd:YAG laser; B = harmonic generator (1064, 532 and 335/266 nm); C = harmonic separator; D = dielectric mirror; E = sample cell; F = cell holder; G = beam dump; H = circulated water supply.

TABLE 3.5. Kinetic parameters for the photopolymerization of vinyl acetate

	Illumination	
	High intensity	*Low intensity*
v_i/mol dm^{-3} s^{-1}	7.29×10^{-9}	1.11×10^{-9}
v_{ps}/mol dm^{-3} s^{-1}	1.19×10^{-4}	0.45×10^{-4}
τ_s/s	1.50	4.00
$(k_p/k_t^{1/2})/(\text{dm}^3\,\text{mol}^{-1}\,\text{s}^{-1})^{1/2}$	0.1826	0.177
(k_p/k_t)	3.3×10^{-5}	3.32×10^{-5}
k_p/dm^3 mol^{-1} s^{-1}	1.01×10^3	0.94×10^3
k_t/dm^3 mol^{-1} s^{-1}	3.06×10^7	2.83×10^7

the polymer chains are initiated, until regenerated by the next pulse. The pulse width is small (15 ns) and the dark time between flashes, t_f can be varied between 0.1 and 10 s. The rate expression for this technique is given as

$$\frac{v_p t_f}{[\text{M}]} + \left(\frac{k_p}{k_t}\right) \ln\left\{1 + \frac{\rho k_t t_f}{2}\left[1 + \left(1 + \frac{4}{\rho k_t t_f}\right)^{1/2}\right]\right\} \tag{3.24}$$

where ρ is the radical concentration. Using the same approach as that used in the rotating sector technique the ratio (k_p/k_t) can be derived, by creating non-steady-state conditions.

If the molar mass distribution of the polymer, resulting from the polymerization, is measured using gel permeation chromatography, then the chain length of the polymer $\bar{v}$ formed between successive pulses can be calculated. This allows the separate evaluation of k_p from

$$\bar{v} = k_p[\text{M}]t_f \tag{3.25}$$

and so can be used to extract k_t.

O'Driscoll *et al.* used this approach to measure average values of k_p for styrene and methyl methacrylate at 298 K. These were found to be 78 and 294 dm^3 mol^{-1} s^{-1} respectively.

KINETIC PARAMETERS

Some typical results for the photo-initiated polymerization of vinyl acetate at two intensities of radiation are given in table 3.5.

3.12 Activation energies and the effect of temperature

The influence of temperature on the course of a polymerization reaction depends on initiator efficiency and decomposition rate, chain transfer, and chain propagation, but it is important to have some knowledge of its effect in order to formulate optimum conditions for a reaction.

The energy of activation of a polymerization reaction is easily determined using an Arrhenius plot when the rate constants have been determined at several temperatures, but even for a simple reaction the overall rate is still a three-stage process and the total activation energy is the sum of the three appropriate contributions for initiation, propagation, and termination.

Remembering that v_p is proportional to $k_p(k_d/k_t)^{1/2}$, the overall activation energy E_a is

$$E_a = \tfrac{1}{2}E_d + (E_p - \tfrac{1}{2}E_t). \tag{3.26}$$

The term $(E_p - \frac{1}{2}E_t)$ provides a measure of the energy required to polymerize a particular monomer and has been estimated for styrene to be $27.2\,\mathrm{kJ\,mol^{-1}}$ and for vinyl acetate $19.7\,\mathrm{kJ\,mol^{-1}}$. Initiators have values of E_d in the range 125 to $170\,\mathrm{kJ\,mol^{-1}}$ and this highlights the controlling role of the initiation step in free radical polymerization. Consequently values of E_a are generally in the range 85 to $150\,\mathrm{kJ\,mol^{-1}}$.

Typical values for these quantities are shown in table 3.6. Since the temperature term in the rate equation is $\exp\{(\frac{1}{2}E_t - \frac{1}{2}E_d - E_p)/RT\}$, the exponent will normally be

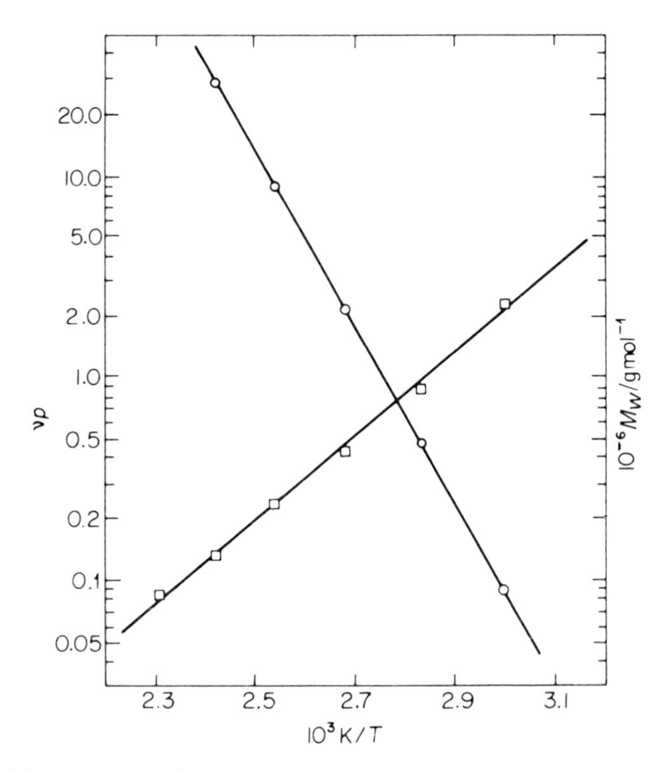

FIGURE 3.9. Dependence of the polymerization rate v_p, expressed as per cent conversion per hour, –○– and polymer molar mass –□– on temperature, for the thermal, self-initiated polymerization of styrene.

TABLE 3.6. Parameters in typical radical chain polymerizations

Monomer	$10^{-3}k_p$ $dm^3\,mol^{-1}\,s^{-1}$	E_p $kJ\,mol^{-1}$	A_p $dm^3\,mol^{-1}\,s^{-1}$	$10^{-7}k_t$ $dm^3\,mol^{-1}\,s^{-1}$	E_t $kJ\,mol^{-1}$	$10^{-9}A_t$ $dm^3\,mol^{-1}\,s^{-1}$
Vinyl chloride	12.3	15.5	0.33	2300	17.6	600
Acrylonitrile	1.96	16.3	—	78.2	15.5	—
Methyl acrylate	2.09	29.7	10	0.95	22.2	15
Methyl methacrylate	0.705	19.7	0.087	2.55	5.0	0.11
1,3-Butadiene	0.100	38.9	12	—	—	—

†Here A is the collision frequency factor in the Arrhenius equation $k = A\exp(-E/RT)$.

negative so that the polymerization rate will increase as the temperature is raised. The change in molar mass can also be examined in this way and now the quantity $\{k_p/(k_d k_t)^{1/2}\}$ is the one of interest. The required energy term is $\exp\{(E_p - \frac{1}{2}E_d - \frac{1}{2}E_t)/RT\}$ and in thermal polymerizations this is negative and usually about $-60\,\mathrm{kJ\,mol^{-1}}$. As the temperature increases, the chain length decreases rapidly and only in pure photochemical reactions, where E_d is zero, is the activation energy slightly positive, leading to a modest increase in x_n as the temperature goes up.

3.13 Thermodynamics of radical polymerization

The conversion of an alkene to a long chain polymer has a negative enthalpy (ΔH_p is negative) because the formation of a σ-bond from a π-bond is an exothermic process. While the enthalpy change favours the polymerization, the change in entropy is unfavourable and negative because the monomer becomes incorporated into a covalently bonded chain structure. However, examination of the relative magnitudes of the two effects shows that whereas $-\Delta S_p$ is in the range 100 to $130\,\mathrm{J\,K^{-1}\,mol^{-1}}$, $-\Delta H_p$ is normally 30 to $150\,\mathrm{kJ\,mol^{-1}}$. The overall Gibbs free energy change $\Delta G_p = \Delta H_p - T\Delta S_p$ is then negative and the polymerization is thermodynamically feasible.

These conditions favour the formation of polymer, but it is obvious from the general treatment of the energetics of the reaction that the chain length decreases as the temperature rises. This can be understood if we postulate the existence of a depolymerization reaction.

When the temperature increases, the depolymerization reaction becomes more important and ΔG_p becomes less negative. Eventually a temperature is reached at which $\Delta G_p = 0$ and the overall rate of polymerization is zero. This temperature is known as the *ceiling temperature* T_c.

If both the forward and reverse reactions are treated as chain reactions then

$$M_n^{\bullet} + M \underset{k_{dp}}{\overset{k_p}{\rightleftharpoons}} \sim M_{n+1}^{\bullet}$$

where k_{dp} is the rate constant for the depropagation. The overall rate expression is then obtained by modifying equation (3.10),

$$v_p = k_p[M^{\bullet}][M] - k_{dp}[M^{\bullet}] \tag{3.27}$$

while the degree of polymerization x is

$$x = (k_p[M] - k_{dp}[M^{\bullet}]/v_t. \tag{3.28}$$

At the ceiling temperature $v_p = 0$ and so

$$K = (k_p/k_{dp}) = 1/[M_e] \tag{3.29}$$

where $[M_e]$ is the equilibrium monomer concentration.

The ceiling temperature, T_c, can be described schematically as the intersection of the propagation and depropagation rate curves (see figure 3.10). Hence above T_c it is impossible for any polymeric material to form.

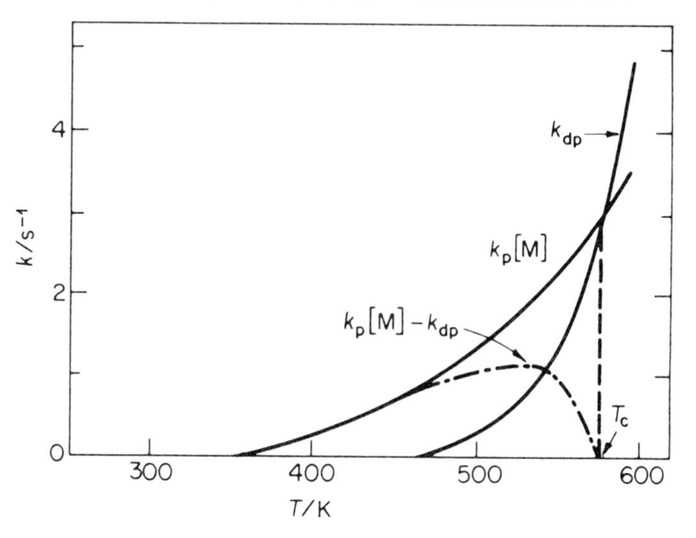

FIGURE 3.10. The temperature dependence of $k_p[M]$ and k_{dp} for styrene. (After Dainton and Ivin.)

The thermodynamic significance of T_c can be related to this kinetic analysis by using the Arrhenius expressions for the rate constants. At equilibrium

$$A_p \exp(-E_p/RT_c)[M_c] = A_{dp} \exp(-E_{dp}/RT_c) \tag{3.30}$$

and it follows that

$$T_c = (E_p - E_{dp})/\{R \ln(A_p/A_{dp}) + R \ln[M_e]\}. \tag{3.31}$$

The difference in activation energies for the forward and reverse reaction is simply the enthalpy change for the polymerization ΔH_p, so equation (3.31) becomes

$$T_c = \Delta H_p/\{R \ln(A_p/A_{dp}) + R \ln[M_e]\}. \tag{3.32}$$

The equilibrium constant can be related to the standard free energy change ΔG°

$$\Delta G^\circ = -RT \ln K = RT \ln[M_e] \tag{3.33}$$

and it follows that $R \ln(A_p/A_{dp}) = \Delta S^\circ$ the standard entropy change and

$$T_c = \Delta H_p/(\Delta S_p^\circ + R \ln[M_e]) \tag{3.34}$$

This shows that the ceiling temperature is a function of the free monomer concentration and that for a given monomer concentration there will be a specific ceiling temperature at which the particular $[M_e]$ will be in equilibrium with polymer chains. Thus for any polymerization temperature an equilibrium monomer concentration can be estimated for a selected temperature of 298 K (table 3.7).

TABLE 3.7. Equilibrium monomer concentrations for 298 K estimated from equation (3.34)

Monomer	$[M_e]$ (mol dm^{-3})
Vinyl acetate	10^{-9}
Styrene	10^{-6}
Methyl methacrylate	10^{-3}
α-Methylstyrene	2.6

As ΔH_p is negative, a rise in temperature will cause $[M_e]$ to increase, thus at 405 K methyl methacrylate has a value of $[M_e] \approx 0.5$ mol dm^{-3}, whereas α-methylstyrene will not polymerize at all. Ceiling temperatures then refer to a given monomer concentration and it is more convenient to refer it to a standard state. This can either be referred to pure liquid monomer or a concentration of 1 mol dm^{-3}; typical examples for pure liquid monomers are given in table 3.8. While the ceiling temperature alters with monomer concentration, it is also sensitive to pressure. As ΔH and ΔS are both negative, an increase in T_c is obtained if $-\Delta S$ can be decreased. This can be achieved either by raising the monomer concentration (in a solution polymerization) or by decreasing the volume change normally observed during polymerization. Experimental data show that there is a linear dependence of log T_c on pressure and the Clapeyron–Clausius equation

$$\frac{dT_c}{dP} = \frac{T_c \Delta V}{\Delta H}$$

is applicable. Typical values for the rate of increase of T_c with pressure are 0.17 K MPa^{-1} for α-methylstyrene and 0.2 K MPa^{-1} for tetrahydrofuran.

It should be noted that whereas the preceding discussion has been cast in terms of free radical polymerizations, the thermodynamic argument is independent of the nature of the active species. Consequently, the analysis is equally valid for ionic polymerizations. A further point to note is that for the concept to apply an active species capable of propagation and depropagation must be present. Thus inactive polymer can be stable above the ceiling temperature for that monomer but the polymer will degrade rapidly by a depolymerization reaction if main chain scission is stimulated above T_c.

TABLE 3.8. Ceiling temperatures based on pure liquid monomer as the standard state

Monomer	T_c *(pure liquid monomer)* (K)
Tetrafluoroethylene	853
Styrene	583
Methyl methacrylate	493
Thioacetone	368
Tetrahydrofuran	353
α-Methylstyrene	334
Acetaldehyde	242

TABLE 3.9 Heats of polymerization for selected monomers

Monomer	$-\Delta H_p{}^a$ (kJ mol^{-1})	$-\Delta H_p{}^b$ (kJ mol^{-1})
α-Methylstyrene	35.2	34.1
Isobutene	52.7	—
Formaldehyde	54.3	58.5
Methyl methacrylate	58.1	56.0
Ethyl methacrylate	58.9	60.2
Styrene	68.5	—
Vinyl acetate	89.0	—
Vinyl chloride	95.7	—
Tetrafluoroethylene	155.5	—

[a]Calorimetric.
[b]Equation (3.34).

3.14 Heats of polymerization

Addition polymerization is an exothermic process and the change in enthalpy is typically in the range 34 to 160 kJ mol^{-1}. The particular values differ for each monomer and are influenced by several factors, *viz.* (a) the energy difference between monomer and polymer resulting from resonance stabilization of the double bond by the substituent or by conjugation; (b) steric strains in the polymer imposed on the new single bonds by substituent interactions; and (c) polar or secondary bonding effects.

The most important factors are (a) and (b) and while the general observation is that the higher the resonance stabilization in the monomer the less exothermic the reaction, it is believed that steric factors have the greatest effect on ΔH_p. Thus the unusually high steric strains generated on forming poly(α-methylstyrene) are caused by interactions between the phenyl rings and α-methyl units and result in low values for ΔH_p and T_c. These also seem to be responsible for the facile unzipping degradation reaction. Polymerization of tetrafluoroethylene, on the other hand, is a highly exothermic reaction, and produces a polymer with little steric strain. Some values for ΔH_p are shown in table 3.9 which have been measured calorimetrically and can be compared favourably with data calculated from equation (3.34).

The large heat of polymerization can have serious practical consequences, especially when polymerizations are rapid, and can even lead to thermal explosions. To avoid these defects the rate of the process must be controlled or other practical expediencies adopted. Heat removal is particularly problematic in bulk polymerizations taken to high conversions as the reaction mixtures become very viscous and efficient stirring becomes difficult.

The generation of dangerous hot spots in the reaction may be avoided by keeping path lengths for heat loss low, by performing polymerizations in solution, or providing a heat sink by carrying out polymerizations in emulsions and dispersions where a large volume of inert liquid phase is present.

3.15 Polymerization processes

Industrial radical initiated polymerizations can be carried out in one of four different ways:

(a) with monomer only – *bulk*;
(b) in a solvent – *solution*;

(c) with monomer dispersed in an aqueous phase – *suspension*;
(d) or as an *emulsion*.

Bulk polymerization is used in the production of polystyrene, poly(methylmethacrylate), and poly(vinyl chloride). The reaction mixture contains only monomer and initiator, but because the reaction is exothermic, hot spots tend to develop when heat removal is inefficient. Auto-acceleration occurs in the highly viscous medium making control difficult and efficient monomer conversion is impeded. To overcome some of the disadvantages low conversions are used, after which the unreacted monomer is stripped off and recycled. The main advantages of the technique lie in the optical clarity of the product and its freedom from contaminations.

Quiescent mass polymerization is an unstirred reaction used for casting sheets of poly(methyl methacrylate). A low molar mass prepolymer is prepared and then the main polymerization is carried out *in situ* making use of the Trommsdorff effect to obtain high molar mass material and tougher sheets. The two stage approach helps to control the heat evolved.

In *solution polymerization*, the presence of the solvent facilitates heat transfer and reduces the viscosity of the medium. Unfortunately the additional complication of chain transfer arises and solvents must be selected with care.

Ethylene, vinyl acetate, and acrylonitrile are polymerized in this way. The redox initiated polymerization of acrylonitrile is an example of precipitation polymerization where the polyacrylonitrile formed is insoluble in water and separates as a powder. This can lead to undesirable side reactions known as popcorn polymerizations when tough crosslinked nodules of polymer grow rapidly and foul the feed lines in industrial plants.

Suspension polymerization counteracts the heat problem by suspending droplets of water-insoluble monomer in an aqueous phase. The droplets are obtained by vigorous agitation of the system and are in the size range 0.01 to 0.5 cm diameter. The method is in effect a bulk polymerization which avoids the complications of heat and viscosity build-up.

Emulsion polymerization is an important technological process widely used to prepare acrylic polymers, poly(vinyl chloride), poly(vinyl acetate), and a large number of copolymers. The technique differs from the suspension method in that the particles in the system are much smaller, 0.05 to 5 μm diameter, and the initiator is soluble in the aqueous phase rather than in the monomer droplets. The process offers the unique opportunity of being able to increase the polymer chain length without altering the reaction rate. This can be achieved by changing either the temperature or the initiator concentration, and the reasons for this will become more obvious when we examine the technique more closely.

The essential ingredients are monomer, emulsifying agent, water, and a water-soluble initiator. The surfactant is normally an amphipathic long chain fatty acid salt with a hydrophilic "head" and a hydrophobic "tail". In aqueous solutions these form aggregates or micelles (0.1 to 0.3 μm long), consisting of 50 to 100 molecules oriented with the tails inwards, thereby creating an interior hydrocarbon environment and a hydrophilic surface of heads in contact with the water. The micelles exist in equilibrium with free molecules in the aqueous phase and the concentration must exceed the "critical micelle concentration" of the emulsifier.

When monomer is added to the dispersion, the bulk of it remains in the aqueous phase as droplets but some dissolves in the micelles, swelling them. Free radicals are generated from a water-soluble redox system such as persulphate + ferrous

$$S_2O_8^{2-} + Fe^{2+} \rightarrow Fe^{3+} + SO_4^{2-} + SO_4^{-\bullet}$$

at a rate of $10^{16}\,dm^{-3}\,s^{-1}$. The radicals diffuse through the aqueous phase and penetrate both the micelles and droplets but as the concentration of micelles (about $10^{21}\,dm^{-3}$) far exceeds that of the droplets (10^{13} to $10^{14}\,dm^{-3}$), polymerization is centred almost exclusively in the micelle interior. After only 2 to 10 per cent conversion the character of the system has changed markedly. Constant replenishment of the polymer swollen micelles takes place by diffusion from the droplets which decrease steadily in size until at about 50 to 80 per cent conversion they have been totally consumed. Polymerization then continues at a steadily decreasing rate until all the remaining monomer in the micelle is converted into polymer.

A schematic diagram of a typical emulsion system is shown in figure 3.11.

The polymerization process can be described using the model developed by Smith and Ewart. This assumes that a micelle-containing monomer is penetrated by a radical diffusing in through the aqueous phase, and this initiates a chain propagation reaction with rate v_p. The chain will continue to grow until another radical reaches the micelle and enters where it will encounter the growing chain end and terminate it. The micelle will then remain quiescent until another chain is initiated by the entry of another radical and the process of chain growth continues until the next radical arrives and once more terminates the chain, *i.e.* the model assumes that only one active radical can be tolerated in a micelle and so this will contain either one or zero radicals at any one time. The result is that the polymerization process within any micelle in the system comprises a series of start-stop reactions, and the rate of the on-off switching is controlled by both the rate of radical production and the number of micelles in the reaction medium. As the entry of a radical into a micellar particle is random, the chances of a chain growing in a micelle at any particular time are 50:50. This means

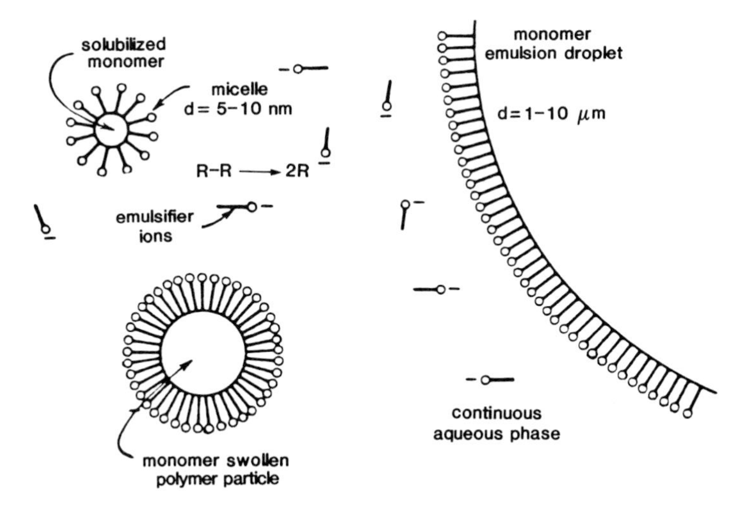

FIGURE 3.11. Schematic representation of an emulsion polymerization system.

that if there are N^* micellar particles containing monomer and polymer in the system, then on average only $N^*/2$ will be active at any period during the course of the complete polymerization reaction so the rate v_p will be proportional to the concentration of the monomer in the micelle [M*] and to the number of active micelles, ($N^*/2$) *i.e.*

$$v_p = k_p[M^*][N^*/2] \tag{3.35}$$

where k_p [M*] is the rate of polymerization within a single micelle. If the rate of radical production is v_i then the rate at which they enter a micelle is (v_i/N^*) which is the rate of initiation (or termination) in the micelle. The kinetic chain length in a micellar particle is then

$$\bar{v} = \frac{k_p[M^*]}{(v_i/N^*)} = \frac{k_p[M^*][N^*]}{2fk_d[I]} \tag{3.36}$$

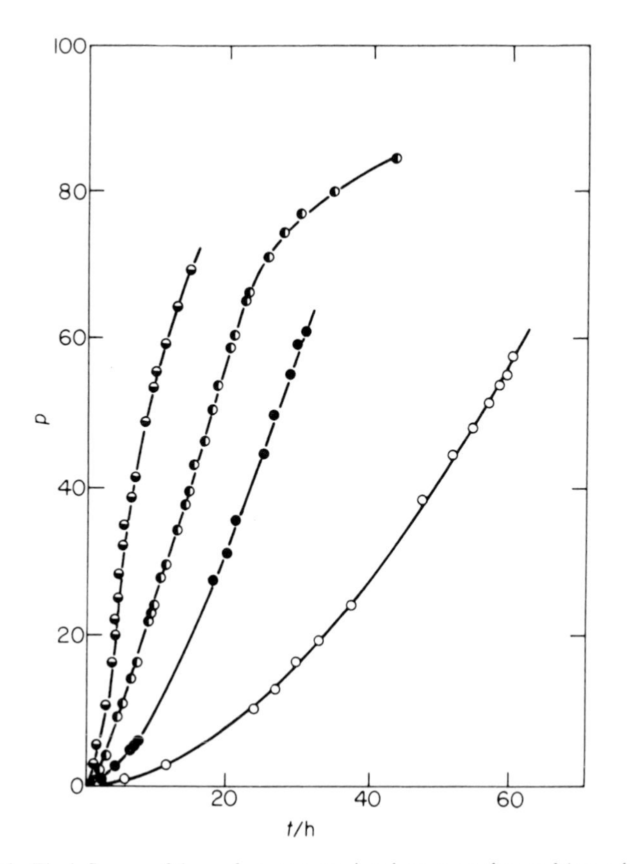

FIGURE 3.12. The influence of the surfactant potassium laurate on the emulsion polymerization of isoprene at 323 K. The percentage polymerization p is shown as a function of time t for four concentration of potassium laurate: ○, 0.01 mol dm^{-3}; ●, 0.04 mol dm^{-3}; ◐, 0.10 mol dm^{-3}; and ◒, 0.50 mol dm^{-3}.

The consequences of this analysis are (i) an increase in the initiator concentration decreases the polymer chain length while leaving the rate of polymerization unaffected, and more surprisingly, (ii) for a fixed initiator concentration both v_p and the chain length are a function of the number of micelles in the system. Thus an increase in the surfactant concentration alone is sufficient to increase the polymerization rate and the molar mass of the product. This can be understood on the basis of the model, as increasing the number of micelles while holding the rate of radical production constant, means that the time between the penetration of the micelle by successive radicals will be increased thereby allowing the chain propagation to continue for longer periods before termination.

This is illustrated in figure 3.12 for the polymerization of isoprene at 323 K using four concentrations of potassium laurate. Thus a combination of high rates and large x_n can be obtained without temperature variation and this provides the system with its particular appeal. Control of the chain length can be achieved, when desired, by adding a chain transfer agent such as dodecyl mercaptan.

3.16 Features of free radical polymerization

The main features of a radical polymerization can now be summarized and contrasted with the corresponding step-growth reactions (section 2.9).

(1) A high molar mass polymer is formed immediately the reaction begins and the average chain length shows little variation throughout the course of the polymerization.

(2) The monomer concentration decreases steadily throughout the reaction.

(3) Only the active centre can react with monomer and add units onto the chain one after the other.

(4) Long reaction times increase the polymer yield, but not the molar mass of the polymer.

(5) An increase in temperature increases the rate of the reaction but decreases the molar mass.

General Reading

H. Alter and A. D. Jenkins, "Chain-reaction polymerization" in *Encyclopaedia of Polymer Science and Technology*. Interscience Publishers Inc. (1965).

H. R. Allcock and F. W. Lampe, *Contemporary Polymer Chemistry*, Prentice-Hall (1981).

G. Allen and J. C. Bevington, Eds, *Comprehensive Polymer Science*, Vols. 3 and 4. Pergamon Press (1989).

C. H. Bamford and C. F. H. Tipper, Eds, *Free Radical Polymerization*, Vol. 14A of *Comprehensive Chemical Kinetics*. Elsevier, Scientific Publishing Company (1976).

F. W. Billmeyer, *Textbook of Polymer Science*. John Wiley and Sons, 3rd Ed. (1984).

D. C. Blackley, *Emulsion Polymerization*. Wiley, N.Y. (1975).

T. C. Bouton, J. N. Henderson and J. C. Bevington, Eds, *Polymerization Reactors and Processes*. ACS Symposium Series 104 (1979).

P. J. Flory, *Principles of Polymer Chemistry*, Chapter 4. Cornell University Press, Ithaca, N.Y. (1953).

G. E. Ham, *Vinyl Polymerization*, Vol. I. Marcel Dekker (1967).

A. D. Jenkins, "The reactivity of polymer radicals" in *Adv. in Free Radical Chemistry*, Vol. 2. Logos Press Ltd. (1967).

R. W. Lenz, *Organic Chemistry of Synthetic High Polymers*, Chapters 9–11. Interscience (1967).

D. Margerison and G. C. East, *Introduction to Polymer Chemistry*, Chapter 4. Pergamon Press (1967).

D. H. Napper, *Polymer Stabilization of Colloidal Dispersions*. Academic Press, London (1984).

G. Odian, *Principles of Polymerization*, John Wiley and Sons, 2nd Ed. (1981).

P. Rempp and E. W. Merrill, *Polymer Synthesis*. Hüthig and Wepf Verlag Basel (1986).

D. A. Smith, *Addition Polymers*, Chapter 2. Butterworths (1968).

References

1. J. N. Cardenas and K. F. O'Driscoll, *J. Polym. Sci. Polym. Chem. Ed.*, **14**, 883 (1976).
2. F. S. Dainton and K. J. Ivin, *Quarterly Reviews*, **12**, 61 (1958).
3. T. P. Davis, K. F. O'Driscoll, M. C. Piton, and M. A. Winnik, *Macromolecules*, **22**, 2785 (1989).
4. H. W. Melville, *J. Chem. Soc.*, **247** (1947).
5. O. F. Olaj, I. Bitai, and F. Hinkelmann, *Makromol. Chem.*, **188**, 1689 (1987).
6. G. V. Schulz, *Ber.*, **80**, 232 (1947).

CHAPTER 4

Ionic Polymerization

4.1 General characteristics

Radical-initiated polymerizations are generally non-specific, but this is not true for ionic initiators, since the formation and stabilization of a carbonium ion or carbanion depends largely on the nature of the group R in the vinyl monomer $CH_2{=}CHR$. For this reason cationic initiation is usually limited to monomers with electron-donating groups which help to stabilize the delocalization of the positive charge in the π-orbitals of the double bond. Anionic initiators require electron withdrawing substituents (—CN, —COOH, $—CH{=}CH_2$, *etc.*) to promote the formation of a stable carbanion, and when there is a combination of both mesomeric and inductive effects the stability is greatly enhanced.

As these ions are associated with a counter-ion or gegen-ion the solvent has a profound influence. Chain propagation will depend significantly on the separation of the two ions and this separation will also control the mode of entry of an adding monomer. Also, the gegen-ion itself can influence both the rate and stereochemical course of the reaction. While polar and highly solvating media are obvious choices for ionic polymerizations, many cannot be used because they react with and negate the ionic initiators. This is true of the hydroxyl solvents, and even ketones will form stable complexes with the initiator to the detriment of the reaction. As solvents of much lower dielectric constant have to be used, the resulting close proximity of the gegenion to the chain end requires that one must treat the propagating species as an *ion pair*, but even in low polarity media such as methylene chloride, ether, THF, nitrobenzene, *etc.*, the ion pair separation can vary sufficiently for the effects to be distinguishable.

Ionic-initiated polymerizations are much more complex than radical reactions. When the chain carrier is ionic, the reaction rates are rapid, difficult to reproduce, and yield high molar mass material at low temperatures by mechanisms which are often difficult to define.

Complications in the kinetic analysis can arise from co-catalyst effects where small quantities of an inorganic compound, such as water, will have an unexpectedly large influence on the polymerization rate.

Initiation of an ionic polymerization can occur in one of four ways involving essentially the loss or gain of an electron e^- by the monomer to produce an ion or radical ion.

(a) $M + I^+ \rightarrow MI^+$	Cationic
(b) $M + I^- \rightarrow MI^-$	Anionic
(c) $M + e^- \rightarrow {}^{\bullet}M^-$	Anionic
(d) $M - e^- \rightarrow {}^{\bullet}M^+$	Cationic (charge transfer)

4.2 Cationic polymerization

Ionic polymerizations proceed by a chain mechanism and can be dealt with under the general headings that were used for the radical reactions: initiation, propagation, and termination. A common type of cationic initiation reaction is that represented in (a) where I^+ is typically a strong Lewis acid. These electrophilic initiators are classed in three groups: (1) classical protonic acids or acid surfaces – HCl, H_2SO_4, $HClO_4$; (2) Lewis acids or Friedel-Crafts catalysts – BF_3, $AlCl_3$, $TiCl_4$, $SnCl_4$; (3) carbenium ion salts.

The most important initiators are the Lewis acids MX_n, but they are not particularly active alone and require a co-catalyst SH to act as a proton donor. In general we have first an ionization process

$$MX_n + SH \rightleftharpoons [SMX_n]^- H^+$$

followed by an initiation mechanism which is probably the two-step process

$$\text{(i)}\quad \mathrm{H_2C{=}CR_1R_2} + H^+[SMX_n]^- \rightarrow \left[\mathrm{H_2C \overset{H}{\vdots} {=} CR_1R_2}\right]^+ [SMX_n]^-$$

$$\text{(ii)}\quad \rightleftharpoons \left[\mathrm{H{-}CH_2{-}C^+R_1R_2}\right] [SMX_n]^-$$

where step one is the rapid formation of a π-complex and step two is a slow intramolecular rearrangement. While the need for a co-catalyst is recognized it is often difficult to demonstrate, and a useful reaction which serves this purpose is the polymerization of isobutylene. This reaction proceeds rapidly when trace quantities of water are present but remains dormant under anhydrous conditions. The active catalyst–co-catalyst species required to promote this reaction is

$$BF_3 + H_2O \rightleftharpoons H^+[BF_3OH]^-,$$

and the complex reacts with the monomer to produce a carbenium ion chain carrier which exists as an ion pair with $[BF_3OH]^-$.

$$H^+[BF_3OH]^- + (CH_3)_2C{=}CH_2 \rightarrow (CH_3)_3C^+[BF_3OH]^-$$

The type of co-catalyst also influences the polymerization rate because the activity of the initiator complex depends on how readily it can transfer a proton to the monomer.

If the polymerization of isobutylene is initiated by $SnCl_4$, the acid strength of the co-catalyst governs the rate, which decreases in the co-catalyst order: acetic acid > nitroethane > phenol > water.

Other types of initiator are less important; thus strong acids protonate the double bond of a vinyl monomer

$$HA + CH_2{=}CR_1R_2 \rightarrow A^- CH_3C^+R_1R_2,$$

while iodine initiates polymerization with the ion pair

$$2I_2 \rightarrow I^+I_3^-$$

which forms a stable π-complex with olefins such as styrene and vinyl ethers.

$$CH_2{=}CHR + I^+I_3^- \rightarrow ICH_2{-}\overset{+}{\underset{\displaystyle R}{\underset{|}{C}}}HI_3^-$$

A recent suggestion that it may be a charge transfer mechanism has not been fully substantiated.

High energy radiation is also thought to produce cationic initiation, but this may lead to fragmentation and a mixture of free radical and cationic centres.

4.3 Propagation by cationic chain carriers

Chain growth takes place through the repeated addition of a monomer in a head-to-tail manner, to the carbenium ion, with retention of the ionic character throughout.

$$CH_3{-}\overset{\displaystyle R_1}{\overset{|}{\underset{\displaystyle R_2}{\underset{|}{C}}}}{}^+[SMX_n]^- + nCH_2{=}CR_1R_2 \xrightarrow{k_p} CH_3[CR_1R_2CH_2]_n\overset{\displaystyle R_1}{\overset{|}{\underset{\displaystyle R_2}{\underset{|}{C}}}}{}^+[SMX_n]^-$$

TABLE 4.1. Cationic polymerization of styrene in media of varying dielectric constant ε

Solvent	ε	Catalyst	k_p $dm^3\,mol^{-1}\,s^{-1}$
CCl_4	2.3	$HClO_4$	0.0012
$CCl_4 + (CH_2Cl)_2 (40/60)$	5.16	$HClO_4$	0.40
$CCl_4 + (CH_2Cl)_2 (20/80)$	7.0	$HClO_4$	3.20
$(CH_2Cl)_2$	9.72	$HClO_4$	17.0
$(CH_2Cl)_2$	9.72	$TiCl_4/H_2O$	6.0
$(CH_2Cl)_2$	9.72	I_2	0.003

The mechanism depends on the *counterion*, the *solvent*, the *temperature*, and the *type of monomer*. Reactions can be extremely rapid when strong acid initiators such as BF_3 are used, and produce long chain polymer at low temperatures. Rates tend to be slower when the weaker acid initiators are used and a polymerization with $SnCl_4$ may take several days. Useful reaction temperatures are in the range 170 to 190 K and both molar mass and reaction rate decrease as the temperature is raised.

Propagation also depends greatly on the position and type of the gegen-ion associated with the chain carrier. The position of the gegen-ion can be altered by varying the dielectric constant of the solvent and large changes in k_p can be obtained as shown in table 4.1 for a perchloric acid initiated polymerization of styrene in several media.

It has been suggested that the various stages of the ionization producing carbenium ions can be represented as

$$\underset{\text{covalent}}{RX} \rightleftharpoons \underset{\text{intimate ion pair}}{R^+X^-} \rightleftharpoons \underset{\text{solvent separated ion pair}}{R^+/\!/X^-} \rightleftharpoons \underset{\text{free ions}}{R^+ + X^-}.$$

The increasing polarity of the solvent alters the distance between the ions from an intimate pair, through a solvent separated pair to a state of complete dissociation. As free ions propagate faster than a tight ion pair, the increase in free ion concentration with change in dielectric constant is reflected in an increase in k_p. The separation of the ions also lowers the steric restrictions to the incoming monomer, so that free ions exert little stereo-regulation on the propagation and too great a separation may even hinder reactions which are assisted by co-ordination of the monomer with the metal in the gegen-ion. To the first approximation it can be stated that as the dielectric constant of the medium increases, there is a linear increase in the polymer chain length and an exponential increase in the reaction rate, but in some cases the bulk dielectric of the medium may not determine the effect of the solvent on an ion in its immediate environment. This leads to deviations from the simple picture. The nature of the gegen-ion affects the polymerization rate. Larger and less tightly bound ions lead to larger values of k_p, hence a decrease in k_p is observed as the initiator changes from $HClO_4$ to $TiCl_4.H_2O$ to I_2, for the reaction of styrene in 1,2-dichloroethane.

4.4 Termination

The termination reaction in a cationic polymerization is less well defined than for the radical reactions, but is thought to take place either by a unimolecular rearrangement of the ion pair

$$\sim CH_2\overset{\displaystyle R_1}{\underset{\displaystyle R_2}{\overset{|}{\underset{|}{C^+}}}}[SMX_n]^- \rightarrow \sim CH{=}CR_1R_2 + H^+[SMX_n]^-,$$

or through a bimolecular transfer reaction with a monomer

$$\sim CH_2\overset{\overset{\displaystyle R}{|}}{\underset{\underset{\displaystyle R_2}{|}}{C^+}}[SMX_n]^- + CH_2{=}\overset{\overset{\displaystyle R_1}{|}}{\underset{\underset{\displaystyle R_2}{|}}{C}} \longrightarrow \sim CH{=}CR_1R_2 + CH_3\overset{\overset{\displaystyle R_1}{|}}{\underset{\underset{\displaystyle R_2}{|}}{C^+}}[SMX_n]^-$$

The first involves hydrogen abstraction from the growing chain to regenerate the catalyst–co-catalyst complex, while the second re-forms a monomer-initiator complex, thereby ensuring that the kinetic chain is not terminated by the reaction. In the unimolecular process, actual covalent combination of the active centre with a catalyst–co-catalyst complex fragment may occur giving two inactive species. This serves to terminate the kinetic chain and reduce the initiator complex and, as such, is a more effective route to reaction termination.

4.5 General kinetic scheme

Many cationic polymerizations are both rapid and heterogeneous which makes the formulation of a rigorous kinetic scheme extremely difficult. At best, one of general validity can be deduced, but this should not be applied indiscriminately. Following the steady-state approach outlined for radical reactions, the rate of initiation v_i of a cationic reaction is proportional to the catalyst–co-catalyst concentration c and the monomer concentration [M].

$$v_i = k_i c[M] \tag{4.1}$$

Termination can be taken as a first-order process in contrast to the free radical mechanisms, and

$$v_t = k_t[M^+] \tag{4.2}$$

Under steady-state conditions, $v_i = v_t$ and

$$[M^+] = k_i c[M]/k_t. \tag{4.3}$$

This gives an overall polymerization rate v_p of

$$v_p = k_p[M][M^+] = (k_p k_i/k_t)c[M]^2 \tag{4.4}$$

and a chain length of

$$x_n = v_p/v_t = (k_p/k_t)[M], \tag{4.5}$$

if termination, rather than transfer, is the dominant process. When chain transfer is significant,

$$x_n = k_p/k_{tr}. \tag{4.6}$$

Although not universally applicable, this scheme gives an adequate description of the polymerization of styrene by $SnCl_4$ in ethylene dichloride at 298 K.

4.6 Energetics of cationic polymerization

Having established a kinetic scheme, some explanation for the increase in overall rate with decreasing temperature may be forthcoming. The rate is proportional to $(k_i k_p/k_t)$ so the overall activation energy E is given by

$$E = E_i + E_p - E_t, \tag{4.7}$$

and for the chain length

$$E_x = E_p - E_t. \tag{4.8}$$

Propagation in a cationic polymerization requires the approach of an ion to an uncharged molecule in a relatively non-polar medium and as this is an operation with a low activation energy, E_p is much less than E_i, E_t, or E_{tr}. Consequently E is normally in the range -40 to $+60\,\text{kJ mol}^{-1}$, and when it is negative, the rather unexpected increase in k_p is obtained with decreasing temperature. It should be noted, however, that not all cationic polymerizations have negative activation energies; the polymerizations of styrene by trichloroacetic acid in nitromethane and by 1,2-dichloroethylene have E equal to $+57.8\,\text{kJ mol}^{-1}$ and $+33.6\,\text{kJ mol}^{-1}$ respectively.

The chain length, on the other hand, will always decrease as the reaction temperature rises because E_t is always $> E_p$.

4.7 Telechelic polymers via cationic polymerization

A telechelic polymer is defined as a relatively low molar mass species ($M_n \leqslant 20\,000$), with functional end groups that can be used for further reaction to synthesize block copolymers or for network formation. Cationic polymerization methods can be used to prepare these functionalized polymers using the *initiator-transfer* or "Inifer" technique perfected by Kennedy. If the initiating catalyst–co-catalyst system is prepared from a Lewis acid and an alkyl or aryl halide *i.e.*

$$BCl_3 + RCl \rightleftharpoons [R^{\oplus} BCl_4^{\ominus}]$$

and a monomer such as isobutene is added, then the resulting polymer can undergo a transfer reaction with RCl to produce a halogen terminated chain and a regenerated initiating species.

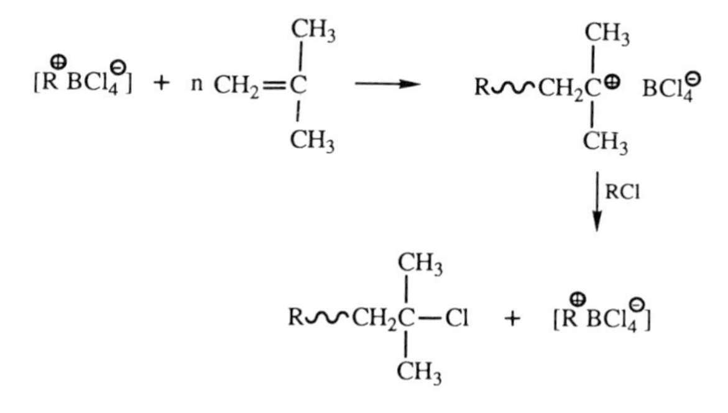

The terminal chlorine can then be converted by subsequent reactions into other useful functional groups.

$$>C{=}CH_2;\quad -CH_2OH;\quad -NCO;\quad -\underset{\displaystyle CH_2SO_3H}{\underset{|}{C}}{=}CH_2$$

Bifunctional telomers can also be prepared and these functional polymers can be used to form block copolymers using coupling reactions.

4.8 Cationic ring opening polymerization

Cyclic monomers such as lactones, lactams, cyclic amines and cyclic ethers can be encouraged to undergo ring opening reactions under the influence of cationic initiators to form linear polymers. The tendency to ring open depends on the ring size, and the primary driving force in small rings, is the relief of ring strain. The main reasons for the presence of ring strain are bond angle distortion, conformational strain and non bonded interactions, in the ring. The heat of polymerization, ΔH_p, is a good indication of the magnitude of this quantity. Table 4.2 lists some representative values for cyclic ethers, while table 4.3 shows a range of cyclic monomers that are susceptible to cationic ring opening polymerization reactions. The data in table 4.2 indicate that oxane and 1,4 dioxane do not polymerize, presumably because of the stability of the six-membered ring, but it should be noted that six-membered lactones, lactams and trioxane can form polymers. Monomers with rings larger than six tend to have lower ring strains and if they can polymerize they may also have low ceiling temperatures.

Two general mechanisms for the chain propagation in cationic ring opening polymerizations have been suggested.

One mechanism involves the primary interaction of the catalyst system with the monomer, to form an onium ion which acts as the initiating species. Propagation is then predominantly an S_N2 type substitution reaction.

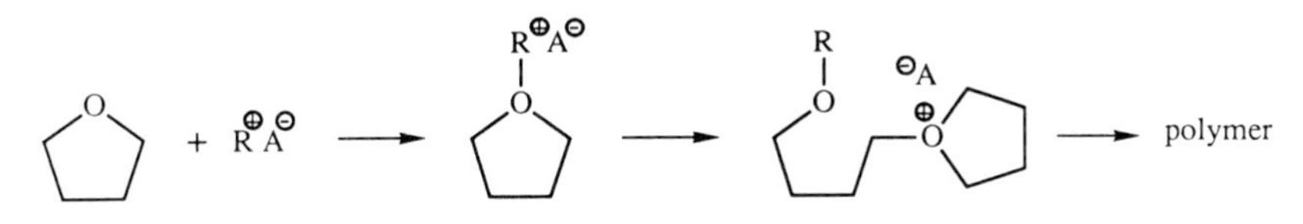

TABLE 4.2. Dependence of ΔH_p on the ring size for cyclic ethers[a]

Monomer	*Ring size*	$-\Delta H_p$(kJ mol^{-1})
Ethylene oxide (oxirane)	3	94.5
Trimethylene oxide (oxetane)	4	81
Tetrahydrofuran (oxolane)	5	15
Tetrahydropyran (oxane)	6	~0
1,4-Dioxane	6	~0
Hexamethylene oxide (oxepane)	7	33.5[b]

[a] H. Sawada, *J. Macromol. Sci. Rev. Macromol. Chem.*, **C5** (1), 151 (1970).
[b] W. K. Busfield, R. M. Lee and D. Merigold, *Makromol. Chem.*, **156**, 183 (1972).

TABLE 4.3. Heterocycles which can be polymerized cationically

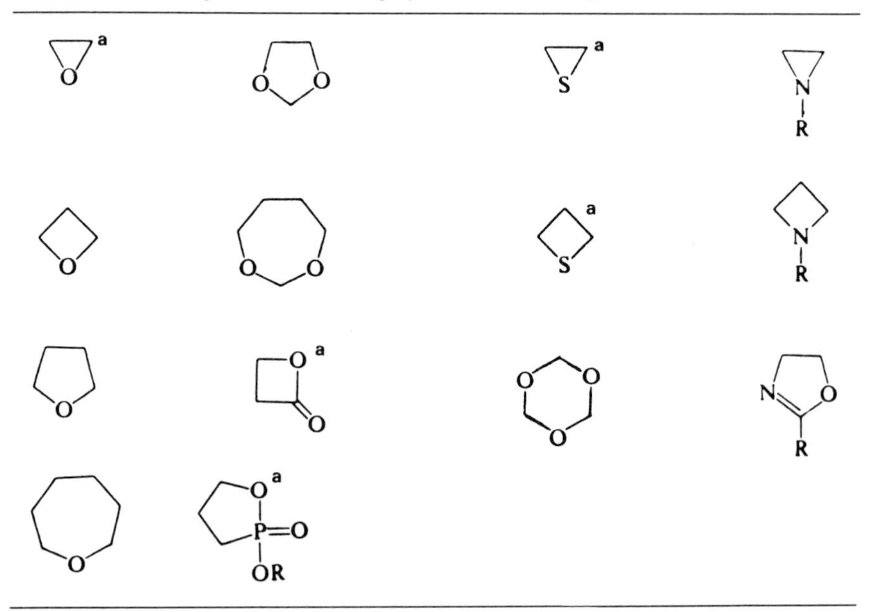

[a]These compounds can also be polymerized anionically.

An alternative mehcanism requires ring cleavage by the catalyst to form an ionic species. This is followed by attack of another monomer, with ring opening and regeneration of the active site.

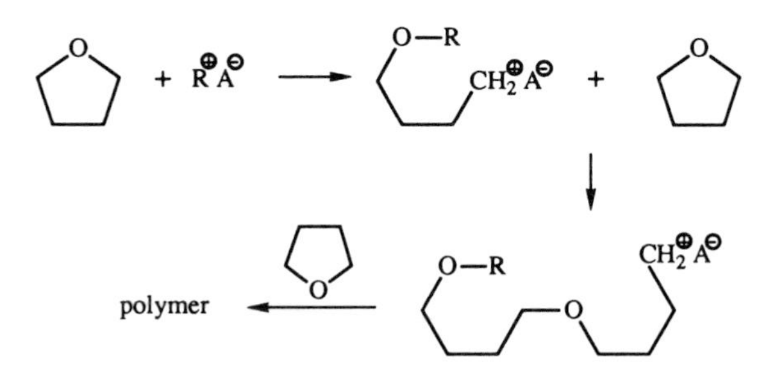

The reactions can produce commercially useful polymers but because of the low ceiling temperatures, and the tendency to degrade by an "unzipping" mechanism, end-capping of the molecules may be desirable. Thus polyformaldehyde is an engineering plastic, prepared from 1,3,5-trioxane using boron trifluoride etherate as the initiator and can be stabilized by acetylating the terminal hydroxyl group.

$$\xrightarrow{BF_3O(C_2H_5)_2} HO\!+\!CH_2O\!+_{n}\!H \xrightarrow{(CH_3COO)_2} CH_3COO\!+\!CH_2O\!+_{n}\!H$$

Cyclic ethers are not the only heterocycles that can be polymerized. Poly(ethylene imine) can be prepared by the ring opening of aziridine, which can be initiated by protonic acids followed by nucleophilic attack of the monomer to give the dimer. Further addition of the monomer can lead to linear polymer formation, but the dimer

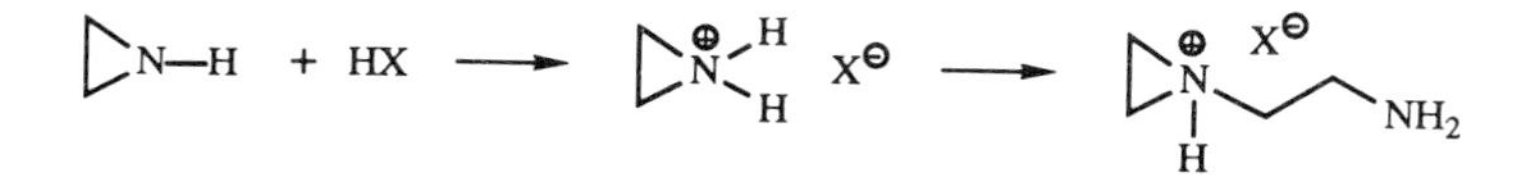

can also transfer a proton to another amino group and form an uncharged dimer. Continued chain growth leads to the formation of secondary amine functions that can react with an arizidinium ion followed by proton transfer to form branched points in the chain.

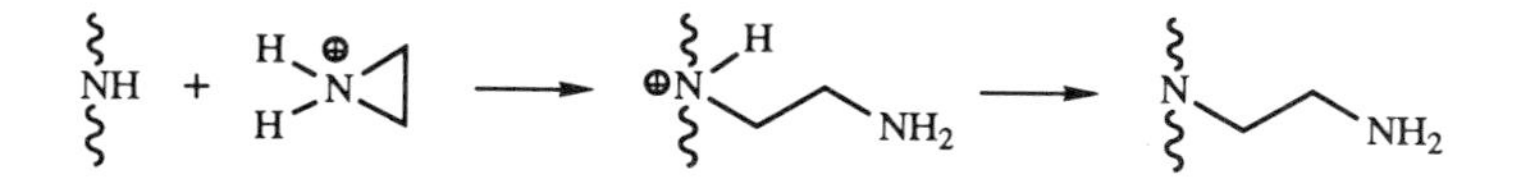

Consequently 20–30 per cent branching can be present in the final product.

A linear form of poly(ethylene imine) LPEI can be prepared by using a protecting group.

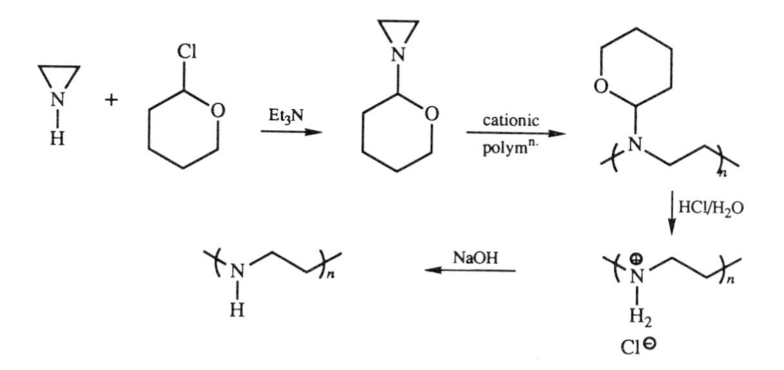

N-(α-tetrahydropyranyl)aziridine is prepared and polymerized cationically to give the poly(iminoether) which is then hydrolysed with aqueous acid to give the polymer salt. On neutralizing, LPEI is formed.

Other N-substituted aziridines can be polymerized but tend to give low molar mass products. The nature of the substituent can also have an important influence on the termination reaction, which is reflected in the magnitude of the (k_p/k_t) ratio. For N-ethyl aziridine $(k_p/k_t) \approx 6\ \mathrm{dm^3\,mol^{-1}}$, but for N-t-butyl(aziridine) it is $(k_p/k_t) \approx 12\,000\ \mathrm{dm^3\,mol^{-1}}$. In the latter case the termination reaction is so slow that the polymer produced can be regarded as a "temporary living" entity.

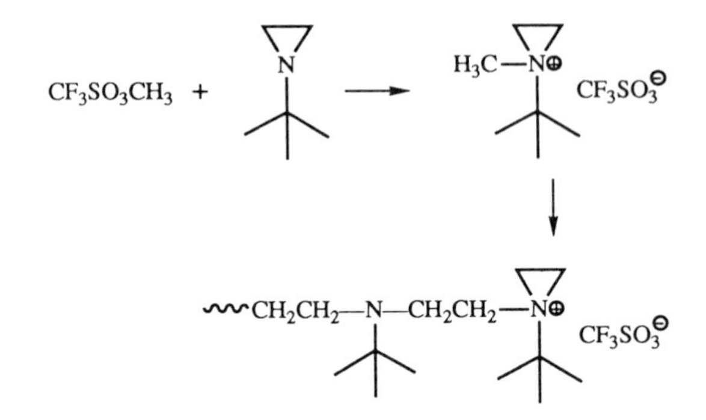

The fact that the polymers produced have a narrow molar mass distribution and can be quantitatively functionalized by a deactivator, bears testament to the living nature of this polymerization.

4.9 Stable carbocations

Many of the uncertainties inherent in Friedel-Crafts catalyst–co-catalyst systems can be removed if stable, well defined, initiators are used. Bawn and his co-workers have made use of triphenyl methyl and tropylium salts of the general formula $Ph_3C^+X^-$ and $C_7H_7^+X^-$ where X^- is a stable anion such as ClO_4^-, $SbCl_6^-$, and PF_6^-.

Initiation occurs by one of three mechanisms:

(i) Direct addition: $I^+ + CH_2{=}CHR \rightarrow ICH_2{-}\overset{+}{C}HR$
(ii) Hydride extraction: $I^+ + CH_2{=}CHR \rightarrow IH + CH{\cdots}\overset{+}{C}H{-}R$
(iii) Electron transfer: $I^+ + CH_2{=}CHR \rightarrow I^{\bullet} + {}^{\bullet}[CH_2{=}CHR]^+$

The reaction of trityl hexafluorophosphate and tetrahydrofuran (THF) has been shown to proceed without evidence of a termination reaction and a "living" cationic system can be obtained. The reaction takes place below room temperature.

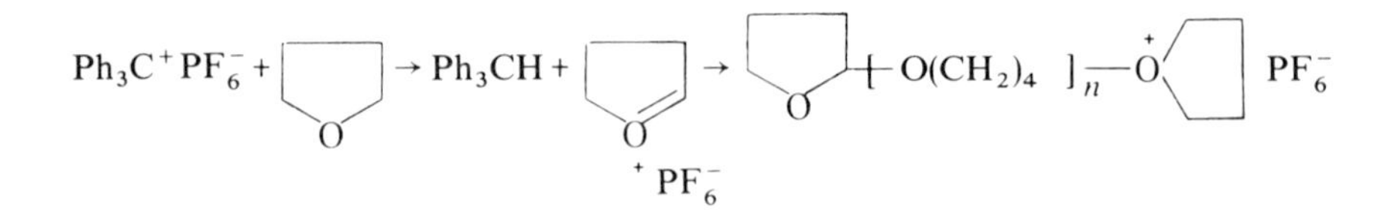

The effect of the counter-ion is a noticeable factor in the elimination of the termination reaction and neither $SbCl_6^-$ nor any other anion studied has proved as good as PF_6^-.

It has been reported that when alkyl vinyl ethers are polymerized using an (HI/I_2) initiating system, then "living" polymers are produced. The reaction involves the addition of HI to a solution of the monomer in a non polar solvent at low temperatures. This produces an inert adduct, but no polymer formation. However, a rapid reaction takes place when I_2 is introduced. This is much faster than is usually observed by simple I_2 initiation, and polymers with narrow molar mass distribution can be prepared.

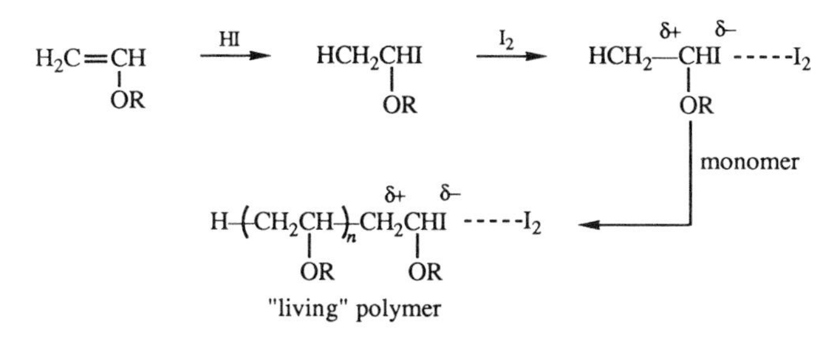

The "living" character of the system is thought to arise from the stability of the propagating species that suppresses any tendency towards termination or chain transfer reactions.

Other similar non-terminating systems have been identified, but the influence of the anion on the efficiency of the system to produce "living" polymers varies from monomer to monomer.

4.10 Anionic polymerization

The polymerization of monomers with strong electronegative groups – acrylonitrile, vinyl chloride, styrene, and methyl methacrylate – can be initiated by either mechanism (b) or (c) of section 4.1.

In (b) an ionic or ionogenic molecule is required, capable of adding the anion to the vinyl double bond and so create a carbanion.

$$CX \rightarrow C^+ + X^-$$
$$X^- + M \rightarrow MX^-$$

The gegen-ion C^+ may be inorganic or organic and typical initiators include KNH_2, *n*-butyl lithium, and Grignard reagents (alkyl magnesium bromides).

If the monomer has a strong electron withdrawing group, then only a weakly positive initiator (Grignard) will be required for polymerization, but when the side group is phenyl or the electronegativity is low, a highly electropositive metal initiator, such as a lithium compound, is needed.

Mechanism (c) is the direct transfer of an electron from a donor to the monomer to form a radical anion. This can be accomplished by means of an alkali metal, and Na or K can initiate the polymerization of butadiene and methacrylonitrile; the latter reaction is carried out in liquid ammonia at 198 K.

$$Na + CH_2{=}\underset{\displaystyle CN}{\overset{\displaystyle CH_3}{\underset{|}{\overset{|}{C}}}} \rightarrow Na^+ + {}^{\bullet}\left[CH_2{-}\underset{\displaystyle CN}{\overset{\displaystyle CH_3}{\underset{|}{\overset{|}{C^-}}}}\right]$$

The anionic reactions have characteristics similar in many ways to the cationic polymerizations. In general they are rapid at low temperatures but are slower and less sensitive to changes in temperature than the cationic reactions. Reaction rates depend on the dielectric constant of the solvent, the resonance stability of the carbanion, the electronegativity of the initiator, and the degree of solvation of the gegen-ion. Many anionic polymerizations have no formal termination step but are sensitive to traces of impurities and as carbanions are quickly neutralized by small quantities of water, alcohol, carbon dioxide, and oxygen these are effective terminating agents. This imposes the need for rather rigorous experimental procedures to exclude impurities when anionic polymerizations are being studied and a few of these procedures will be mentioned later.

4.11 Polymerization of styrene by KNH_2

One of the first anionic reactions studied in detail was the polymerization of styrene in liquid ammonia, with potassium amide as initiator, reported by Higginson and Wooding. This serves to illustrate the general mechanism encountered and has the added interest that it is one of the few reactions involving free ions rather than ion pairs. Polymerizations were performed at 240 K in a highly polar medium, liquid ammonia.

Initiation is a two-step process; the dissociation of the potassium amide, first into its constituent ions, followed by addition of the anion to the monomer to create an active chain carrier.

$$KNH_2 \rightleftharpoons K^+ + :NH_2^-$$

$$:NH_2^- + CH_2{=}CHC_6H_5 \xrightarrow{k_1} H_2NCH_2{-}\ddot{C}HC_6H_5$$

The second step is rate determining so that

$$v_i = k_i c[M], \tag{4.9}$$

where c is the concentration of the ion $:NH_2^-$.

Propagation is then the usual addition of monomer to the carbanion and the rate is given by

$$v_p = k_p[M][M^-]. \tag{4.10}$$

Termination of the growing chain occurs when there is transfer to the solvent with regeneration of the amide ion, which is usually capable of initiating another chain.

$$H_2N{+}\underset{\substack{|\\ Ph}}{CH_2CH}{)_n}CH_2{-}\overset{\substack{H\\ |}}{\underset{\substack{|\\ Ph}}{C^-}} + NH_3 \xrightarrow{k_{tr}} NH_2{+}CH_2{-}CHPh)_nCH_2CH_2Ph + :NH_2^-$$

The rate of termination is then

$$v_t = k_{tr}[M^-][NH_3]. \tag{4.11}$$

The assumption of steady-state conditions gives an expression for the concentration of propagating polycarbanions,

$$[M^-] = (k_i/k_{tr})c[M]/[NH_3], \tag{4.12}$$

giving

$$v_p = (k_p k_i/k_{tr})c[M]^2/[NH_3] \tag{4.13}$$

and

$$x_n = (k_p/k_{tr})[M]/[NH_3]. \tag{4.14}$$

The activation energy for the transfer process is larger than that for propagation and so the chain length decreases with increasing temperature, but as the overall activation energy for the reaction is positive, $+38\,\text{kJ mol}^{-1}$, the reaction rate decreases with decreasing temperature.

4.12 "Living" polymers

The reaction scheme proposed for the initiation with potassium amide contains no formal termination step and if all the impurities which are liable to react with the carbanions are excluded from the system, propagation should continue until all monomer has been consumed, leaving the carbanion intact and still active. This means that if more monomer could be introduced, the active end would continue growing unless inadvertently terminated. These active polycarbanions were first referred to as "living" polymers by Szwarc.

One of the first "living" polymer systems studied was the polymerization of styrene initiated by sodium naphthalene. The initiator is formed by adding sodium to a solution of naphthalene in an inert solvent, tetrahydrofuran.

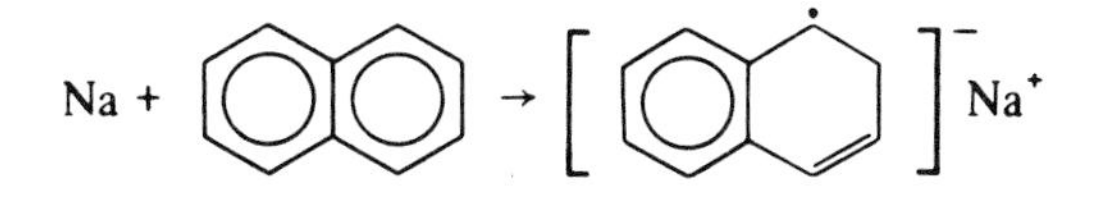

The sodium dissolves to form an addition compound and, by transferring an electron, produces the green naphthalene anion radical. Addition of styrene to the system leads to electron transfer from the naphthyl radical to the monomer to form a red styryl radical anion.

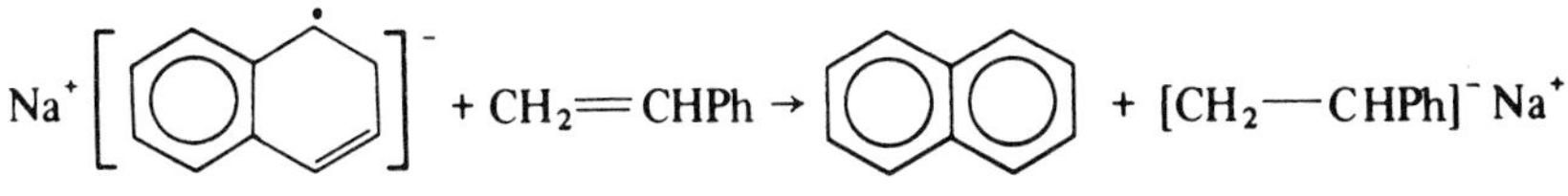

Na^+ [naphthalene radical anion]$^-$ + $CH_2{=}CHPh$ → naphthalene + $[CH_2{-}CHPh]^- Na^+$

It is thought, a dianion is finally formed capable of propagating from both ends.

$$Na^{+\,-}[PhCHCH_2CH_2CHPh]^- Na^+$$

Note that the absence of both a termination and a transfer reaction means that if no accidental termination by impurity occurs the chains will remain active indefinitely. The validity of this assumption has been demonstrated (i) by adding more styrene to

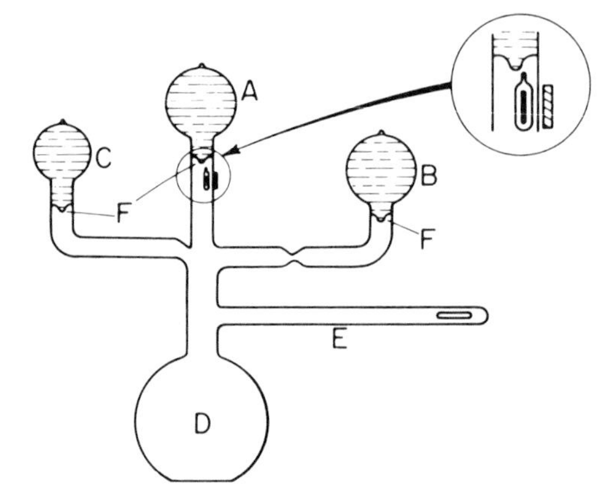

FIGURE 4.1. Apparatus similar to that used by Szwarc to demonstrate the existence of "living" polymers. The insert shows the arrangement of the internal and external magnets in relation to a break seal.

the "living" polystyryl carbanions, and (ii) by adding another monomer such as isoprene, to form a block copolymer.

The existence of "living" polymers was originally demonstrated by Szwarc using an all-glass apparatus of the type shown in figure 4.1.

The components of the reaction were subjected to stringent purification procedures and sealed in the apparatus under vacuum. The green solution of the initiator, sodium naphthalene, in THF was contained in B and introduced into D by rupturing the break-seal using a glass-encased magnet with a sharp tip, contained in the apparatus. The magnet can be held in position by a second magnet taped in position on the exterior surface of the glass tubing. Pure styrene from C was then admitted to the reactor and an immediate colour change from green to red was observed which persisted after the rapid reaction was complete. The viscosity of the reaction mixture was tested by tipping it into the side arm E, after rotation to the vertical position, and timing the fall of a piece of metal encased in glass, through the medium. The apparatus was returned through 90° to its original position and a fresh solution of styrene in THF, having the same concentration as the reaction mixture, was added from bulb A. A marked increase in viscosity indicated further growth of the existing chains (rather than new ones being formed) and the red colour of the polystyryl ion was retained. In a second experiment, isoprene was contained in bulb A and when added, formed a block copolymer with the styrene. Analysis of the product showed that no polyisoprene was formed, again substantiating the concept of a "living" polymer. The lack of a formal termination reaction in "living" anionic systems can be put to good use, particularly in the preparation of block copolymers as well as the more unusual star-shaped and combshaped structures, as will be seen in chapter 5. Block copolymer synthesis can be effected in several ways, either by direct initiation of a new block sequence on addition of a second monomer, or by coupling reactions involving the living anionic chain ends. As there are monomers that are not reactive under anionic conditions, use of a coupling reaction between preformed blocks with functional terminal units is an alternative method. These functionalized blocks can be prepared by deliberately adding a compound at the end of the reaction that terminates the chain and is itself incorporated to form a useful end group. Thus hydroxyl groups can be inserted by reaction with a lactone or oxirane, and the addition of water will protonate the anion to form the primary alcohol group.

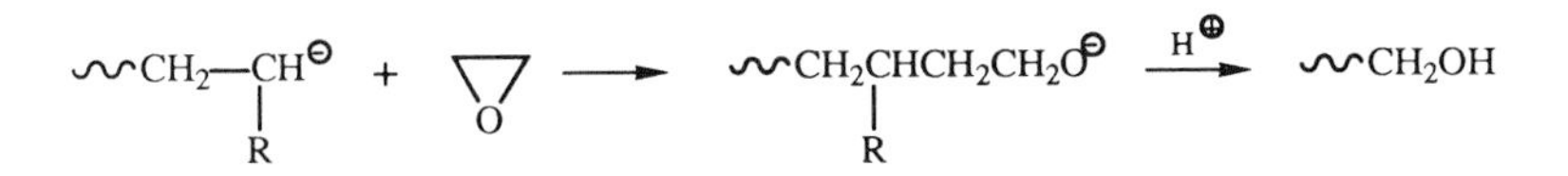

Carboxylic acid functions can be formed using carbon dioxide,

$$\sim\!\!\sim CH_2{-}\underset{\displaystyle R}{\underset{|}{C}}H^{\ominus} + CO_2 \longrightarrow \sim\!\!\sim\underset{\displaystyle R}{\underset{|}{C}}H{-}C(=O)O^{\ominus} \xrightarrow{H^{\oplus}} \sim\!\!\sim\underset{\displaystyle R}{\underset{|}{C}}HCOOH$$

while phosgene in excess produces a terminal acid chloride

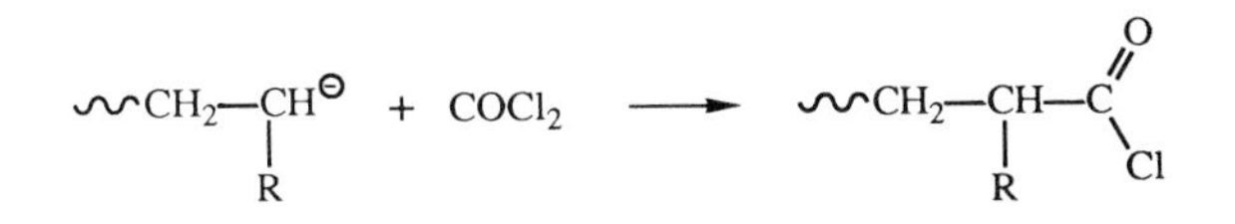

Ketones and amide functional groups can also be introduced.

4.13 Kinetics and molar mass distribution in "living" anionic systems

It can be assumed that in a system designed specifically to produce a "living" polymer, the initiator is completely dissociated in the medium before monomer is added. Under those circumstances free ion propagation of the chain should occur, when monomer is brought in contact with the initiator, and all chains will begin to grow at approximately the same time. This should lead to a polymer sample with a very narrow distribution of molar masses and, as we shall see, a Poisson type distribution is expected. If the initial concentration of the initiator is [GA] then the initiating steps are

$$\begin{gathered} GA \rightarrow G^{\oplus} + A^{\ominus} \\ G^{\oplus} + A^{\ominus} + M \rightarrow AM_1^{\ominus} + G^{\oplus} \end{gathered} \tag{4.15}$$

and chain growth will start at $[AM_1^{\ominus}] = [GA]$ centres. The rate of propagation is then

$$-\frac{d[M]}{dt} = v_p = k_p[AM^{\ominus}][M] = k_p[GA][M] \tag{4.16}$$

and as this is a first order rate equation with respect to the monomer concentration it can be integrated to give

$$[M] = [M]_0 \exp(-k_p[GA]t) \tag{4.17}$$

where $[M]_0$ is the monomer concentration at $t = 0$.

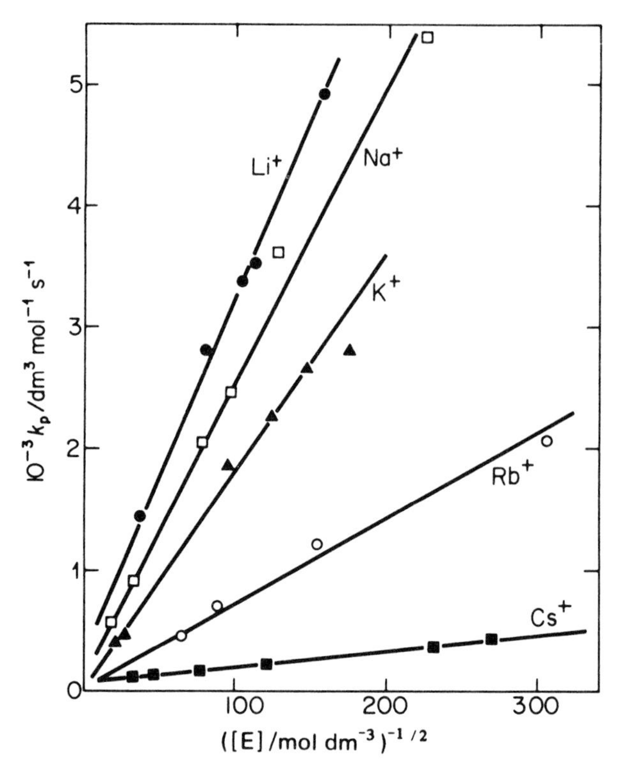

FIGURE 4.2. Behaviour of the experimental propagation rate constant k_p as a function of the concentration [E] of "living ends" for various salts of "living" polystyrene in tetrahydrofuran at 298 K. (From data by Szwarc.)

The kinetic chain length ($\bar{\nu}$) at any time during the reaction is then

$$\bar{\nu} = ([M]_0 - [M])/[GA] \tag{4.18}$$

and substituting equation (4.17) gives

$$\bar{\nu} = \frac{[M]_0}{[GA]}\{1 - \exp(-k_p[GA]t)\} \tag{4.19}$$

demonstrating that after all the monomer is consumed, *i.e.* ($t \to \infty$), then

$$\bar{\nu} = [M]_0/[GA] = x_n$$

when there is no termination reaction.

It can be shown that this type of polymerization leads to a Poisson distribution of chain lengths by considering the following steps.

The rate of addition of the second monomer to the active centre

$$AM_1^{\ominus} + M \rightarrow AM_2^{\ominus}$$

is (always assuming the presence of $G^{\oplus}$)

$$-\frac{d[AM_1^{\ominus}]}{dt} = k_p[AM_1^{\ominus}][M] = k_p[AM_1^{\ominus}][M]_0 \exp(-k_p[GA])t) \tag{4.20}$$

which on integration, remembering that $[AM_1^{\ominus}] = [GA]$ at $t = 0$, is

$$[AM_1^{\ominus}] = [GA] \exp\left\{\frac{-[M]_0}{[GA]}[1 - \exp(k_p[GA]t)]\right\} \tag{4.21}$$

or in a simplified form using equation (4.19), this becomes

$$[AM_1^{\ominus}] = [GA] \exp(-\bar{v}) \tag{4.22}$$

Consider next the addition of the third monomer

$$AM_2^{\ominus} + M \rightarrow AM_3^{\ominus}$$

and the rate of change in concentration of species $[AM_2^{\ominus}]$,

$$\frac{d[AM_2^{\ominus}]}{dt} = k_p[AM_1^{\ominus}][M] - k_p[AM_2^{\ominus}][M]$$

or, again using equation (4.17) and equation (4.22),

$$\frac{d[AM_2^{\ominus}]}{dt} = k_p[M]_0 \exp(-k_p[GA]t)\{[GA] \exp(-\bar{v}) - [AM_2^{\ominus}]\} \tag{4.23}$$

This can be simplified by differentiating equation (4.19) with respect to t

$$d\bar{v} = k_p[M]_0 \exp(-k_p[GA]t)\, dt \tag{4.24}$$

and substituting to give

$$\frac{d[AM_2^{\ominus}]}{d\bar{v}} = [GA] \cdot \bar{v} \cdot \exp(-\bar{v}) \tag{4.25}$$

Generalization of this analysis for $n - 1$ additions of monomer to the first active site $[AM^{\ominus}]$ to give an n-mer chain leads to

$$[AM_n^{\ominus}] = [GA] \cdot \bar{v}^{n-1} \frac{\exp(-\bar{v})}{(n-1)!} \tag{4.26}$$

This can be expressed as the number fraction N_n/N where N_n is the number of chains

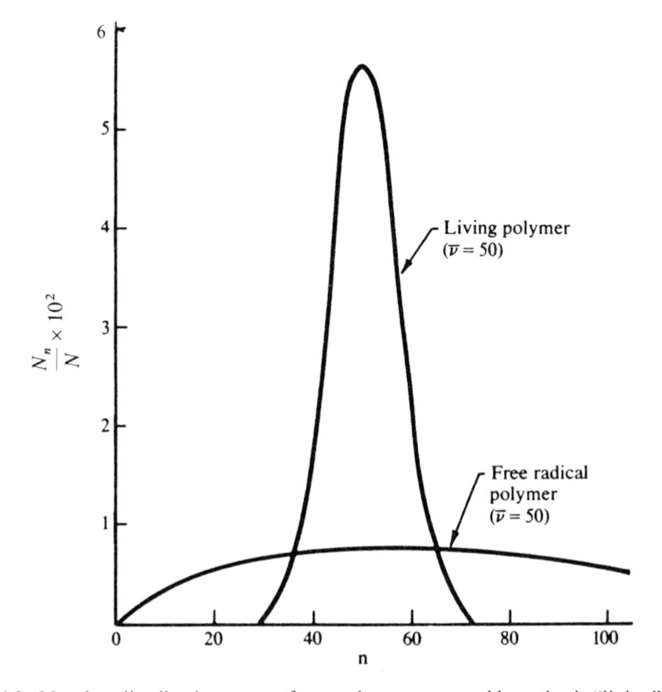

FIGURE 4.3. Number distribution curves for a polymer prepared by anionic "living" polymerization and, for comparison, that generated by free radical polymerization with termination by combination. Both curves calculated for $\bar{\nu} = 50$.

with degree of polymerization n, and N is the total number of chains

$$\frac{N_n}{N} = \frac{[AM_n^-]}{[GA]} = \frac{\bar{\nu}^{n-1}\exp(-\bar{\nu})}{(n-1)!} \tag{4.27}$$

which is the form of a Poisson distribution.

The number fraction distribution is shown in figure 4.3 for $\bar{\nu} = 50$, for which the heterogeneity index

$$\frac{M_w}{M_n} = 1 + \frac{\bar{\nu}}{(\bar{\nu}+1)^2} \tag{4.28}$$

can be calculated from equation (4.28) to be 1.02.

Also shown for comparison is the distribution curve for a radical initiated polymerization for a similar kinetic chain length calculated for termination by combination.

4.14 Metal alkyl initiators

The organo-lithium derivatives, such as n-butyl lithium, are particular members of this group of electron deficient initiators. In general, the initiation involves addition to the double bond of the monomer

$$\mathrm{RLi + CH_2{=}CHR_1 \rightarrow RCH_2{-}\overset{\displaystyle H}{\underset{\displaystyle R_1}{\overset{|}{\underset{|}{C}}}}{}^{-}\,Li^+}$$

and propagation is then

$$\mathrm{RCH_2\overset{\displaystyle H}{\underset{\displaystyle R_1}{\overset{|}{\underset{|}{C}}}}{}^{-}\,Li^+ + \mathit{n}CH_2{=}CHR_1 \rightarrow R{+}CH_2{-}CHR_1)_\mathit{n}CH_2\overset{\displaystyle H}{\underset{\displaystyle R_1}{\overset{|}{\underset{|}{C}}}}{}^{-}\,Li^+}$$

Kinetic analysis of the reactions shows that the initiation is not a simple function of the basicity of R, however, owing to the characteristic tendency for organo-lithium compounds to associate and form tetramers or hexamers. The kinetics are usually complicated by this feature, which is solvent dependent, and consequently fractional reaction orders are commonplace.

The alkyl lithiums have proved commercially useful in diene polymerization and some steric control over the polymerization can be obtained.

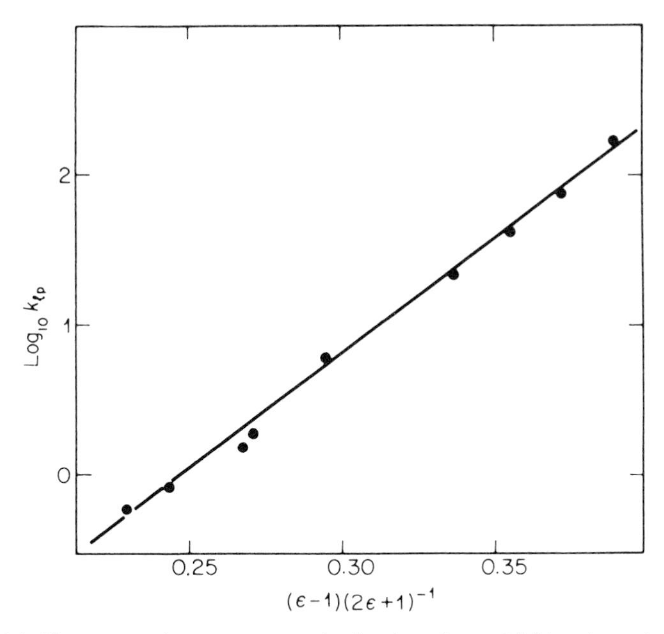

FIGURE 4.4. The propagation rate constant k_{1p} for the polystyryl-lithium ion pair plotted as a function of the dielectric constant of the reaction medium. (Adapted from Bywater and Worsfold.)

4.15 Solvent and gegen-ion effects

Both the solvent and the gegen-ion have a pronounced influence on the rates of anionic polymerizations. The polymerization rate generally increases with increasing polarity of the solvent, for example, $k_p = 2.0\,\text{dm}^3\,\text{mol}^{-1}\,\text{s}^{-1}$ for the anionic polymerization of styrene in benzene, but $k_p = 3800\,\text{dm}^3\,\text{mol}^{-1}\,\text{s}^{-1}$ when the solvent is 1,2-dimethoxyethane. Unfortunately the dielectric constant is not a useful guide to polarity or solvating power in these systems as $k_p = 550\,\text{dm}^3\,\text{mol}^{-1}\,\text{s}^{-1}$ when the solvent is changed to THF whose dielectric constant ε is higher than ε for 1,2-dimethoxyethane.

The influence of the gegen-ion on the polymerization of styrene in THF at 298 K is shown in figure 4.2 compiled from data obtained by Szwarc. Clearly the smaller Li^+ ions can be solvated to a greater extent than the larger ions and the decreasing rate reflects the increasing tendency for ion pairs to be the active species, rather than free ions, as the solvating power of the solvent deteriorates.

The effect of increasing the dielectric constant of the solvent on the propagation rate constant for an ion pair k_{lp}, has been demonstrated by Bywater and Worsfold (see figure 4.4). They plotted $\log_{10}k_{\text{lp}}$ against $(\varepsilon - 1)/(2\varepsilon + 1)$ for polystyryllithium in several THF + benzene mixtures and found that k_{lp} increased as the solvation increased. The solvation was measured by an increase in ε.

4.16 Anionic ring opening polymerization

The ring opening polymerization of oxiranes, thiiranes and thietanes can be initiated by both cationic and anionic methods, but there are some heterocyclic compounds such as lactones and lactams that are more suited to the anionic technique.

Polyethylene oxide is readily prepared by the reaction of ethylene oxide with the potassium salt of an alcohol, and the chain is terminated by a transfer reaction to excess alcohol present.

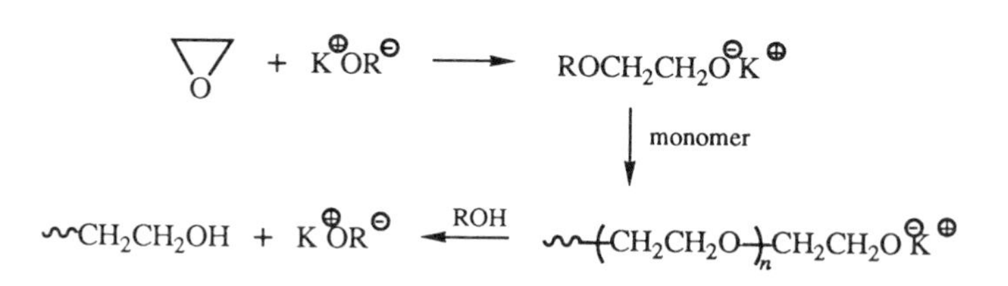

Lactones can be used to prepare polyesters, but the ring size is an important consideration *e.g.* the five membered ring γ-butyrolactone will not polymerize, whereas the six membered ring δ-valerolactone reacts.

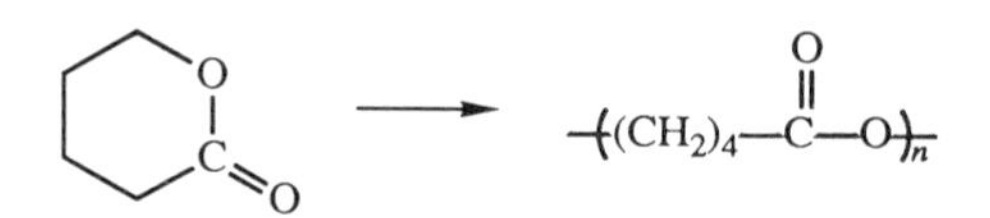

Certain lactams will undergo ring opening polymerizations to give polyamides, and nylon-6 can be prepared from the water catalysed reaction of caprolactam.

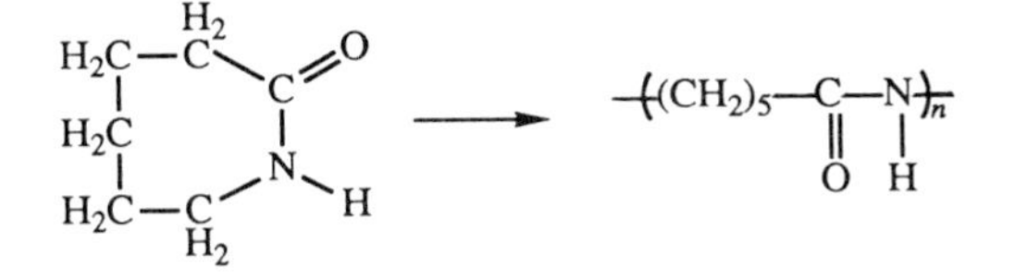

Care must be taken with this reaction and an alternative procedure is to use a two component catalyst system by reacting the lactam with a base to produce an activated monomer. This then reacts with a promotor such as the acyl lactam which initiates the ring opening growth of the linear polymer. For a series of cyclic lactams the reaction rates are a function of ring size and are in the order $8 > 7 > 11 \gg 5$ or 6 membered rings. These are important commercial processes and nylon-4 is also prepared using this type of reaction.

General Reading

G. Allen and J. C. Bevington, Eds, *Comprehensive Polymer Science*. Vol. 3. Pergamon Press (1989).

A. M. Eastham, "Cationic polymerization" in *Encyclopaedia of Polymer Science and Technology*. Interscience Publishers Inc. (1965).

E. J. Goethals, Ed., *Cationic polymerization and Related Processes*. Academic Press (1984).

T. E. Hogen-Esch and J. Smid, Eds, *Recent Advances in Anionic Polymerization*. Elsevier Science Publishing Co. Ltd (1987).

J. P. Kennedy and E. Marechal, *Carbocationic Polymerization*. John Wiley and Sons Ltd (1982).

K. J. Ivin and T. Saegusa, Eds, *Ring Opening Polymerization*. Elsevier Applied Science Publishers (1984).

R. W. Lenz, *Organic Chemistry of Synthetic High Polymers*, Chapters 13 and 14. Interscience Publishers Inc. (1967).

D. Margerison and G. C. East, *Introduction to Polymer Chemistry*, Chapter 5. Pergamon Press (1967).

M. Morton, *Anionic Polymerization: Principles and Practice*. Academic Press (1983).

P. H. Plesch, *The Chemistry of Cationic Polymerization*. Pergamon Press (1963).

D. A. Smith, *Addition Polymers*, Chapter 3. Butterworths (1968).

M. Szwarc, *Carbanions, Living Polymers and Electron Transfer Processes*. Interscience Publishers Inc. (1968).

References

1. S. Bywater, "Polymerization initiated by lithium and its compounds", *Adv. in Polymer Science*, **4**, 66 (1965).
2. S. Bywater and D. J. Worsfold, *J. Phys. Chem.*, **70**, 162 (1966).
3. D. N. Bhattacharyya, C. L. Lee, J. Smid, and M. Szwarc, *J. Phys. Chem.*, **69**, 612, (1965).
4. J. P. Kennedy and A. W. Langer, "Recent advances in cationic polymerization", *Adv. in Polymer Science*, **3**, 508 (1964).

CHAPTER 5

Copolymerization

5.1 General characteristics

In the addition reactions considered in the previous chapters, the emphasis has been on the formation of a polymer from only one type of monomer. Often it is found that these homopolymers have widely differing properties and one might think that by using physical mixtures of various types, a combination of all the desirable properties would be obtained in the resulting material. Unfortunately this is not always so, and instead it is more likely that the poorer qualities of each become exaggerated in the mixture.

An alternative approach is to try to synthesize chains containing more than one monomer and examine the behaviour of the product. By choosing two (or perhaps more) suitable monomers, A and B, chains incorporating both can be prepared using free radical or ionic initiators, and many of the products exhibit the better qualities of the parent homopolymers. This is known as *copolymerization*.

Even in the simplest case, that of copolymerization involving two monomers, a variety of structures can be obtained, and five important types exist:

(i) *Statistical copolymers* are formed when irregular propagation occurs and the two units enter the chain in a statistical fashion, *i.e.* ⌇ ABBAAAABAABBBA ⌇. This is the most commonly encountered structure.

(ii) *Alternating copolymers* are obtained when equimolar quantities of two monomers are distributed in a regular alternating fashion in the chain ⌇ ABABABA ⌇. Many of the step-growth polymers formed by the condensation of two (A—A), (B—B) type monomers could be considered as alternating copolymers but these are commonly treated as homopolymers with the repeat unit corresponding to the dimeric residue.

(iii) *Block copolymers.* Instead of having a mixed distribution of the two units, the copolymer may contain long sequences of one monomer joined to another sequence or block of the second. This produces a linear copolymer of the form AA ~ AABBB ~ B, *i.e.* an {A} {B} block, but other combinations are possible.

(iv) *Graft copolymers.* A non-linear or branched block copolymer is formed by attaching chains of one monomer to the main chain of another homopolymer.

(v) *Stereoblock copolymers.* Finally a very special structure can be formed from one monomer where now the distinguishing feature is the tacticity of each block, *i.e.*

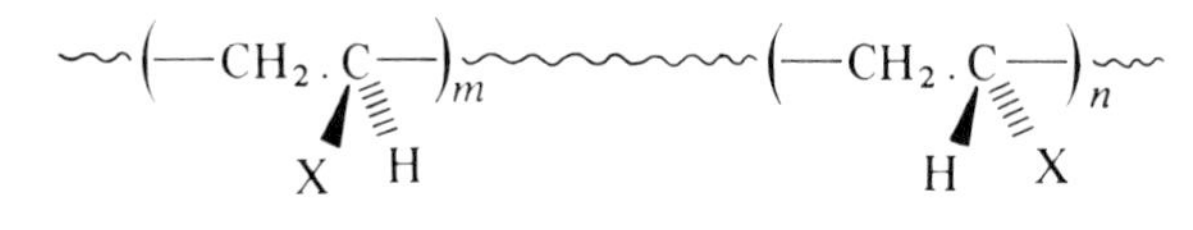

In general block and graft copolymers possess the properties of both homopolymers, whereas the random and alternating structures have characteristics which are more of a compromise between the extremes.

It soon becomes obvious that the factors influencing the course of even simple copolymerizations are much more complex than those in a homopolymerization. For example, attempts to polymerize styrene and vinyl acetate result in copolymers containing only 1 to 2 per cent of vinyl acetate while a small quantity of styrene will tend to inhibit the free radical polymerization of vinyl acetate. At the other extreme, two monomers like maleic anhydride and stilbene are extremely difficult to polymerize separately, but form copolymers with relative ease.

5.2 Composition drift

It was realized by Staudinger, as early as 1930, that when two monomers copolymerize the tendency of each monomer to enter the chain can differ markedly. He found that if an equimolar mixture of vinyl acetate and vinyl chloride were copolymerized, the chemical composition of the product varied throughout the reaction, and that the ratio of chloride to acetate in the copolymers changed from 9:3 to 7:3 to 5:3 to 5:7.

This phenomenon, known as *composition drift*, is a feature of many copolymerizations and has been attributed to the greater reactivity of one of the monomers in the mixture. Consequently, in a copolymerization, it is necessary to distinguish between the composition of a copolymer being formed at any one time in the reaction and the overall composition of all the polymer formed at a given degree of conversion.

Two major questions arise which must be answered if the criteria controlling copolymerizations are to be formulated.

(1) Can the composition of the copolymer be predicted when it is prepared from the restricted conversion of a mixture of two monomers?

(2) Can one predict the behaviour of two monomers which have never reacted before?

To answer the first question, we must explore the relative reactivity of one monomer to another, while an attempt to answer the second is embodied in the *Q–e* scheme.

5.3 The copolymer equation

To begin to answer question (1) we must establish a suitable kinetic scheme. The following group of homo- and hetero-polymerization reactions were proposed by Dostal in 1936 for a radical copolymerization between two monomers M_1 and M_2, and ultimately extended and formalized by a number of workers who established a practical equation from the reactions:

$$\sim M_1^{\bullet} + M_1 \xrightarrow{k_{11}} \sim M_1^{\bullet} \qquad (5.1a)$$

$$\sim M_1^{\bullet} + M_2 \xrightarrow{k_{12}} \sim M_2^{\bullet} \qquad (5.1b)$$

$$\sim M_2^{\bullet} + M_2 \xrightarrow{k_{22}} \sim M_2^{\bullet} \qquad (5.1c)$$

$$\sim M_2^{\bullet} + M_1 \xrightarrow{k_{21}} \sim M_1^{\bullet} \qquad (5.1d)$$

where k_{11} and k_{22} are the rate constants for the *self-propagating* reactions and k_{12} and k_{21} are the corresponding *cross-propagation* rate constants.

Under steady-state conditions, and assuming that the radical reactivity is independent of chain length and depends only on the nature of the terminal unit, the rate of consumption of M_1 from the initial reaction mixture is then

$$-\mathrm{d}[M_1]/\mathrm{d}t = k_{11}[M_1][M_1^{\bullet}] + k_{21}[M_1][M_2^{\bullet}], \tag{5.2}$$

and M_2 by

$$-\mathrm{d}[M_2]/\mathrm{d}t = k_{22}[M_2][M_2^{\bullet}] + k_{12}[M_2][M_1^{\bullet}]. \tag{5.3}$$

The *copolymer equation* can then be obtained by dividing equation (5.2) by (5.3) and assuming that $k_{21}[M_2^{\bullet}][M_1] = k_{12}[M_1^{\bullet}][M_2]$ for steady-state conditions, so that

$$\mathrm{d}[M_1]/\mathrm{d}[M_2] = ([M_1]/[M_2])\{(r_1[M_1] + [M_2])/([M_1] + r_2[M_2])\}, \tag{5.4}$$

where $k_{11}/k_{12} = r_1$, and $k_{22}/k_{21} = r_2$.

The quantities r_1 and r_2 are the relative reactivity ratios defined more generally as the ratio of the reactivity of the propagating species with its own monomer to the reactivity of the propagating species with the other monomer.

5.4 Monomer reactivity ratios

The copolymer equation provides a means of calculating the amount of each monomer incorporated in the chain from a given reaction mixture or feed, when the reactivity ratios are known. It shows that if monomer M_1 is more reactive than M_2, then M_1 will enter the copolymer more rapidly, consequently the feed becomes progressively poorer in M_1 and composition drift occurs. The equation is then an "instantaneous" expression which relates only to the feed composition at any given time.

As r_1 and r_2 are obviously the factors which control the composition of the copolymer, one must obtain reliable values of r for each pair of monomers (comonomers) if the copolymerization is to be completely understood and controlled. This can be achieved by analysing the composition of the copolymer formed from a number of comonomer mixtures with various $[M_1]/[M_2]$ ratios, at low (5 to 10 per cent) conversion (where monomer reactivities do not differ greatly).

If we now define F_1 and F_2 as the mole fractions of monomers M_1 and M_2 being added to the growing chain at any given time, and f_1 and f_2 as the corresponding mole fractions of the monomers in the feed mixture, then the copolymer equation can be written as

$$F_1 = (r_1 f_1^2 + f_1 f_2)/(r_1 f_1^2 + 2 f_1 f_2 + r_2 f_2^2) \tag{5.5}$$

This can be rearranged and simplified further if

$$F = (F_1/F_2) \text{ and } f = (f_1/f_2)$$

TABLE 5.1. Some reactivity ratios r_1 and r_2 for free radical initiated copolymerizations

M_1	M_2	r_1	r_2	r_1r_2
Acrylonitrile	Acrylamide	0.87	1.37	1.17
	Butadiene	2.0	0.1	0.2
	Methyl acrylate	0.84	0.83	0.70
	Styrene	0.01	0.40	0.004
	Vinyl acetate	6.0	0.07	0.42
Butadiene	Methyl methacrylate	0.70	0.32	0.22
	Styrene	1.40	0.78	1.1
Ethylene	Propylene	17.8	0.065	1.17
Maleic anhydride	Acrylonitrile	0	6	0
	Methyl acrylate	0.02	3.5	0.07
Methyl methacrylate	Vinyl acetate	22.2	0.07	1.55
	Vinyl chloride	10	0.1	1.0
Styrene	*p*-Fluorostyrene	1.5	0.7	1.05
	α-Methylstyrene	2.3	0.38	0.87
	Vinyl acetate	55	0.01	0.55
	2-Vinyl pyridine	0.55	1.14	0.63
Tetrafluoroethylene	Monochlorotrifluoroethylene	1.0	1.0	1.0
Vinyl chloride	Vinyl acetate	1.35	0.65	0.88
	Vinylidene chloride	0.5	0.001	0.0005

to give

$$\{f(1-F)/F\} = r_2 - (f^2/F)r_1$$

which is a linear form of equation (5.5) proposed by Finemann and Ross.

A plot of $\{f(1-F)/F\}$ against (f^2/F) should then be linear and yield r_1 from the slope and r_2 from the intercept. Several other linear forms have been suggested for the determination of the reactivity ratios but it is now much easier to estimate r_1 and r_2 from a non linear least squares fit to the composition data.

Some representative values of r_1 and r_2 for a number of comonomers, are shown in table 5.1 These are seen to differ widely.

5.5 Reactivity ratios and copolymer structure

It is obvious, from the wide ranging values of reactivity ratios shown in table 5.1, that the structure of the copolymer will also be a function of r_1 and r_2.

Several types of copolymeric structure can be obtained as we saw in section 5.1 and the influence of monomer reactivity ratios can be illustrated by examining plots of the "instantaneous" copolymer composition F_1 against the "instantaneous" monomer composition in the feed f_1, for various combinations of r_1 and r_2.

Consider first the unusual case when $r_1 \approx r_2 \approx 1$. This situation arises when little or no preference for either monomer is shown by the polymer radical, *i.e.* $k_{11} \approx k_{12}$ and $k_{22} \approx k_{21}$, and copolymerization is then entirely random. Under these conditions $F_1 = f_1$ and this is represented by curve I in figure 5.1. As this plot is reminiscent of corresponding plots for an ideal system of two liquids, the copolymers formed under

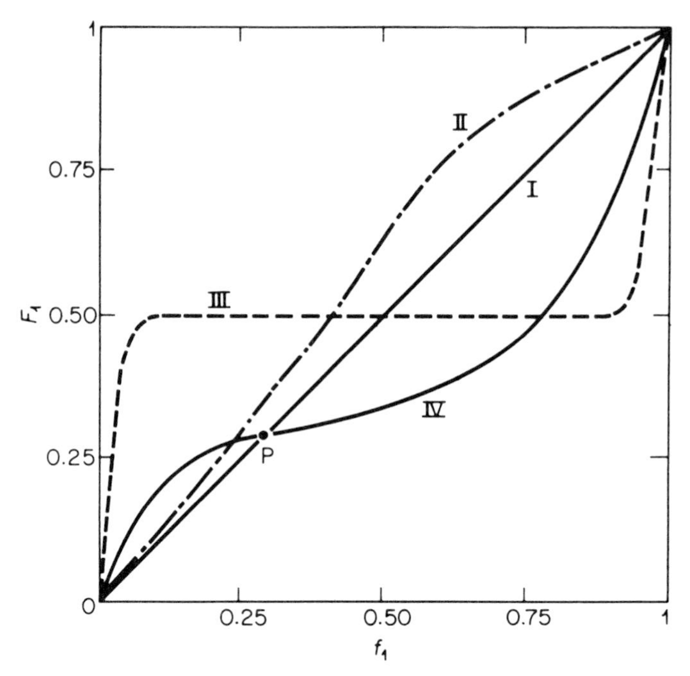

FIGURE 5.1. Variation of F_1 with f_1 for copolymerizations which are: I, completely random; II, almost ideal (i.e. $r_1r_2 = 1.17$); III, regular alternating; VI, intermediate between alternating and random (i.e. $0 < r_1r_2 < 1$) viz. statistical.

these conditions, and indeed any copolymer where the product (r_1r_2) is unity, are called IDEAL copolymers. Completely random copolymers are formed from the comonomer pairs: tetrafluoroethylene + monochlorotrifluoroethylene; isoprene + butadiene; vinyl acetate + isopropenyl acetate. However, if $r_1 > 1$ and $r_2 < 1$ or *vice versa*, but $r_1r_2 = 1$, there will be composition drift of the kind shown in figure 5.2, and when the differences between r_1 and r_2 become large, departure from ideal conditions is significant. The curve for $r_1 = 5.0$ and $r_2 = 0.2$ clearly shows that M_1 enters the copolymer more frequently than M_2 and random copolymers become increasingly difficult to prepare.

Values of (r_1r_2) are, however, more likely to be above or below unity and curve II, of figure 5.1 represents the nearly ideal pair acrylamide + acrylonitrile, for which $r_1r_2 = 1.17$. This shows the slight deviation from ideal copolymerization and illustrates the use of the curve as a guide to the composition drift which can be expected when $r_1 \neq r_2$.

In systems where r_1 and r_2 are both less than unity, copolymerization is favoured and only short sequences of M_1 and M_2 tend to form. In the extreme case when k_{11} and k_{22} are zero, $r_1 = r_2 = 0$ and a regular alternating (1:1) copolymer is formed; this is represented by curve III of figure 5.1. Strictly alternating copolymers can be prepared from the comonomers maleic anhydride + styrene, fumaronitrile + α-methyl styrene and others; however, these are rather special cases and more generally there is a greater likelihood for systems lying in the range $0 < r_1r_2 < 1$. Thus the closer the product (r_1r_2) is to zero, the greater is the tendency for M_1 and M_2 to alternate in the chain. The

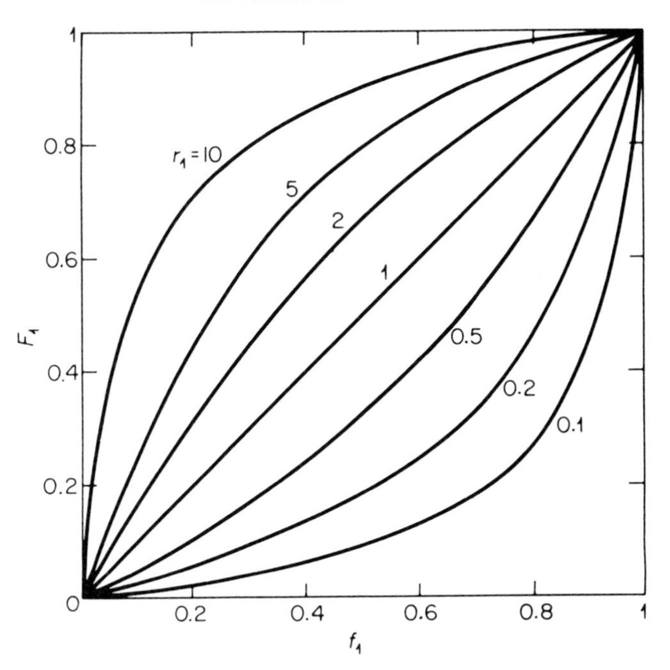

FIGURE 5.2. Plot of mole fraction F_1 of comonomer 1 in the copolymer as a function of the mole fraction f_1 of comonomer 1 in the feed, for copolymerizations in which $r_1 r_2 = 1$, showing compositional variations for several indicated values of r_1.

copolymer composition plots for these types of system are sigmoidal (curve IV) and cross the ideal line at a point P. At this point $F_1 = f_1$, and P indicates the *azeotropic copolymer composition*. This is an important feature of the system, as it represents a feed composition which will produce a copolymer of constant composition throughout the whole reaction, without having to make adjustment to the feed. This type of copolymerization, where no composition drift is observed, is known as *azeotropic* copolymerization and the critical composition f_{1C} required to obain the necessary conditions can be calculated from

$$f_{1C} = (1 - r_2)/\{2 - (r_1 + r_2)\}. \tag{5.6}$$

When r_1 and r_2 are greater than unity, *i.e.* $r_1 r_2 \gg 1$, conditions favouring long sequences or blocks of each monomer in the copolymer are obtained, and, in extreme cases, homopolymer formation may predominate.

5.6 Monomer reactivities and chain initiation

Monomer reactivities have been found to be essentially independent of the free radical process used (*e.g.* bulk, emulsion) but can be affected tremendously, for the same pair of monomers, if the chain carrier is changed.

For example, monomer reactivity ratios for styrene and methyl methacrylate in a

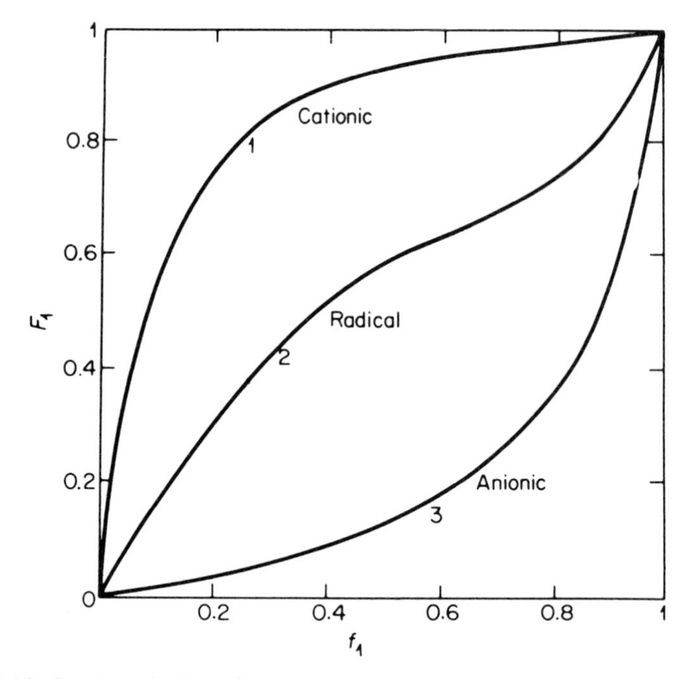

FIGURE 5.3. Copolymerization of styrene and methyl methacrylate initiated by 1, $SnCl_4$; 2, benzoyl peroxide; and 3, sodium in liquid ammonia, showing the vast differences in the dependence of F_1 on f_1 for the various types of initiator, where component 1 is styrene. (After Pepper.)

free radical copolymerization are $r_1 = 0.5$, $r_2 = 0.44$. This represents a statistical copolymerization. Contrast this with the anionic reaction, where $r_1 = 0.12$ and $r_2 = 6.4$, or the cationic reaction where $r_1 = 10.5$ and $r_2 = 0.1$. Obviously the propagation rates are no longer similar and this is represented in figure 5.3 where it can be seen that the anionic technique produces a copolymer rich in methyl methacrylate while the cationic system leads to a copolymer with a high styrene content.

This illustration merely accentuates the need to answer the questions, why do the values of r_1 and r_2 differ so widely and why does r for a given monomer change when the comonomer is changed?

5.7 Influence of structural effects on monomer reactivity ratios

The propagation rates in ionic polymerizations are influenced by the polarity of the monomers; in free radical reactions the relative reactivity of the monomers can be correlated with resonance stability, polarity, and steric effects; we shall consider only radical copolymerizations.

Resonance effects. The reactivity of a free radical is known to depend on the nature of the groups in the vicinity of the radical. If, in a vinyl monomer ($CH_2{=}CHR$), the group R is capable of aiding delocalization of the radical, the radical stability will increase, and some of the more common substituents can be arranged in order of increasing electron withdrawal:

$$C_6H_5 > -CH{=}CH_2 > -\underset{\underset{O}{\|}}{C}-CH_3 > C{\equiv}N > -\underset{\underset{O}{\|}}{C}-OR > Cl > R > -O-\underset{\underset{O}{\|}}{C}-CH_3$$

Thus styrene ($R = C_6H_5$) has a radical whose resonance stabilization is high ($84\ kJ\ mol^{-1}$) whereas vinyl acetate

$$(R = O-\underset{\underset{O}{\|}}{C}-CH_3)$$

has a very unstable radical.

As a reactive monomer forms a stable free radical, the radical reactivity will be the reverse order of the groups above. This means that monomers containing conjugated systems (styrene, butadiene, acrylates, acrylonitriles, *etc.*) will be highly reactive monomers but will form stable and so relatively unreactive radicals. Conversely, unconjugated monomers (ethylene, vinyl halides, vinyl acetate, *etc.*) are relatively unreactive towards free radicals but will form unstable and highly reactive adducts.

The suppression of radical reactivity towards a monomer is also found to be a stronger effect than the corresponding enhancement of monomer reactivity. This is true for styrene whose radical is about 10^3 times less reactive towards a given monomer than the vinyl acetate radical, but the styrene monomer is only 50 times more reactive towards a given radical than the vinyl acetate monomer.

We can now see why styrene and vinyl acetate are such a poor comonomer pair. The copolymerization requires that the stable styrene radical reacts with the unreactive vinyl acetate monomer, but this is such a slow process that the styrene tends to homopolymerize.

Broadly speaking, an efficient copolymerization tends to take place when the comonomers are either both reactive or both relatively unreactive, but not when one is reactive and the other unreactive. As with most generalizations, this is rather an extreme statement and cannot be treated too rigorously, especially when one realizes that resonance is not the only factor contributing to copolymerization behaviour, and that both steric and polar effects have to be considered.

Polar effects. It has been observed that strongly alternating copolymers are formed when comonomers with widely differing polarities are reacted together. The polarity is again determined by the side group. Thus electron withdrawing substituents, *e.g.* $-COOR$, $-CN$, $-COCH_3$, all decrease the electron density of the double bond in a vinyl monomer relative to ethylene, whereas electron donating groups, *e.g.* $-CH_3$, $-OR$, $-OCOCH_3$ increase the electron density. Hence acrylonitrile forms statistical copolymers with methyl vinyl ketone ($r_1 r_2 = 1.1$), while copolymerization of acrylonitrile with vinyl ethers leads to alternating structures ($r_1 r_2 \approx 0.0004$).

Polar forces also help to overcome steric hindrance. Neither maleic anhydride nor diethyl fumarate will form homopolymers, but both will react with styrene, stilbene, and vinyl ethers to form alternating copolymers because of the strong polar interaction. For example, the reaction between stilbene and maleic anhydride is

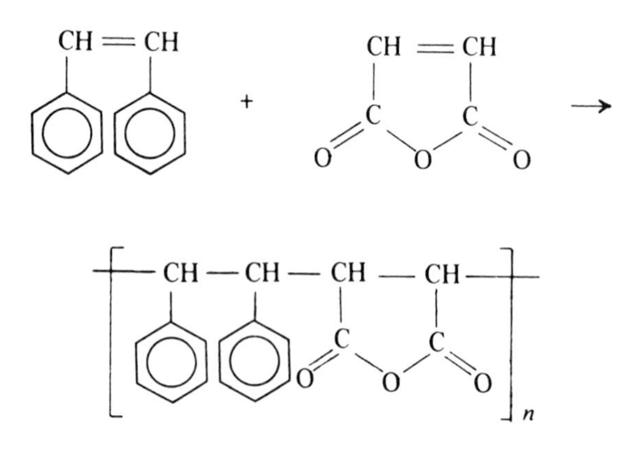

5.8 The *Q-e* scheme

All these factors contribute to the rate of copolymerization, but in a manner which makes it difficult to distinguish the magnitude of each effect.

Attempts to correlate copolymerization tendencies are thus mainly on a semi-empirical footing and must be treated as useful approximations rather than rigorous relations. A generally useful scheme was proposed by Alfrey and Price who denoted the reactivities or resonance effects of monomers by a quantity Q and radicals by P, while the polar properties were assigned a factor e which is assumed to be the same for both a monomer and its radical.

An expression for the rate constant of the cross-propagation reaction can then be derived as

$$k_{12} = P_1 Q_2 \exp(-e_1 e_2), \tag{5.7}$$

where P_1 relates to the radical $M_1^{\bullet}$ and Q_2 to the monomer M_2. This has been called the $Q-e$ scheme and can be used to predict monomer reactivity ratios by extending the treatment to give the relations for r_1 and r_2:

$$r_1 = (k_{11}/k_{12}) = (Q_1/Q_2)\exp\{-e_1(e_1 - e_2)\}, \tag{5.8}$$

$$r_2 = (k_{22}/k_{21}) = (Q_2/Q_1)\exp\{-e_2(e_2 - e_1)\}, \tag{5.9}$$

and

$$r_1 r_2 = \exp\{-(e_1 - e_2)^2\}. \tag{5.10}$$

By choosing arbitrary reference values for styrene of $Q = 1.0$ and $e = -0.8$, a table of relative values of Q and e for monomers can be compiled.

On doing this one finds that for substituents capable of conjugating with the double

TABLE 5.2 Selected values of Q and e for monomers

Monomer	Q	e
Styrene (reference)	1.0	−0.8
Acrylonitrile	0.60	1.20
1,3-Butadiene	2.39	−1.05
Isobutylene	0.033	−0.96
Ethylene	0.015	−0.20
Isoprene	3.33	−1.22
Maleic anhydride	0.23	2.25
Methyl methacrylate	0.74	0.40
α-Methyl styrene	0.98	−1.27
Propylene	0.002	−0.78
Vinyl acetate	0.026	−0.25
Vinyl chloride	0.044	0.20

bond $Q > 0.5$, whereas for groups such as Cl, OR, and alkyl, $Q < 0.1$, thereby reflecting the assumption that Q is a measure of resonance stabilization.

The values of e are also informative; for instance, maleic anhydride with two strong electron attracting side groups has $e = +1.5$ indicating an electropositive double bond. This leads to a repulsion of other maleic anhydride molecules and so no homopolymerization takes place. Similarly isobutylene has $e = -1.1$, and repulsion of like monomers is again a strong possibility. Copolymerization of oppositely charged monomers, however, should take place readily.

Although the scheme suffers from the disadvantages that steric effects are ignored, that the use of the same value of e for both monomer and radical is a doubtful assumption, and that monomers other than monosubstituted ethylenes do not fit in very well, it has proved useful in a qualitative way and should be accepted for what it is – a useful approximation.

The equation is similar to the Hammett equation which correlates monomer reactivity with structure, but the Hammett treatment is limited to substituted aromatic compounds.

5.9 Alternating copolymers

The factors that control the entry of monomers into a chain in a strictly alternating sequence are combinations of strong polar and steric effects. Thus a powerful electron donor such as SO_2 can react "spontaneously" with an electron acceptor like bicyclo(2.2.1)hept-2-ene even at temperatures as low as 230 K to form a (1:1) alternating copolymer, *i.e.*

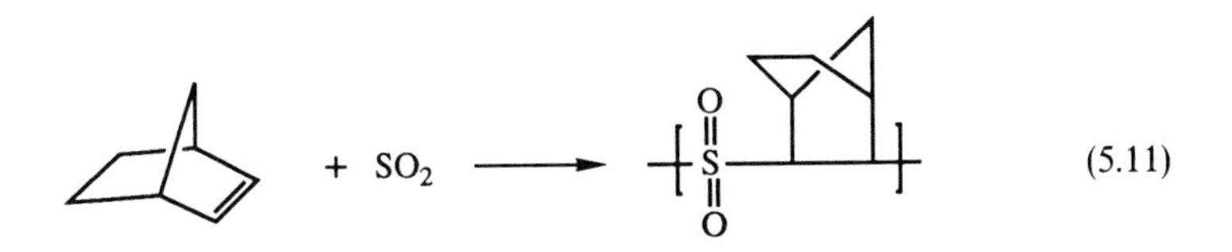

(5.11)

In most cases the spontaneity is absent but in the presence of a radical initiator, maleic anhydride, which is a powerful electron acceptor, reacts readily with a wide range of donor molecules (*e.g.* styrene, vinyl acetate, vinyl ethers) to produce copolymers with a strong tendency to form alternating structures. The perfection of the alternating sequence will depend on the relative strengths of the donor-acceptor pairs, and as this becomes weaker, statistical copolymer formation becomes more likely.

Strong acceptor molecules are usually vinyl compounds with a cyano or a carbonyl group conjugated with the double bond. The alternating tendency in systems involving this type of monomer can be enhanced by adding a Lewis acid that can complex with the acceptor and in so doing reduce the electron density on the double bond. The formation of these complexes can alter the characteristics of the acceptor molecule quite markedly; thus uncomplexed methyl methacrylate has $Q = 0.74$ and $e = 0.4$ but when complexed with $ZnCl_2$ these change to $Q = 26.2$ and $e = 4.2$.

The acceptor–Lewis acid complex reacts with conjugated donor molecules (*e.g.* styrene) under quite mild conditions to produce highly alternating structures. However, much stronger conditions are required for non-conjugated donor molecules (ethylene, propylene, vinyl acetate) where it is necessary to use alkyl aluminium sesquichloride as the Lewis acid at a temperature of 195 K. It has been postulated that a ternary molecular complex is formed

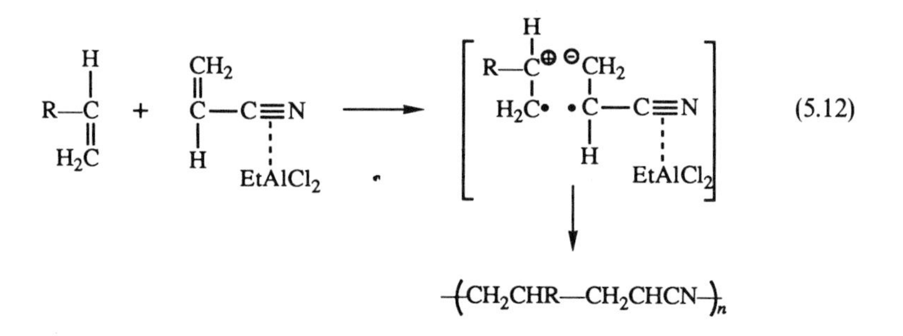

(5.12)

that can then undergo polymerization to form the alternating copolymer. This can be an oversimplification and not necessarily the mechanism for every system.

A more specialized reaction involves the use of a Ziegler catalyst, formed from vanadium and titanium halides complexed with alkyl aluminium compounds, to synthesize alternating copolymers from propylene and dienes, but the mechanism is now different.

A novel type of spontaneous alternating copolymerization, developed by Saegusa, leads to copolymer formation *via* a zwitterion, where both the propagating ion and the gegen-ion are situated at opposite ends of the chain. In general, an electrophilic monomer (M_E) interacts with a nucleophilic monomer (M_N) in the absence of a catalyst to form a dimeric dipolar species.

$$M_N + M_E \longrightarrow {}^{\oplus}M_N - M_E^{\ominus} \qquad (5.13)$$

Polymerization then proceeds with the retention of the separated charges to form a "living" polymer.

$$n(^{\oplus}M_N - M_E^{\ominus}) \longrightarrow {}^{\oplus}M_N - (M_E - M_N)_{n-1} - M_E^{\ominus} \qquad (5.14)$$

The majority of the M_N monomers are heterocyclic

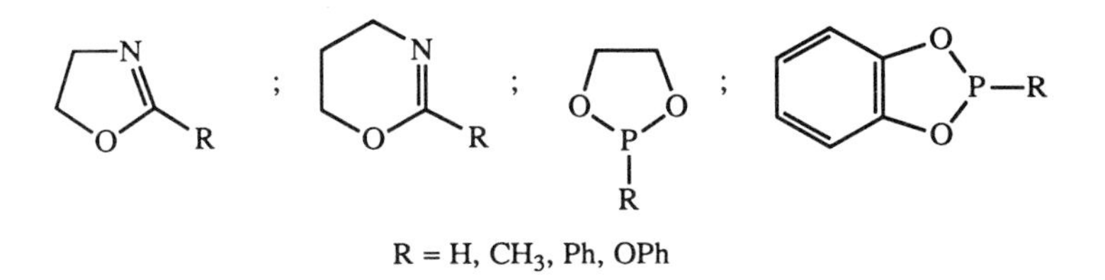

whereas M_E monomers exist in a greater variety of forms including both heterocyclic and acrylic types, *e.g.*

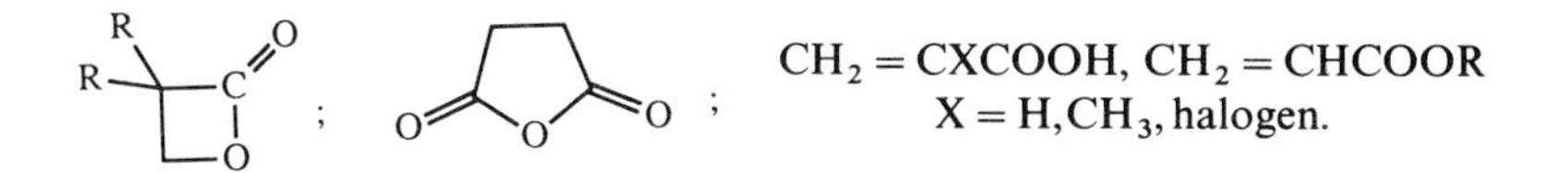

Typical reactions occurring between M_N and M_E monomers are

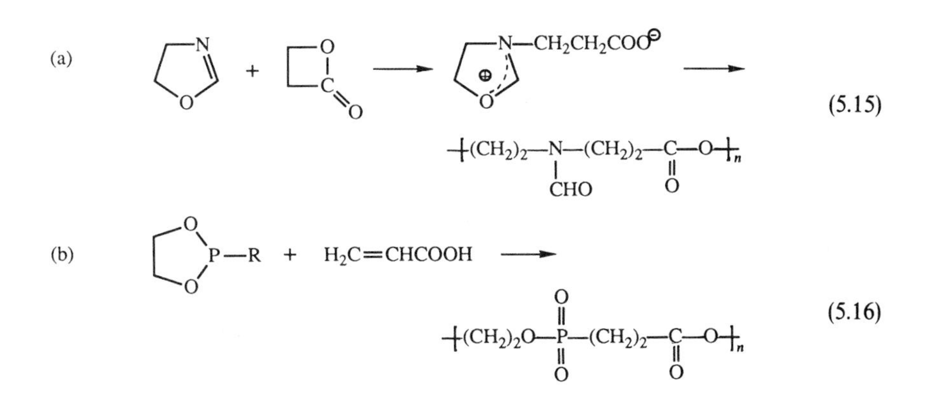

but in the majority of cases only low molecular weight polymers ($< 10\,000$) are produced.

5.10 Block copolymer synthesis

Block copolymers (figure 5.4) can be synthesized by sequential addition reactions using: (i) ionic initiators where an active site is kept "alive" on the end of the initial block, which is then capable of initiating chain growth of a second monomer on the end of the first chain; (ii) coupling of different blocks with functional terminal units, either

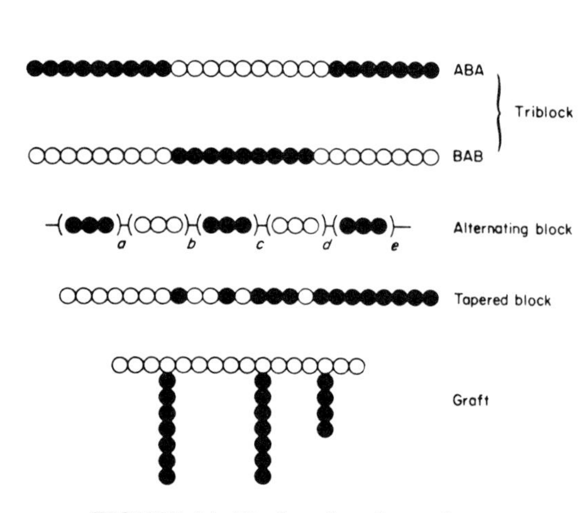

FIGURE 5.4. Block and graft copolymers.

directly or through a reaction involving a small intermediate molecule; and (iii) bifunctional radical initiators where a second potentially active site is incorporated at one end of the first chain grown, which can initiate at a later stage a new chain from the macroradical produced.

Ionic reactions are particularly successful in preparing well-defined block copolymers by making use of the observation that there is no easily discernible termination step and, if kept free from impurities, the "living" carbanionic end groups can be used to initiate the polymerization of a second monomer.

The main limitation to the method is that the anion of one monomer must be able to initiate the polymerization of a second monomer and this may not always be the case. Thus polystyryl lithium can initiate the polymerization of methyl methacrylate to give an (A – B) di-block but, because of its relatively low nucleophilicity, the methyl methacrylate anion cannot initiate styrene propagation. Best results are achieved when two monomers of high electrophilicity are used, *e.g.* styrene (St) with butadiene (Bd) or isoprene and (A – B – A) tri-blocks can be formed as shown in equations (5.17a and b).

$$(\mathrm{PSt})_x^- \mathrm{Li}^+ + y\mathrm{Bd} \longrightarrow (\mathrm{PSt}\!\!+_x\!\!(\mathrm{PBd})_y^- \mathrm{Li}^+ \qquad (5.17\mathrm{a})$$

$$(\mathrm{PSt}\!\!+_x\!\!(\mathrm{PBd})_y^- \mathrm{Li}^+ + z\mathrm{St} \longrightarrow (\mathrm{PSt}\!\!+_x\!\!(\mathrm{PBd}\!\!+_y\!\!(\mathrm{PSt})_z^- \mathrm{Li}^+ \qquad (5.17\mathrm{b})$$

$$(\mathrm{PSt}\!\!+_w\!\!(\mathrm{PBd})_x^- \mathrm{Li}^+ + \mathrm{ClCH_2Cl} + \mathrm{Li}^{+\,-}(\mathrm{PBd}\!\!+_y\!\!(\mathrm{PSt})_z$$

$$\longrightarrow (\mathrm{PSt}\!\!+_w\!\!(\mathrm{PBD}\!\!+_{x+y}\!\!(\mathrm{PSt})_z + 2\mathrm{LiCl} \qquad (5.17\mathrm{c})$$

The tri-block can also be prepared by coupling the two carbanions using an organic dihalide (equation 5.17c) and other coupling agents such as phosgene or dichlorodimethylsilane are equally effective. This method can also be used to prepare radial blocks with multifunctional compounds as illustrated with silicon tetrachloride (equation (5.18)).

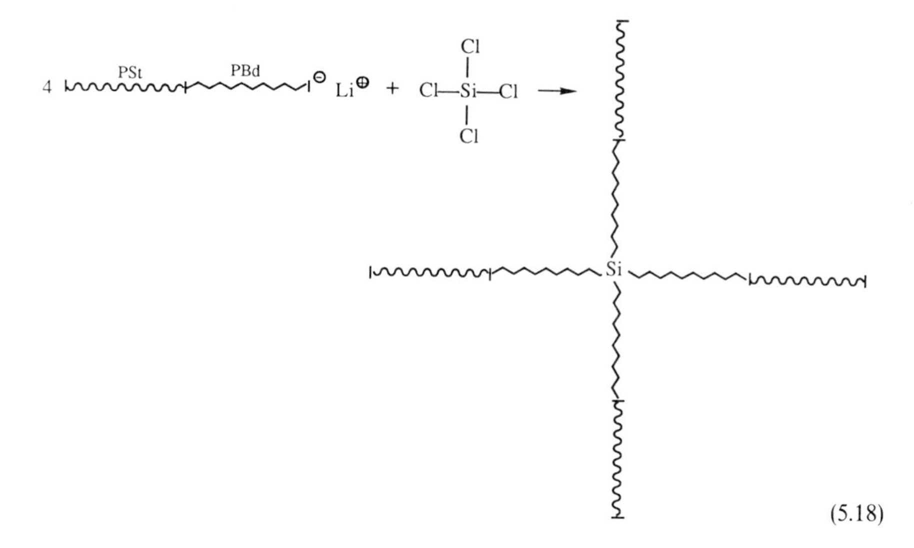

An interesting consequence of the marked differences in reactivity ratios found in some of the anionic systems is that in a mixture of monomers pure blocks of one can be obtained without incorporation of the second monomer. In styrene/butadiene mixtures the latter reacts most rapidly and can be almost completely polymerized before the styrene begins to react. As the butadiene becomes depleted, styrene is incorporated progressively until it is the only monomer left and the remaining chain grown is purely polystyrene. This produces a "tapered" di-block copolymer.

Tri-block copolymers can be constructed if a bifunctional initiator is generated as when sodium naphthalene is used with styrene or α-methylstyrene. Radical anions are formed, which combine to give a dianion and growth can then take place from both ends. Addition of a second monomer then yields a tri-block structure.

TRANSFORMATION REACTIONS

Potentially there are greater numbers of monomers that are suitable for cationic polymerizations than for anionic but the cationic method is less successful in block copolymer synthesis because, in many systems, the existence of a living carbocationic species is doubtful. Consequently, the involvement of carbocations in block copolymer synthesis tends to be limited to mixed reactions, *e.g.* the coupling of poly(tetrahydrofuran) cations with polystyryl anions to give an (A – B) di-block (equation (5.19)).

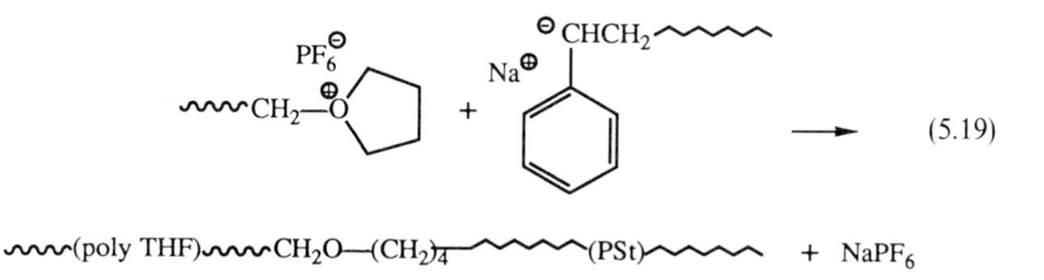

A more versatile approach is to use a transformation reaction in which one type of active terminal species is converted into a second type. Two general reactions have been identified: (i) a terminal unit anion–cation transformation by a two-electron oxidation process and (ii) carbanion to free radical conversion, which is a one-electron oxidation step

In the anion–cation transformation reaction, the anionically generated "living" polymer chain is end-capped with a halide, producing a chain which can be isolated for subsequent reaction. This can be used to initiate a cationic polymerization of a suitable monomer by activating the end with a silver or lithium salt according to the general scheme shown in equations (5.20a to c).

$$\sim M_1^- Li^+ + BrRBr \longrightarrow \sim M_1RBr + LiBr \tag{5.20a}$$

$$\sim M_1RBr + Ag^+Y^- \longrightarrow \sim M_1R^+Y^- + AgBr \tag{5.20b}$$

$$\sim M_1R^+Y^- + nM_2 \longrightarrow \sim M_1 \sim M_2^+Y^- \tag{5.20c}$$

Halides may not always be the best terminating agents and Grignard reagents have been used for this purpose with much greater success.

The reverse cation–anion transformation is also feasible and involves the end capping of the carbenium ion with a species capable of further reaction with an alkyllithium.

$$\sim M_1^+Y^- + RNH_2 \longrightarrow \sim M_1NRH + HY \tag{5.21a}$$

$$\sim M_1NRH + R'Li \longrightarrow \sim M_1NR^-Li^+ + R'H \tag{5.21b}$$

$$\sim M_1NR^-Li^+ + nM_2 \longrightarrow \sim M_1N \sim M_2^-Li^+ \tag{5.21c}$$

Anion-radical transformations can be effected in a number of ways but one must always begin with the carbanion-terminated chain.

(a) This chain can be end capped with a halide perester (equation (5.22)), which provides a chain with a potential radical-forming site at one end. Thermal decomposition of this group in a second-stage reaction, in the presence of another monomer, generates an alkoxy macroradical from which to grow the second block but also produces a second radical fragment likely to produce some homopolymer as a contaminant.

$$\sim\!\sim M_1^{\ominus}Li^{\oplus} + XR{-}\underset{\underset{O}{\|}}{C}{-}O{-}O{-}\underset{\underset{O}{\|}}{C}{-}RX \longrightarrow$$

$$\sim\!\sim M_1\,R{-}\underset{\underset{O}{\|}}{C}{-}O{-}O{-}\underset{\underset{O}{\|}}{C}{-}RX \xrightarrow[nM_2]{\Delta}$$

$$\sim\!\sim M_1R{-}\underset{\underset{O}{\|}}{C}{-}O\sim\!\sim M_2^{\bullet} + XR{-}\underset{\underset{O}{\|}}{C}{-}O\sim\!\sim M_2^{\bullet} \tag{5.22}$$

(b) An alternative route involves end capping to produce a terminal hydroxyl followed by reaction with trichloroacetyl isocyanate (equation (5.23)). This new, reactive end group can be used to initiate the growth of a second block *via* the photoreduction method proposed by Bamford where magnesium or rhenium carbonyls are excited by UV or visible radiation and extract a chlorine atom from the terminal unit, thereby creating a radical site. As only one radical is formed, this is a much "cleaner" reaction compared with (a); however, block lengths are more difficult to control in both these radical reactions and the exact structure of the product formed can depend on the mechanism of the termination reaction.

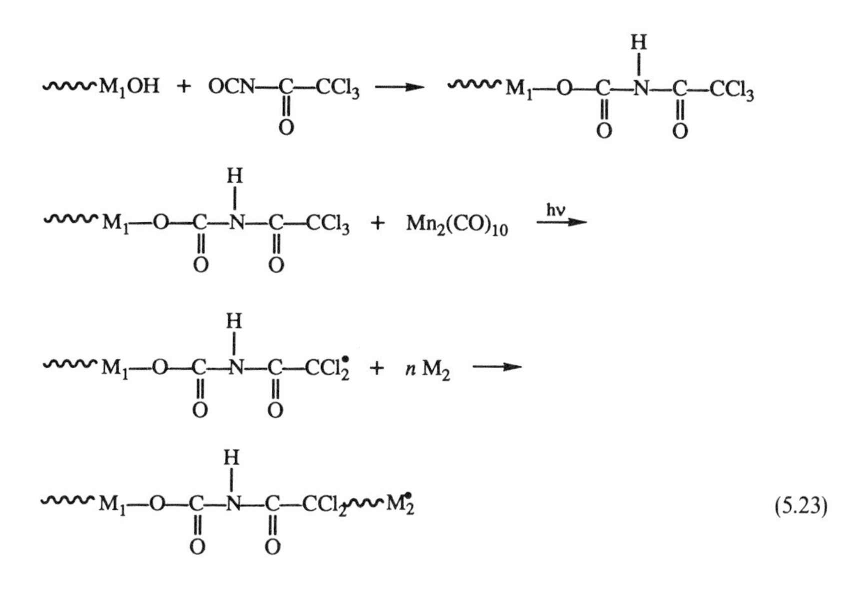

(5.23)

COUPLING REACTIONS

It is clear from the foregoing that polymer chains can be synthesized with functional groups in the α or the ω position, or both. If two different types of block are functionalized, they can be linked together to form copolymers.

Anionic polymerizations can be terminated by addition of another molecule which will introduce an ω-functional group in the chain. Excess carbon dioxide or cyclic anhydrides lead to terminal carboxylic groups, while addition of excess phosgene produces an acid chloride function. Similarly isocyanates generate ω-amide functions and lactones yield ω-hydroxyl groups.

The "Inifer" process developed by Kennedy can be used to functionalize vinyl monomers *via* a cationic route by initiating a polymerization with an alkyl halide/boron trichloride mixture $\{R^+ BCl_4^-\}$. The termination by transfer to an alkyl halide leaves a halide-terminated polymer. This can be transformed to a hydroxyl terminal unit *via* the sequence (i) dehydrohalogenation, (ii) hydroboration, (iii) oxidation and hydrolysis (equation (5.24)). These ω-functional blocks may be coupled to form di-block copolymers using standard reaction techniques, *e.g.* diisocyanates will couple ω-hydroxy and/or ω-amine blocks together. Direct reactions can also occur and ω-acid chlorides combine readily with ω-hydroxy units.

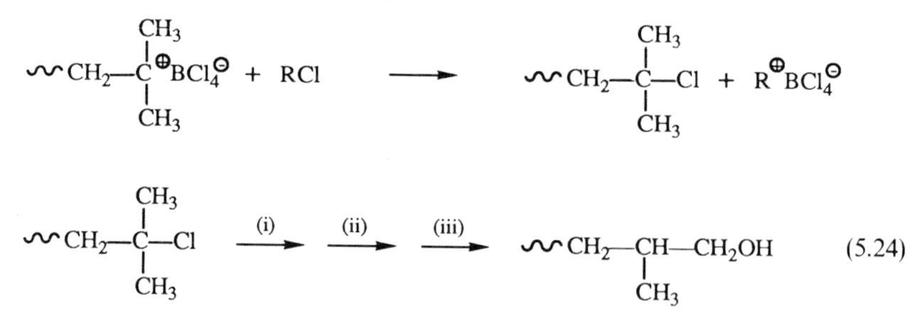

Macroazonitriles can be employed and structures based on (**1**) in equation (5.25) with either diol, acid, or acid chloride terminal functions, are preferred. Functionalized chains can be linked to the azo compounds, then the azo group can be decomposed thermally to produce radical sites for further chain growth.

$$\underset{(\mathbf{1})}{Cl{-}\overset{\displaystyle O}{\overset{\|}{C}}{-}R{-}N{=}N{-}R{-}\overset{\displaystyle O}{\overset{\|}{C}}{-}Cl} + 2OH{\sim\sim\sim} \longrightarrow$$

$$\sim\sim\sim O{-}\overset{\displaystyle O}{\overset{\|}{C}}{-}R{-}N{=}N{-}R{-}\overset{\displaystyle O}{\overset{\|}{C}}{-}O\sim\sim\sim \qquad (5.25)$$

5.11 Graft copolymer synthesis

In the synthesis of graft copolymers (figure 5.4) the sites at which the second and subsequent blocks are attached to the first are no longer terminal units but are at positions along the backbone of the first chain. There are three main techniques for preparing graft copolymers: (a) grafting "from"; (b) grafting "onto"; and (c) *via* macromonomers.

The grafting "from" procedure requires active sites to be created on the polymer chain capable of initiating the growth of other chain branches comprising a second monomer. Free radical sites can be formed by direct or mutual radiation with γ-rays of a polymer in the presence of the second monomer. This is a simple method but can also lead to homopolymer formation.

Preirradiation of a polymer in the presence of oxygen leads to formation of peroxy groups on the polymer which are relatively stable and this allows the polymer to be isolated and stored for further reaction. Polymers prepared in this way can be heated in the presence of a second monomer: the peroxy groups decompose to produce radical sites and the grafting process can take place. This method has been used to prepare poly(styrene-*graft*-acrylonitrile). The free radical approach can be used in other ways. Graft copolymers are formed when a chain transfer to preformed polymer can be effected with a second monomer present in the reaction mixture, but this depends on the radical source, *e.g.* methyl acrylate can be grafted to natural rubber when benzoyl peroxide is the initiator but is much less likely to do so with azobisisobutyronitrile. The effectiveness of this grafting "from" technique is a function of the reactivity and polarity of the radical site and the monomer. Alternatively, radical-forming sites can be introduced into the chain backbone by *in situ* modification of some monomer units or by copolymerization. Thus, a polymer with trihalide groups pendant to the chain can be activated in the presence of a second monomer thereby forming a graft rather than a block copolymer. This type of reaction may also lead to a crosslinked structure if termination of the radical by combination predominates.

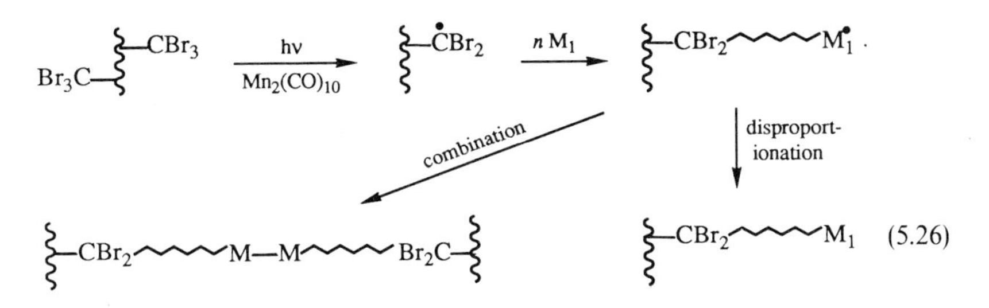

(5.26)

By altering the number of active sites on the backbone the number and distribution of grafted chains can be controlled. The length of each graft will depend on both the rate of initiation and the monomer concentration, but the mechanism by which termination of the growing radical takes place will determine the ratio of branches to crosslinks in the system. If this is exclusively by combination, then the occurrence of crosslinking will be high. However, the network formation can be modified by addition of a chain transfer agent which will produce a mixture of branches and crosslinks but also some homopolymer. In such systems the amount of chain transfer agent added will determine the ratio of branches to crosslinks.

A similar mixture of structures, but little of the contaminating homopolymer, will be obtained when a second monomer is used whose radicals terminate partly by disproportionation and partly by combination.

Finally, photodegradation of pendant ketones results in the formation of radical sites capable of initiating a graft, but like many of the other radical techniques there is also a tendency for homopolymerization to occur.

Anionic sites suitable for grafting "from" reactions can be introduced by metallation, involving the complexation of a hydrocarbon polymer by organolithium compounds. The reaction is assisted by complexing the lithium first with tetramethylenediamine, which acts as a solvating base. Aromatic chlorine is readily exchanged for lithium, which then acts as an initiator for the anionic polymerization of suitable monomers.

Grafting "onto" methods involve having sites on the main chain which can be

$$2\ \sim\!\sim M_1^{\ominus} + COCl_2 \longrightarrow \sim\!\sim(M_1)\sim C(=O)\sim(M_1)\sim\!\sim \qquad (5.27a)$$

$$\sim\!\sim(M_1)\sim C(=O)\sim(M_1)\sim\!\sim + \sim\!\sim M_2^{\ominus} \longrightarrow \sim\!\sim(M_1)\sim C(OH)(M_2\sim\!\sim)\sim(M_1)\sim\!\sim \qquad (5.27b)$$

attacked by a growing second chain, thereby linking the two by covalent bonding. In anionic polymerizations the linking of two chains by phosgene creates a polymeric ketone which can react with a second chain. Other electrophilic functional groups are effective in this reaction. *e.g.* ester, nitrile, anhydride, *etc.*, and can be used as grafting sites by growing carbanions such as the polystyryl ion. Many of these grafting techniques produce random branching along the primary chain, also a distribution of branch lengths. A more controlled grafting procedure can be used to produce regular comb-branch structures and this is achieved either by polymerization of macromonomers or by using a polymer analogous reaction on a suitable backbone.

In the first case macromonomers can be prepared by functionalizing a short chain with a vinyl unit. A typical reaction is shown in equation (5.28) and other methods have been reported. Polymerization of these monomers produces a well-defined graft structure with branches located at regular intervals along the chain. If the starting macromonomers have a uniform length, then the branches will also be regular, but mixed lengths can also be prepared. Copolymerization with another monomer will alter the regularity of the branching points but will maintain the uniformity of branch length.

Poly(acid chloride)s have been used for polymer analogous reactions and ω-functionalized units can be condensed at these sites to produce structures similar to those obtained using macromonomers.

General Reading

G. Allen and J. C. Bevington Eds, *Comprehensive Polymer Science.* Vols 3 and 4. Pergamon Press (1989).

D. G. Allport and W. H. Janes, Eds, *Block Copolymers.* Applied Science Publishers (1973).

J. M. G. Cowie, *Alternating Copolymers.* Plenum Press (1985).

A. Noshay and J. E. McGrath, *Block Copolymers: Overview and Critical Survey.* Academic Press Inc. (1977).

G. Odian, *Principles of Polymerization.* John Wiley and Sons Ltd (1981).

M. J. Folkes, Ed., *Processing, Structure and Properties of Block Copolymers.* Elsevier Applied Science Publishers (1985).

P. Rempp and E. W. Merrill, *Polymer Synthesis.* Hüthig and Wepf, Basel (1986).

References

1. G. M. Estes, S. L. Cooper and A. B. Tobolsky, "Block copolymers", *Reviews in Macromolecular Chemistry*, 5–2, 167 (1970).
2. D. C. Pepper, *Quarterly Reviews*, **8**, 88 (1954).

CHAPTER 6

Polymer Stereochemistry

The physical behaviour of a polymer depends not only on the general chemical composition but also on the more subtle differences in microstructure which are known to exist. As it is now possible to exercise a large degree of control over the synthesis of specific structures it is prudent at this point to elaborate on the types of microstructural variations encountered before discussing how each can be produced. Several kinds of isomerism or microstructural variations can be identified and these are grouped under four main headings: architectural, orientational, configurational, and geometric.

6.1 Architecture

Differences here include branching, network formation, and polymers derived from isomeric monomers, for example, poly(ethylene oxide), I, poly(vinyl alcohol), II, and polyacetaldehyde, III where the chemical composition of the monomer units is the same but the atomic arrangement is different in each case. This makes a considerable difference to the physical properties of the polymers, *e.g.* the glass transition temperature T_g of structure I is 206 K, for II $T_g = 358$ K, and for III $T_g = 243$ K.

I: $-\!\!\left(CH_2CH_2-O\right)\!\!-_n$

II: $-\!\!\left(CH_2-\underset{\displaystyle OH}{\underset{|}{CH}}\right)\!\!-_n$

III: $-\!\!\left(\underset{\displaystyle CH_3}{\underset{|}{CH}}-O\right)\!\!-_n$

6.2 Orientation

When a radical attacks an asymmetric vinyl monomer two modes of addition are possible

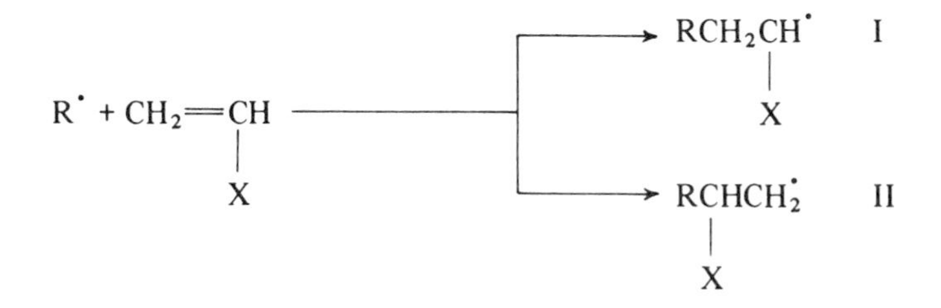

This leads to the configuration of the monomer unit in the chain being either head-to-tail

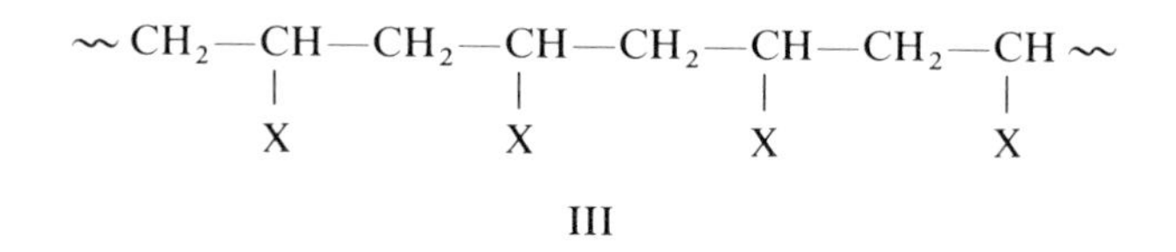

III

if route I is favoured, or a chain containing a proportion of head-to-head, tail-to-tail structure IV if route II is followed.

$$\sim CH_2 \vdots - \underset{\displaystyle X}{\underset{|}{CH}} - \underset{\displaystyle X}{\underset{|}{CH}} \vdots - CH_2 - CH_2 - \underset{\displaystyle X}{\underset{|}{CH}} - \underset{\displaystyle X}{\underset{|}{CH}} \vdots - CH_2 - CH_2 \vdots - \underset{\displaystyle X}{\underset{|}{CH}} \sim$$

head-head IV tail-tail

The actual mode of addition depends on two factors: the stability of the product and the possible steric hindrance to the approach of $R^{\bullet}$ caused by a large group X in the molecule. The reaction in route I is highly favoured, because firstly there is a greater possibility of resonance stabilization of this structure by interaction between group X and the unpaired electron on the adjacent α-carbon atom, and secondly this direction of radical attack is least impeded by the substituent X. The preferred structure is then the head-to-tail orientation (III) and while the alternative structure IV may occur occasionally in the chain, especially when termination by combination predominates, the existence of an exclusively head-to-head orientation is unlikely unless synthesized by a special route.

Experimental evidence supports the predominance of structure III in the majority of polymers; the most notable exceptions are poly(vinylidene fluoride) with 4 to 6 per cent and poly(vinyl fluoride) with 25 to 32 per cent head-to-head links detected by n.m.r. studies. The presence of head-to-tail structures can be demonstrated in a number of ways and the general principle is illustrated using poly(vinyl chloride) as an example. Treatment of this polymer with zinc dust in dioxan solution leads to elimination of chlorine which can proceed by two mechanisms.

$$\text{(a)} \quad \sim CH_2 - \underset{\displaystyle Cl}{\underset{|}{CH}} - CH_2 - \underset{\displaystyle Cl}{\underset{|}{CH}} - CH_2 \sim \;\longrightarrow\; \sim CH_2 - \underbrace{CH - CH}_{CH_2} - CH_2 \sim + ZnCl_2$$

$$\text{(b)} \quad \sim CH_2 - \underset{\displaystyle Cl}{\underset{|}{CH}} - \underset{\displaystyle Cl}{\underset{|}{CH}} - CH_2 \sim \;\longrightarrow\; \sim CH_2 - CH{=}CH - CH_2 \sim + ZnCl_2$$

Statistical analysis of chlorine loss via route (a) indicates that only 86.4 per cent of the chlorine will react due to the fact that, as elimination is a random process, about 13.6 per cent of the chlorine atoms become isolated during the reaction and will remain in the chain. Elimination by mechanism (b) results in total removal of chlorine. Analysis of poly(vinyl chloride) after treatment with zinc dust showed 84 to 86 per cent chlorine elimination and this figure remained constant even after prolonged heating of the reaction mixture. This leads one to the conclusion that the polymer is almost entirely in the head-to-tail orientation.

6.3 Configuration

It has long since been recognized that when an asymmetric vinyl monomer $CH_2=CHX$ is polymerized, every tertiary carbon atom in the chain can be regarded as a chiral centre by virtue of the fact that *m* and *n* are not normally equal in any chain. Under these circumstances the two possible configurations shown (i) and (ii) can only be interconverted by breaking a bond.

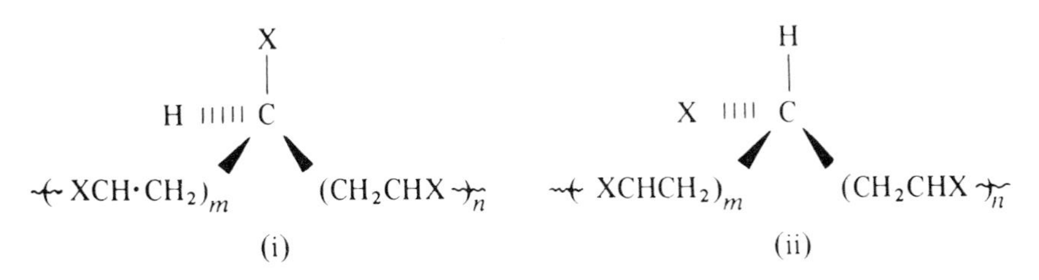

No real progress in the preparation of distinguishable stereoisomers was made until the advent of the Ziegler-Natta catalysts which will be discussed later. Since then the study of stereoregular polymers has expanded rapidly in a vigorous and exciting manner, helped greatly by the application of n.m.r. to the accurate characterization of the microstructure, but before examining these topics, a brief outline of the nomenclature is required.

If every tertiary carbon atom in the chain is asymmetric one might expect the polymer to exhibit optical activity. Normally homoatomic carbon chains show no optical activity because two long chains constitute part of the group variations and as these become longer (and more alike) in relation to the chiral centre, the optical activity decreases to a vanishingly small value. Vinyl polymers derived from $(CH_2=CXY)$ monomers fall into this category as they are centrosymmetric relative to the main chain, and the tertiary carbons are then only pseudo-asymmetric.

This is not true for heteroatomic chains such as $-(CH_2C^*HX{\cdot}O)-$ where C* is a true asymmetric centre, and these polymers are optically active. In this case an absolute configuration can be assigned, using preferably the Cahn-Ingold-Prelog system, referring either to the R- (rectus) or to the S- (sinister) form.

The two forms (i) and (ii) can be distinguished by arbitrarily assigning them *d*- or *l*-configurations, which have nothing to do with optical activity and merely refer to the group X being positioned either below or above the chain in a planar projection.

There are then three distinctive distributions of the *d*- and *l*-forms among the units in a chain and these decide the chain tacticity.

MONOTACTIC POLYMERS

(a) The *isotactic* form. When a polymer, in the all *trans* zig-zag conformation, is viewed along the bonds comprising the chain backbone, then if each asymmetric chain atom has its substituents in the same steric order the polymer is said to be *isotactic*. In other words the arrangement of the substituent groups is either all *d* or all *l*.

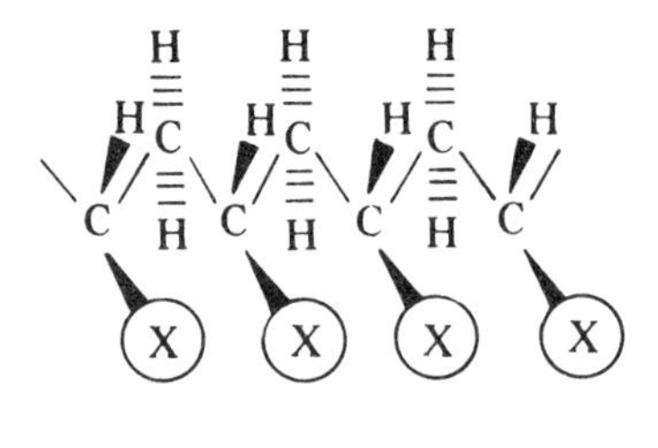

dddd

(b) *Syndiotactic*. A chain is termed syndiotactic when observation along the main chain shows the opposite configuration around each successive asymmetric centre in the chain.

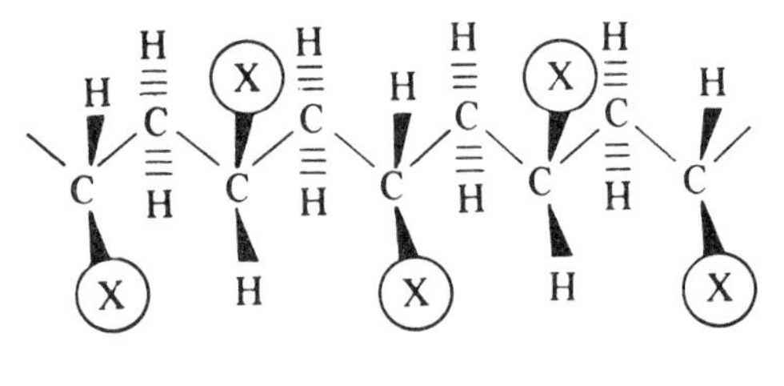

dldld

(c) *Atactic*. When the stereochemistry of the tertiary carbons in the chain is random the polymer is said to be atactic or heterotactic.

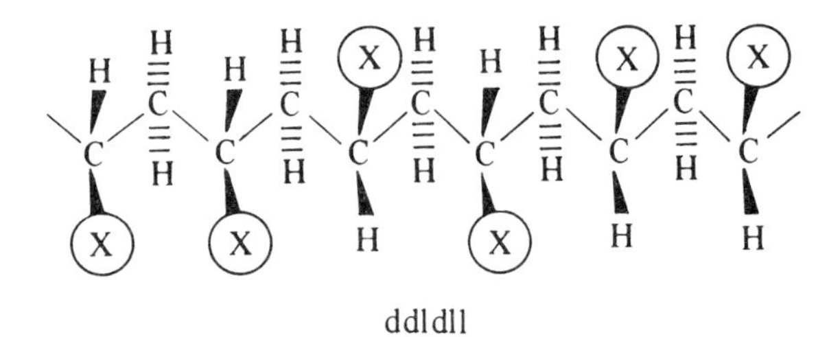

ddldll

It is often easier to obtain a clear picture of the spatial arrangement of the chains by referring to a Newman projection, and these are shown for comparison in figure 6.1a.

DITACTIC POLYMERS

A more complicated picture emerges when the polymerization of 1,2-disubstituted ethylenes (CHR=CHR′) is considered, because now each carbon atom in the chain becomes a chiral centre. The resulting ditactic structure are illustrated in figure 6.1b. Two isotactic structures are obtained, the *erythro*, where all the carbon atoms have the

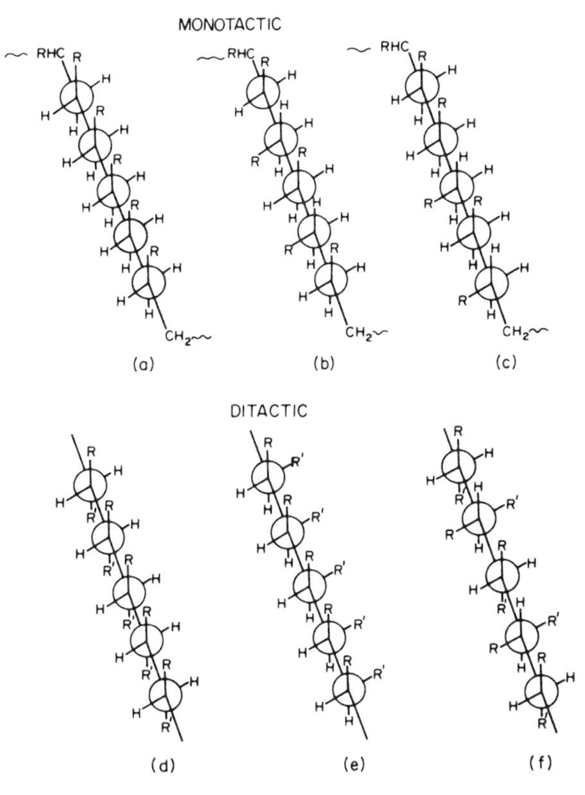

FIGURE 6.1. Newman projections of various stereoregular forms, (a) Isotactic, (b) Syndiotactic, (c) Atactic, (d) Erythro-di-isotactic, (e) Threo-di-isotactic, and (f) Di-syndiotactic.

same configuration and the *threo*, in which the configuration alternates. Only one di-syndiotactic structure is possible. The differences arise from the stereochemistry of the starting material; if the monomer is *cis*-substituted the *threo* form is obtained whereas a *trans* monomer leads to the *erythro* structure.

POLYETHERS

When the spacing between the asymmetric centres increases, as in the heteroatomic polymer poly(propylene oxide), the isotactic and syndiotactic structures become less easily recognized in planar projection. Using an extended zig-zag structure, the *isotactic* form now has its substituents alternating across the plane containing the main chain bonds.

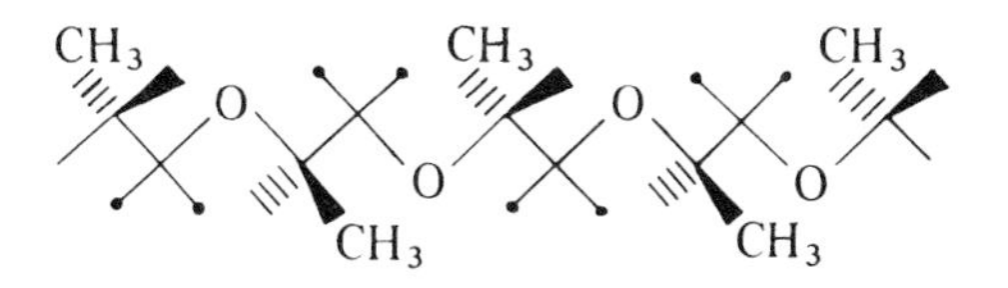

The reverse is true for the *syndiotactic* chain where the substituents are all located on one side.

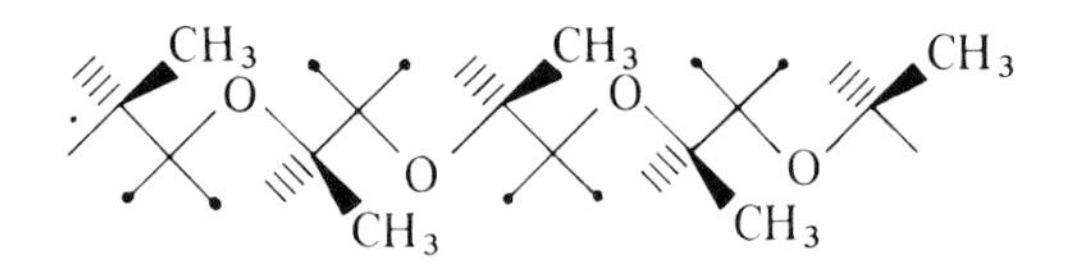

6.4 Geometric isomerism

In addition to the configurational isomerism encountered in polymers derived from asymmetric olefins, geometric isomerism is obtained when conjugated dienes are polymerized, *e.g.* ($CH_2{=}CX{-}CH{=}CH_2$). Chain growth from monomers of this type can proceed in a number of ways, illustrated conveniently by 2-methyl-1,3-butadiene (isoprene). Addition can take place either through a 1,2-mechanism or a 3,4-mechanism, both of which could lead to isotactic, syndiotactic, or atactic structures, or by a 1,4-mode leaving the site of unsaturation in the chain.

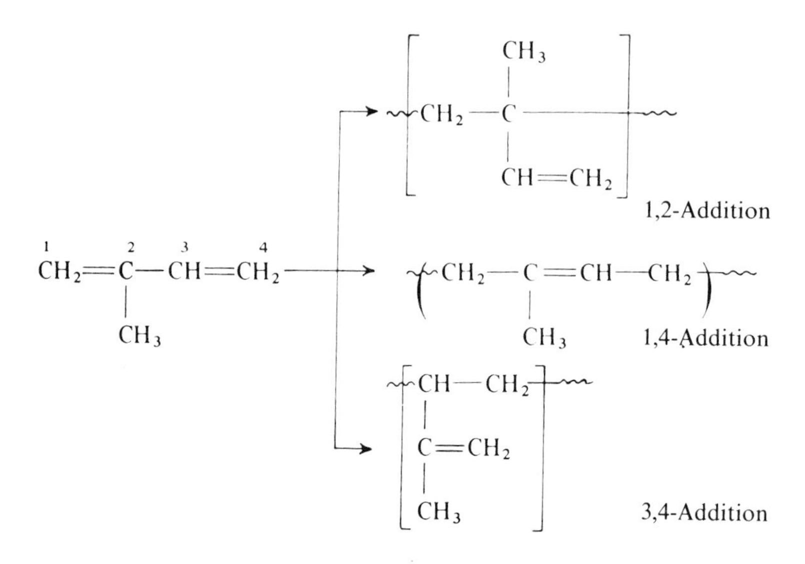

This means that the 1,4 polymer can exist in the *cis* or *trans* form or a mixture of both.

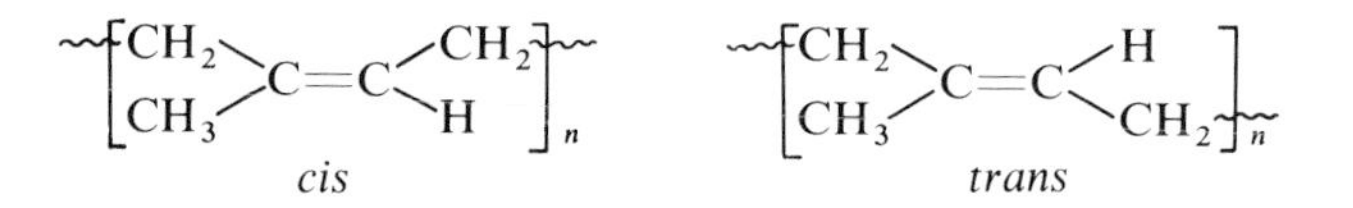

In theory it is possible to synthesize eight distinguishable stereochemical forms or mixtures of these. For a symmetrical monomer such as 1,3-butadiene ($CH_2{=}CH{-}CH{=}CH_2$) the 1,2- and 3,4-additions are indistinguishable and the possible number of stereoforms diminishes accordingly.

Additional variations are possible when 1,4-disubstituted dienes are considered, XCH=CH—CH=CHY. For Y = H

$$\mathrm{XCH{=}CH{-}CH{=}CH_2} \rightarrow \left[\begin{matrix}\mathrm{CH} \\ | \\ \mathrm{X}\end{matrix}\mathrm{{-}CH{=}CH{-}CH_2}\right]_n$$

both *cis* and *trans* isomerism is possible together with isotactic, syndiotactic, and atactic placements for the group X when the addition is 1,4. When Y ≠ H then *threo* and *erythro* forms are also possible in the 1,4-polymer. The name *tritactic* has been suggested for polymers prepared from monomers with different X and Y groups.

6.5 Conformation of stereoregular polymers

Many of the stereoregular polymers prepared are highly crystalline, and the tendency to form ordered structures increases as the stereoregularity becomes more pronounced. We shall see later that crystalline order is usually associated with regular symmetrical polymer structures, whereas the asymmetric monomers form highly unsymmetrical chains. Some other factors must aid crystallite formation.

A stable form of polyethylene is the all *trans* zig-zag form in which it crystallizes. An extended zig-zag pattern becomes untenable, however, for an isotactic polymer with a bulky substituent because the distance between the substituent centres in this conformation is only 0.254 nm. Obviously, the low energy form for an isotactic species

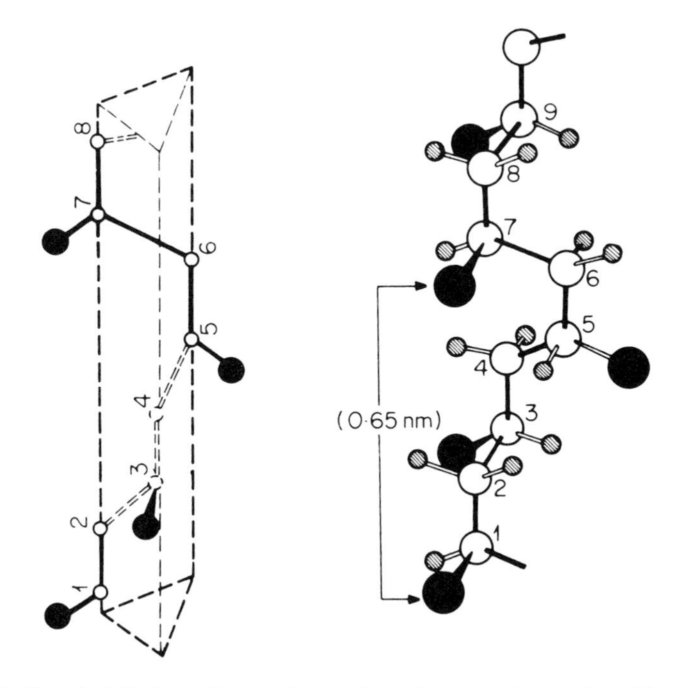

FIGURE 6.2. A 3_1-helix formed by a poly α-olefin, in this case polypropylene. This structure is also seen to fit a triangular template.

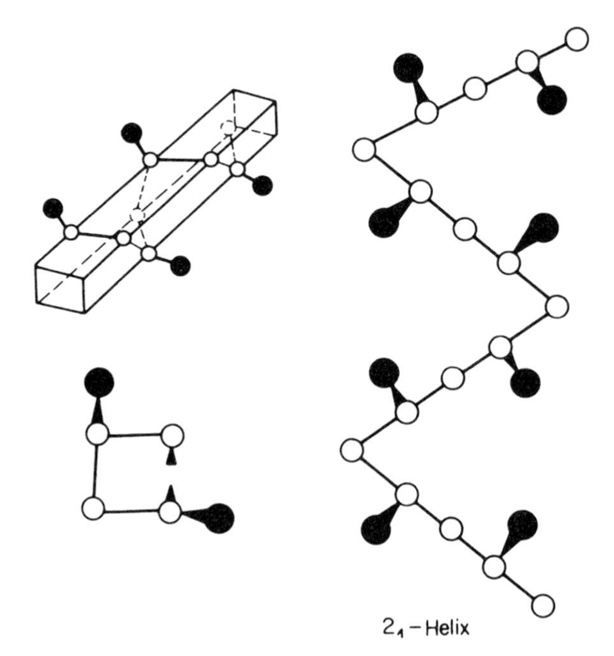

FIGURE 6.3. A poly α-olefin in the syndiotactic configuration showing the *ttgg* sequence along the chain, and the two fold helix which fits a square template.

must be attained by placing the substituents in staggered positions of maximum separation, and this is achieved when bond rotation generates a helix. One particular helical form is shown for polypropylene in figure 6.2. Working from carbon 1 we have the following sequence: 1 and 4 are *trans* to each other (t) carbons 2 and 5 are *gauche* (g), 3 and 6 are *trans*, 4 and 7 are *gauche*, and so on. Carbon 1 repeats at carbon 7, hence the helix is three fold with three monomer residues constituting one complete turn. In a shorthand notation this is a 3_1-helix with a *tgtgtg* conformation, and departure from this pattern to a *ttgg* sequence would simply lead the chain back on itself. This type of helix can also be built up on a triangular template which can be used as a simple model to demonstrate the structure. A helix generated in this way using rotations of 120° should result in an identity period of 0.62 nm. Polypropylene has a period of 0.65 nm and the structure may be generated with equal ease either as a left- or right-handed helix. A helix turning in a clockwise direction, when the chain is viewed along its axis is said to be a right-handed helix; if anti-clockwise, it is left-handed.

The syndiotactic configuration is much more suitable for the extended zig-zag form as the substituents are already staggered for convenient packing on either side of the chain, but a two-fold helix can also be generated by adopting a *ttgg* sequence which has been identified in syndiotactic polypropylene. A square mandrel can be used to demonstrate this structure (figure 6.3). The type of helix formed depends largely on the size of the substituent, and a number of these are shown in figure 6.4. As the helix is a regular ordered structure, it can be arranged compactly in a three-dimensional close structure with relative ease which explains how the unsymmetrical chain monomer can be

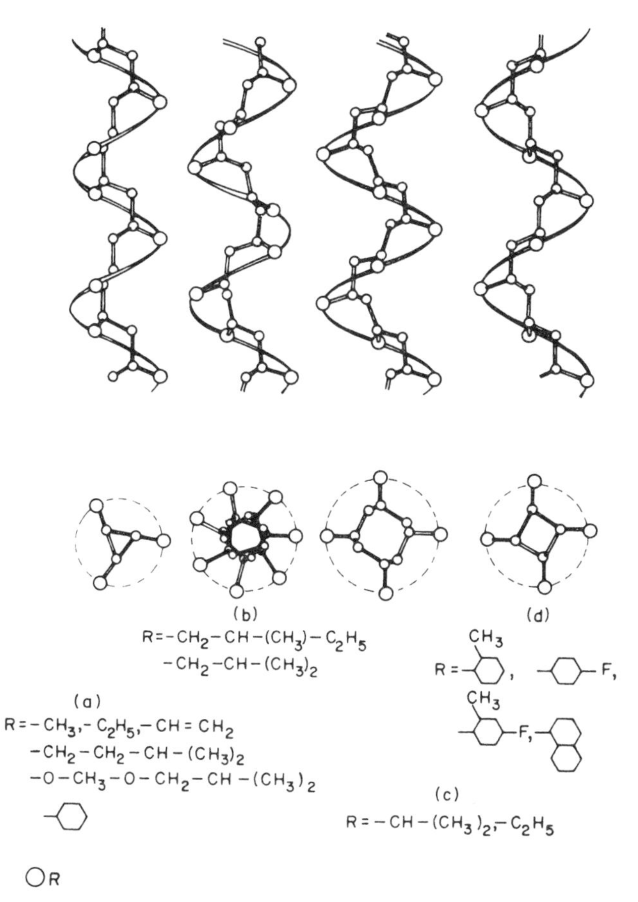

FIGURE 6.4. A diagrammatic representation of various other ordered helical structures adopted by isotactic polymers. (After Natta and Corradini.)

accommodated in a crystalline polymer structure. Highly crystalline samples are obtained when the polymer is sufficiently stereoregular to enable it to form significantly long helical or regular zig-zag sections for ordered chain arrangement to take place.

The automatic identification of crystallinity with stereoregularity should be avoided, however, as they are not necessarily synonymous, and while highly stereoregular polymers tend to be crystalline, the existence of any polymer in a crystalline state does not automatically mean the sample is markedly stereoregular.

6.6 Factors influencing stereoregulation

The low pressure polymerization of ethylene, reported by Ziegler in 1955, signalled the emergence of a new phase in polymer science. The catalyst, prepared from $TiCl_4$ and $(C_2H_5)_3Al$, was heterogeneous and it was subsequently demonstrated by Natta that co-ordination catalysts of this type could be used to exercise control over the stereoregular structure of the polymer. Initially it was thought that only heterogeneous catalysis would lead to stereoregular polymers, but we now know that this is untrue and that stereoregulation can be effected under specific rigorously defined conditions, regardless of the solubility of the catalyst system.

If stereoregulation is simply the control of the mode of entry of a monomer unit to a growing chain, examination of the factors influencing this addition should provide an understanding of how to exert such control.

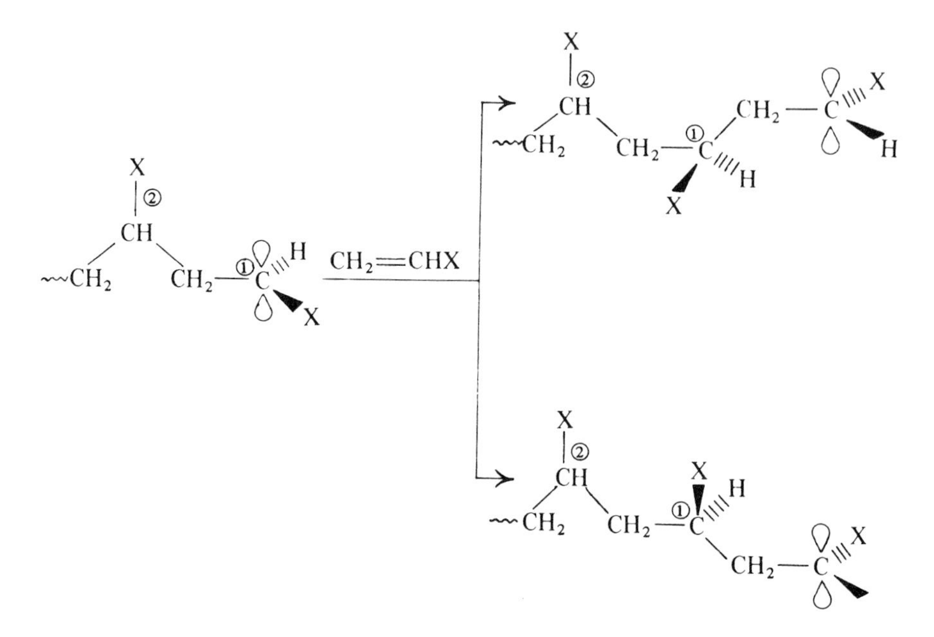

Free radical initiation can usually be thought of as generating a chain by a Bernoulli-trial propagation, in which the orientation of the incoming monomer is unaffected by the stereostructure of the polymer. It can then add on in one of two ways where the active end is assumed to be a planar sp^2 hybrid, and the configuration of the adding monomer is finally determined only when another monomer adds on to it in the next step. In other words this addition leads to an isotactic or syndiotactic placement of the pseudo-asymmetric centre 1 with respect to 2. When the chain carrier is a free species, *i.e.* a radical, the stereoregularity of the polymer is a function of the relative rates of the two methods of addition and this is governed by the temperature. Consideration of the relative magnitudes of the enthalpy and entropy of activation for isotactic and syndiotactic placements shows that while the differences are small, the syndiotactic structure is favoured. This is, of course, aided by the greater steric hindrance and repulsions experienced by the substituents in the isotactic configuration and will vary in extent with the nature of the group X. Thus for a free radical polymerization at 373 K, the fraction of syndiotactic placements is 0.73 for methyl methacrylate monomer but only 0.51 when vinyl chloride is used. A decrease in the polymerization temperature increases the tendency towards syndiotactic placements, but as radical reactions are normally high temperature processes atactic structures predominate. Low temperature free radical propagation has been found

TABLE 6.1. Polymerizations using co-ordination catalysts where quoted tacticities are > 90 per cent

Monomer	*Catalyst*	*Structure*
Isobutyl vinyl ether	$BF_3(C_2H_5)_2O$ in propane at 213 K	Isotactic
Methyl acrylate	C_6H_5MgBr or n-C_4H_9Li in toluene at 253 K	Isotactic
Propylene	$TiCl_4 + (C_2H_5)_3Al$ in heptane at 323 K	Isotactic
Propylene	$VCl_4 + Al(i\text{-}C_4H_9)_2Cl$ in anisole or toluene at 195 K	Syndiotactic

to produce syndiotactic polymers from the polar monomers, isopropyl and cyclohexyl acrylate, and methyl methacrylate.

The same general principles apply for freely propagating ionic chain carriers, but if co-ordination between the monomer and the active end takes place, the stereoregulation is altered. The configuration of the monomer is then influenced by the stereochemistry of the growing end, and the possible number of ways the monomer can join the chain is in excess of two. These co-ordination catalysts include the Ziegler-Natta type as the largest group, and others such as butyl lithium, phenyl magnesium bromide, and boron trifluoride etherate. The resulting polymer is normally isotactic, although some cases exist where highly syndiotactic polymers are obtained.

The orienting stage in co-ordination polymerization can be pictured as being multicentred, with the monomer position governed by co-ordination with the gegen-ion and the propagating chain end. As the gegen-ion will tend to repel the substituent X on the incoming monomer, it is forced to approach in a way that leads to predominantly isotactic placement. If co-ordination plays a major role in determining the configuration of the incoming monomer, then the greater the co-ordinating power the more regular the resulting polymer should be, but the nature of the monomer is also important. Polar monomers (the acrylates and vinyl ethers) are capable of taking an active part in the co-ordination process, and will only require catalysts with moderate powers of orientation, but non-polar monomers such as the α-olefins will require stronger co-ordinating catalysts to maintain the required degree of stereoregulation in the addition process. In extreme cases the heterogeneous Ziegler-Natta catalysts are required, where severe restrictions are imposed on the method of monomer approach to the growing chain end and these must be used for the non-polar monomers which yield only atactic polymers with homogeneous catalyst systems.

6.7 Homogeneous stereospecific cationic polymerizations

An example of this type of reaction is provided by the alkyl vinyl ethers ($CH_2{=}CHOR$). Isobutyl vinyl ether was the first monomer studied which produced a stereoregular polymer using a $BF_3 + (C_2H_5)_2O$ catalyst and will be used as the illustrative monomer. A homogeneous stereospecific polymerization can be carried out in toluene at 195 K using such soluble complexes as $(C_2H_5)_2TiCl_2AlCl_2$ or $(C_2H_5)_2TiCl_2Al(C_2H_5)Cl$, or, if a suitable choice of mixed solvents is made, a homogeneous system with $BF_3 + (C_2H_5)_2O$ can be obtained which is capable of producing the isotactic polymer.

The mechanism proposed by Bawn and Ledwith, postulates the existence of an sp^3 configuration for the terminal carbon in the growing chain due to the attendant gegen-ion, and especially in low dielectric solvents. They also point out that the structure of the alkyl vinyl ethers, with the exception of the ethyl and isopropyl members, will be subject to steric shielding of one side of the double bond, *i.e.*

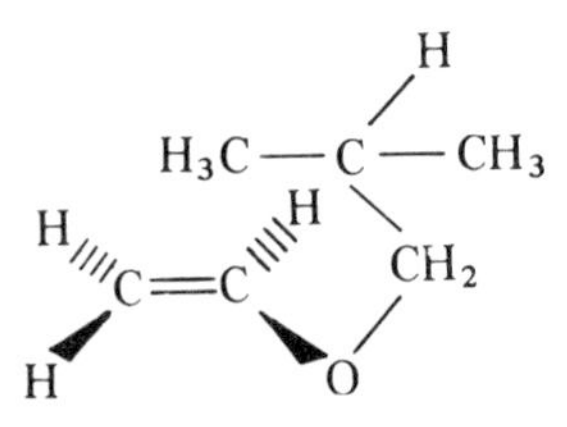

This blocks one mode of double-bond opening and assists stereoregulation. This conclusion is supported by the lack of any crystalline polymer in the product when the ethyl and isopropyl groups are used, where no blocking is possible.

The formation of a loose six-membered ring is thought to stabilize the growing carbonium ion in the reaction so that the only route for monomer approach is past the counter-ion.

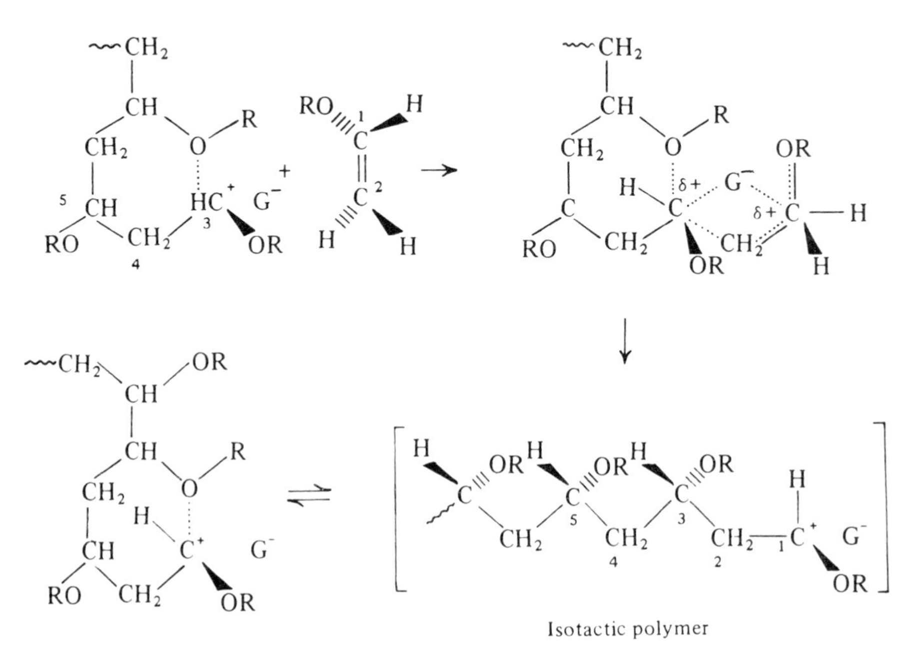

Isotactic polymer

A four-centred cyclic transition state is involved in the propagation stage leading to the insertion of a monomer unit between the catalyst and the chain end, with subsequent regeneration of the cyclic structure. An alternative transition state, proposed by Cram and Kopecky, has a similar but more rigid structure.

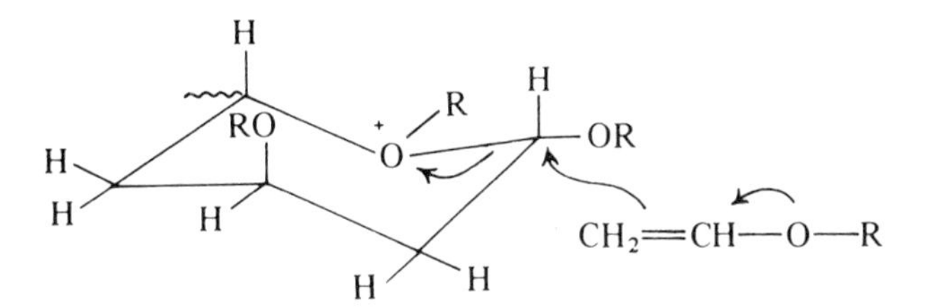

Both mechanisms ignore the nature of the catalyst forming the gegen-ion, but obviously as this will act as a template for the attacking monomer it will exert an influence on the rate of reaction and the type of stereoregularity imposed. The most probable configuration is isotactic because of the tendency for the gegen-ion to repel the substituent group of the incoming monomer.

6.8 Homogeneous stereoselective anionic polymerizations

The various factors influencing the stereoregularity, when the propagating chain end is a carbanion, are conveniently highlighted in a study of the polymerization of methyl methacrylate by organo-lithium catalysts.

The propagating chain end in an anionic reaction initiated by a reagent such as *n*-butyl lithium can be thought of as existing in one of the following states, analogous to carbonium ion formation.

RLi	$\rightleftharpoons$ R^-Li^+	$\rightleftharpoons$ $R^-//Li^+$	$\rightleftharpoons$ $R^\ominus + Li^+$
covalent	contact ion pair	solvent separated ion pair	free ions

The extent of the separation will depend on the polarity of the reaction medium and in non-polar hydrocarbon solvents, such as toluene, covalent molecules or contact ion pairs, are most likely to exist. With increasing solvent polarity there is a greater tendency to solvate the ions, eventually producing free ions for strictly anionic polymerizations. These lead to conditions similar to a free radical polymerization where the stereoregulation is reduced and syndiotactic placements are favoured at low temperatures.

The effects of solvent and temperature are manifest in the polymerization of methyl methacrylate with *n*-butyl lithium at 243 K in a series of mixed solvents prepared from toluene and dimethoxyethane (DME). The n.m.r. spectra of the products indicate the compositions in table 6.2 and reveal that a predominantly isotactic material is produced in a low polarity medium, but that this becomes highly syndiotactic as the solvating power of the medium increases.

An additional point emerges from this; higher syndiotactic contents are obtained when the Lewis base strength of the solvent increases and this factor probably accounts for the efficiency of the ether in this system. When the catalyst is 9-fluorenyl lithium, the reaction of methyl methacrylate at 195 K in toluene leads to isotactic polymer, whilst a change of solvent to tetrahydrofuran results in a syndiotactic product.

TABLE 6.2 The effect of mixed solvent composition on the tacticity of poly(methyl methacrylate) initiated by *n*-butyl lithium at 243 K; the mole fractions of the various configurations are given

Toluene/DME	*Isotactic*	*Heterotactic*	*Syndiotactic*
100/0	0.59	0.23	0.18
64/36	0.38	0.27	0.35
38/62	0.24	0.32	0.44
2/98	0.16	0.29	0.55
0/100*	0.07	0.24	0.69

*Measured at 203 K.

Stereoregulation is also altered by the nature of the solvent when Grignard reagents and alkali metal alkyls are used as initiators. In toluene, for example, the isotactic placements in the chain decrease as reagents change from Li to Na to K.

If general conclusions can be drawn from the behaviour of methyl methacrylate, it appears that stereoregulation in anionic polymerizations, involving either polar monomers or monomers with bulky substituents, will lead to predominantly syndiotactic polymers when a free, dissociated ion, occurs at the propagating end. This is because it is the most stable form arising from a minimization of steric and repulsive forces. If, however, some strongly regulating kinetic mechanism is available, for instance monomer + gegen-ion co-ordination, then the less favourable isotactic placement occurs.

In an attempt to explain the mechanism of stereoregulation in systems catalysed by lithium alkyls, Bawn and Ledwith proposed that the penultimate residue plays an important role in the addition. A loose cyclic intermediate is formed when the Li^+ counter-ion co-ordinates with the carbonyl of the penultimate unit and with the terminal unit in a resonance enolic structure V.

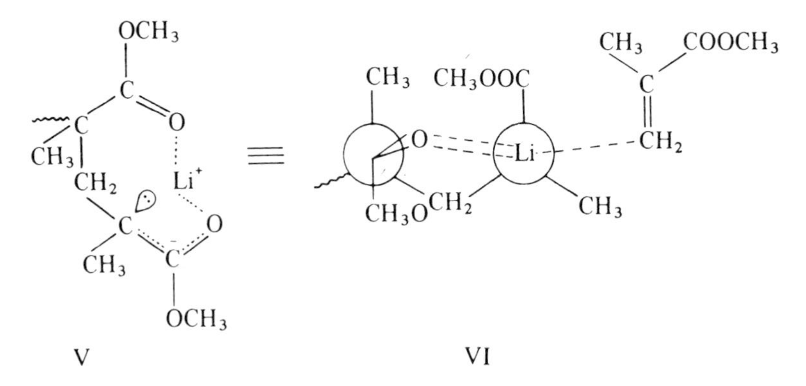

This can be represented alternatively as a transition state VI similar to that in an S_N2 reaction. With one side of the Li^+ shielded, monomer approach is restricted, and the path of least resistance places the α-methyl group on the incoming monomer in a *trans* position, relative to the α-methyl group on the carbanion, during the π-complex formation. Addition then proceeds through a series of bond exchanges as the carbanion joins the monomer methylene group. The carbonyl group of the monomer co-ordinates with the ion by replacing the interaction with the previous penultimate group and the cyclic intermediate is regenerated.

The steric restriction imposed by the α-methyl group aids the formation of an isotactic polymer and in its absence (*i.e.* methyl acrylate) there is a reduced probability of isotactic placements. A compensating feature in the higher acrylates arises from the shielding of one side of the monomer by the bulkiness of the ester group. In the branched homologues, isopropyl and *t*-butyl acrylate, π-bonding with the Li^+ ion is forced to take place on one side of the monomer only, thereby enhancing the formation of isotactic polymer quite markedly.

As this and other mechanisms all postulate the existence of structures stabilized by intramolecular solvation, the addition of Lewis bases or polar solvents should disrupt the required template and encourage conventional anionic propagation by free ions. This automatically reduces the probability of an isotactic placement occurring.

6.9 Homogeneous diene polymerization

The principles applied in the previous section to essentially polar monomers can be extended to the stereoregular polymerization of dienes by alkali metals and metal alkyls. We have already seen that the *cis-trans* isomerism presents a variety of possible structures for the polydiene to adopt and complicates the preparation of a sample containing only one form rather than a mixture. Thus polyisoprene may contain units in the 1,2 or 3,4 or *cis*-1,4 or *trans*-1,4 configuration without even considering the tacticity of the 1,2 or 3,4 monomer sequences in the chain.

Most work has centred on the preparation of a particular form of geometric isomer because the type and distribution of each isomeric form in the chain has a profound influence on the mechanical and physical properties of the sample. The original discovery that metallic lithium in a hydrocarbon solvent catalysed the production of an all *cis*-1,4 polyisoprene stimulated interest in this area and quickly raised two points which must be satisfied if a suitable mechanism is to be postulated.

(i) Lithium and lithium alkyl catalysts produce highly specific stereostructures, but when replaced by Na or K this effect diminishes.

(ii) Stereospecific polymerization takes place in the bulk state or in hydrocarbon solvents, but the addition of a polar solvent leads to drastic changes.

To explain these features the following mechanism has been put forward. Initiation produces a "Schlenk" adduct VII.

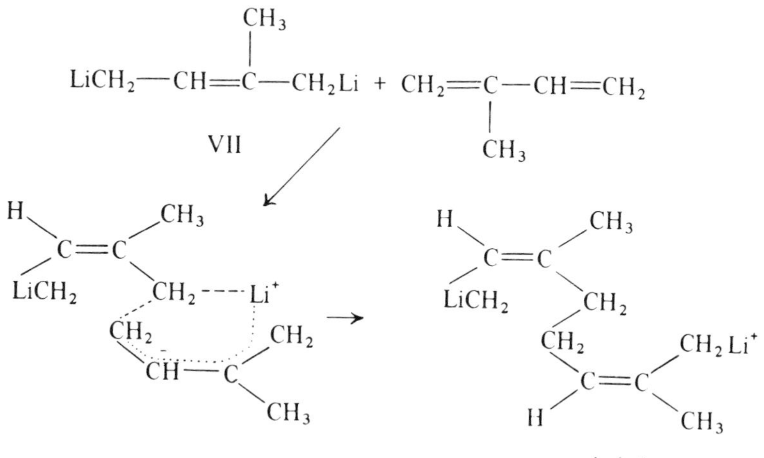

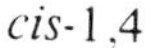
cis-1,4

The lithium ion then forms a chelate complex with the isoprene monomer locking it into a *cis*-configuration which is maintained during the addition reaction. This type of complex is suitable when a small ion like Li^+ is used but will be disrupted by the larger gegen-ions Na^+ and K^+, thereby allowing freer approach of the reactants. The presence of ethers also alters the stereospecificity by competing for the Li^+ and altering the spatial arrangement of the chelating pattern.

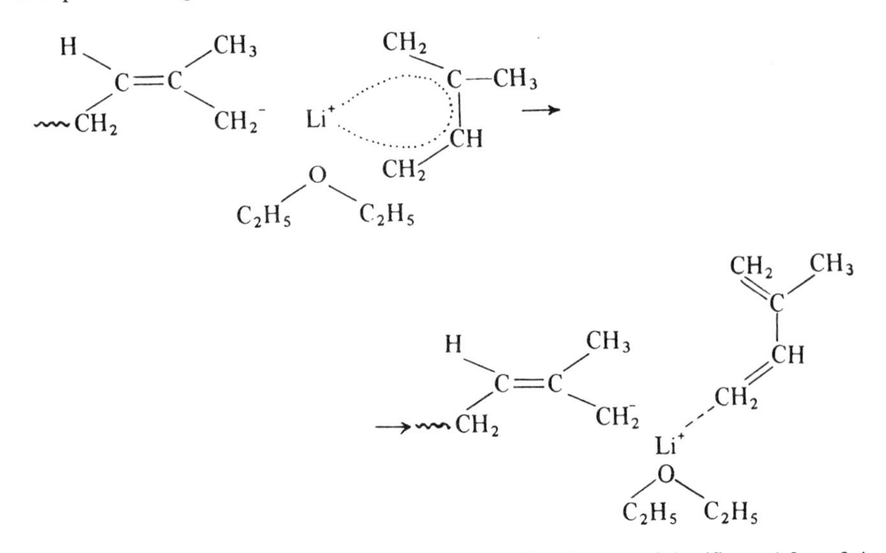

The monomer can then enter in a random fashion. The absence of significant 1,2- or 3,4-addition is thought to be caused by the shielding of carbon 3 in the transition state. However, all such proposals remain speculative.

6.10 Summary

We can now summarize a few important points dealt with so far.

Three factors influence stereoregularity during chain propagation:

(1) *Steric factors* which force the unit into a spatial arrangement determined by the size and position of the substituents already in the chain;

(2) *Polar factors* because solvents which allow contact ion pairs favour isotactic placements, but pairs separated by heavy solvation (free ions) lead to syndiotactic structures;

(3) *Co-ordination* because if the end group of a growing chain has a planar (sp^2) configuration with no established parity, such as that found in free radical or free ion propagation, then the configuration of this unit is established only during the course of addition of an incoming monomer. Normally this will result in a syndiotactic placement with respect to the penultimate unit. Otherwise, co-ordination occurs between the gegen-ion, the incoming polar monomer, and the end or penultimate unit.

For polar monomers the soluble catalysts can produce isotactic structures but for non-polar monomers the homogeneous catalysts lead mainly to atactic or syndiotactic polymers and a heterogeneous catalyst is required for isotactic placements to occur. These will be discussed in the next chapter.

General Reading

W. Cooper, "Stereospecific polymerization", in *Progress in High Polymers*, Vol. I, Academic Press (1961).

M. Goodman, "Concepts of polymer stereochemistry", *Topics in Stereochemistry*, Vol. 2. Wiley-Interscience (1967).

A. D. Ketley, *The Stereochemistry of Macromolecules*, Vols. I–III. Edward Arnold (1968).

G. Natta, "Precisely constructed polymers", *Scientific American*, **205**, 33 (1961).

G. E. Schildknecht, "Stereoregular polymers" in *Encyclopaedia of Chemistry*. Reinhold Publishing Corp. (1966).

R. B. Seymour, *Introduction to Polymer Chemistry*, Chapter 6. McGraw-Hill (1971).

References

1. C. E. H. Bawn and A. Ledwith, *Quarterly Reviews*, **16**, 361 (1962).
2. G. Natta and P. Corradini, *Rubber Chem. Technol*, **33**, 703 (1960).

CHAPTER 7

Polymerization Reactions Initiated by Metal Catalysts and Transfer Reactions

7.1 Polymerization using Ziegler-Natta catalysts

Stereoregular polymerizations carried out in homogeneous systems, using essentially polar monomers whose ability to co-ordinate with the catalyst-complex imposes a stereospecific mechanism on the addition, have been dealt with in chapter 6. As the polarity of the monomer decreases, however, the ability to control the configuration of the incoming monomer decreases and atactic polymers result.

One of the most significant advances in synthetic polymer science in the 1950s was the discovery by Ziegler that, in the presence of a catalyst prepared from aluminium alkyl compounds used in conjunction with a transition metal halide, ethylene could be polymerized at ambient temperatures and atmospheric pressure. Ziegler found that this reaction produced a linear, highly crystalline polymer (now manufactured and marketed as high density polyethylene, HDPE) as opposed to the extensively branched, less crystalline polyethylene produced by radical initiated, high pressure processes (low density polyethylene, LDPE). The work was developed further by Natta and his co-workers, who found that semi-crystalline linear polymers from propylene, 1-butene and several other α-olefins could be prepared using similar catalyst systems. It was also demonstrated that the crystallinity in these polymers was enhanced by their highly stereoregular structure, and this opened the way for the preparation of stereoregular α-olefins.

The systems were in all cases heterogeneous and the active initiators are now known by the general name *Ziegler-Natta* catalysts. This encompasses a vast number of substances prepared from different combinations of organometallic compounds where the metal comes from the main Groups I, II, or III and is combined with the halide or ester of a transition metal (Groups IV to VIII). Table 7.1 contains a number of the common components of the Ziegler-Natta catalysts but this list is far from exhaustive.

These catalysts tend to control two features, (a) the rate and (b) the specificity of the reaction, but this varies from reaction to reaction and only a judicious choice of catalyst can effect control over both of these aspects.

Unfortunately the insolubility of the catalyst poses the problems that the kinetics are hard to reproduce and the reaction mechanisms are difficult to formulate with real confidence. This means that the choice of a suitable catalyst for a system is somewhat empirical and very much trial and error, until optimum conditions are established.

It is useful to remember that both heterogeneous and homogeneous catalysts exist in

TABLE 7.1 Components of Ziegler-Natta Catalysts

Metal Alkyl or Aryl	*Transition metal compounds*
$(C_2H_5)_3Al$	$TiCl_4$; $TiBr_3$
$(C_2H_5)_2AlCl$	$TiCl_3$; VCl_3
$(C_2H_5)AlCl_2$	VCl_4; $(C_5H_5)_2TiCl_2$
$(i\text{-}C_4H_9)_3Al$	$(CH_3COCHCOCH_3)_3V$
$(C_2H_5)_2Be$	$Ti(OC_4H_9)_4$
$(C_2H_5)_2Mg$	$Ti(OH)_4$; $VOCl_3$
$(C_4H_9)Li$	$MoCl_5$; $CrCl_3$
$(C_2H_5)_2Zn$	$ZrCl_4$
$(C_2H_5)_4Pb$	$CuCl$
$((C_6H_5)_2N)_3Al$	WCl_6
C_6H_5MgBr	$MnCl_2$
$(C_2H_5)_4AlLi$	NiO

the Ziegler-Natta group but the latter only yield atactic or occasionally syndiotactic polymers from non-polar monomers. As only the heterogeneous Ziegler-Natta catalysts produce isotactic poly-α-olefins, these have received most attention. The interest in this type of system has been immense, as evidenced by the vast quantity of published material, and it was most fitting that both Ziegler and Natta were recognized for their work by being awarded jointly the Nobel Prize for chemistry in 1963.

EXPERIMENTAL DEMONSTRATION

Before dealing with the mechanism of catalysis and the nature of the catalyst in greater depth, a description of a laboratory preparation of polyethylene will be given to demonstrate the use of such systems.

Preparation of the catalyst. This is prepared from either aluminium triethyl or aluminium diethyl chloride in combination with titanium tetrachloride. The main disadvantage in using aluminium alkyls is that they ignite spontaneously in air and, to avoid this hazard, must be handled in an inert atmosphere.

A safer procedure is to use amyl lithium which can be prepared from lithium wire and amyl chloride. Petroleum ether (50 cm^3) is stirred in a three-necked flask and degassed under a stream of nitrogen, which is first purified by passing through a pyrogallol and sodium hydroxide train to remove oxygen. Lithium wire (3 g) is added, followed by 2 cm^3 of a solution of amyl chloride (20.7 cm^3) in petroleum ether (25 cm^3). This is stirred vigorously until the solution becomes turbid (LiCl) and then the remaining amyl chloride solution is added slowly over a period of 20 min to the reaction flask now being cooled in an ice-bath. The reaction mixture turns blue-brown and after 2.5 h it is filtered under nitrogen through glass wool into a graduated flask to remove unreacted Li. The filtrate is allowed to settle and the supernatant liquid is analysed by hydrolysing an aliquot with water and titrating the LiOH formed with 0.1 M HCl. The amyl lithium solution can be stored for some days at 273 K.

Reaction. The catalyst is prepared *in situ*. An apparatus similar to that in figure 7.1 is used. The reaction kettle (1 dm^3) is charged with 400 cm^3 petroleum ether and 0.05 mol

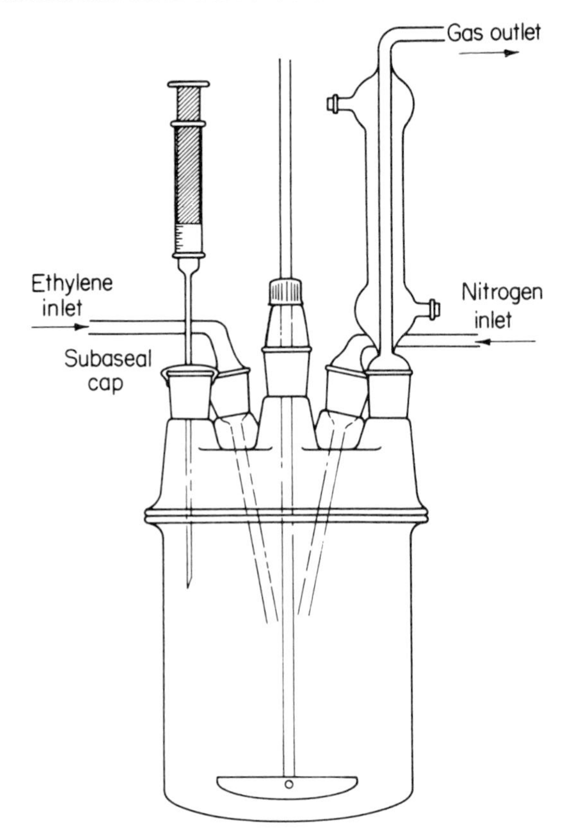

FIGURE 7.1. Apparatus for polymerization of ethylene.

lithium amyl. Anhydrous $TiCl_4$ (2 cm^3) is added and the formation of the catalyst as a brownish-black precipitate is complete in 20 min. The formation is accompanied by a rise in temperature of about 10 K. Ethylene is then passed into the stirred mixture and polyethylene forms immediately. The reaction can be allowed to continue for 30 min, then the catalyst is destroyed by the addition of butanol (40 cm^3). The polymer is filtered, washed with a 1:1 mixture of HCl and methanol, and dried at 350 K. It has a high degree of crystallinity, a higher density, and a melting temperature some 20 to 30 K higher than samples prepared using high pressure techniques.

The apparatus shows a syringe in position which may be used if the more inflammable aluminium alkyl catalysts are used, as these are often handled in hydrocarbon solvents.

7.2 Nature of the catalyst

Frequently the product of a Ziegler-Natta polymerization is sterically impure and can be preferentially extracted to give two products – a highly crystalline stereoregular fraction, and an amorphous atactic one. This may be attributed to the size of the catalyst particles as stereoregularity is enhanced by having large particles, whereas a finely divided catalyst tends to produce an amorphous polymer.

The crystal form of the catalyst is also important and the violet α, γ, and δ forms of $TiCl_3$ produce a greater quantity of isotactic polypropylene when combined with an aluminium alkyl, than the brown β-structure. As the active sites for heterogeneous polymerizations are believed to be situated on the crystal surfaces, the structure is all important. In the layered structure of α-$TiCl_3$, where every third Ti^{3+} ion in the lattice is missing, a number of Cl vacancies occur on the surface to maintain electrical neutrality in the crystal. The Ti^{3+} on the surface is then only 5-co-ordinated leaving a vacant d-orbital, □, and an active site is created when an alkyl group replaces a chloride ion to form $TiRCl_4$ □. (See diagrams page 144).

In β-$TiCl_3$, the linear chains form bundles in which some Ti ions are surrounded by five Cl^- ions and some by only four Cl^-. This means that the steric control at the sites with two vacancies is now less rigid and stereoregulation is much poorer.

Catalyst composition affects both stereoregulation and polymer yield. Thus Ti^{3+} is a more active producer of isotactic polypropylene than Ti^{4+} or Ti^{2+}, while an increase in the length of the associating alkyl group decreases the efficiency of stereoregular placements. Varying the transition metal and the associated aluminium compounds in the catalysts also influences the nature of the product.

7.3 Nature of active centres

Most of the experimental evidence points to propagation taking place at a carbon to transition metal bond with the active centre being anionic in character. Free radical reactions are considered to be non-existent in the Ziegler-Natta systems because neither (i) chain transfer nor (ii) catalyst consumption occurs. The active centres also live longer than radicals and resemble "living" polymer systems in many ways, one being that block copolymers can be produced by feeding two monomers alternatively into the system.

While a number of reaction mechanisms have been suggested, two are worth considering in detail. These are based on the view that the active centres are localized rather than migrating and that the α-olefin is complexed at the transition metal centre prior to incorporation into the chan, *i.e.* growth is always from the metal end of the growing chain.

The active species are then considered to be either *bimetallic* or *monometallic*.

7.4 Bimetallic mechanism

Natta and his associates have postulated a mechanism involving chain propagation from an active centre formed by the chemisorption of an electropositive metal alkyl of small ionic radius on the co-catalyst surface. This yields an electron deficient bridge complex such as I and chain growth then emanates from the C–Al bond.

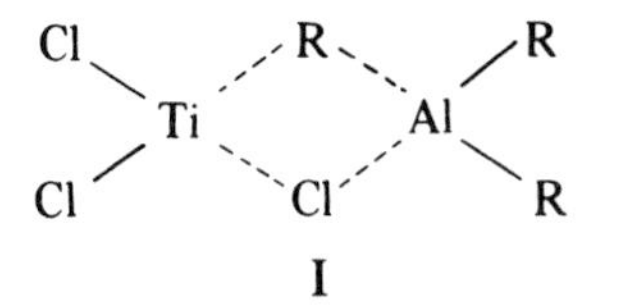

It is suggested that the nucleophilic olefin forms a π-complex with the ion of the transition metal and, following a partial ionization of the alkyl bridge, the monomer is included in a six-membered ring transition state. The monomer is then incorporated into the growing chain between the Al and the C allowing regeneration of the complex.

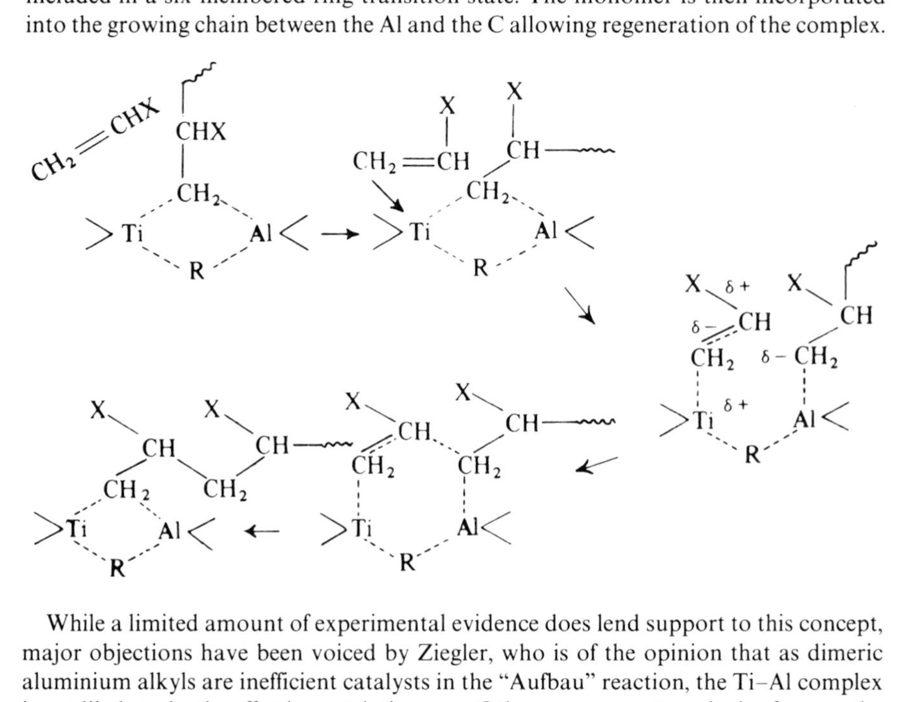

While a limited amount of experimental evidence does lend support to this concept, major objections have been voiced by Ziegler, who is of the opinion that as dimeric aluminium alkyls are inefficient catalysts in the "Aufbau" reaction, the Ti–Al complex is not likely to be the effective catalytic agent. Other more recent work also favours the second and simpler alternative, the monometallic mechanism.

7.5 Monometallic mechanism

Majority opinion now favours the concept that the d-orbitals in the transition element are the main source of catalytic activity and that chain growth occurs at the titanium-alkyl bond. The ideas now presented are predominantly those of Cossee and Arlman, and will be developed using propylene as the monomer.

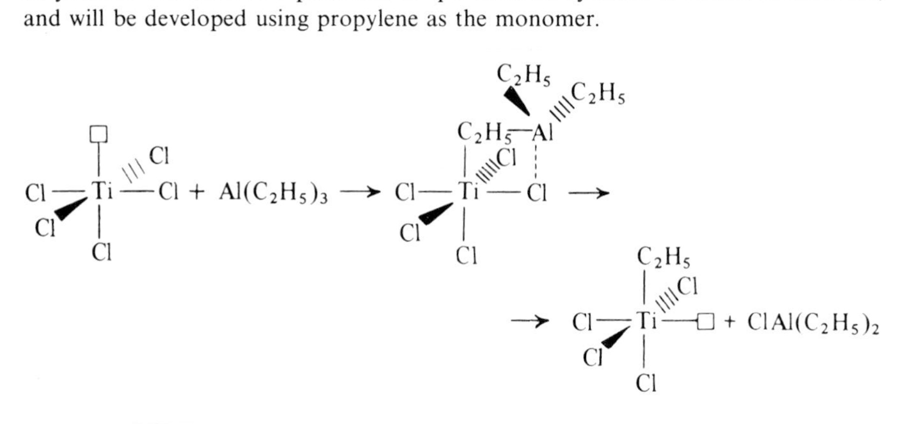

The first stage is the formation of the active centre illustrated here using α-$TiCl_3$ as catalyst. The suggestion is that alkylation of the 5-co-ordinated Ti^{3+} ion takes place by an exchange mechanism after chemisorption of the aluminium alkyl on the surface of the $TiCl_3$ crystal. The four chloride ions remaining are the ones firmly embedded on the lattice and the vacant site is now ready to accommodate the incoming monomer unit. The reaction is confined to the crystal surface and the active complex is purely a surface phenomenon in heterogeneous systems.

The attacking monomer is essentially non-polar but forms a π-complex with the titanium at the vacant d-orbital. A diagram of a section of the complex shows that the propylene molecule is not much bigger than a chloride ion and consequently the double bond can be placed adjacent to the Ti ion and practically as close as the halide. After insertion of the monomer between the Ti–C bond, the polymer chain then migrates back into its original position ready for a further complexing reaction.

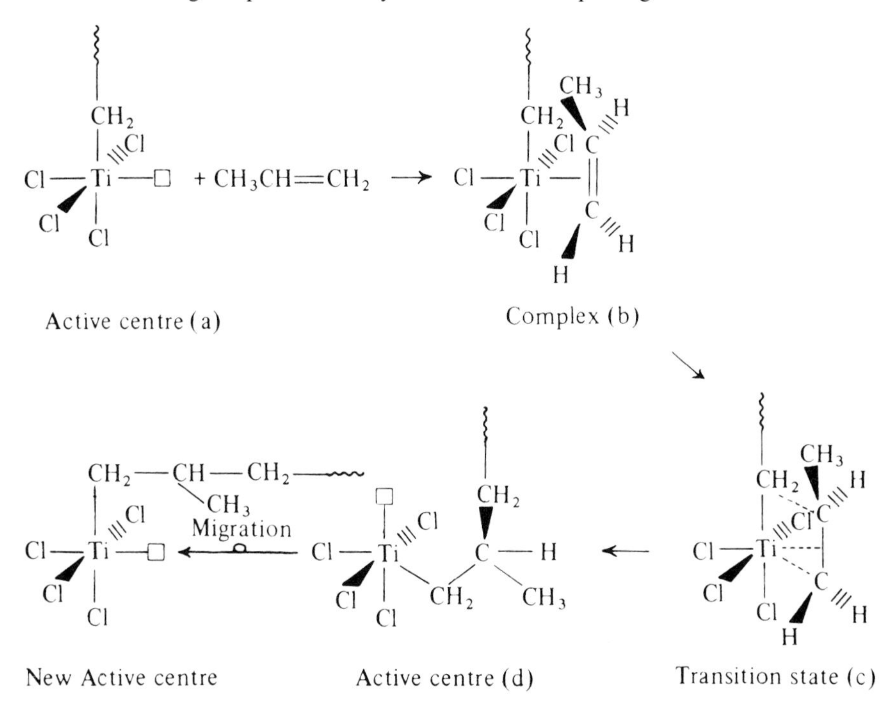

The reactivity of the active centre is attributed primarily to the presence of d-orbitals in the transition metal. The initial state of the active centre shows that the $=CH_2$ group will be capable of considerable distortion from its equilibrium position, because of the availability of adjacent d-orbitals. Complexing (b) takes place when the π-bonding orbitals of the olefin overlap with the vacant $d_{x^2-y^2}$ orbital of the Ti^{3+} while at the same time the π^*-antibonding orbitals can overlap the d_{yz} orbitals of the Ti^{3+}. Formation of the transition state is aided by the ability of the $=CH_2$ group to migrate by partial overlap with the d_{z^2}, d_{yz}, and π^*-orbitals.

The main features of the monometallic mechanism are: (1) an octahedral vacancy on the Ti^{3+} is available to complex the olefin; (2) the presence of an alkyl to transition

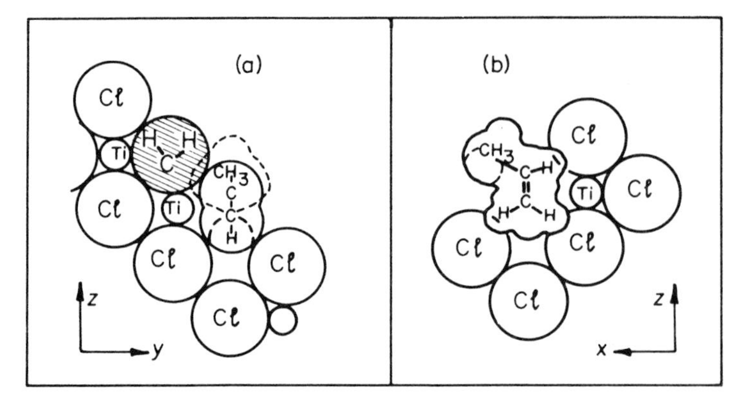

Figure 7.2. Cross-sectional diagrams of the propylene-catalyst complex through (a) the y–z plane and (b) the x–z plane of the octahedral structure. (Adapted from Bawn and Ledwith.)

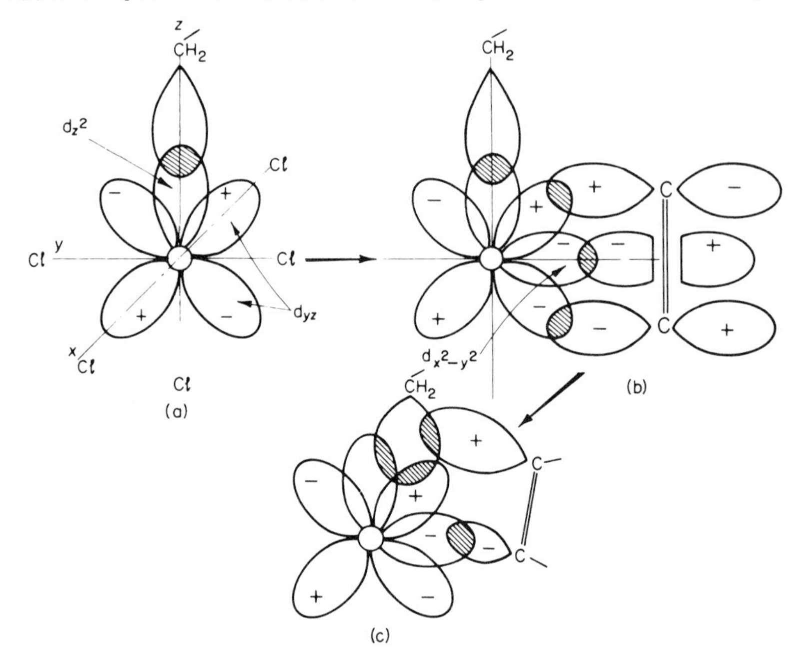

FIGURE 7.3. Representation of the relevant orbital overlap in (a) the active centre, (b) the titanium-olefin complex, and (c) the transition state.

metal bond at this site is required; and (3) the growing polymer chain is always attached to the transition metal.

7.6 Stereoregulation

To obtain a stereoregular polymer, the chemisorption of the monomer on the catalyst surface must be controlled so that the orientation of the incoming monomer is always the same. Examination of models shows that a molecule such as propylene will fit into the catalyst surface in only one way if a position of closest approach of the double bond to the Ti^{3+} ion is to be achieved. This places the $=CH_2$ group pointing into the lattice and for steric reasons the orientation of the $-CH_3$ group to one side is preferred. This determines the configuration of the monomer during the complexing stage and is always the same. Repeated absorption of the monomer in this orientation, prior to reaction, leads to an isotactic polymer.

For the Cossee-Arlman mechanism to operate, migration of the vacant site back to its original position is necessary, else an alternating position in offered to the chemisorbed monomer and a syndiotactic polymer would result. This implies that the tacticity of the polymer formed depends essentially on the rates of both the alkyl shift and the migration. As both of these will slow down when the temperature is decreased, formation of syndiotactic polymer should be favoured at low temperatures, and syndiotactic polypropylene can in fact be obtained at 203 K.

7.7 Natural synthetic rubber – "Natsyn" (IR)

The original goal of most chemists interested in elastomers was the synthesis of *cis*-polyisoprene, natural synthetic rubber, or "Natsyn" as it is sometimes called. The problem was complicated by the ability of the starting material, isoprene, to polymerize in four different ways and form mixtures of the various combinations. With the advent of stereospecific Ziegler-Natta catalysts the "impossible" was achieved in 1956.

In the preparation of *cis*-polyisoprene, great care must be taken in the purification of reagents. Various catalysts are available, but the trialkylaluminium + titanium tetrachloride combination has been found to produce a *cis*-content in excess of 94 per cent.

A 15 per cent solution of high purity isoprene in dry pentane (99 per cent) is prepared and 55 g, drained directly from a cooled silica gel column, are added to a clean dry screw cap bottle. The contents of the bottle are reduced to about 50 g by heating the open bottle on a sand bath before adding by syringe 0.200 mmol of aluminium tri-isobutyl and 0.185 mmol of titanium tetrachloride. Both of these compounds react violently with oxygen or water and are conveniently handled by first preparing 0.2 to 0.5 mol dm^{-3} solutions in pure dry heptane which are stored under nitrogen in bottles sealed by serum caps. These can be kept in a dry-box in an inert atmosphere, and, when required, a quantity is withdrawn by syringe.

After the reagents have been added, the bottle is sealed with a teflon lined cap and rotated for 16 h in a thermostat bath at 323 K. When the reaction is complete and cool, an antioxidant is added (2 per cent solution of di-*tert*-amyl hydroquinone in benzene is suitable). The mixture is then poured into isopropyl alcohol (0.2 dm^3) containing 2 g of antioxidant. The precipitated polymer is separated and dried at 310 K under vacuum.

The *trans*-1,4-polyisoprene (99 per cent) can be prepared using an aluminium + titanium + vanadium catalyst. "Natsyn" has a very small share of world markets at present, but this may change as demands increase.

7.8 Ring opening metathesis polymerizations (ROMP)

The Ziegler catalysts are not the only group of complexes capable of promoting polymerizations. There is a growing interest in "Olefin metathesis reactions", particularly those involving the transition metal catalysed, ring opening polymerizations of cycloalkenes and bicycloalkenes. The reactions can be used to produce linear chains containing unsaturated sites, and in the case of bicycloalkanes, ring structures are also incorporated. The development of catalysts capable of sustaining living polymer systems has led to the ability to prepare block copolymer structures or functionalized telomers, and statistical copolymers have also been synthesized by these routes.

The majority of catalyst systems are based on transition metal compounds of tungsten, molybdenum, rhenium, rhuthenium and titanium carbene complexes. Among the most widely used are the tungsten halides, WCl_6, WF_6 and $WOCl_4$, but in general these and other catalysts may also require activation by a co-catalyst that is an organometallic compound or a Lewis acid. Thus [WCl_6:$(C_2H_5)_2AlCl$], [$TiCl_4$:$(C_2H_5)_3Al$], [$RuCl_3$(hydrate):C_2H_5OH], and the general group M(CHR)(NAr)$(OR)_2$ where M = Mo or W, R = alkyl, Ar = 2,6-diisopropylphenyl, are all active catalyst–co-catalyst combinations. Many more exist, and the structure of the polymer formed can depend on the ratio of the catalyst to co-catalyst in the mixture. Metal carbene and metallacyclic complexes have now been discovered that have extended the range of monomer derivatives capable of undergoing ROMP. For example, using titanocene compounds such as Cp_2TiCl_2 ($Cp = \eta^5 - C_5H_5$), active metallacyclobutane derivatives can be generated and used to form stable, living polymer systems that can be stored for some considerable time while retaining their activity.

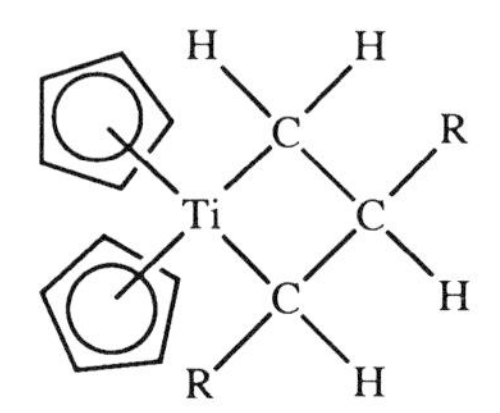

The types of monomer susceptible to metathesis polymerization reactions are limited, however, and those that are most suitable are strained ring structures. Thermodynamic considerations account for the lack of reaction with six-membered rings, and strain-free cyclohexene derivatives, or conjugated cyclo dienes, tend to be excluded. Cyclic monomers containing functional groups, —OH, —COOH, —COOR, —$CONH_2$, and —NH_2 are also regarded as unlikely candidates, at least with the catalytic systems known at present.

7.9 Monocyclic monomers

Rapid polymerization reactions are observed when highly strained cycloalkene rings are used, an example being the conversion of cyclopentene into polypentenamer using (WCl_6:$Al_2Et_3Cl_3$) as catalyst.

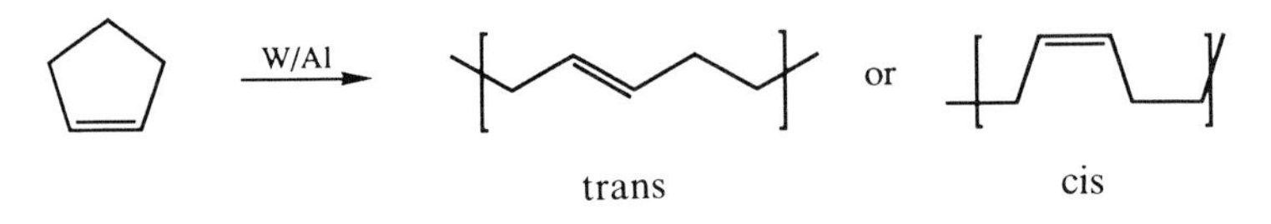

The material is a linear polymer retaining the double bond in the chain and this means that there is the possibility of *cis-trans* geometric isomerism which will affect the properties of the product. Polypentenamer is a useful elastomer and can be prepared with either a high *trans* content ($T_g = 183$ K; $T_m = 293$ K), or a high *cis* content ($T_g = 159$ K; $T_m = 232$ K), *e.g.* 99 per cent *cis*-content is obtained with $MoCl_5/AlEt_3$, the catalyst where there are equimolar ratios of Mo:Al.

Both structures have a glass transition temperature T_g, that is lower than natural rubber, whilst the T_g for the *cis*-polymer is the lowest recorded so far for a hydrocarbon elastomer. Variation of the catalyst composition can be used to control the *cis-trans* content of the chain. For (Al:W) molar ratios of up to (2:1), a predominantly *cis*-polypentenamer is formed but as the molar ratio is raised, increasing amounts of the *trans* placements are found in the chain and at (6:1) this amounts to ≈ 90 per cent *trans* content.

Chain propagation is thought to proceed by co-ordination of the carbon-carbon double bond of the monomer at the vacant site of a transition metal carbene to form a π-complex. The unstable metallocyclobutane produced is then opened to reform the vacant site, and in so doing incorporates the monomer in the chain.

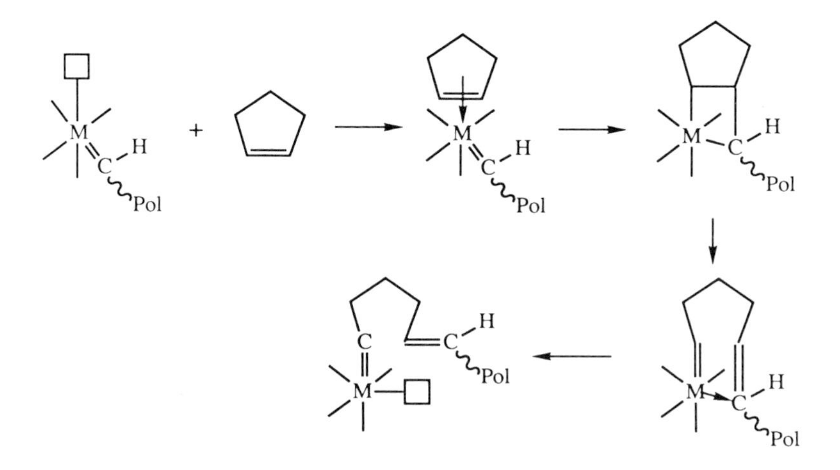

This is known as the "non-pairwise mechanism".

Other monocyclic monomers will undergo ring opening reactions; cyclobutene and 1-methyl cyclobutene produce polybutadiene and *cis*-polyisoprene respectively, but this synthetic route cannot compete with established methods for preparing synthetic elastomers. Reactions are slower with the larger rings, as the strain decreases, but a useful elastomer with a high *trans* content and well developed crystallinity (sold under the trade name Vestenamer) can be prepared from *cis*-cyclooctene (the *trans*-compound is unreactive).

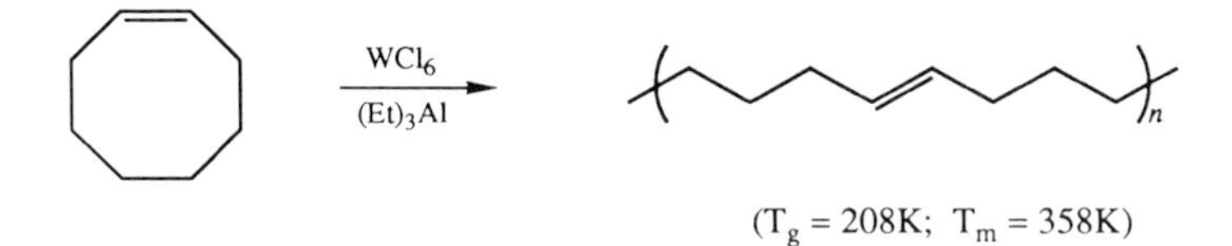

($T_g = 208$K; $T_m = 358$K)

7.10 Bicyclo- and tricyclo-monomers

In general bi- and tricyclic monomers can be polymerized more easily than the monocyclic monomers, and this is reflected in the wider range of catalytic systems that can be used.

Norbornene, or bicyclo[2.2.1]hept-2-ene, is one of the better known monomers that can be subjected to ring opening polymerization by metathesis catalysts. This monomer, or its derivatives, produce polymers with both a ring, and a site of unsaturation in the main chain repeat unit.

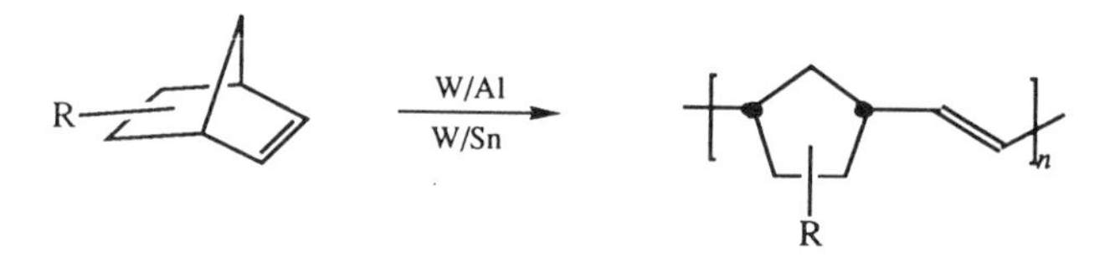

Such polymers can have both *cis* and *trans* structures, and a polynorbornene with a high *trans* content ($T_g = 308$ K; $T_m > 440$ K), prepared using $MoCl_5$ or $RuCl_3$ catalysts, is sold as the product Norsorex. In addition, the chain contains the chiral centres (shown as —•—), and other stereoregular structures are now possible. If the chirality of the centres on either side of the double bond is the same then a *racemic* diad is formed, whereas if they have opposite chiralities a *meso* diad results. The structures with *meso* diads are called isotactic placements and those with *racemic* diads are called syndiotactic placements.

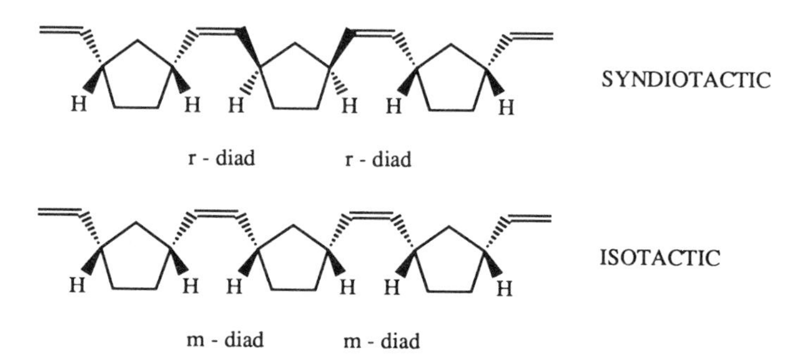

Successful polymerization of heterobicyclic monomers depends on careful selection of the catalyst system, and only a few examples exist at present. Thus 7-oxa-norbornene derivatives can ring open when reacted with osmium or rhuthenium catalysts.

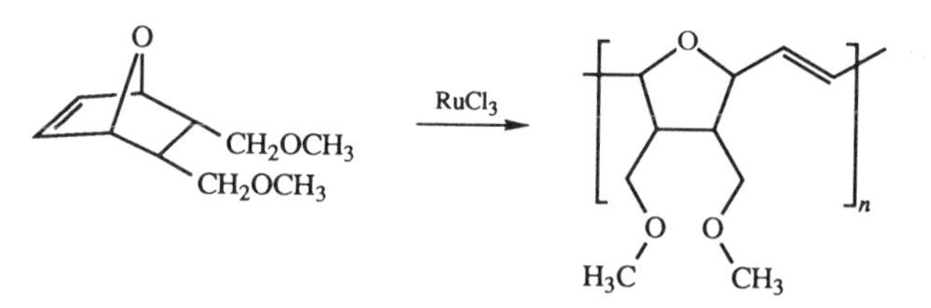

Tricyclo[5.2.1.0]dec-8-ene can be polymerized by several catalytic combinations to produce highly stereospecific polymers.

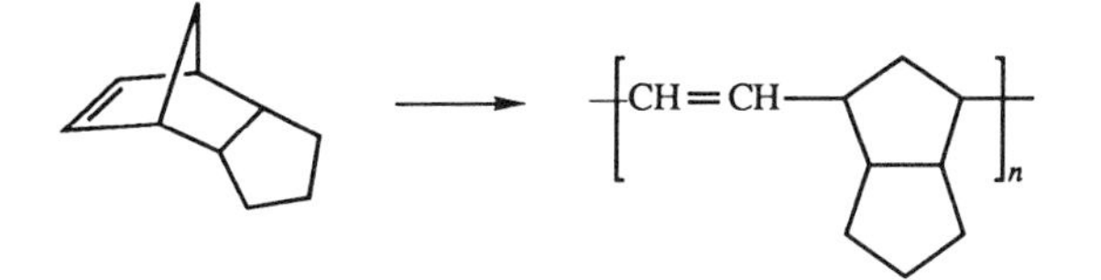

If $MoCl_5$ is used a *trans* polyalkenamer is formed, when $ReCl_5$ is the catalyst the *cis*-polymer is produced and, if the monomer is reacted in the presence of WCl_6, a mixed *cis-trans* structure is obtained.

7.11 Copolyalkenamers

The copolymerization of various pairs of cycloalkenes can be carried out in the presence of ROMP catalytic systems. The structures obtained are predominantly statistical because the differences in ring size or ring substituents, that will always exist, ensure that the reactivities of the monomers are sufficiently dissimilar to make uniform alternating copolymers virtually impossible to prepare.

The copolymerization of cyclopentene and cycloheptene in the presence of (WCl_6:Et_2AlCl) and benzoyl peroxide leads to a copolyalkenamer with 75% *trans* placements and approximately 80% of the pentenamer units in the chain. Mono- and bicyclo-alkenes can also copolymerize when tungsten or molybdenum catalysts are used, hence cyclopentene and norbornene form the elastomeric structure shown:

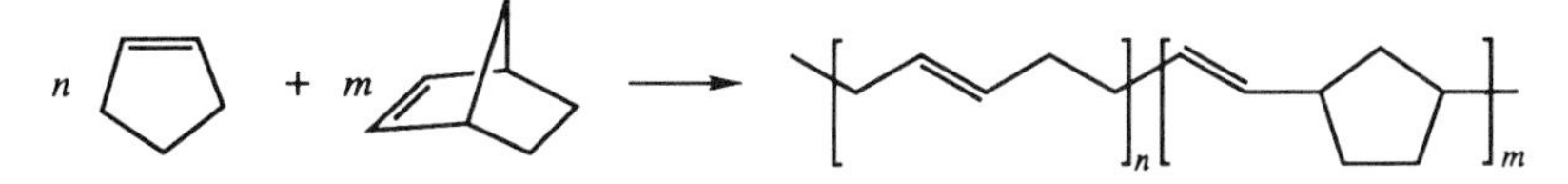

Similarly, substituted norbornenes can be used to produce copolymers with differing pendant groups.

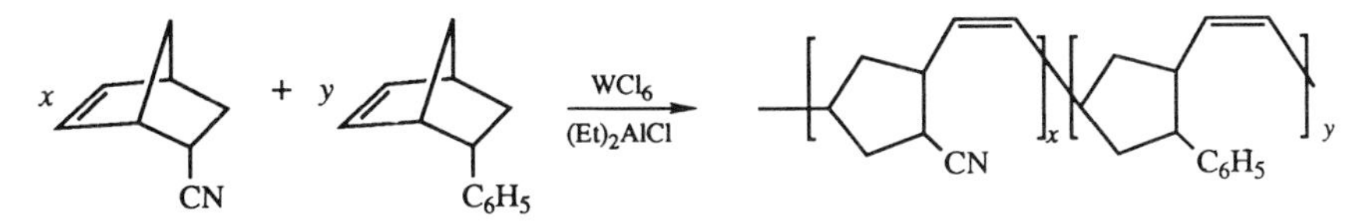

7.12 Living systems

The development of novel titanium carbene complexes by Grubbs has opened up a route to "living polymer" systems, using co-ordinating polymerizations as opposed to those derived from ionic initiators, that can be used to form block copolymers or to produce chains with a functionalized end group. The initiating species are formed by the reaction of norbornene with a titanocyclobutane derived from 3,3-dimethyl cycloprene

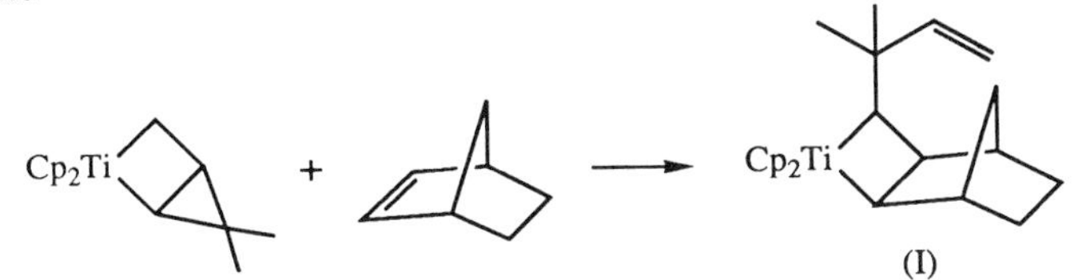

or from

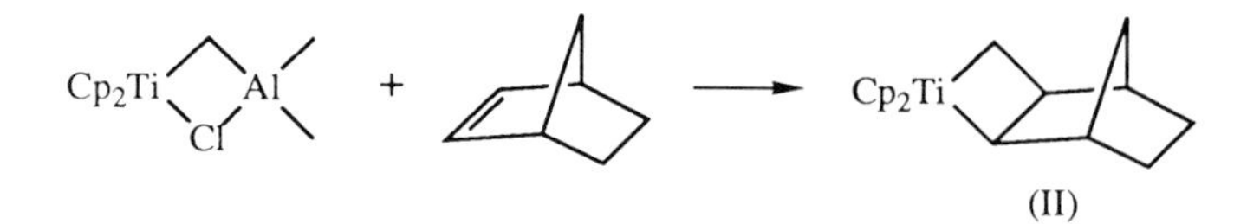

(II)

These can be used to prepare narrow molar mass distribution polymers, and as no termination reaction can be detected, they have all the features of a "living" polymer system, *e.g.* a plot of per cent conversion of monomer against the molar masses of the polymers formed is a straight line passing through zero, which is a good test of a "living" system.

The stable "living polymer" chain derived from norbornene is:

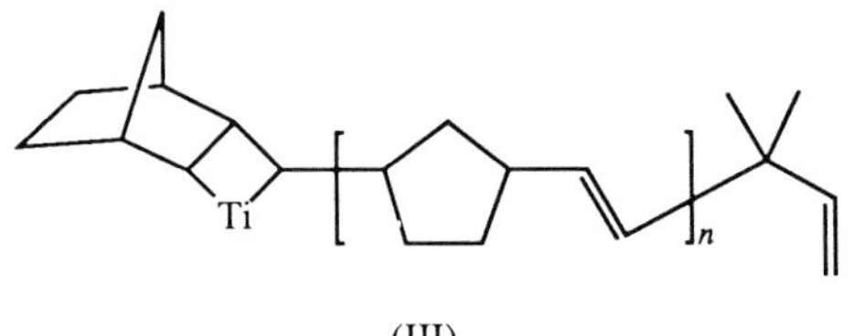

(III)

Heating (III) in the presence of another monomer leads to di-block copolymer formation

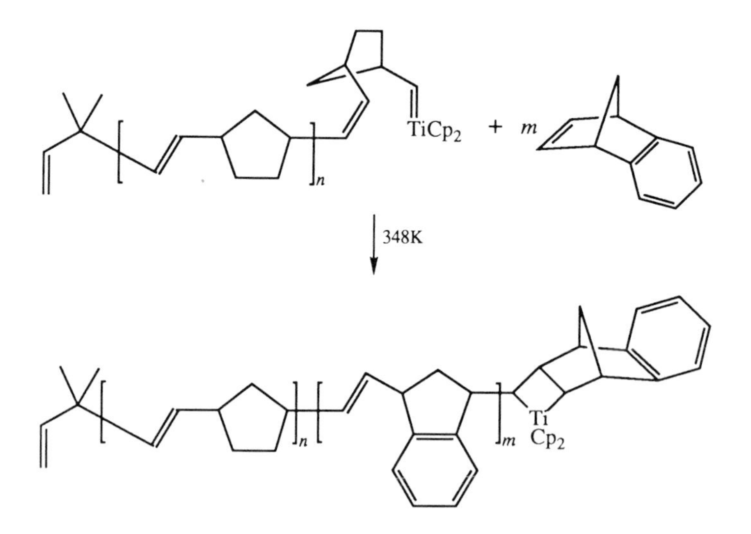

and by making a third addition to the reaction mixture, of another monomer (or norbornene again), a tri-block will be formed.

The chains can be terminated by heating with a reagent that will react with the carbene, and aldehydes or ketones will undergo a Wittig-type reaction to form a terminal olefin unit.

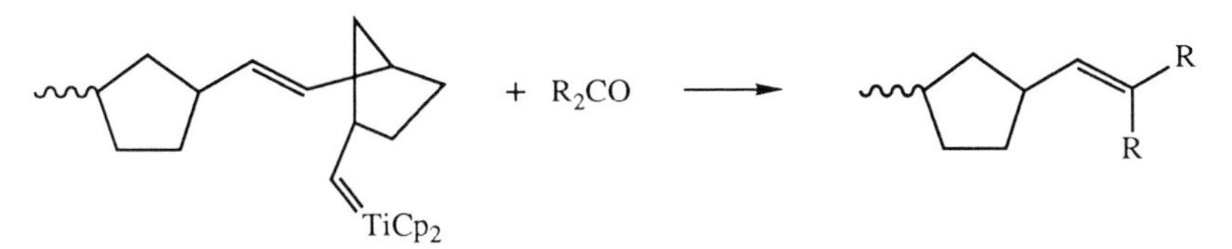

The end-capping reaction can be used to functionalize the terminal group and provide a telomer for further graft or block copolymer synthesis by coupling. Alternatively, it can be designed to give a group capable of initiating a different polymerization technique by a transfer reaction. For example, end-capping with phthalaldehyde gives a chain with a terminal aldehyde group that can be used to initiate an aldol GTP reaction (see section 7.14) and form a di-block copolymer.

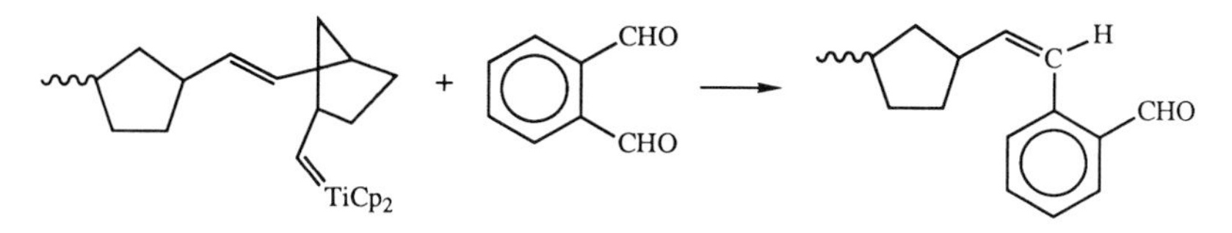

7.13 Group transfer polymerization (GTP)

The use of conventional anionic polymerization methods to produce "living" polymers from acrylic or methacrylic monomers has not met with much success. In 1983, however, the discovery of group transfer polymerization by workers at Du Pont, rectified the situation. The method developed is a new type of reaction leading to "living" polymers, that can be used for polar monomers particularly derivatives of acrylic and methacrylic acids.

The reaction is a sequential Michael addition of an organosilicon compound to α,β-unsaturated esters, ketones, nitriles, and carboxamides. Chain propagation proceeds by transfer of the silyl group, from the silyl ketene acetal catalyst, to the monomer with the generation of a new ketene acetal, and, if inadvertent termination is avoided, repeated addition of the monomer leads to a "living" polymer*. The typical reaction shown on page 154 uses the catalyst 1-methoxy-1-trimethylsiloxy-2-methyl prop-1-ene, but a co-catalyst is required and this is either an anionic species or a Lewis acid. A catalytic complex forms first, to which the monomer is added sequentially, with the trimethyl silyl group transferring to the incoming monomer. A chain with the 'living" ketene silyl acetal group is then produced, which is capable of further reaction if more monomer is made available.

Effective anionic co-catalysts are bifluoride ions *e.g.* tris(dimethyl amino) sulphonium bifluoride $[(Me_2N)_3S{\cdot}HF_2]$ or azides (TAS N_3) and cyanides (TASCN) where TAS $= (Me_2N)_3S$. Lewis acids, zinc halides and dialkyl aluminium chloride can also act as co-catalysts.

Monomers containing active hydrogen (*e.g.* acids and hydroxy compounds) are not suitable for GTP, but acrylates, acrylonitrile, N,N-dimethyl acrylamide can be polymerized quite readily. The reaction is very susceptible to impurities, and all

*Recent work suggests GTP is not truly a "living" polymerization reaction and that there is a slow termination reaction involving cyclization. Consequently, block copolymer formation may depend on the rate of addition of a second monomer and the rate of polymerization.

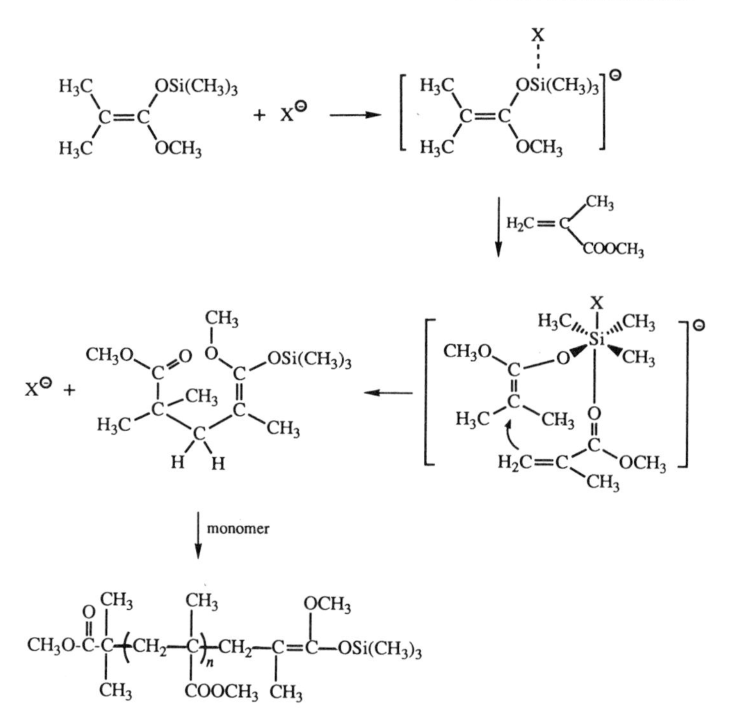

reagents and solvents must be scrupulously dried. While both low and high temperature polymerization reactions are possible, a range of 270 K to 320 K is preferred.

As with other "living" systems, GTP lends itself to the preparation of block copolymers using, *e.g.*, monomers such as methylmethacrylate (MMA) and butyl methacrylate (BMA).

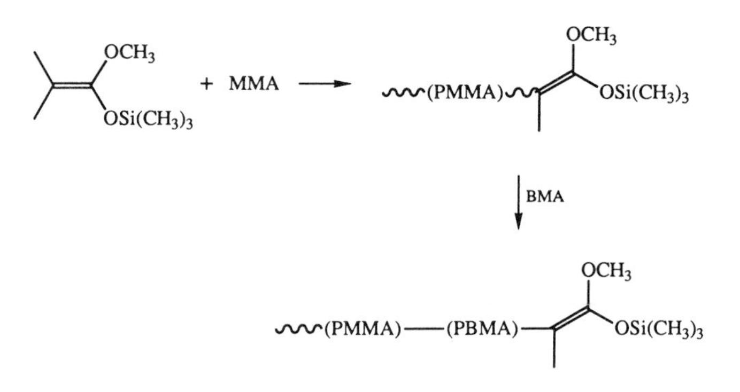

The block lengths can be controlled by altering the monomer to initiator molar ratio.

The formation of telechelic polymers is also possible. Modification of the initiator with a trimethyl silyl group, that can subsequently be removed by treatment with a methanol/Bu_4NF mixture, gives a hydroxy terminated chain.

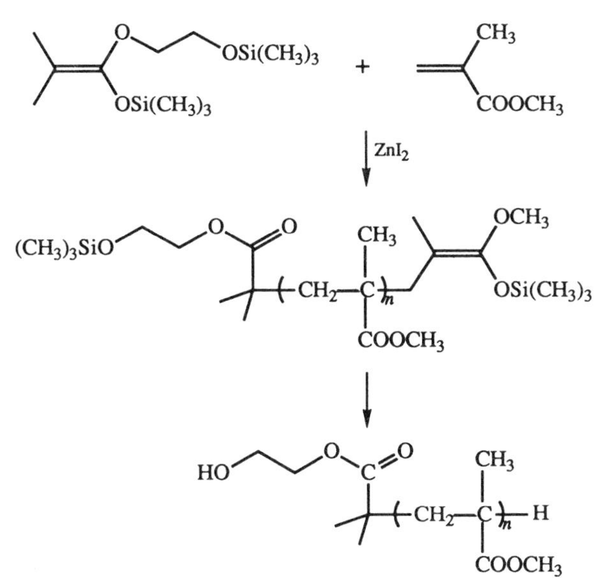

If $(Me)_2C{=}C(OSiMe_3)_2$ is used as the initiating species, a carboxyl terminated chain can be obtained. In either case, coupling reactions between these functional chains will lead to block copolymer formation.

7.14 Aldol group transfer polymerization

A related technique called aldol GTP makes use of the reaction of an aldehyde with a silyl vinyl ether. The co-catalysts used are again Lewis acids, and if used with *t*-butyl dimethyl silyl vinyl ether, the combination will polymerize aldehydes:

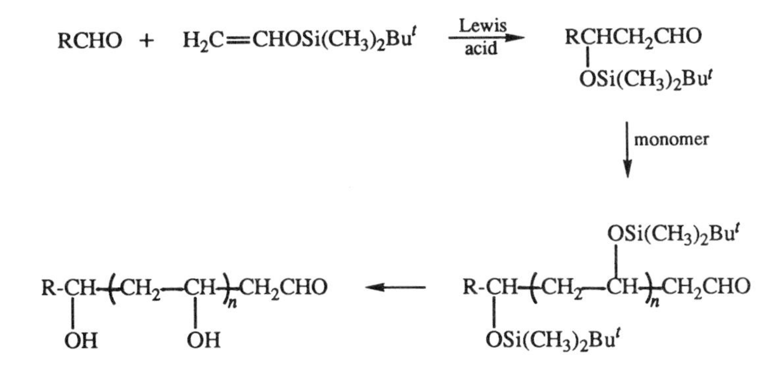

Hydrolysis of the trialkyl silyl groups gives poly(vinyl alcohol). Block copolymers in which one block is poly(vinyl alcohol) can be synthesised if a telechelic with an aldehyde terminal group is used to initiate the aldol GTP, and structures such as poly(styrene-*block*-vinyl alcohol) can be prepared in this way. Alternatively, as silyl ketene acetals can react with aldehydes, block structures can be formed by a coupling process.

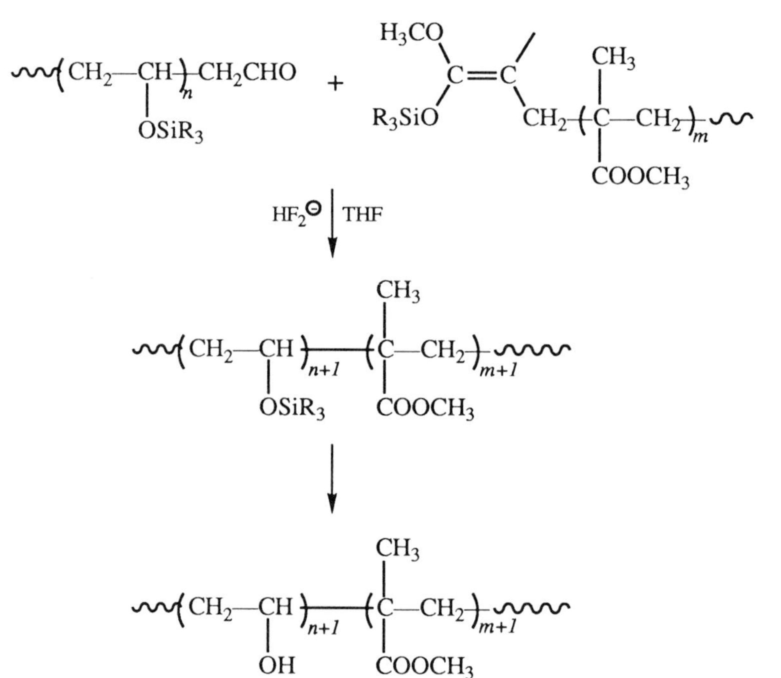

During this reaction, cleavage of the silyl groups, by fluoride ion in the presence of methanol, forms the vinyl alcohol block.

General Reading

G. Allen and J. C. Bevington, Eds, *Comprehensive Polymer Science*, Vol. 4. Pergamon Press (1989).

J. Boor, *Ziegler-Natta Catalysis and Polymerization*. Academic Press (1979).

V. Dragutan, A. T. Balaban and M. Dimonie, *Olefin Metathesis and Ring Opening Polymerization of Cyclo Olefins*. John Wiley, N. Y. (1985).

K. J. Ivin, *Olefin Metathesis*. Academic Press, N. Y. (1983).

W. Kaminsky and H. Sinn, Eds, *Transition Metals and Organometallics as Catalysts for Olefin Polymerization*. Springer-Verlag, Berlin (1988).

T. Keii, *Kinetics of Ziegler-Natta Polymerization*. Chapman and Hall, London (1972).

R. P. Quirk, Ed., *Transition Metal Catalysed Polymerizations*. Harwood Academic Press, N. Y. (1983).

References

G. Bazan, R. R. Schrock, E. Khosravi, W. J. Feast, and V. C. Gibson, *Polymer Commun.*, **9**, 258 (1989).

L. R. Gilliom and R. H. Grubbs, *J. Amer. Chem. Soc.*, **108**, 733 (1986).

O. W. Webster, W. R. Hertler, D. Y. Sogah, W. B. Farnham, and T. V. Rajanbabu, *J. Macromol. Sci.-Chem.*, **A21**, 943 (1984).

CHAPTER 8

Polymers in Solution

8.1 Thermodynamics of polymer solutions

The interaction of long chain molecules with liquids is of considerable interest from both a practical and theoretical view point. For linear and branched polymers, liquids can usually be found which will dissolve the polymer completely to form a homogeneous solution, whereas cross-linked networks will only swell when in contact with compatible liquids. In this chapter we shall deal with linear or branched polymers and treat the swelling of networks in chapter 14.

When an amorphous polymer is mixed with a suitable solvent, it disperses in the solvent and behaves as though it too is a liquid. In a good solvent, classed as one which is highly compatible with the polymer, the liquid-polymer interactions expand the polymer coil, from its unperturbed dimensions, in proportion to the extent of these interactions. In a "poor" solvent, the interactions are fewer and coil expansion or perturbation is restricted.

The fundamental thermodynamic equation used to describe these systems relates the Gibbs free energy function G to the enthalpy H and entropy S, *i.e.* $G = H - TS$. A homogeneous solution is obtained when the Gibbs free energy of mixing $\Delta G^M \leqslant 0$, *i.e.* when the Gibbs free energy of the solution G_{12} is lower than the Gibbs functions of the components of the mixture G_1 and G_2.

$$\Delta G^M = G_{12} - (G_1 + G_2) \tag{8.1}$$

8.2 Ideal mixtures of small molecules

To understand the behaviour of polymers in solution more fully, a knowledge of the enthalpic and entropic contributions to ΔG^M is essential, and it is instructive to consider first mixtures of small molecules, to establish some fundamental rules concerning ideal and non-ideal behaviour. Raoult's law is a useful starting point and defines an ideal solution as one in which the activity of each component in a mixture a_i is equal to its mole fraction x_i. This is valid only for components of comparable size, and where the intermolecular forces acting between both like and unlike molecules are equal. The latter requirement means that component molecules of each species can interchange positions without altering the total energy of the system, *i.e.* $\Delta H^M = 0$ and consequently it only remains for the entropy contribution ΔS^M to be calculated.

For a system in a given state, the entropy is related to the number of distinguishable arrangements the components in that state can adopt, and can be calculated from the

Boltzmann law $S = k \ln W$, where W is the number of statistical microstates available to the system. We can begin by considering the mixing of N_1 molecules of component (1) with N_2 molecules of component (2) and this can be assumed to take place on a hypothetical lattice containing $(N_1 + N_2) = N_0$ cells of equal size. Although this formalism is not strictly necessary for the analysis, the arrangement of spherical molecules of equal size in the liquid state will, to the first near neighbour approximation, be similar to a regular lattice structure and so it is a useful structure to use as a framework for the mixing process.

The total number of possible ways in which the component molecules can be arranged on the lattice increases when mixing takes place and is equal to $(N_1 + N_2)! = N_0!$, but as the interchanging of a molecule of (1) with another molecule (1), or (2) with (2) will be an indistinguishable process, the net number of distinguishable arrangements will be

$$W = \frac{(N_1 + N_2)!}{N_1!N_2!} = \frac{N_0!}{\Pi N_i!} \tag{8.2}$$

The configurational (or combinatorial) entropy S_c can then be derived from the Boltzmann law and

$$S_c = k \ln \frac{N_0!}{N_1!\,N_2!} \tag{8.3}$$

For large values of N_i, Stirling's approximation can be used to deal with the factorials *viz.* $\ln N! = N \ln N - N$, and equation (8.3) becomes

$$S_c = k(N_0 \ln N_0 - N_0 - N_1 \ln N_1 + N_1 - N_2 \ln N_2 + N_2) \tag{8.4}$$

which on dividing by N_0 gives

$$S_c = -k\left[N_1 \ln \frac{N_1}{N_0} + N_2 \ln \frac{N_2}{N_0}\right] \tag{8.5}$$

If $x_i = (N_i/N_0)$, the mole fraction of component i, then

$$S_c = -k[N_1 \ln x_1 + N_2 \ln x_2] \tag{8.6}$$

For the pure components, $x_i = 1$, and as ΔS^M, the change in entropy on mixing, is given by $(S_c - S_1 - S_2)$ then we can write

$$S_c = \Delta S^M_{id} = -k\Sigma N_i \ln x_i$$

so for a two component mixture

$$\Delta S^M_{id} = -k[N_1 \ln x_1 + N_2 \ln x_2] \tag{8.7}$$

This expression is derived assuming (a) the volume change on mixing $\Delta V^M = 0$, (b) the molecules are all of equal size, (c) all possible arrangements have the same energy,

$\Delta H^M = 0$, and (d) the motion of the components about their equilibrium positions remains unchanged on mixing. Thus the free energy of mixing, ΔG^M is

$$\Delta G^M = -T\Delta S^M = -kT(N_1 \ln x_1 + N_2 \ln x_2) \tag{8.8}$$

which shows that mixing in ideal systems is an entropically driven, spontaneous process.

8.3 Non-ideal solutions

Any deviations from assumptions (a) to (d) will constitute a deviation from ideality (an ideal solution is a rare occurrence) and several more realistic types of solution can be identified:

(i) *Athermal* solutions; where $\Delta H^M = 0$ but ΔS^M is not ideal
(ii) *Regular* solutions; where ΔS^M is ideal but $\Delta H^M \neq 0$,
(iii) *Irregular* solutions; in which both ΔS^M and ΔH^M deviate from their ideal values.

Polymer solutions tend to fall into category (iii) and the non-ideal behaviour can be attributed not only to the existence of a finite heat of mixing but also to the large difference in size between the polymer and solvent molecules. The polymer chain can be regarded as a series of small segments covalently bonded together and it is the effect of this chain connectivity which leads to deviations from an ideal entropy of mixing. The effect of connectivity can be assessed by calculating the entropy change associated with the different number of ways of arranging polymer chains and solvent molecules on a lattice and, as it will be demonstrated, this differs from that calculated for the ideal solution. This is embodied in the theory developed by Flory and Huggins, but still represents only the combinatorial contribution, whereas there are other (non-combinatorial) contributions to the entropy which come from the interaction of the polymer with the solvent and are much harder to quantify. Nevertheless, the Flory-Huggins theory forms the cornerstone of polymer solution thermodynamics and is worth considering further.

8.4 Flory-Huggins theory

The dissolution of a polymer in a solvent can be regarded as a two stage process. The polymer exists initially in the solid state where it is restricted to only one of the many conformations which are available to it as a free isolated molecule. On passing into the liquid solution the chain achieves relative freedom and can now change rapidly among a multitude of possible equi-energetic conformations, dictated partly by the chain flexibility and partly by the interactions with the solvent.

Flory and Huggins considered that formation of the solution depends on (a) the transfer of the polymer chain from a pure, perfectly ordered state to a state of disorder which has the necessary freedom to allow the chain to be placed randomly on a lattice, and (b) the mixing process of the flexible chains with solvent molecules.

The formalism of the lattice was used, for convenience, to calculate the combinatorial entropy of mixing following the method outlined in section 8.2 for small molecules, including the same starting assumptions and restrictions.

ENTROPY OF MIXING FOR ATHERMAL POLYMER SOLUTIONS

Consider a polymer chain consisting of r covalently bonded segments whose size is the same as the solvent molecules, *i.e.* $r = (V_2/V_1)$ where V_i is the molar volume of component i. To calculate the number of ways this chain can be added to a lattice, the necessary restriction imposed is that the segments must occupy r contiguous sites on the lattice because of the connectivity. The problem is to examine the mixing of N_1 solvent molecules with N_2 monodisperse polymer molecules comprising r segments and we can begin by adding i polymer molecules to an empty lattice with a total number of cells N_0

$$N_0 = (N_1 + rN_2) \tag{8.9}$$

Thus the number of vacant cells left which can accommodate the next (i + 1) molecule will be

$$(N_0 - ri) \tag{8.10}$$

The (i + 1) molecule can now be placed on the lattice, segment by segment, bearing in mind the restrictions imposed, *viz.* the connectivity of the segments, which requires the placing of each segment in a cell adjoining the preceding one. This in turn will depend on the availability of a suitable vacancy. The first segment can be placed in any empty cell but the second segment is restricted to the immediate near neighbours surrounding the first. This can be given by the co-ordination number of the lattice z but we must also know if a cell in the co-ordination shell is empty. If we let p_i be the probability that an adjacent cell is vacant, then to a reasonable approximation this can be equated with the fraction of cells occupied by i polymer chains on the lattice *i.e.*

$$p_i = (N_0 - ri)/N_0 \tag{8.11}$$

which is valid for large values of z. So the expected number of empty cells available for the secong segment is zp_i, and having removed one more vacant cell from the immediate

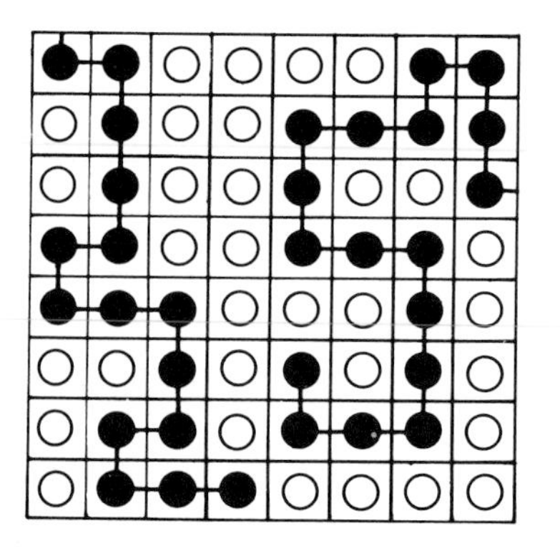

FIGURE 8.1. Placement of polymer chains and solvent molecules on a lattice as required by the Flory-Huggins theory.

vicinity, the third and each succeeding segment will have $(z-1)\,p_i$ empty cells to choose from. The total number of ways in which the (i + 1) molecule can be placed on the lattice is then

$$W_{(i+1)} = (N_0 - ri)z(z-1)^{r-2}[(N_0 - ri)/N_0]^{r-1} \tag{8.12a}$$

$$= (N_0 - ri)^r\{(z-1)/N_0)^{r-1} \tag{8.12b}$$

This gives the set of possible ways in which the (i + 1) molecule can be accommodated on the lattice. The total number of ways for all N_2 molecules to be placed can then be obtained from the product of all possible ways, *i.e.*

$$W_1 W_2 \ldots W_i \ldots W_{N_2} = \prod_{i=1}^{N_2} W_i$$

The polymer molecules are all identical and so by analogy with equation (8.2) the total number of distinguishable ways of adding N_2 polymer molecules is

$$W_p = \prod_{i=1}^{N_2} W_i/N_2! \tag{8.13}$$

Substituting for W_i gives

$$W_p = \frac{1}{N_2!}\prod_{i=1}^{N_2}\{[N_0 - r(i-1)]^r[(z-1)/N_0]^{r-1}\}$$

$$= \left(\frac{1}{N_2!}\right)\left(\frac{z-1}{N_0}\right)^{N_2(r-1)}\prod_{i=1}^{N_2}\{N_0 - r(i-1)\}^r \tag{8.14}$$

To evaluate the product term we can multiply and divide by r

$$\prod_{i=1}^{N_2}\{N_0 - r(i-1)\}^r = r^{(N_2 r)}\prod_{i=1}^{N_2}\left\{\frac{N_0}{r} - i + 1\right\}^r \tag{8.15}$$

This can be converted into the more convenient factorial form by remembering that the product

$$\left(\frac{N_0}{r}+1-1\right)^r\left(\frac{N_0}{r}+1-2\right)^r\left(\frac{N_0}{r}+1-3\right)^r\cdots\left(\frac{N_0}{r}+1-N_2\right)^r \tag{8.16}$$

is equivalent to

$$\left\{\frac{(N_0/r)!}{(N_0/r - N_2)!}\right\}^r = \left\{\frac{(N_0/r)!}{(N_1/r)!}\right\}^r \tag{8.17}$$

and so equation (8.14) can be written as

$$W_p = \left(\frac{1}{N_2!}\right)\left\{\frac{(N_0/r)!}{(N_1/r)!}\right\}\left[\frac{z-1}{N_0}\right]^{N_2(r-1)} \tag{8.18}$$

The remaining empty cells on the lattice can now be filled by solvent molecules, but as there is only one distinguishable way in which this can be done, $W_s = 1$, there is no further contribution to W_p and the entropy of the system. The latter can now be calculated from the Boltzmann equation. The factorials can again be approximated using Stirling's relation and while this requires considerable manipulation, which will be omitted here, it can eventually be shown that

$$S^{\mathrm{M}}/k = \ln W_p = -N_1 \ln\left(\frac{N_1}{N_0}\right) - N_2 \ln\left(\frac{N_2}{N_0}\right) + N_2\{(\mathrm{r}-1)\ln(z-1) - (\mathrm{r}-1)\} \quad (8.19)$$

To convert this into a form which will allow us to express this in the correct site fraction form we can add and subtract $N_2 \ln \mathrm{r}$ on the r.h.s. of equation (8.19) to give

$$S^{\mathrm{M}}/k = -N_1 \ln\left[\frac{N_1}{N_1 + \mathrm{r}N_2}\right] - N_2 \ln\left[\frac{\mathrm{r}N_2}{N_1 + \mathrm{r}N_2}\right] + N_2\left\{(\mathrm{r}-1)\ln\frac{(z-1)}{\mathrm{e}} + \ln \mathrm{r}\right\} \quad (8.20)$$

For the pure solvent $N_2 = 0$ and the entropy $S_1 = 0$. Similarly the entropy of the pure polymer S_2 can be obtained for $N_1 = 0$, which gives

$$S_2 = N_2\left\{(\mathrm{r}-1)\ln\frac{(z-1)}{\mathrm{e}} + \ln \mathrm{r}\right\} \quad (8.21)$$

Equation 8.21 then represents the entropy associated with the disordered or amorphous polymer on the lattice in the absence of solvent.

It follows that the entropy change on mixing disordered polymer and solvent

$$\Delta S^{\mathrm{M}} = S^{\mathrm{M}} - S_1 - S_2$$

and so

$$\Delta S^{\mathrm{M}} = -k\{N_1 \ln \phi_1 + N_2 \ln \phi_2\} \quad (8.22)$$

where ϕ_{i}, the volume fraction can replace the site fraction if it is considered that the number of sites occupied by the polymer and solvent is proportional to their respective volumes.

Equation (8.22) is the expression for the combinatorial entropy of mixing of an athermal polymer solution and comparison with equation (8.7) shows that they are similar in form except for the fact that now the volume fraction is found to be the most convenient way of expressing the entropy change, rather than the mole fraction used for small molecules. This change arises from the differences in size between the components which would normally mean mole fractions close to unity for the solvent especially when dilute solutions are being studied.

We can gain a further understanding of how the size of the polymer chain affects the magnitude of ΔS^{M} and why it differs from $\Delta S^{\mathrm{M}}_{\mathrm{id}}$ (equation 8.7), by recasting equation (8.22) in the following way. The volume fraction ϕ_{i} can be expressed in terms of the number of moles n_{i}, and the volume V_{i} of component i, as

$$\phi_{\mathrm{i}} = (n_{\mathrm{i}} V_{\mathrm{i}}/V)$$

where V is the total volume and $\Delta V^{M} = 0$. If n_i is converted to molar quantities then

$$\Delta S^{M} = -RV\left[\frac{\phi_1}{V_1}\ln\phi_1 + \frac{\phi_2}{V_2}\ln\phi_2\right] \tag{8.23}$$

As V_i can conveniently be expressed as a function of a reference volume V_0 such that $V_i = r_i V_0$ and assuming that, without introducing significant error, r can be equated with the degree of polymerization for the polymer then

$$\Delta S^{M} = -\frac{RV}{V_0}\left[\frac{\phi_1}{r_1}\ln\phi_1 + \frac{\phi_2}{r_2}\ln\phi_2\right] \tag{8.24}$$

If the volume fraction form is retained, then for a simple liquid mixture $r_1 = r_2 = 1$, but for a polymer solution $r_2 \gg 1$ and the last term in equation (8.24) will be smaller than the equivalent term calculated for small molecules. Consequently ΔS^{M} per mole of lattice sites (or equivalent volume) will be very much less than ΔS^{M}_{id} and the contribution of the combinatorial entropy to the mixing process in a polymer solution is not as large as that for solutions of small molecules when calculated in terms of volume fractions and expressed as per mole of sites.

8.5 Enthalpy change on mixing

The derivation of ΔS^{M} from the lattice theory has been made on the assumption that no heat or energy change occurs on mixing. This is an uncommon situation as experimental experience suggests that the energy change is finite. We can make use of regular solution theory to obtain an expression for ΔH^{M} where this change in energy is assumed to arise from the formation of new solvent-polymer (1–2) contacts on mixing which replace some of the (1–1) and (2–2) contacts present in the pure solvent, and the pure polymer components respectively. This can be represented by a quasi-chemical process

$$\tfrac{1}{2}(1-1) + \tfrac{1}{2}(2-2) \rightarrow (1-2) \tag{8.25}$$

where the formation of a solvent-polymer contact requires first the breaking of (1–1) and (2–2) contacts, and can be expressed as an interchange energy $\Delta\epsilon_{12}$ per contact, given by

$$\Delta U^{M} = \Delta\epsilon_{12} = \epsilon_{12} - \tfrac{1}{2}(\epsilon_{11} + \epsilon_{22}) \tag{8.26}$$

Here ϵ_{ii} and ϵ_{ij} are the contact energies for each species. The energy of mixing ΔU^{M} can be replaced by ΔH^{M} if no volume change takes place on mixing, and for q new contacts formed in solution

$$\Delta H^{M} = q\Delta\epsilon_{12} \tag{8.27}$$

The number of contacts can be estimated from the lattice model by assuming that the probability of having a lattice cell occupied by a solvent molecule is simply the volume fraction ϕ_1. This means that each polymer molecule will be surrounded by $(\phi_1 rz)$

solvent molecules, and for N_2 polymer molecules

$$\Delta H^{\mathrm{M}} = N_2\phi_1 \mathrm{r} z\,\Delta\epsilon_{12} \tag{8.28}$$

From the definition of ϕ_2 we obtain $\mathrm{r}N_2\phi_1 = N_1\phi_2$, hence

$$\Delta H^{\mathrm{M}} = N_1\phi_2 z\Delta\epsilon_{12} \tag{8.29}$$

which is the van Laar expression derived for regular solutions and shows that this approach can be applied to polymer systems. To eliminate z, a dimensionless parameter (χ_1) per solvent molecule, is defined as

$$kT\chi_1 = z\Delta\epsilon_{12} \tag{8.30}$$

which is the difference in energy between a solvent molecule when it is immersed in pure polymer and when in pure solvent. It can also be expressed in the alternative form $RT\chi_1 = BV_1$ where B is now an interaction density.

The final expression is

$$\Delta H^{\mathrm{M}} = kT\chi_1 N_1\phi_2 \tag{8.31}$$

and the interaction parameter χ_1 is an important feature of polymer solution theory which will be met with frequently.

8.6 Free energy of mixing

Having calculated the entropy and enthalpy contributions to mixing, these can now be combined to give the expression for the free energy of mixing, $\Delta G^{\mathrm{M}} = \Delta H^{\mathrm{M}} - T\Delta S^{\mathrm{M}}$ as

$$\Delta G^{\mathrm{M}} = kT[\underbrace{N_1 \ln\phi_1 + N_2 \ln\phi_2}_{\text{Combinatorial term}} + \underbrace{N_1\phi_2\chi_1}_{\text{Contact dissimilarity}}] \tag{8.32}$$

It is more useful to express equation (8.32) in terms of the chemical potentials of the pure solvent (μ_1^0) and the solvent in solution (μ_1), by differentiating the expression with respect to the number of solvent molecules N_1 to obtain the partial molar Gibbs free energy of dilution (after multiplying by Avogadro's number),

$$\frac{\partial\Delta G^{\mathrm{M}}}{\partial N_1} = (\mu_1 - \mu_1^0) = RT\left[\ln(1-\phi_2) + \left(1 - \frac{1}{\mathrm{r}}\right)\phi_2 + \chi_1\phi_2^2\right] \tag{8.33}$$

This could also be carried out for the polymer (N_2), but as it makes no difference which one is taken (both having started from ΔG^{M}), equation (8.33) is more convenient to use. While this expression is not strictly valid for the dilute solution regime it can be converted into a structure which is extremely informative about deviations from ideal solution behaviour encountered when measuring the molar mass by techniques such as osmotic pressure. If the logarithmic term is expanded using a Taylor series:

$$\ln(1-\phi_2) = -\phi_2 - \phi_2^2/2 - \phi_2^3/3\ldots..$$

but truncated after the squared term, assuming ϕ_2 is small, then,

$$(\mu_1 - \mu_1^0) = -RT[(\phi_2/\mathrm{r}) + (\tfrac{1}{2} - \chi_1)\phi_2^2] \tag{8.34}$$

This can be modified by remembering that $\mathrm{r} = (V_2/V_1)$ and $\phi_2 = c_2\bar{v}_2$, where $\bar{v}_2$ is the partial specific volume of the polymer. This can be related to the polymer molecular weight M_2 through

$\bar{v}_2 = (V_2/M_2)$ so that $(\phi_2/\mathrm{r}) = c_2 V_1/M_2$ and finally

$$(\mu_1 - \mu_1^0) = -RT\left[\frac{c_2 V_1}{M_2} + \bar{v}_2^2(\tfrac{1}{2} - \chi_1)c_2^2\right] \tag{8.35}$$

Let us now anticipate the molar mass measurements to be described in chapter 9 and examine the osmotic pressure of a polymer solution in the light of equation (8.35).

8.7 Osmotic pressure

The osmotic pressure π of a solution can be regarded as the pressure which must be exerted on that solution to raise the chemical potential of the solvent in the solution (μ_1) back up to that of the pure solvent (μ_1^0) at a standard pressure P, *i.e.*

$$\mu_1^0 = \mu_1 + \int_P^{P+\pi} (\partial\mu_1/\partial P)_T \mathrm{d}P \tag{8.36}$$

The compressibility of the solvent $(\partial\mu_1/\partial P)_T$ is equal to the molar volume of the solvent in solution, V_1, and can be assumed to be unchanged over a small range of pressures, thus

$$\mu_1^0 = \mu_1 + V_1 \int_P^{P+\pi} \mathrm{d}P \tag{8.37a}$$

$$= \mu_1 + V_1[(P + \pi) - P] \tag{8.37b}$$

giving

$$(\mu_1 - \mu_1^0) = -V_1\pi \tag{8.37c}$$

Substitution in (8.35) gives

$$\pi V_1 = RT\left[\frac{c_2 V_1}{M_2} + \bar{v}_2^2(\tfrac{1}{2} - \chi_1)c_2^2\right] \tag{8.38}$$

or

$$\frac{\pi}{c_2} = \frac{RT}{M_2} + RT\frac{\bar{v}_2^2}{V_1}(\tfrac{1}{2} - \chi_1)c_2 \tag{8.39}$$

This is a limited virial expansion in which the first term is the classical van't Hoff expression for the osmotic pressure at infinite dilution. The second term is related to the deviation from ideal behaviour and gives a relationship between the second virial coefficient B and the interaction parameter χ_1

$$B = RT\frac{\bar{v}_2^2}{V_1}(\tfrac{1}{2} - \chi_1) \tag{8.40}$$

Thus when $\chi_1 = \frac{1}{2}$ then $B = 0$ and the osmotic pressure is given by the ideal solution law.

8.8 Limitations of the Flory-Huggins theory

The simple lattice theory does not describe the behaviour of dilute polymer solutions particularly well because the following simplifications in the theoretical treatment are invalid: (1) it was assumed that the segment-locating process is purely statistical, but this would only be true if $\Delta\epsilon_{12}$ was zero; (2) the treatment assumed that the flexibility of the chain is unaltered on passing into the solution from the solid state – this limits the calculation of ΔS^{M} to the combinatorial contribution only and neglects any contribution from continual flexing of the chain in solution which will contribute to the non-combinatorial or excess entropy of mixing; (3) any possible specific solvent-polymer interactions which might lead to orientation of the solvent molecules in the vicinity of the polymer chain are neglected *i.e.* polar solutions may be inadequately catered for by this theory; (4) a uniform density of lattice site occupation is assumed, but this will only apply to relatively concentrated solutions; (5) The parameter χ_1 is often concentration-dependent but this is ignored. It is now accepted that a non-combinatorial entropy contribution arises from the formation of new (1–2) contacts in the mixture which change the vibrational frequencies of the two components, *i.e.* assumption (d) in section 8.2 must be relaxed. This can be allowed for by recognizing that χ_1 is actually a free energy parameter comprising entropic χ_H and enthalpic χ_S contributions, such that $\chi_1 = \chi_H + \chi_S$. These are defined by

$$\chi_H = -T(\mathrm{d}\chi_1/\mathrm{d}T) \quad \text{and} \quad \chi_S = \mathrm{d}(T\chi_1)/\mathrm{d}T. (= -\Delta S/k)$$

Experiments tend to show that the major contribution comes from the χ_S component, indicating that there is a large decrease in entropy (non-combinatorial) which is acting against the dissolution process of a polymer in a solvent.

In spite of much justifiable criticism, the Flory-Huggins theory can still generate considerable interest because of the limited amount of success which can be claimed for it in relation to phase equilibria studies.

8.9 Phase equilibria

Use can be made of the Flory-Huggins theory to predict the equilibrium behaviour of two liquid phases when both contain amorphous polymer and one or even two solvents.

Consider a two component system consisting of a liquid (1) which is a poor solvent for a polymer (2). Complete miscibility occurs when the Gibbs free energy of mixing is less than the Gibbs free energies of the components, and the solution maintains its homogeneity only as long as ΔG^{M} remains less than the Gibbs free energy of any two possible co-existing phases.

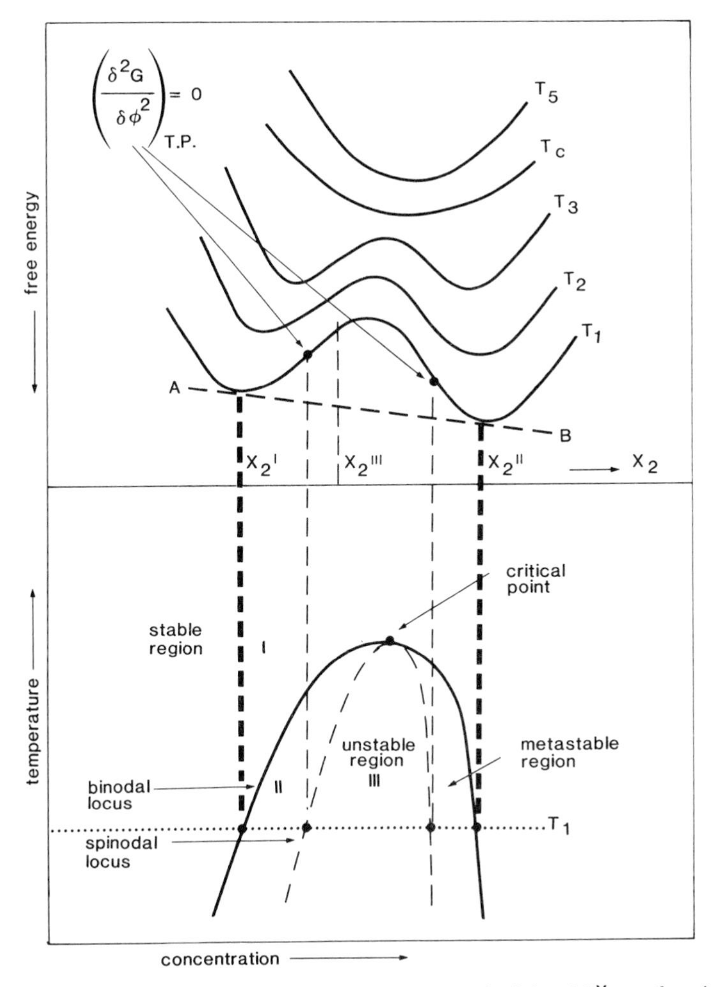

FIGURE 8.2. Schematic diagram of the Gibbs free energy of mixing ΔG^{M} as a function of the mole fraction x_2 of solute (top half) showing the transition from a system miscible in all proportions at a temperature T_5 through the critical temperature T_c, to partially miscible systems at temperatures T_3 to T_1. The contact points for the common tangents drawn to the minima are shown projected onto the Temperature – Concentration plane to form the binodal (cloud-point) curve, whilst projection of the inflexion points forms the spinodal curve. The lower part of the diagram indicates the one phase stable region I, the metastable region II and the unstable region III.

The situation is represented by curve T_5 in figure 8.2. The miscibility of this type of system is observed to be strongly temperature dependent and as T decreases the solution separates into two phases. Thus at any temperature, say T_1, the Gibbs free energy of any mixture, composition x_2''' in the composition range x_2' to x_2'', is higher than either of the two co-existing phases whose compositions are x_2' and x_2'' and phase separation takes place. The compositions of the two phases x_2' and x_2'' do not correspond to the two minima, but are measured from the points of contact of the double tangent AB with the Gibbs free energy curve. The same is true for other temperatures lying below T_c, and the inflexion points can be joined to bound an area representing the heterogeneous two phase system, where there is limited solubility of component 2 in 1 and vice-versa. This is called a *cloud-point curve*.

As the temperature is increased the limits of this two phase co-existence contract, until eventually they coalesce to produce a homogeneous, one phase, mixture at T_c, the *critical solution temperature*. This is sometimes referred to as the *critical consolute point*.

In general, we can say that if the free energy-composition curve has a shape which allows a tangent to touch it at two points, phase separation will occur.

The critical solution temperature is an important quantity and can be accurately defined in terms of the chemical potential. It represents the point at which the inflexion points on the curve merge, and so it is the temperature where the first, second, and third derivatives of the Gibbs free energy with respect to mole fraction are zero.

$$\partial(\Delta G^M)/\partial x_2 = \partial^2(\Delta G)/\partial x_2^2 = \partial^3(\Delta G^M)/\partial x_2^3 = 0. \tag{8.41}$$

It is also true that the partial molar Gibbs free energies of each component are equal at this point and it emerges that the conditions for incipient phase separation are

$$\partial\mu_1/\partial\phi_2 = \partial^2\mu_1/\partial\phi_2^2 = \partial^3\mu_1/\partial\phi_2^3 = 0. \tag{8.42}$$

By remembering that $\Delta G_1 = (\mu_1 - \mu_1^0)$, application of these criteria for equilibrium to equation (8.33) leads to the first derivative of that equation

$$(1-\phi_{2,c})^{-1} - (1 - 1/x_n) - 2\phi_{2,c}\,\chi_{1,c} = 0, \tag{8.43}$$

while the second derivative is

$$(1-\phi_{2,c})^{-2} - 2\chi_{1,c} = 0, \tag{8.44}$$

where the subscript c denotes critical conditions. The critical composition at which phase separation is first detected is then

$$\phi_{2,c} = 1/(1 + x_n^{1/2}) \approx 1/x_n^{1/2}, \tag{8.45}$$

and

$$\chi_{1,c} = \tfrac{1}{2} + 1/x_n^{1/2} + 1/2x_n, \tag{8.46}$$

which indicates that $\chi_{1,c} = 0.5$ at infinitely large chain length.

The interaction parameter χ_1 is a useful measure of the solvent power. Poor solvents have values of χ_1 close to 0.5 while an improvement in solvent power lowers χ_1. Generally, a variation from 0.5 to -1.0 can be observed although for many synthetic polymer solutions the range is 0.6 to 0.3. A linear temperature dependence for χ_1 is also predicted of the general form $\chi_1 = a + b/T$, which suggests that as the temperature increases the solvating power of the liquid should increase. This has implications for polymer fractionation.

8.10 Fractionation

The relations derived in this and other chapters normally assume that the polymer sample has a unique molar mass. This situation is rarely achieved in practice and it is useful to know the form of the molar mass distribution in a sample, as this can have a significant bearing on the physical properties. It is also advantageous to be able to prepare sample fractions, whose homogeneity is considerably better than the parent polymer, especially when testing dilute solution theory.

We have seen that the chain length can be related to the solvent power, expressed as χ_1, by equation (8.46) and this is illustrated in figure 8.3. The implication is that if χ_1 can be carefully controlled, conditions could be attained which would allow a given molecular species to precipitate, while leaving larger or smaller molecules in solution. This process is known as *fractionation*.

Experimentally, a polymer sample can be fractionated in a variety of ways and three in common use are: (1) addition of a non-solvent to a polymer solution; (2) lowering the temperature of the solution; and (3) column chromatography.

In the first method the control of χ_1 is effected by adding a non-solvent to the polymer solution. If the addition is slow, χ_1 increases gradually until the critical value for large molecules is reached. This causes precipitation of the longest chains first and these can be separated from the shorter chains which remain in solution. In practice the polymer solution is held at a constant temperature while precipitant is added to the stirred solution. When the solution becomes turbid the mixture is warmed until the precipitate dissolves. The solution is then returned to the original temperature and the

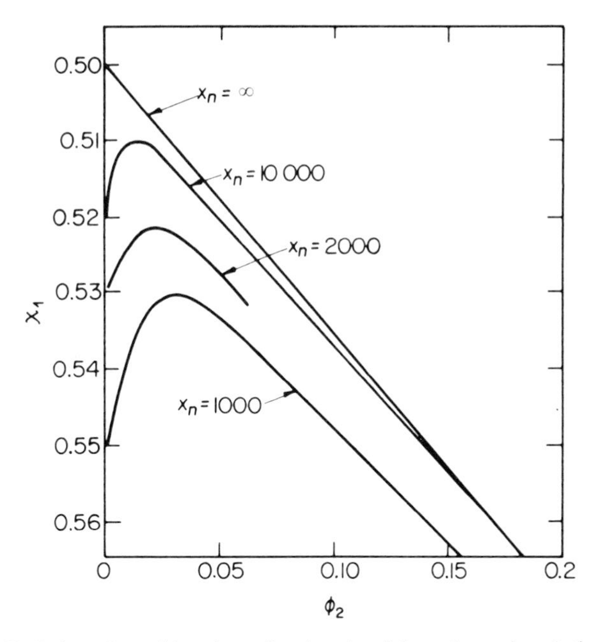

FIGURE 8.3. Variation of χ_1 with volume fraction ϕ_2 of the polymer in solution, showing the effect of changing chain length x_n.

precipitate which reforms is allowed to settle and then separated. This ensures that the precipitated fraction is not broadened by local precipitation during addition of the nonsolvent. Successive additions of small quantities of non-solvent to the solution allow a series of fractions of steadily decreasing molar mass to be separated.

In the second method, χ_1 is varied by altering the temperature, with similar results. For both techniques, it is useful to dissolve the polymer initially in a poor solvent with a large χ_1 value. This ensures that only small quantities of non-solvent are required to precipitate the polymer in method 1, and that the temperature changes required in method 2 are small.

In column chromatography the polymer is precipitated on the inert support medium at the top of a column which has a temperature gradient imposed along its length. The packing is usually glass beads of 0.1 to 0.3 mm diameter. A solvent + non-solvent mixture is used to elute the sample and fractionation is achieved by using a solvent gradient. This is generated in a mixing system, situated above the column, by constantly increasing the solvent to non-solvent ratio and as the mixture is initially a poor solvent which is gradually enriched by the good solvent the low molar mass fractions are eluted first. Fractions of increasing molar mass are collected from the bottom of the column.

In each of the techniques, the mass and molar mass of the fractions are recorded and a distribution curve for the sample can be constructed from the results. However, as the fractions themselves have a molar mass distribution, extensive overlapping of the fractions will occur as shown schematically in figure 8.4. Consequently a simple histogram constructed from the mass and molar mass of each fraction will not provide a good representation of the distribution and a method must be used to compensate for the overlapping.

A useful approach was proposed by Schulz who suggested that a cumulative mass fraction be plotted against the molar mass. The cumulative mass fraction $C(M_i)$ can be calculated by adding half the mass fraction w_i of the ith fraction to the total mass fraction

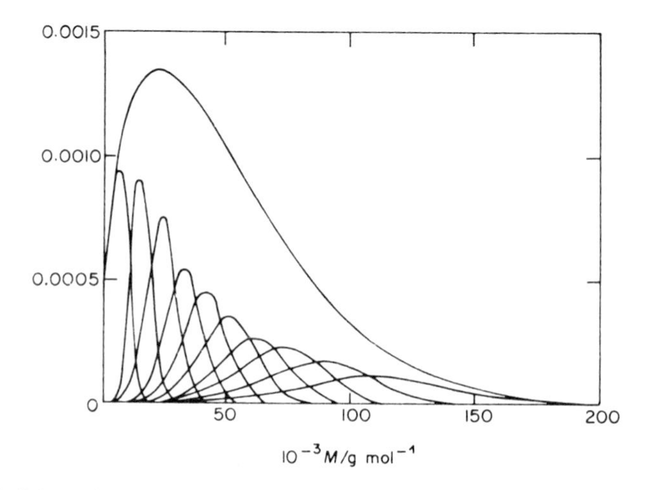

FIGURE 8.4. Schematic representation of the overlapping molar mass distributions of fractions f obtained from the parent sample.

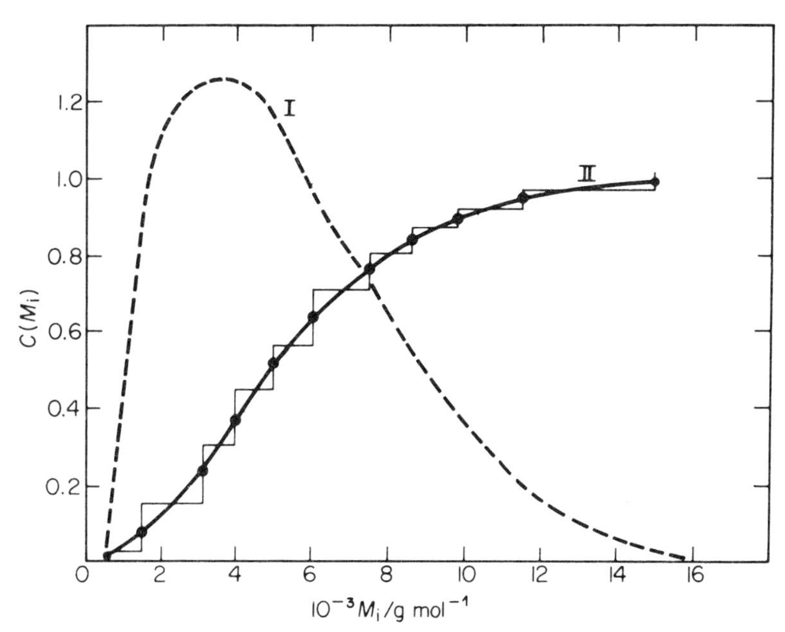

FIGURE 8.5. The differential distribution I, and integral distribution II obtained using equation (8.47).

of those fractions preceding it, *i.e.*

$$C(M_i) = (w_i/2) + \sum_{j=1}^{i-1} w_j. \tag{8.47}$$

The values of $C(M_i)$ are plotted against the corresponding M_i and connected by a smooth curve as shown in figure 8.5, to give the integral distribution curve. The differential curve can be obtained by determining the slope of this curve at selected molar masses and plotting this against the appropriate molar mass.

8.11 Flory-Krigbaum theory

To overcome the limitations of the lattice theory resulting from the discontinuous nature of a dilute polymer solution, Flory and Krigbaum discarded the idea of a uniform distribution of chain segments in the liquid. Instead they considered the solution to be composed of areas containing polymer which were separated by the solvent. In these areas the polymer segments were assumed to possess a Gaussian distribution about the centre of mass, but even with this distribution the chain segments still occupy a finite volume from which all other chain segments are excluded. It is within this excluded volume that the long range interactions originate which are discussed more fully in chapter 10.

Flory and Krigbaum defined an enthalpy (κ_1) parameter and an entropy of dilution (ψ_1) parameter such that the thermodynamic functions used to describe these long range effects are given in terms of the excess partial molar quantities

$$\Delta H_1^{\mathrm{E}} = RT\kappa_1\phi_2^2 \tag{8.48}$$

$$\Delta S_1^{\mathrm{E}} = R\psi_1\phi_2^2 \tag{8.49}$$

From equation (8.33) it can be seen that the excess free energy of dilution is

$$(\mu_1 - \mu_1^0)^{\mathrm{E}} = \Delta G_1^{\mathrm{E}} = -RT(\tfrac{1}{2} - \chi_1)\phi_2^2 \tag{8.50}$$

Combination of these non-ideal terms then yields

$$(\tfrac{1}{2} - \chi_1) = (\psi_1 - \kappa_1) \tag{8.51}$$

As we saw from equation (8.40), when $B = 0$ and $\chi_1 = \frac{1}{2}$ the solution appears to behave as though it were ideal. The point at which this occurs is known as the FLORY or THETA point and is in some ways analogous to the Boyle point for a non-ideal gas. Under these conditions

$$\psi_1 = \kappa \text{ i.e. } \Delta H_1^{\mathrm{E}} = T\Delta S_1^{\mathrm{E}}$$

The temperature at which these conditions are obtained is the FLORY or THETA temperature Θ, conveniently defined as $\Theta = T\kappa_1/\psi_1$. This tells us that Θ will only have a meaningful value when ψ_1 and κ_1 have the same sign.

Substitution in (8.50) followed by rearrangement gives

$$(\mu_1 - \mu_1^0)^{\mathrm{E}} = -RT\psi_1\left(1 - \frac{\Theta}{T}\right)\phi_2^2 \tag{8.52}$$

and shows that deviations from ideal behaviour vanish when $T = \Theta$.

The theta temperature is a well defined state of the polymer solution at which the excluded volume effects are eliminated and the polymer coil is in an unperturbed condition (see chapter 10). Above the theta temperature expansion of the coil takes place, caused by interactions with the solvent, whereas below Θ the polymer segments attract one another, the excluded volume is negative, the coils tend to collapse and eventual phase separation occurs.

8.12 Location of the theta temperature

The theta temperature of a polymer-solvent system can be measured from phase separation studies. The value of $\chi_{1,\mathrm{c}}$ at the critical concentration is related to the chain length of the polymer by equation (8.46), and substitution in (8.52) leads to

$$\psi_1(\Theta/T_{\mathrm{c}} - 1) = 1/x_n^{1/2} + 1/2x_n, \tag{8.53}$$

where now we have replaced r with the equivalent degree of polymerization x_n. Rearrangement gives

$$1/T_{\mathrm{c}} = (1/\Theta)\{1 + (1/\psi_1)(1/x_n^{1/2} + 1/2x_n)\} \tag{8.54}$$

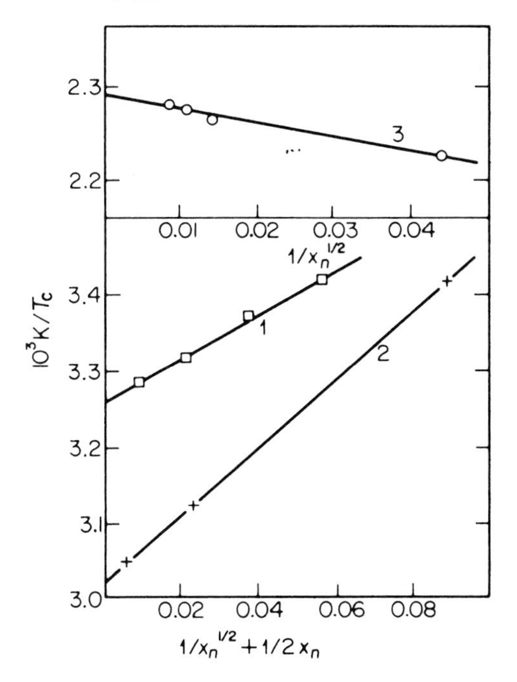

FIGURE 8.6. Chain length x_n dependence of the upper critical consolute temperature T_c for (1) polystyrene in cyclohexane, and (2) polyisobutylene in di-isobutyl ketone (data of Schultz and Flory); and the lower critical solution temperature for (3) polyoctene-1 in *n*-pentane (data of Kinsinger and Ballard.)

Remembering that $x_n = (M\bar{v}_2/V_1)$, where M and $\bar{v}_2$ are the molar mass and partial specific volume of the polymer, and V_1 is the molar volume of the solvent, the equation states that the critical temperature is a function of M and the value of T_c at infinite M is the theta temperature for the system.

Precipitation data for several systems have proved the validity of equation (8.54). Linear plots are obtained with a positive slope from which the entropy parameter ψ_1 can be calculated as shown in figure 8.6. Typical values are shown in table 8.1, but ψ_1

TABLE 8.1. Theta temperatures and entropy parameters for some polymer + solvent systems, derived from equation (8.54)

Polymer	*Solvent*	Θ/K	ψ_1
1. Polystyrene	Cyclohexane	307.2	1.056
2. Polyethylene	Nitrobenzene	503	1.090
3. Polyisobutene	Di-isobutyl Ketone	333.1	0.650
4. Poly(methylmethacrylate)	4-Heptanone	305	0.610
5. Poly(acrylic acid)	Dioxan	302.2	−0.310
6. Polymethacrylonitrile	Butanone	279	−0.630

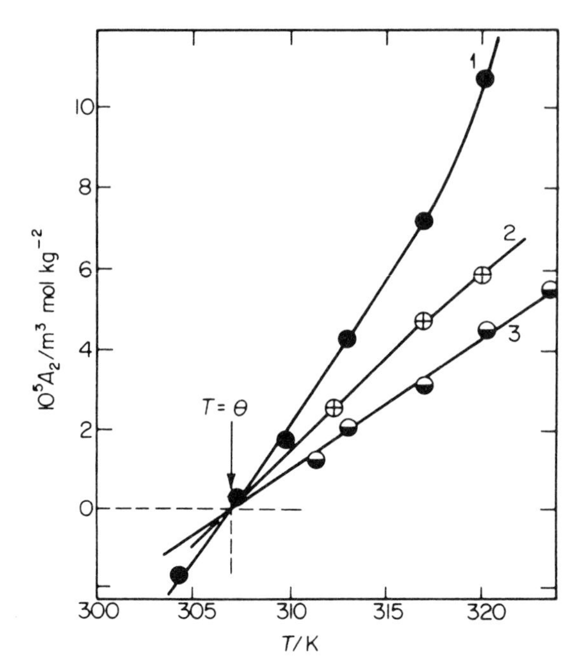

FIGURE 8.7. Location of the theta temperature Θ for poly(α-methyl styrene) in cyclohexane. Values of A_2 are measured for (1) $M_n = 8.6 \times 10^4\,\mathrm{g\,mol^{-1}}$, (2) $M_n = 3.8 \times 10^5\,\mathrm{g\,mol^{-1}}$; and (3) $M_n = 1.5 \times 10^6\,\mathrm{g\,mol^{-1}}$.

values measured for systems such as polystyrene + cyclohexane have been found to be almost ten times larger than those derived from other methods of measurement. This appears to arise from the assumption in the Flory-Huggins theory that χ_1 is concentration independent and improved values of ψ_1 are obtained when this is rectified.

The theta temperature, calculated from equation (8.54) for each system is in good agreement with that measured from the temperature variation of $A_2(=B/RT$ see chapter 9). Curves of A_2, measured at various temperatures in the vicinity of Θ, are constructed as a function of temperature for one or more molar masses as shown in figure 8.7. Intersection of the curves with the T-axis occurs when $A_2 = 0$ and $T = \Theta$. The curves for each molar mass of the same polymer should all intersect at $T = \Theta$.

8.13 Lower critical solution temperatures

So far we have been concerned with non-polar solutions of amorphous polymers, whose solubility is increased with rising temperature, because the additional thermal motion helps to decrease attractive forces between like molecules, and encourages energetically less favourable contacts. The phase diagram for such systems, when the solvent is poor, is depicted by area A in figure 8.8, where the critical temperature T_c occurs near the maximum of the cloud-point curve and is often referred to as the *upper*

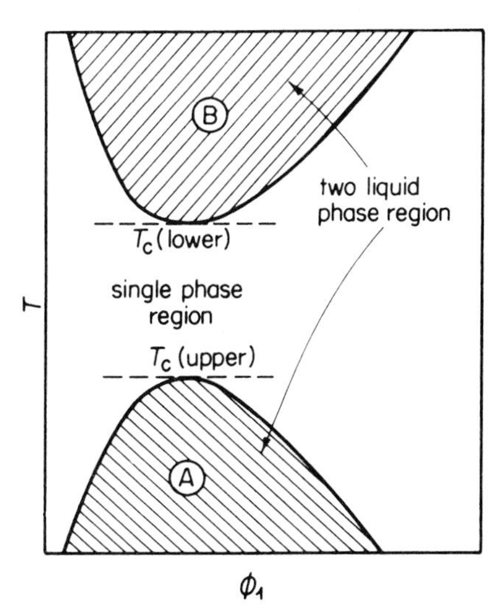

FIGURE 8.8. Schematic diagram of the two types of phase boundaries commonly encountered in polymer solutions; A, the two-phase region characterized by the upper critical solution temperature; and B the two-phase region giving the lower critical solution temperature with a single phase region, lying between the two.

critical solution temperature (UCST). This behaviour follows from that depicted in figure 8.2.

For non-polar systems ΔS^M is normally positive but weighted heavily by T and so solubility depends mainly on the magnitude of ΔH^M, which is normally endothermic (positive). Consequently as T decreases ΔG^M eventually becomes positive and phase separation takes place.

Values of Θ and ψ_1, in table 8.1, show that for systems 1 to 4 the entropy parameter is positive, as expected, but for poly(acrylic acid) in dioxan and polymethacrylonitrile in butanone, ψ_1 is negative at the theta temperature. As $\psi_1 = \kappa_1$ when $T = \Theta$, the enthalpy is also negative for these systems. This means that systems 5 and 6 exhibit an unusual decrease in solubility as the temperature rises, and the cloud-point curve is now inverted as in area B. The corresponding critical temperature is located at the minimum of the miscibility curve and is known as the *lower critical solution temperature* (LCST).

In systems 5 and 6 this phenomenon is a result of hydrogen-bond formation between the polymer and solvent, which enhances the solubility. As hydrogen bonds are thermally labile a rise in T reduces the number of bonds and causes eventual phase separation. In solutions, which are stabilized in this way by secondary bonding, the LCST usually appears below the boiling temperature of the solvent but it has been found experimentally that an LCST can be detected in non-polar systems when these are examined at temperatures approaching the critical temperature of the solvent. Polyisobutylene in a series of n-alkanes, polystyrene in methyl acetate and cyclohexane, and cellulose acetate in acetone all exhibit LCSTs.

The LCST is located by heating the solutions, in sealed tubes, up to temperatures approaching the gas-liquid critical point of the solvent. As the temperature rises, the liquid expands much more rapidly than the polymer, which is restrained by the covalent bonding between its segments. At high temperatures, the spaces between the solvent molecules have to be reduced if mixing is to take place and when this eventually results in too great a loss of entropy, phase separation occurs.

The separation of polymer/solvent systems into two phases as the temperature increases is now recognized to be a characteristic feature of all polymer solutions. This presents a problem of interpretation within the framework of regular solution theory, as the accepted form of χ_1 predicts a monotonic change with temperature and is incapable of dealing with two critical consolute points.

The problem of how to accommodate, in a theoretical framework, the existence of two miscibility gaps requires a new approach, and a more elaborate treatment by Prigogine and co-workers encompasses the difference in size between the components of a mixture, which cannot be ignored for polymer solutions. They replaced the rigid lattice model used by Flory and Huggins, which is valid only at absolute zero, with a flexible lattice whose cells change volume, with temperature and pressure. This allowed them to include in their theory dissimilarities in free volume between polymer and solvent, together with the corresponding interactions. The same approach was extended by both Patterson and Flory to deal specifically with polymer systems.

The most important of the new parameters is the so-called "structural effect" which is related to the number of degrees of freedom "$3c$" which a molecule possesses, divided by the number of external contacts q. This structural factor (c/q) is a measure of the number of external degrees of freedom per segment and changes with the length of the component. Thus the ratio decreases as a liquid becomes increasingly polymeric.

The expansion and free volume can then be characterized by the ratio of the thermal energy arising from the external degrees of freedom available to the component, U_{thermal}, and the interaction energy between neighbouring non-bonded segments, U_{cohesive}, which will oppose the thermal energy effects, *i.e.*

$$\frac{U_{\text{thermal}}}{U_{\text{cohesive}}} = \left(\frac{ckT}{q}\right) \cdot \frac{1}{\epsilon^*} \tag{8.55}$$

where ϵ^* is the characteristic cohesive energy per contact.

For convenience q may be replaced by r, the number of chain segments, although q will actually be less than r because some of the external contacts are used in forming the covalent bonds in the chain.

Free volume dissimilarities become increasingly important as the size of one component increases with respect to the second, as in polymer solutions, and when these differences are sufficiently large, phase separation can be observed at the LCST. The differences in expansivity can be accounted for if the interaction parameter is now expressed as

$$\chi = -(U_1/RT)\nu^2 + (C_{p1}/2R)\tau^2 \tag{8.56}$$

where the first term reflects the interchange energy on forming contacts of unlike type and includes segment size differences, while the second term is the new "structural" contribution coming from free volume changes on mixing a dense polymer with an expanded solvent. This can be represented schematically in figure 8.9.

The first term in (8.56), shown by curve 1, is merely an expression of the Flory-Huggins theory where χ decreases constantly with rising temperature, but now

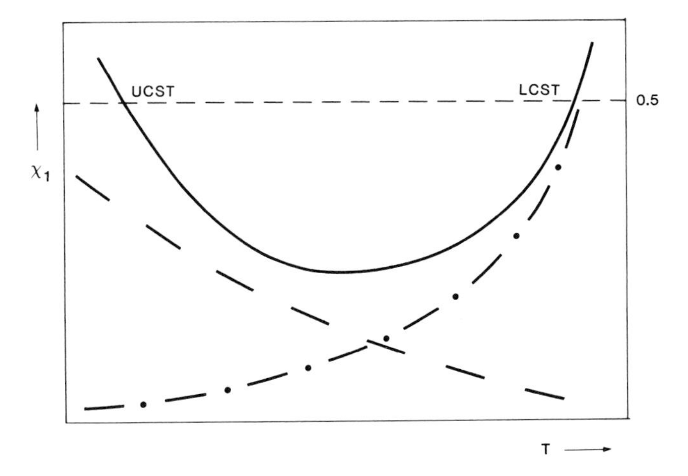

FIGURE 8.9. Schematic diagram of χ_1 as a function of temperature, showing the composite curve of 1 (– –), first term in equation (8.56); and 2 (–·–), the free volume contribution from the second term in equation (8.56) which results in the observation of the LCST.

inclusion of the new free volume term, shown by curve 2, modifies the behaviour of χ. The second term gains in importance as the expansivities of the two components become increasingly divergent with temperature and the net effect is to increase χ again until it once more attains its critical value at high temperature. The LCST which results, is then a consequence of these free volume differences and is an entropically controlled phenomenon.

This can be illustrated in the following ways. In terms of the flexible lattice model, one can imagine the polymer and liquid lattices expanding at different rates until a temperature is reached at which the highly expanded liquid lattice can no longer be distorted sufficiently to accommodate the less expanded polymer lattice and form a solution, *i.e.* the loss in entropy during the distortion becomes so large and unfavourable that phase separation (LCST) takes place. Alternatively, a polymer solution can be thought of as a system formed by the condensation of solvent into a polymer. As the temperature increases, the entropy loss incurred during condensation becomes greater until eventually it is so unfavourable that condensation in the polymer is impossible, and phase separation takes place. Neither picture is particularly rigorous but they serve to emphasize the fact that the LCST is an entropically controlled phenomenon.

8.14 Solubility and the cohesive energy density

Solvent-polymer compatibility problems are repeatedly encountered in industry. For example, in situations requiring the selection of elastomers for use as hose-pipes or gaskets, the correct choice of elastomer is of prime importance, as contact with highly compatible fluids may cause serious swelling and impair the operation of the system. The wrong selection can have far reaching consequences; the initial choice of an elastomer for the seals in the landing gear of the DC-8 aircraft resulted in serious jamming because the seals become swollen when in contact with the hydraulic fluid.

TABLE 8.2. Average solubility parameters for some common polymers

Polymer	$\delta/(\mathrm{J\,cm^{-3}})^{1/2}$	*Polymer*	$\delta/(\mathrm{J\,cm^{-3}})^{1/2}$
Polytetrafluoroethylene	12.7	Polystyrene	18.7
Polyisobutylene	16.3	Poly(methylmethacrylate)	19.0
Polyethylene	16.4	Poly(vinyl acetate)	19.2
Polyisoprene	16.7	Poly(vinyl chloride)	20.7
Polybutadiene	17.1	Nylon 6.6	27.8
Polypropylene	17.8	Poly(acrylonitrile)	28.8

This almost led to grounding of the plane but replacement with an incompatible elastomer made from ethylene-propylene copolymer rectified the fault.

To avoid such problems a technologist may wish to have at his disposal a rough guide to aid the selection of solvents for a polymer or to assess the extent of polymer-liquid interaction other than those already described. Here use can be made of a semi-empirical approach suggested by Hildebrand and based on the premise that "like dissolves like". The treatment involves relating the enthalpy of mixing to the cohesive energy density (E/V) and defines a solubility parameter $\delta = (E/V)^{1/2}$, where E is the molar energy of vaporization and V is the molar volume of the component. The

TABLE 8.3. Group contributions to F

Group	*Small*	*Hoy*
$-CH_3$	438	303.4
$-CH_2$	272	269.0
$-\overset{|}{\underset{|}{C}}H-$	57	176.0
$-\overset{|}{\underset{|}{C}}-$	−190	65.5
$-CH(CH_3)$	495	(479.4)
$-C(CH_3)_2$	686	(672.3)
$-CH{=}CH-$	454	497.4
$-\overset{|}{C}{=}CH-$	266	421.5
$-C(CH_3){=}CH-$	(704)	(724.9)

TABLE 8.3 Group contributions to F (*continued*)

Group	*Small*	*Hoy*
cyclopentyl	–	1295.1
cyclohexyl	–	1473.3
phenyl	1504	1398.4
p-phenylene	1346	1442.2
—F	(250)	84.5
—Cl	552	419.6
—Br	696	527.7
—I	870	–
—CN	839	725.5
—CHCN—	(896)	(901.5)
—OH	–	462.0
—O—	143	235.3
—CO—	563	538.1
—COOH	–	(1000.1)
—COO—	634	668.2
—O—C(=O)—O—	–	(903.5)
—C(=O)—O—C(=O)—	–	1160.7
—C(=O)—N(H)—	–	(906.4)
—O—C(=O)—N(H)—	–	(1036.5)
—S	460	428.4

proposed relation for the heat of mixing of two non-polar components

$$\Delta H^{\mathrm{M}} = V_{\mathrm{M}}(\delta_1 - \delta_2)^2 \phi_1 \phi_2 \tag{8.57}$$

shows that ΔH^{M} is small for mixtures with similar solubility parameters and this indicates compatibility.

Values of the solubility parameter for simple liquids can be readily calculated from the enthalpy of vaporization. The same method cannot be used for a polymer and one must resort to comparative techniques. Usually δ for a polymer is established by finding the solvent which will produce maximum swelling of a network or the largest value of the limiting viscosity number, as both indicate maximum compatibility. The polymer is then assigned a similar value of δ. Alternatively, Small and Hoy have tabulated a series of group molar attraction constants from which a good estimate of δ for most polymers can be made.

The suggested group contributions are shown in table 8.3 and the solubility parameter for a polymer can be estimated from the sum of the various molar attraction constants F for the groups which make up the repeat unit *i.e.*

$$\delta = (\sum F)/V = (\sum F)\rho/M_0.$$

Here V is the molar volume of the repeat unit whose molar mass is M_0 and ρ is the polymer density.

Thus for poly(methyl methacrylate) with $M_0 = 100.1$ and $\rho = 1.19\,\mathrm{g\,cm^{-3}}$, we have using the Hoy values

$$-\!\!\left[CH_2-\overset{\displaystyle CH_3}{\underset{\displaystyle COOCH_3}{C}} \right]\!\!-_n$$

Group	F
$2(-CH_3)$	2(303.4)
$>CH_2$	269
$-COO-$	668.2
$>C<$	65.5
	$\sum 1609.5$

Therefore $\delta = (1609.5)(1.19)/100.1$
$= 19.13\ (\mathrm{J\,cm^{-3}})^{1/2}$

For a more complex polyhydroxyether of Bisphenol A structure and with $\rho = 1.15\,\mathrm{g\,cm^{-3}}$

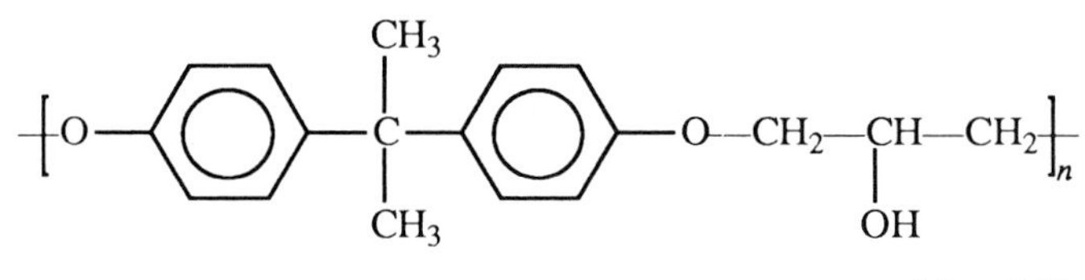

$M_0 = 268$

Group	F
2(p-phenylene)	2(1442)
$2(-CH_3)$	2(303)
$2(>CH_2)$	2(269)
2(Ether oxygen)	2(235.3)
$-OH$	462
$>CH-$	176
$>C<$	65.5
	$\sum 5202$

$\delta = (5202)(1.15)/268 = 22.32\ (\mathrm{J\,cm^{-3}})^{1/2}$

Both estimates are within 10 per cent of experimentally determined values.

Attempts to correlate δ with χ_1 from the Flory-Huggins equation have met with limited success because of the unjustifiable assumptions made in the derivation. It is now believed, however, that χ_1 is not an enthalpy parameter but a free energy parameter and a relation of the form c.f. section 8.8

$$\chi_1 = 1/z + (V_1/RT)(\delta_1 - \delta_2)^2, \qquad (8.58)$$

has improved the correlation. Here $1/z = \chi_s$ is supposed to compensate for the lack of a non-combinatorial entropy contribution in the Flory-Huggins treatment.

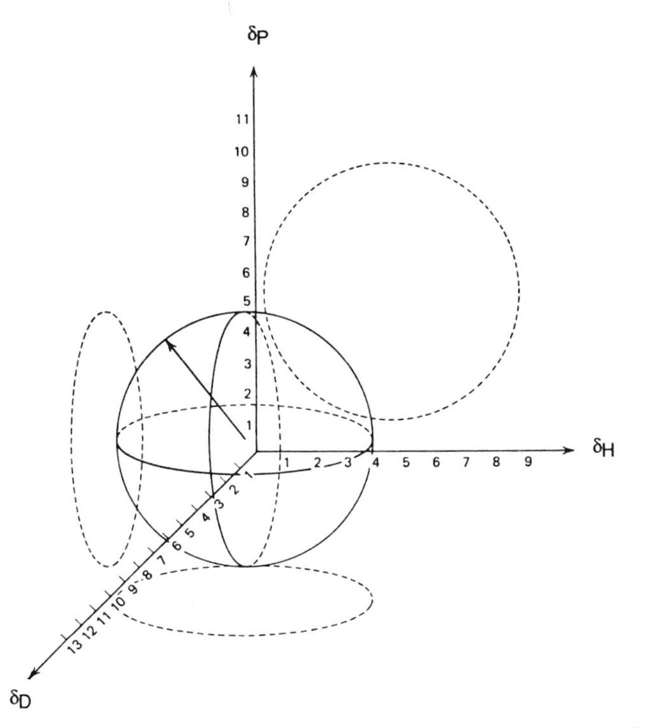

FIGURE 8.10. Schematic representation of a "three dimensional, solubility volume" estimated from the component parts of the solubility parameter: δ_H = hydrogen bonding; δ_P = dipole-dipole; and δ_D = dispersion forces.

Unfortunately, solubility is not a simple process and secondary bonding may play an important role in determining component interactions.

More detailed approaches have been suggested, which introduce a three-dimensional δ composed of contributions from van der Waals dispersion forces, dipole-dipole interaction, and hydrogen bonding.

The overall solubility parameter is then the sum of the various contributions

$$\delta = (\delta_D^2 + \delta_P^2 + \delta_H^2)^{1/2}.$$

Usually two dimensional plots are constructed first, before the three dimensional "solubility volume" is established, as shown in figure 8.10. This is not a convenient construction and often a plot of $\delta_V = (\delta_D^2 + \delta_P^2)^{1/2}$ versus δ_H is considered to be sufficiently accurate as δ_D and δ_P are usually similar and the main polar contribution comes from the hydrogen bonding factor δ_H.

8.15 Polymer-polymer mixtures

In the constant search for new materials with improved performance, the idea of mixing two or more different polymers to form new substances having a combination of all the attributes of the components, is deceptively attractive. Deceptively, because in practice it is rarely accomplished and only in a few cases have polymer blends or mixtures achieved industrial importance. The main reason is that most common polymers do not mix with one another to form homogeneous, one phase solutions or blends, and an explanation for this is to be found in the thermodynamics of solutions which have been outlined in the previous sections.

As we have seen, when two liquids, or a liquid and a polymer are mixed, the formation of a homogeneous, one phase solution is assisted mainly by the large favourable gain in combinatorial entropy. This entropic contribution is progressively reduced when one or both components increase in size, and the reason for this becomes obvious on inspection of equation (8.24). When r_1 and r_2 both increase, then ΔS^M becomes smaller; consequently attempts to mix two high molar mass polymer samples will receive little assistance from this function and must depend increasingly on a favourable (negative) heat of mixing embodied in the χ parameter. This loss of entropy can be conveniently illustrated using the simple lattice model shown in figure 8.11. Here a 10×10 lattice, containing 50 white and 50 black units randomly mixed (a), will result in approximately 10^{30} possible different arrangements of the units on mixing. If these white units are now connected to other white units, and black to black (b), to form five equal chains of each colour with $r_1 = r_2 = 10$, then the number of possible arrangements of these chains decreases to about 10^3. Thus as r_1 and r_2 approach infinity ΔS^M will become negligible and the free energy of mixing will become essentially dependent on ΔH^M which now has to be either very small or negative.

The heat of mixing for the majority of polymer (1)- polymer (2) pairs tends to be endothermic and can be approximated by reference to the solubility parameters using equation (8.57). This can be written as

$$\chi_{12} = \frac{V_0}{RT}(\delta_1 - \delta_2)^2 \tag{8.59}$$

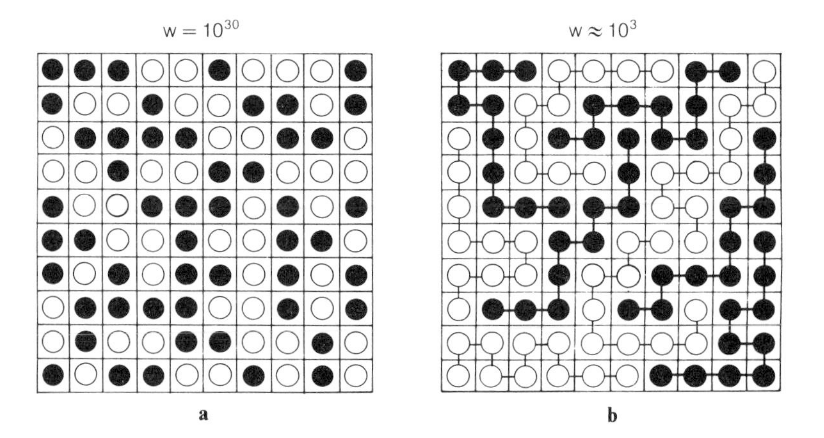

FIGURE 8.11. (a) A ten × ten lattice containing 50 black and 50 white spheres distributed randomly. (b) The same lattice with 5 white and 5 black polymer chains placed on the lattice. Each chain has $x_n = 10$. The number of possible arrangements falls from $W = 10^{30}$ for (a) to $W \approx 10^3$ for (b). Hence $S = k \ln W$ will also decrease dramatically.

where the reference volume normally assumes a value of $100\,cm^3\,mol^{-1}$. The critical value for χ_{12} can be estimated from

$$(\chi_{12})_c = \frac{1}{2}\left[\frac{1}{x_1^{1/2}} + \frac{1}{x_2^{1/2}}\right]^2 \tag{8.60}$$

where x_i is the degree of polymerization, related to the actual degree of polymerization x_n and the reference volume by

$$x_i = x_n(V_R/V_0)$$

with V_R the molar volume of the repeat unit. The critical values for χ_{12}, above which the two polymers will phase separate, calculated for various mixtures with $x_1 = x_2$, are shown in table 8.4 along with the corresponding differences in δ. This shows that for mixing to take place between high molecular weight components the solubility parameters would have to be virtually identical. This limits the number of possible combinations such that only a few examples exist in this category. These include polystyrene/poly(α-methyl styrene) below $M \approx 70\,000$, and the polyacrylates mixed with the corresponding poly vinyl esters, *e.g.*

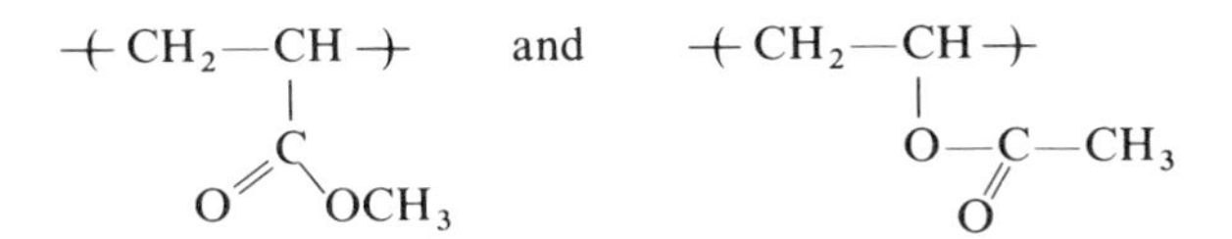

The situation changes if ΔH^M is negative as this will encourage mixing, and the search for binary polymer blends which are miscible has focussed on combinations in which specific intermolecular interactions, such as hydrogen bonds, dipole-dipole interactions, ion-dipole interactions, or charge transfer complex formation, can exist between the component polymers. A substantial number of miscible blends have now been discovered using this principle and it is possible to identify certain groups or repeat units, which when incorporated in polymer chains tend to enter into these intermolecular interactions and enhance the miscibility.

A short selection of some of these complementary groups is given in table 8.5, where

TABLE 8.4. Critical values for χ_{12} derived for various chain lengths, showing also the necessary differences in solubility parameters

$x_1 = x_2$	$(\chi_{12})_c$	$(\delta_1 - \delta_2)_c$
50	0.0400	0.49
100	0.0200	0.35
200	0.0100	0.25
500	0.0040	0.15
700	0.0028	0.13
1000	0.0020	0.11
2000	0.0010	0.08
5000	0.0005	0.05

TABLE 8.5. Complementary groups and repeat units found in miscible binary polymer blends

	Group 1	*Group* 2
1.	$-(CH_2-CH(C_6H_5))-$	$-(CH_2-CH(O-CH_3))-$
2.	$-(CH_2-CH(C_6H_5))-$	$-(C_6H_2R_2-O)-$
3.	$-(CH_2-CR(C(=O)-OCH_3))-$	$-(CH_2-CF_2)-$
4.	$-(CH_2-CR(C(=O)-OCH_3))-$	$-(CH_2-CH(Cl))-$
5.	$-(R_1-O-C(=O)-R_2-C(=O)-O)-$	$-(CH_2-CH(Cl))-$
6.	$-(CH_2-CH(O-C(=O)-CH_3))-$	$\sim\sim(ONO_2)$
7.	$-(CH_2-CH(C(=O)-O-CH_3))-$	$\sim\sim(ONO_2)$
8.	$-C_6H_4-O-C_6H_4-SO_2-$	$-(CH_2-CH_2-O)-$

a polymer containing groups or composed of units from column 1 will tend to form miscible blends with polymers containing groups or composed of units from column 2. Thus it is believed that polystyrene forms miscible blends with poly(vinyl methylether), and with poly(phenylene oxide)s (examples 1 and 2, respectively) because of interactions between the π-electrons of the phenyl rings and the lone pairs of the ether oxygens. Similarly it has been suggested that a weak hydrogn bond, which is strong enough to induce miscibility can form between the carbonyl unit of poly(methyl methacrylate) and the α-hydrogen of poly(vinyl chloride) (example 3 where $R = CH_3$). *i.e.*

$$>C{=}O\cdots H-C-Cl.$$

Much stronger hydrogen bonding interactions can be obtained if units such as

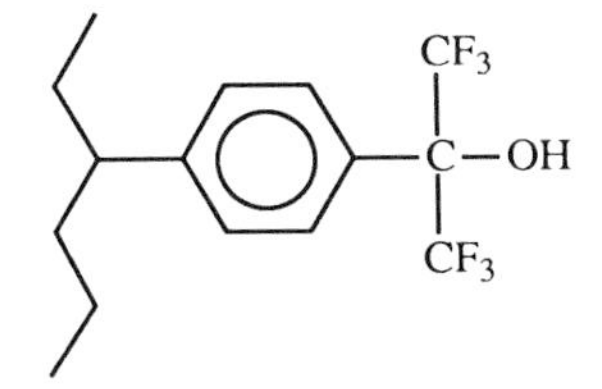

or sites for ion-dipole interactions such as

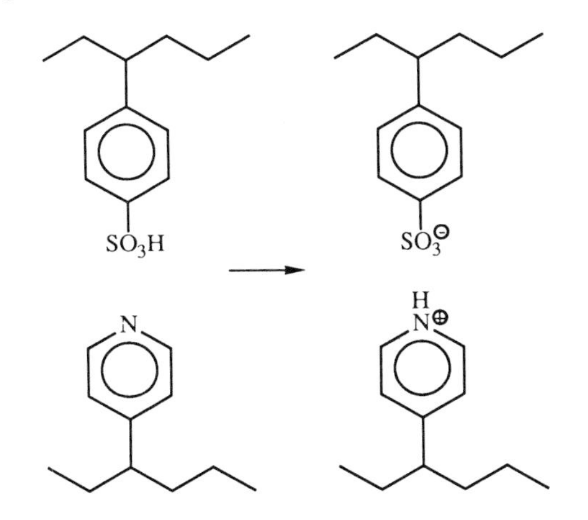

can be built into chains, and even in relatively small amounts these can transform immiscible pairs into totally miscible blends.

Many of these blends undergo quite rapid demixing as the temperature is raised and an LCST phase boundary can be located above the glass transition temperature of the blend. The origins of the lower critical phase separation phenomenon in polymer blends are not yet clearly understood and three possible causes have been proposed.

(i) Free volume dissimilarities may become unfavourable to mixing on increasing the temperature.

(ii) There may be unfavourable entropy contributions arising from non-random mixing.

(iii) A temperature dependent heat of mixing may result if specific intermolecular interactions, which dissociate on heating, are responsible for miscibility at lower temperatures.

While the latter seems the most likely cause in many blends where specific interactions have been identified, miscible blends can also be obtained when certain statistical copolymers are mixed with either a homopolymer, or another copolymer, in which no such interactions have been located. Thus poly(styrene-*stat*-acrylonitrile) will form miscible blends with poly(methyl methacrylate) if the composition of the copolymer lies in the range 10–39 wt% acrylonitrile. This range of compositions is called the "miscibility window" and has been reported to be present in other systems. The drive towards formation of a miscible solid solution in these cases is believed to arise when large repulsive interactions exist between the monomer units (A) and (B) comprising the copolymer; on mixing with a polymer (C), the number of these unfavourable (A–B) contacts are reduced by forming less repulsive (A–C) or (B–C) contacts and a miscible blend results. Many of these blends also exhibit an LCST. Thus the driving force towards lower critical phase separation in polymer-polymer solutions may depend on the system or may be a combination of the effects (i)–(iii).

General Reading

G. Allen and J. C. Bevington, Eds, *Comprehensive Polymer Science*. Vol. 2. Pergamon Press (1989).
F. W. Billmeyer, *Textbook of Polymer Science*. John Wiley and Sons Ltd (1985).
P. J. Flory, *Principles of Polymer Chemistry*, Chapters 12 and 13. Cornell University Press, Ithaca, N.Y. (1953).
J. H. Hildebrand and R. L. Scott, *Regular Solutions*. Prentice-Hall (1962).
R. Koningsveld, M. H. Onclin and L. A. Kleintjens, in *Polymer Compatibility and Incompatibility*. Harwood Academic Publishers (1982).
D. W. Van Krevelen and P. J. Hoftyzer, *Properties of Polymers*. Elsevier Scientific Publishing Co. (1976).
M. Kurata *Thermodynamics of Polymer Solutions*. Gordon and Breach (1982).
H. Morawetz, *Macromolecules in Solution*, 2nd Ed. John Wiley and Sons Ltd (1975).
O. Olabisi, L. M. Robeson, and M. T. Shaw, *Polymer-Polymer Miscibility*. Academic Press, N.Y. (1969).
D. R. Paul and S. Newman, Eds, *Polymer Blends*. Vols. 1 and 2. Academic Press Inc. (1978).
H. Tompa, *Polymer Solutions*. Butterworths (1956).
L. H. Tung Ed., *Fractionation of Synthetic Polymers*. Marcel Dekker Inc. (1977).
H. Yamakawa, *Modern Theory of Polymer Solutions*. Gordon and Breach (1982).

References

1. J. B. Kinsinger and L. E. Ballard, *Polymer Letters*, **2**, 879 (1964).
2. A. R. Schultz and P. J. Flory, *J. Amer. Chem. Soc.*, **74**, 4760 (1952).

CHAPTER 9

Polymer Characterization—Molar Masses

9.1 Introduction

Many of the distinctive properties of polymers are a consequence of the long chain lengths, which are reflected in the large molar masses of these substances. While such large molar masses are now taken for granted, it was difficult in 1920 to believe and accept that these values were real and not just caused by the aggregation of much smaller molecules. Values of the order of $10^6\,g\,mol^{-1}$ are now accepted without question, but the accuracy of the measurements is much lower than for simple molecules. This is not surprising, especially when polymer samples exhibit polydispersity, and the molar mass is, at best, an average dependent on the particular method of measurement used. Estimation of the molar mass of a polymer is of considerable importance, as the chain length can be a controlling factor in determining solubility, elasticity, fibre forming capacity, tear strength, and impact strength in many polymers.

The methods used to determine the molar mass M are either relative or absolute. Relative methods require calibration with samples of known M and include viscosity and vapour pressure osmometry. The absolute methods are often classified by the type of average they yield, *i.e.* colligative techniques yield number averages, light scattering and the ultracentrifuge yield higher averages, *e.g.* weight and z-average.

9.2 Molar masses, molecular weights, and SI units

The dimensionless quantity "the relative molecular mass" (molecular weight) defined as the average mass of the molecule divided by 1/12th the mass of an atom of the nuclide C^{12}, is often used in polymer chemistry, and called the molecular weight. In this book the quantity molar mass is used and appropriate SI units are given.

9.3 Number average molar mass M_n

Determination of the number average molar mass M_n involves counting the total number of molecules, regardless of their shape or size, present in a unit mass of the polymer. The methods are conveniently grouped into three categories: end-group assay, thermodynamic, and transport methods.

9.4 End-group assay

The technique is of limited value and can only be used when the polymer has an end group amenable to analysis. It can be used to follow the progress of linear condensation reactions when an end group, such as a carboxyl, is present which can be titrated. It is used to detect amino end groups in nylons dissolved in *m*-cresol, by titration with methanolic perchloric acid solution, and can be applied to vinyl polymers if an initiator fragment, perhaps containing halogen, is attached to the end of the chain.

The sensitivity of the method decreases rapidly as the chain length increases and the number of end groups drops. A practical upper limit might reach an M_n of about $15\,000\,\mathrm{g\,mol^{-1}}$.

9.5 Colligative properties of solutions: thermodynamic considerations

Because chemical methods are rather limited, the most widely used techniques for measuring the molar mass of a polymer are physical. Among the more common methods are those which depend on the colligative properties of dilute solutions. These are

(a) lowering of the vapour pressure
(b) elevation of the boiling point
(c) depression of the freezing point
(d) osmotic pressure.

A colligative property is defined as one which is a function of the number of solute molecules present per unit volume of solution and is unaffected by the chemical nature of the solute. Thus if Y represents any of the above colligative properties then

$$Y = K\frac{\sum N_i}{V} \tag{9.1}$$

where N_i is the number of particles of each solute component i, and K is a proportionality constant. The concentration of a solution per unit volume of solution V is

$$c = \frac{\sum w_i}{V} = \frac{\sum N_i M_i}{N_A V} \tag{9.2}$$

where w_i is the mass of the component and N_A is the Avogadro constant. The colligative property can be expressed in the reduced form Y/c so that

$$\frac{Y}{c} = K\frac{\sum N_i}{V}\cdot\frac{N_A V}{\sum N_i M_i} = \frac{K N_A}{M_n} \tag{9.3}$$

Hence any colligative method should yield the number average molar mass M_n of a polydisperse polymer.

When a solute, such as a polymer (component 2), is dissolved in a solvent (component

1) to form a homogeneous solution, there is a change in the chemical potential which can be expressed in terms of the activity of the solvent a_1.

$$(\mu_1 - \mu_1^0) = RT \ln a_1 \tag{9.4}$$

During the measurement of the molar mass of the polymer using a colligative method, an equilibrium is established when the chemical potential of the solvent in the solution (μ_1) is equal to that of the pure solvent (μ_1^0), where the pure solvent is either in another phase or separated from the solution by a semi-permeable membrane. The equilibration procedure can be achieved either by changing the temperature or the pressure of the system, and the amount of this change is then a measure of the activity of the solvent in solution. This tells us nothing immediately about the solute but, if very dilute solutions are used, the following useful approximations can be made.

The activity of the solvent can be considered to be equal to the mole fraction of the solvent x_1.

$$\ln a_1 = \ln x_1 = \ln(1 - x_2) \tag{9.5}$$

By expanding the logarithmic term and assuming that in dilute solutions this can be restricted to the first expansion term, ln a_1 can be related to the mole fraction of the solute x_2.

$$\ln(1 - x_2) \approx -x_2 \tag{9.6}$$

We can now make use of these approximations to calculate M_n.

9.6 Ebullioscopy and cryoscopy

In principle these two methods can be treated together and the relevant expressions are derived from the Clausius-Clapeyron equation describing the temperature dependence of the vapour pressure of a liquid

$$\mathrm{d}P/\mathrm{d}T = P\Delta H_1/RT^2 \tag{9.7}$$

where ΔH_1 is the latent heat of vapourization. If P is taken as the vapour pressure of a solution whose pure solvent vapour pressure is P_0, then for solutions containing an involatile solute

$$\int_{P_0}^{P} \frac{\mathrm{d}P}{P} = \frac{\Delta H_1}{R} \int_{T_1}^{T_2} \frac{\mathrm{d}T}{T^2} \tag{9.8}$$

which gives

$$\ln(P/P_0) = \ln a_1 = -\frac{\Delta H_1}{R} \cdot \frac{T_2 - T_1}{T_1 T_2} \tag{9.9}$$

If the solution is very dilute the change in temperature ΔT can be related to the solute mole fraction by

$$\Delta T = \frac{RT^2}{\Delta H_1} \cdot x_2 \tag{9.10}$$

and substituting for $x_2 = (n_2 V_1/V) = (c_2 V_1/M_2)$ gives

$$\Delta T = \frac{RT^2 V_1}{\Delta H_1} \frac{c_2}{M_2} \tag{9.11}$$

where c_2 is the solute concentration (mass per unit volume solution).

Polymer solutions do not behave in this ideal manner even in the dilute solution regime and for accurate molar mass measurements, deviations from ideality must be eliminated. A more accurate representation of the behaviour of the polymer solution can be obtained using equation (8.35) where

$$\frac{(\mu_1 - \mu_1^0)}{RT} = \ln a_1 = -\left[\frac{c_2 V_1}{M_2} + \bar{v}_2^2(\tfrac{1}{2} - \chi)c_2^2 + \bar{v}_2^3 c_2^3\right] \tag{9.12}$$

Substitution in equation (9.9) yields

$$\Delta T = \frac{RT^2}{\Delta H_1}\left[\frac{c_2 V_1}{M_2} + \bar{v}_2^2(\tfrac{1}{2} - \chi)c_2^2 + \bar{v}_2^3 c_2^3\right] \tag{9.13}$$

and rearrangement eliminating higher terms gives

$$\frac{\Delta T}{c_2} = \frac{RT^2 V_1}{\Delta H_1} \cdot \frac{1}{M_2} + \frac{RT^2 \bar{v}_2^2}{\Delta H_1}(\tfrac{1}{2} - \chi)c_2 \tag{9.14}$$

The non-ideal behaviour can then be eliminated by extrapolating the experimental data $(\Delta T/c_2)$ to $c_2 \to 0$ where equation (9.14) reduces to equation (9.11) and $M_2 = M_n$ can be calculated. For ebulliometry, T, ΔH, and ΔT are the boiling temperature of the solvent, the enthalpy of vaporization of the solvent, and the elevation of the boiling temperature respectively, while for cryoscopy they represent the freezing temperature of the solvent, the enthalpy of fusion of the solvent, and the depression of the freezing temperature. The equation represents the limiting case at infinite dilution and it is necessary to extrapolate $(\Delta T/c)$ for a series of solutions to $c = 0$ in order to calculate M_n.

The measurements are limited by the sensitivity of the thermometer used to obtain ΔT. At present this can rarely detect a ΔT of less than 1×10^{-3} K with any precision, and the limit of accurate measurement of M_n is in the region of 25 000 to 30 000 g mol^{-1}.

9.7 Osmotic pressure

Measurement of the osmotic pressure π of a polymer solution can be carried out in the type of cell represented schematically in figure 9.1. The polymer solution is separated

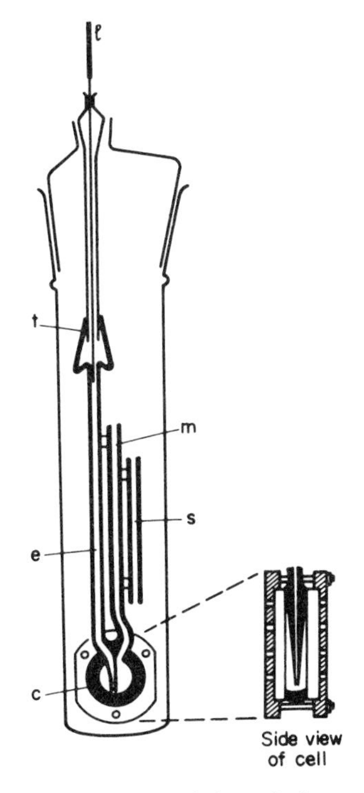

FIGURE 9.1. Pinner Stabin osmometer, consisting of glass cell c, measuring capillary m, reference capillary s, filling tube e, levelling rod 1, and mercury cup t.

from the pure solvent by a membrane, permeable only to solvent molecules. Initially, the chemical potential μ_1, of the solvent in the solution, is lower than that of the pure solvent μ_1^0, and solvent molecules tend to pass through the membrane into the solution in order to attain equilibrium. This causes a build up of pressure in the solution compartment until, at equilibrium, the pressure exactly counteracts the tendency for further solvent flow. This pressure is the osmotic pressure.

The expression for the reduced osmotic pressure (π/c_2) has already been derived in section 8.7 and has the form

$$\frac{\pi}{c_2} = \frac{RT}{M_2} + \frac{RT\bar{v}_2^2}{V_1}(\tfrac{1}{2} - \chi_1)c_2 + RT\bar{v}_2^3 c_2^2 \ldots \tag{9.15}$$

where the limiting form, valid only at infinite dilution, is

$$(\pi/c_2)_{c \to 0} = RT/M_n \tag{9.16}$$

Only under special conditions, when the polymer is dissolved in a theta-solvent, will (π/c) be independent of concentration. Experimentally, a series of concentrations is studied and the results treated according to one or other of the following virial expansions. McMillan and Meyer suggested,

$$\pi/c = RT/M_n + Bc + B_3c^2 + \ldots, \tag{9.17}$$

while alternative forms are also used:

$$\pi/c = RT(1/M_n + A_2c + A_3c^2 + \ldots); \tag{9.18}$$

and

$$\pi/c = (\pi/c)_0(1 + \Gamma_2c + \Gamma_3c^2 + \ldots). \tag{9.19}$$

The coefficients B, A_2, Γ_2 and B_3, A_3, Γ_3, are the second and third virial coefficients. When solutions are sufficiently dilute a plot of (π/c) against c is linear and the third virial coefficients (B_3, A_3, Γ_3) can be neglected. The various forms of the second virial

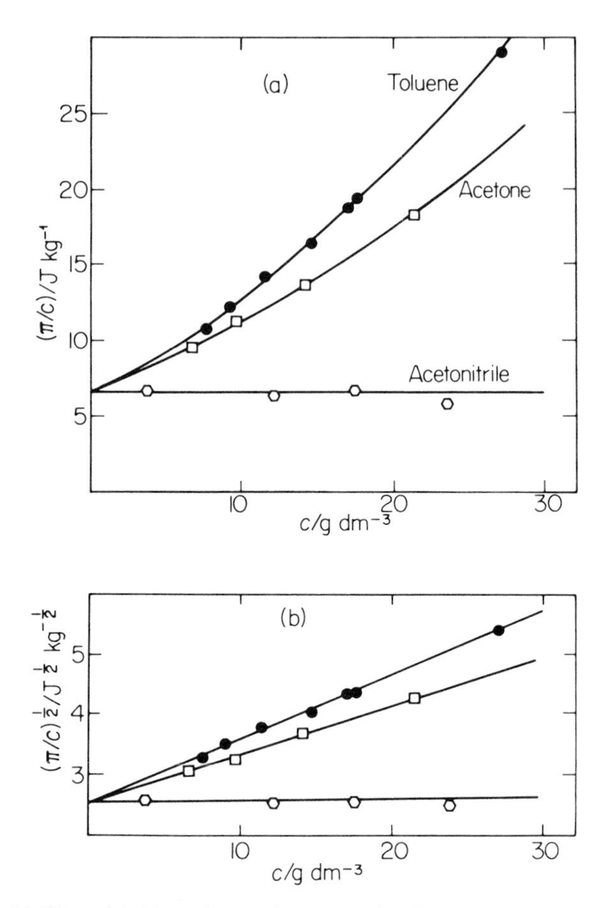

FIGURE 9.2. (a) Plot of (π/c) against c for a sample of poly(methyl methacrylate) $M_n = 382\,000\,\mathrm{g\,mol^{-1}}$ in three solvents. (b) Plot $(\pi/c)^{1/2}$ against c for the same data as in (a). (From data by Fox *et al.*, 1962.)

TABLE 9.1. Values of B, A_2 and Γ_2 for poly(methyl methacrylate) dissolved in three solvents

Solvent	$B/\mathrm{J\,m^3\,kg^{-2}}$	$A_2/\mathrm{m^3\,mol\,kg^{-2}}$	$\Gamma_2/\mathrm{m^3\,kg^{-1}}$
Toluene	0.525	2.08×10^{-4}	8.2×10^{-2}
Acetone	0.410	1.63×10^{-4}	6.4×10^{-2}
Acetonitrile	0	0	0

coefficient are interrelated by

$$B = RTA_2 = RT\Gamma_2/M_n. \tag{9.20}$$

Although not normally detected, the third virial coefficient occasionally contributes to the non-ideal behaviour in dilute solutions and a curved plot is obtained (figure 9.2a.) This increases the uncertainty of the extrapolation, but can be overcome by recasting equation (9.19) and introducing a polymer-solvent interaction parameter g

$$\pi/c = (RT/M_n)(1 + \Gamma_2 c + g\Gamma_2^2 c^2). \tag{9.21}$$

It has been found that $g = 0.25$ in good solvents so that equation (9.21) becomes

$$\pi/c = (RT/M_n)(1 + \tfrac{1}{2}\Gamma_2 c)^2. \tag{9.22}$$

A plot of $(\pi/c)^{1/2}$ against c is now linear and this extrapolation is illustrated in figure 9.2b.

This example (figure 9.2) also shows the differing solubility of poly(methyl methacrylate) in the three solvents. In a good solvent, toluene, the slope or A_2 is large, but as the solvent becomes poorer (acetone) A_2 decreases, until it is zero in the theta-solvent acetonitrile. Thus A_2 provides a useful measure of the thermodynamic quality of the solvent and measures the deviation from ideality of the polymer solution.

The value of M_n is calculated from the intercept $(\pi/c)_0 = 6.4\,\mathrm{J\,kg^{-1}}$ using equation (9.16).

$$M_n = RT/(\pi/c)_0 = 8.314\,\mathrm{J\,K^{-1}\,mol^{-1}} \times 303\,\mathrm{K}/6.4\,\mathrm{J\,kg^{-1}} = 393.62\,\mathrm{kg\,mol^{-1}}.$$

The corresponding values of the second virial coefficient are obtained from the slopes of the plots (table 9.1).

PRACTICAL OSMOMETRY

The static method of determining the osmotic pressure of a polymer solution, using volumes of 3 to 20 cm^3 of solution, is a relatively slow process which requires about 24 h to equilibrate at each concentration. Several designs, suitable for this type of measurement, are typified by the Pinner Stabin instrument shown schematically in figure 9.1. The osmometer is assembled, under a layer of solvent, by clamping two membranes (kept continually moist with solvent) on either side of the glass cell c. These are retained in position by two metal plates perforated and grooved to allow contact between the membrane and solvent which is in the outer container.

The preparation of the membranes is very important and must be carefully carried out. They are normally made of cellulose or a cellulose derivative and should be slowly conditioned from the storage liquid to the solvent in use. This is done by transferring the membrane to mixtures progressively richer in the solvent, allowing them time to equilibrate with the mixture, then transferring again until pure solvent is reached. Equilibration in each mixture usually takes a few hours.

When assembled, the osmometer is placed in a jacket containing enough solvent to cover the bottom part of the reference capillary s. Solvent is then withdrawn from the cell c and a solution of polymer added by means of a syringe. Care is taken during the filling stage to avoid trapping bubbles in the cell. The level of the solution is then adjusted to a few centimetres above the level of solvent in s by means of a levelling rod l. Mercury is added to the cup t, to ensure a leak free system, and the osmometer is left undisturbed in a thermostat bath controlled to ± 0.01 K to reach equilibrium.

The osmotic pressure can be calculated from the difference in heights h between the solvent and solution in s and m respectively and π is measured from $\pi = h\rho g$, for each concentration where ρ is the density of the solution and g the acceleration of free fall. Results are plotted as (π/c) against c as described and M_n is calculated from the intercept. The method suffers from the disadvantage that it is slow and consequently diffusion of low molar mass material could be large enough to introduce serious error in the measurement.

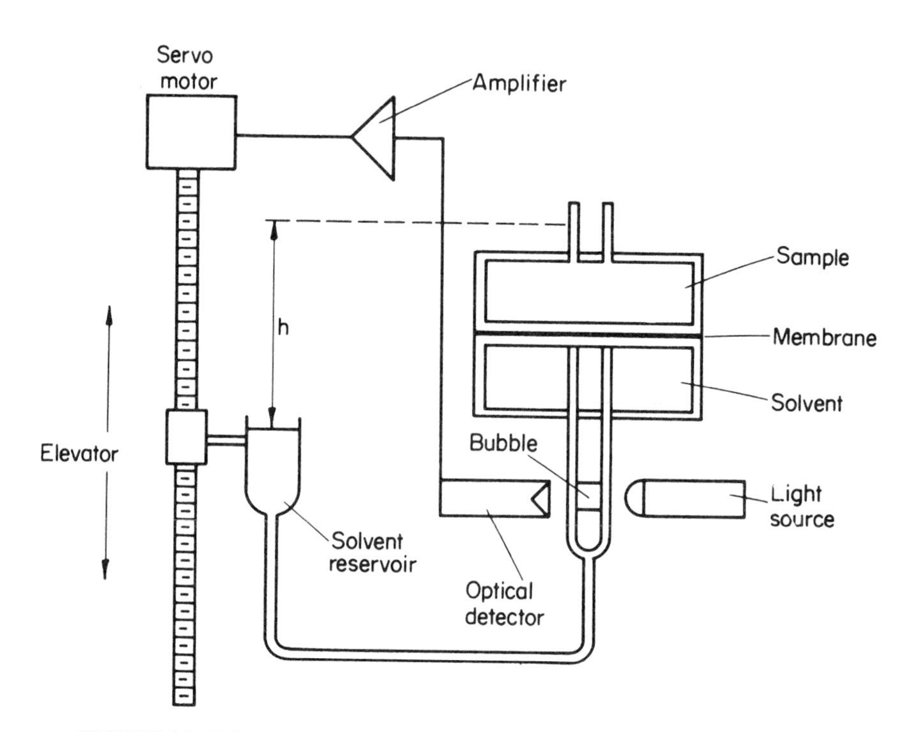

FIGURE 9.3. Schematic diagram of the Mechrolab rapid membrane osmometer.

Two or three high-speed automatic membrane-osmometers have now been designed to reduce these drawbacks and are commercially available. The Mechrolab osmometer, shown schematically in figure 9.3, consists of a solution + solvent cell of volume approximately 1 cm^3, with the solvent side connected to a reservoir attached to a servo-driven elevator. When solution is added to the top-half of the cell, solvent in the lower-half tends to flow into the upper section to equalize the chemical potentials. The flow is detected optically by the movement of a bubble in a capillary below the cell. The movement activates the servo-motor, which alters the hydrostatic head thereby counteracting the flow. The movement of the solvent reservoir is then a measure of the osmotic pressure of the solution. Equalization is rapid (5 to 30 min) and permeation is readily detected, if present, by following the change of head as a function of time on a recorder. There is no actual flow of solvent in the Mechrolab instrument.

A slightly different principle, which allows solvent flow to take place, forms the basis of the Melab and Knauer models. The Melab osmometer has a stainless steel cell (volume 0.5 cm^3), with solution and solvent compartments separated by the membrane. One wall of the cell is a flexible stainless steel diaphragm connected through a strain gauge to a recorder. As solvent diffuses through the membrane, the increase in volume causes the diaphragm to move. The motion is detected by the guage and translated into a pressure. The design has the advantage that both solvent and solution compartments are easily rinsed out and the cell does not have to be dismantled if contamination by permeation of low molar mass solute occurs.

All osmotic pressure measurements are extremely sensitive to temperature and must be carried out under rigorously controlled temperature conditions. This is allowed for in each instrument and in addition, measurements can be made over a range of temperatures (278 to 373 K). Solvents should be chosen which are chemically stable and have a low to medium vapour pressure at the temperature of operation, as this avoids problems of bubble formation in the measuring chamber.

9.8 Transport methods – vapour pressure osmometer

In conventional osmometry, the membrane permeability imposes a lower limit of about $M_n = 15\,000 \text{ g mol}^{-1}$. A technique, based on the lowering of the vapour pressure, called vapour pressure osmometry is a useful method of measuring values of M_n from 50 to 20 000 g mol^{-1}. It is a relative method and is calibrated using such low molar mass standards as benzil, methyl stearate, or glucose penta-acetate.

The apparatus consists of a thermostatted chamber, saturated with solvent vapour at the temperature of measurement, and containing two differential matched thermistors which are capable of detecting temperature differences as low as 10^{-4} K. Two syringes, one for solvent and one for solution, are used to apply a drop of solution to one thermistor, and a drop of solvent to the other. As there is a difference in vapour pressure between the solution and the solvent drops, solvent from the vapour phase will condense on the solution drop causing its temperature to rise. Because of the large excess of solvent present, evaporation, and hence cooling of the solvent drop, is negligible. When equilibrium is attained, the temperature difference between the two drops ΔT is a measure of the extent of the vapour pressure lowering by the solute. The thermistors form part of a Wheatstone bridge, and ΔT is recorded as a difference in resistance ΔR. The molar mass can then be calculated from

$$\Delta R/K^*c = (1/M_n)(1 + \tfrac{1}{2}\Gamma_2 c)^2, \tag{9.23}$$

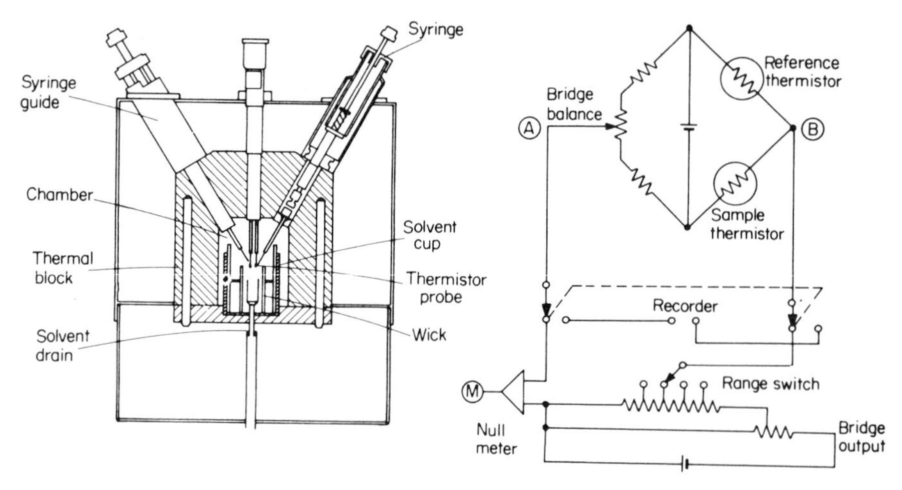

FIGURE 9.4. Sample chamber and circuit diagram for a typical vapour pressure osmometer.

where K^* is the calibration constant. As with other methods M_n is obtained by extrapolating the data to $c = 0$. The calibration constant is estimated by measuring ΔR for solutions of known concentration prepared from standard compounds of known molar mass M_k then

$$K^* = M_k(\Delta R/c)_{c \to 0}. \tag{9.24}$$

In some instances an additional correction for the dilution of the drop of solution may be necessary.

9.9 Light scattering

Light scattering is one of the most popular methods for determining the weight average molar mass M_w. The phenomenon of light scattering by small particles is familiar to us all; the blue colour of the sky or the varied colours of a sunset, the poor penetration of car headlights in a fog is caused by water droplets scattering the light, and the obvious presence of dust in a sunbeam or the Tyndall effect in an irradiated colloidal solution are further examples of this effect.

The fundamentals of light scattering were expounded by Lord Rayleigh in 1871 during his studies on gases, where the particle is small compared with the wavelength of the incident radiation. Light is an electromagnetic wave, produced by the interaction of a magnetic and electric field, both oscillating at right angles to one another in the direction of propagation. When a beam of light strikes the atoms or molecules of the medium, the electrons are perturbed or displaced and oscillate about their equilibrium positions with the same frequency as the exciting beam. This induces transient dipoles in the atoms or molecules, which act as secondary scattering centres by re-emitting the absorbed energy in all directions, *i.e.* scattering takes place.

For gases, Rayleigh showed that the reduced intensity of the scattered light R_θ at any angle θ to the incident beam, of wavelength λ could be related to the molar mass of the gas M, its concentration c, and the refractive index increment $(\mathrm{d}\tilde{n}/\mathrm{d}c)$ by

$$R_\theta = (2\pi^2/N_A \lambda^4)(\mathrm{d}\tilde{n}/\mathrm{d}c)^2(1+\cos^2\theta)M_c. \tag{9.25}$$

The quantity R_θ is often referred to as the Rayleigh ratio and is equal to $(i_\theta r^2/I_0)$ where I_0 is the intensity of the incident beam, i_θ is the quantity of light scattered per unit volume by one centre at an angle θ to the incident beam, and r is the distance of the centre from the observer. This is valid for a gas, where all the particles are considered to be independent scattering centres and the addition of more centres, which increases $\tilde{n}$, increases the scattering. The situation changes when dealing with a liquid as $(\mathrm{d}\tilde{n}/\mathrm{d}c)$ remains unaffected by the addition of molecules and can be expected to be zero. This conceptual difficulty was overcome in the fluctuation theories of Smoluchowski and Einstein; they postulated that optical discontinuities exist in the liquid arising from the creation and destruction of holes during Brownian motion. Scattering emanates from these centres, created by local density fluctuations, which produce changes in $(\mathrm{d}\tilde{n}/\mathrm{d}c)$ in any volume element.

When a solute is dissolved in a liquid, scattering from a volume element again arises from liquid inhomogeneities, but now an additional contribution from fluctuations in the solute concentration is present and for polymer solutions the problem is to isolate and measure these additional effects. This was achieved by Debye in 1944, who showed that for a solute whose molecules are small compared with the wavelength of the light used, the reduced angular scattering intensity of the solute is

$$R_\theta = R_\theta(\text{solution}) - R_\theta(\text{solvent}), \tag{9.26}$$

and that this is related to the change in Gibbs free energy with concentration of the solute. As ΔG is related to the osmotic pressure π, we have

$$R_\theta = (2\pi^2\tilde{n}_0^2/\lambda^4)(1+\cos^2\theta)(\mathrm{d}\tilde{n}/\mathrm{d}c)^2(NM/N_A)\{RT/(\mathrm{d}\pi/\mathrm{d}c)_T\}. \tag{9.27}$$

Here $\tilde{n}_0$ and $\tilde{n}$ are the refractive indices of solvent and solution respectively, and N is the number of polymer molecules. Differentiation of the virial expansion for π with respect to c, followed by substitution in equation (9.27) and rearrangement leads to

$$K'(1+\cos^2\theta)c/R_\theta = 1/M_w + 2A_2c, \tag{9.28}$$

where

$$K' = \{2\pi^2\tilde{n}_0^2(\mathrm{d}\tilde{n}/\mathrm{d}c)^2/\lambda^4 N_A)\}. \tag{9.29}$$

Alternatively, the scattering can be expressed as a turbidity τ where

$$\tau = (16\pi/3)R_\theta, \tag{9.30}$$

and the equation becomes

$$Hc/\tau = 1/M_w + 2A_2c + \ldots \tag{9.31}$$

The new constant is $H = \{(16\pi/3)K'(1+\cos^2\theta)\}$. Both equations are valid for molecules smaller than $(\lambda'/20)$ when the angular scattering is symmetrical. Here λ' is the wavelength of light in solution $\lambda' = (\lambda/\tilde{n}_0)$.

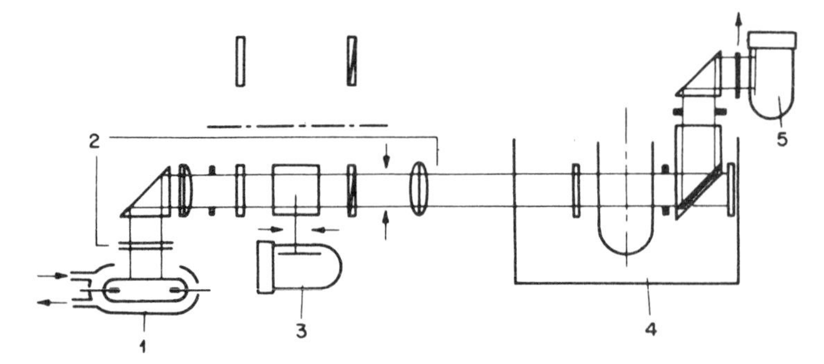

FIGURE 9.5. Schematic representation of the optics of a SOFICA light scattering instrument: 1, the light source, a water cooled mercury vapour lamp; 2, path of incident beam through a system of filters, polarizers, and a variable slit; 3, reference photomultiplier; 4, thermostat; and 5, photomultiplier.

For small particles, M_w can be calculated from either equation (9.28) or (9.31). The important experimental point to remember is that dust will also scatter light and contribute to the scattering intensity. Great care must be taken to ensure that solutions are clean and free of extraneous matter. Solutions of the polymer are prepared in a concentration series and clarified either by centrifugation for a few hours at about 25 000 g, or filtered through a grade 5 sinter glass filter. Alternatively, a millipore filter, porosity 0.45 μm can be used.

A number of instruments are available commercially; only one is described here and the schematic diagram 9.5 provides the main features of the model. Light is obtained from a water-cooled mercury vapour lamp and one of three wavelengths 365, 436, or 546 nm can be selected by means of an appropriate filter. As the scattering intensity is a function of λ^{-4}, use of a lower wavelength enhances the scattering, but the choice is left to the operator. The light beam, which can be polarized, or left unpolarized, is collimated before passing through the cell. The measuring cell is immersed in a vat of liquid, usually benzene or xylene which can be thermostatted at temperatures between 273 and 400 K. Scattering is detected by a photomultiplier, capable of revolving round the cell and the intensity is recorded on a galvanometer. The 90° scattering (R_{90}) is plotted as $(K'c/R_{90})$ against c and linear extrapolation of the results leads to M_w as the intercept at $c = 0$.

Typical results are shown in table 9.2 for a polystyrene sample dissolved in benzene. The relevant constants are $(d\tilde{n}/dc) = 0.112 \times 10^{-3}\,m^3\,kg^{-1}$, $K' = 2.5888 \times$

TABLE 9.2. Polystyrene in benzene at 298 K

$c/g\,dm^{-3}$	$10^3 R_{90}/m^{-1}$	$(K'c/R_{90})/mol\,g^{-1}$
1.760	5.31	8.56
3.708	8.43	11.36
6.244	11.24	14.35
7.736	12.43	16.07
10.230	13.80	19.15

10^{-5} m^2 mol kg^{-2}, the intercept $(K'c/R_{90})_{c=0} = 6.9 \times 10^{-3}$ mol kg^{-1} and $M_w =$ 148 kg mol^{-1}.

SCATTERING FROM LARGE PARTICLES

When polymer dimensions are greater than $\lambda'/20$, intraparticle interference causes the scattered light from two or more centres to arrive considerably out of phase at the observation point, and the scattering envelope becomes dependent on the molecular shape. This attenuation, produced by destructive interference, is zero in the direction of the incident beam, but increases as θ increases because the path length difference $\Delta\lambda_f$ in the forward direction is less than $\Delta\lambda_b$ in the backward (see figure 9.6). This difference can be measured from the dissymmetry coefficient Z

$$Z = R_\theta / R_{\pi-\theta} \tag{9.32}$$

which is unity for small particles, but greater than unity for large particles. The scattering envelope reflects the scattering attenuation and is compared with that for small particles in figure 9.7. The angular attenuation of scattering is measured by the particle scattering factor $P(\theta)$ which is simply the ratio of the scattering intensity to the intensity in the absence of interference, measured at the same angle θ.

Gunier showed that a characteristic shape-independent geometric function, called the radius of gyration $\langle \bar{S}^2 \rangle^{1/2}$ can be measured from large particle scattering. It is defined as an average distance from the centre of gravity of a polymer coil to the chain end.

The function $P(\theta)$ is size dependent and can be related to the polymer coil size by

$$P(\theta) = (2/u^2)\{e^{-u} - (1-u)\}, \tag{9.33}$$

where $u = \{(4\pi/\lambda)\sin(\theta/2)\}^2 \langle \bar{S}^2 \rangle$, for monodisperse randomly coiling polymers. In the

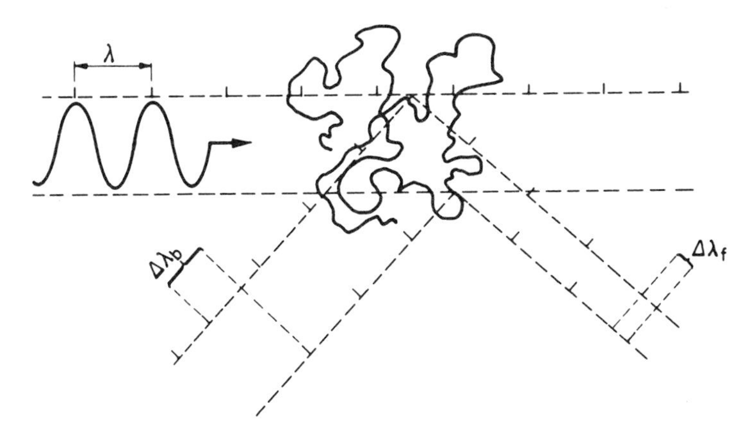

FIGURE 9.6. Destructive interference of light scattered by large particles. Waves arriving at the forward observation point are $\Delta\lambda_f$ out of phase and those arriving at the backward point are $\Delta\lambda_b$ out of phase. For large polymer molecules $\Delta\lambda_f < \Delta\lambda_b$ as shown also in figure 9.7.

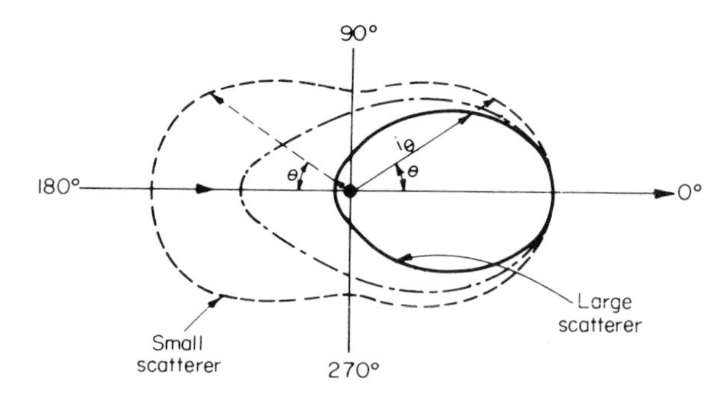

FIGURE 9.7. Intensity distribution of light scattered at various angles. The symmetrical envelope is obtained for small isotropic scatterers in dilute solution, the two asymmetric envelopes are for much larger scattering particles. The solid line represents the scattering from spheres whose diameters are approximately one half of the wavelength of the incident light.

limit of small θ the expansion

$$P(\theta)^{-1} = 1 + u/3 - \dots \tag{9.34}$$

can be used, and the coil size can be estimated from $P(\theta)$ without assuming a particular model. Specific shapes can be related to $P(\theta)$ if desired, as shown in figure 9.8a and b.

Two methods can be used to calculate M_w and the particle size for large molecules. (i) *Dissymmetry method*. If $Z = \{P(\theta)/P(\pi - \theta)\}$ is not too large, one need only measure the scattering intensity at 90° and two angles symmetrical about 90°, usually 45° and 135°. As Z is normally concentration dependent, the value at $c = 0$ is obtained by plotting $(Z - 1)^{-1}$ against c. From published tables $Z_{c=0}$ can be related to $P(90)$, and M_w is calculated from the 90° scattering then corrected by multiplication with $P(90)$. Also available in table form is the ratio $(\langle \bar{r}^2 \rangle^{1/2}/\lambda')$ presented as a function of Z, where $\langle \bar{r}^2 \rangle^{1/2}$ is the root mean square distance between the ends of the polymer coil. The

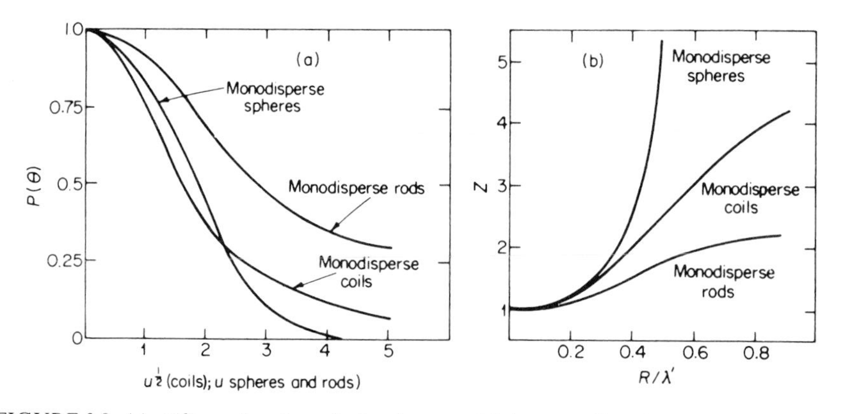

FIGURE 9.8. (a) $P(\theta)$ as a function of u for three model shapes; coils, spheres, and rods. (b) The dissymmetry Z as a function of R/λ', where R is the characteristic linear dimension of either a rod, a sphere, or a coil.

corresponding functions for a rod and a sphere have different forms (figure 9.8b). Polymer dimensions can be calculated in this way if an assumption is made about the best model. A much more satisfying treatment of the data uses the double extrapolation method proposed by Zimm, which leads to the shape independent parameter $\langle \bar{S}^2 \rangle^{1/2}$.

(ii) *Zimm plots.* This is based on the knowledge that, as the scattering at zero angle is independent of size, $P(\theta)$ is unity when $\theta = 0$. Experimentally this is difficult to measure, and an extrapolation procedure has been devised which makes use of a modified form of equation (9.28) for large particles,

$$Kc/R_\theta = 1/M_w P(\theta) + 2A_2 c + \ldots \tag{9.35}$$

Substituting for $P(\theta)$ leads to

$$Kc/R_\theta = 1/M_w + (1/M_w)\{(16\pi^2/3\lambda'^2)\sin^2(\theta/2)\langle \bar{S} \rangle_Z^2\} + 2A_2 c + \ldots \tag{9.36}$$

If the scattering intensity for each concentration in a dilution series is measured over an angular range 35° to 145°, the data can be plotted as (Kc/R_θ) against $\{\sin^2(\theta/2) + k'c\}$, where k' is an arbitrary constant chosen to provide a convenient spread of the data in the grid-like graph which is obtained. A double extrapolation is then carried out, as shown in figure 9.9, by joining all points of equal concentration and

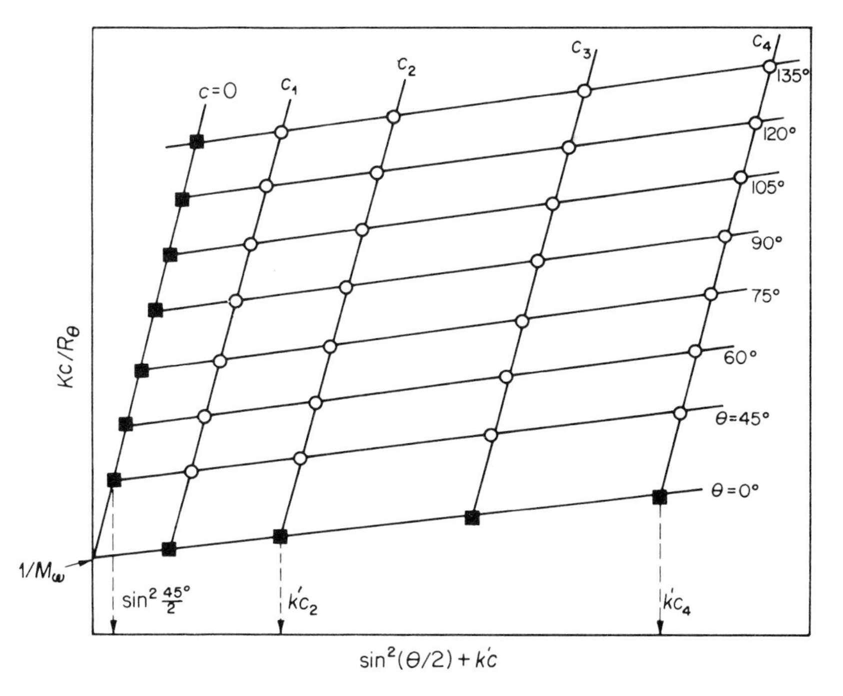

Figure 9.9. Typical Zimm plot showing the double extrapolation technique, where –○– are the experimental points and –■– represent the extrapolated points.

extrapolating to zero angle, and then all points measured at equal angles and extrapolating these to zero concentration. For example, on the diagram the points corresponding to concentration c_3 are joined and extrapolated to intersect with an imaginary line corresponding to the value of $k'c_3$ on the abscissa; similarly all points measured at 90° are joined and extrapolated until the point corresponding to $\sin^2(90/2)$ is reached. This is done for each concentration and each angle and the extrapolated points are then lines of $\theta = 0$ and $c = 0$. Both lines, on extrapolation to the axis, should intersect at the same point. The intercept is then $(M_w)^{-1}$, the slope of the $\theta = 0$ line yields A_2, whereas $\langle \bar{S}^2 \rangle$ is obtained from the initial slope s_i of the $c = 0$ line *i.e.*

$$\langle \bar{S}^2 \rangle_z = s_i M_w (3\lambda'^2/16\pi^2). \tag{9.37}$$

The radius of gyration calculated in this way for a polydisperse sample is a z-average.

9.10 Refractive index increment

Before M_w can be calculated from light scattering measurements, the specific refractive index increment $(d\tilde{n}/dc)$ must be known for the particular polymer + solvent system under examination. It is defined as $(\tilde{n} - \tilde{n}_0)/c$ where $\tilde{n}$ and $\tilde{n}_0$ are the refractive indices of the solution and the solvent and c is the concentration. Measurements of $\Delta\tilde{n} = (\tilde{n} - \tilde{n}_0)$ are made using a differential refractometer employing the same wavelength of light as used in the light scattering. The monochromatic beam (selected by filter) from a mercury vapour lamp is directed through a differential cell, consisting of a solution and solvent compartment separated by a diagonal glass wall. The deflection of the light beam is measured, first with solvent in the forward compartment and solution in the rear, giving deflection d_1; the position is reversed and deflection d_2 measured. If similar readings for solvent alone, d_1^0 and d_2^0, are obtained, then the total displacement Δd is

$$\Delta d = (d_1 - d_2) - (d_1^0 - d_2^0). \tag{9.38}$$

If the instrument is calibrated with aqueous KCl solutions of known $\Delta\tilde{n}$ a relation,

$$\Delta\tilde{n} = c'\Delta d, \tag{9.39}$$

can be obtained where c' is the calibration constant. By measuring Δd for a number of concentrations of polymer, $\Delta\tilde{n}$ is obtained from a knowledge of c', and $(d\tilde{n}/dc)$ from the slope of the plot of $\Delta\tilde{n}$ against c.

9.11 Small angle X-ray scattering

The theoretical outline presented for light scattering studies is valid for electromagnetic radiation of all wavelengths. For X-rays, λ is as low as 0.154 nm, and as this is much smaller than typical polymer dimensions structural information over small distances should be available from X-ray scattering. The intensity of scattering is a function of the electron density and therefore of the refractive index. The molar mass is then related to the excess electron density $\Delta\rho_e$ of solute over solvent for $\lambda = 0.154$ nm by

$$R_0 = (4.8\,\text{cm}^{-1}) M_w (\Delta\rho_e)^2 c, \tag{9.40}$$

where R_0 is the Rayleigh ratio at $\theta = 0$. Experimental techniques are difficult because of the weak scattering, but the method has provided useful information on macromolecules with dimensions in the range 1 to 100 nm and, as such, is complementary to light scattering.

9.12 Ultracentrifuge

When macroscopic particles are allowed to settle in a liquid under gravity it is possible to determine their size and weight. Macromolecules in solution are usually much smaller and it would take years for them to overcome the Brownian motion and form a sediment. This problem can be overcome by subjecting them to an external force, strong enough to alter their spatial distribution by a significant amount in a short time. In 1925, Svedberg first achieved this by subjecting polymer solutions to large force fields, generated at high speeds of rotation.

The technique is now a well established method for measuring M_w and M_z for both synthetic and biological macromolecules and has the added advantage that measurements require only small quantities of material. The dilute solution of polymer is placed in a cell with a sector shaped centre piece in the form of a truncated cone, whose peak is located at the centre of the rotation. The shape ensures that convective disturbances are minimized during the transportation of molecules to the cell bottom. The cells are supported in a rotor of either titanium or aluminium alloy, which is attached to the drive motor by a fine steel wire, thereby allowing limited self-balancing to take place. The rotor is spun in a vacuum chamber to minimize frictional heating during high

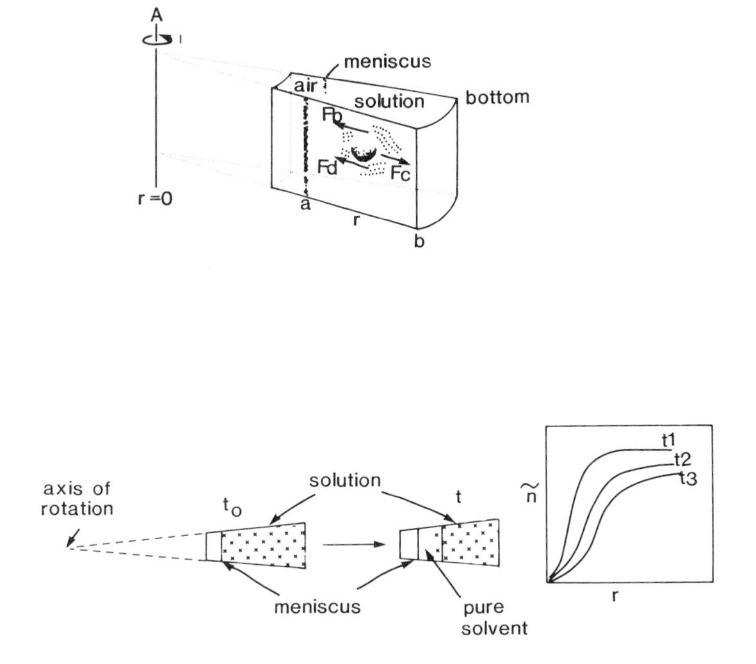

FIGURE 9.10. Schematic diagram of an ultracentrifuge cell showing boundary movement during a sedimentation run. The boundary movement can be followed by measuring the rapid change in refractive index $\tilde{n}$ on passing from solvent to solution.

speed rotations, as speeds of up to 68 000 r.p.m., capable of producing 372 000 g can be generated. During rotation the cell passes through a collimated beam of light from a high pressure mercury lamp and the emergent beam then travels through the optical system to be recorded photographically. Three types of optical system are available, schlieren, interference, and UV absorption. Solvents, having densities and refractive indices sufficiently different from the polymer, are chosen to ensure movement of the polymer chains in the medium and the optical detection of this motion.

Most commercial instruments are extremely versatile, with an extensive choice of rotor speeds and a temperature control system. Molar masses from 10^2 to 10^6 g mol^{-1} can be measured and this range is much wider than any other existing technique.

Two general methods are used to measure M, (1) sedimentation velocity and (2) sedimentation equilibrium.

SEDIMENTATION VELOCITY

The centrifuge is operated at high speeds to transport the polymer molecules through the solvent to the cell bottom if the solvent density is less than the polymer, or to the top (flotation) if the reverse is true. The rate of movement can be measured by following the change in refractive index $\tilde{n}$ through the boundary region. As the molecules sediment, a layer of pure solvent is left whose refractive index differs from the solution. The boundary is located by the sharp change in $\tilde{n}$ and its movement is followed as a function of time using one or other of the optical methods available.

When moving through the solution, the polymer will experience a centrifugal force which is $\boldsymbol{F}' = \omega^2 rm$, but as the molecule displaces a mass of solution $m_0 = m\bar{v}_2\rho$ it will be subject to a *buoyancy* effect and an opposing force $F'' = -\omega^2 rm\bar{v}_2\rho$. The net force is then

$$\boldsymbol{F} = m\omega^2 r - m\omega^2 r\bar{v}_2\rho = m(1-\bar{v}_2\rho)r\omega^2 \tag{9.41}$$

where r is the distance between the boundary and the centre of rotation, $\bar{v}_2$ is the partial specific volume of the polymer, ω is the angular velocity, the mass of the molecule $m = M/N_A$ and ρ is the density of the solution. This force will be balanced by the frictional resistance of the medium F for a particular velocity $(\mathrm{d}r/\mathrm{d}t)$ and

$$F = 6\pi\eta R_s(\mathrm{d}r/\mathrm{d}t) \tag{9.42}$$

where R_s is the spherical radius of the polymer particle and η is the viscosity of the medium. These two forces are in equilibrium when a uniform particle velocity is attained and

$$(M/N_A)(1-\bar{v}_2\rho)r\omega^2 = 6\pi\eta R_s(\mathrm{d}r/\mathrm{d}t). \tag{9.43}$$

The steady-state velocity in a unit gravitational field can then be defined as the sedimentation constant S,

$$S = (1/\omega^2 r)(\mathrm{d}r/\mathrm{d}t), \tag{9.44}$$

and

$$S = (M/N_A)(1-\bar{v}_2\rho)/6\pi\eta R_s = (M/N_A)(1-\bar{v}_2\rho)/f \tag{9.45}$$

where f is the frictional coefficient of the molecule and is related to the diffusion constant D by

$$D = kT/f \tag{9.46}$$

Substitution gives the Svedberg equation,

$$M_{SD} = \{RT/(1 - \bar{v}_2\rho)\}(S/D) \tag{9.47}$$

From this a molar mass M_{SD}, is calculated if both S and D are known. This average is close to M_w but is usually smaller and depends on the method used to measure D.

The term $(1 - \bar{v}_2\rho)$ is called the *buoyancy factor* and determines the direction of macromolecular transport in the cell. If the factor is positive, the polymer chains sediment away from the centre of rotation to the cell bottom, if negative, they move in the opposite direction and float to the top. The determination of M is absolute when S and D are known, but more commonly a relation of the form

$$S = K_s M^b, \tag{9.48}$$

is established, using polymer fractions of known M, for a given solvent + polymer system. This approach is similar to that used for the limiting viscosity number, which is a non-absolute method.

SEDIMENTATION EQUILIBRIUM

In the sedimentation equilibrium experiments the condition for equilibrium requires that the *total* potential, $\tilde{\mu}$, must be constant in all parts of the system. For a polymer of molar mass M_2 dissolved in a solvent (1) and placed in a centrifugal field of angular velocity ω, its potential energy at a distance r from the centre of rotation is $(-M_2\omega^2r^2/2)$ and its chemical potential is μ. The total potential $\tilde{\mu}$ then becomes $\mu - (M_2\omega^2r^2/2)$ and the conditions for equilibrium are

$$\frac{\partial\tilde{\mu}}{\partial r} = \left[\left(\frac{\partial\mu}{\partial r}\right) - M_2\omega^2 r\right] = 0$$

Consider now the transportation of J moles of polymer across unit cross sectional area in unit time; the transport equation for this is

$$J = -L\left[\left(\frac{\partial\mu}{\partial r}\right) - M\omega^2 r\right] \tag{9.49}$$

where $L = c_2/N_A f$, is the mass conductivity which is proportional to the concentration of the substance and inversely proportional to the resistance offered by the medium to transport *i.e.* the frictional coefficient per mole f.

Now μ is a function of T, P and c_2 but if T is constant then

$$\left(\frac{\partial\mu}{\partial r}\right) = \left(\frac{\partial\mu}{\partial P}\right)_{T,C} \cdot \left(\frac{\partial P}{\partial r}\right) + \left(\frac{\partial\mu}{\partial c_2}\right)_{T,P} \cdot \left(\frac{\partial c_2}{\partial r}\right)$$

and as

$$\left(\frac{\partial \mu}{\partial P}\right) = \bar{V}_2 = \bar{v}_2 M_2; \left(\frac{\partial P}{\partial r}\right) = \rho\omega^2 r;$$

$$\left(\frac{\partial \mu}{\partial c_2}\right) = \frac{RT}{c_2}\left[1 + c_2\left(\frac{\partial \gamma_2}{\partial c_2}\right)\right]$$

where γ_2 is the activity coefficient. We have for an ideal solution

$$\left(\frac{\partial \mu}{\partial r}\right) = M_2 \bar{v}_2 \rho\omega^2 r + \frac{RT}{c_2}\left(\frac{\partial c_2}{\partial r}\right)$$

Substitution in equation (9.49) yields

$$J = -L\left[M_2 \bar{v}_2 \rho\omega^2 r - M_2 \omega^2 r + \frac{RT}{c_2}\left(\frac{\partial c_2}{\partial r}\right)\right]$$

or

$$J = L\left[M_2 \omega^2 r(1 - \bar{v}_2 \rho) - \frac{RT}{c_2}\left(\frac{\partial c_2}{\partial r}\right)\right] \tag{9.50}$$

Using the equations (9.45), (9.46) and the definition of L, equation (9.50) can be written as

$$J = S\omega^2 r c_2 - D(\partial c_2/\partial r)$$

This shows that the flux J is then a net result of the sedimentation rate and the back diffusion of the molecules. At equilibrium these balance, and the flow vanishes so that $J = 0$. It follows that

$$\frac{1}{c_2}\left(\frac{\mathrm{d}c_2}{\mathrm{d}r}\right) = \frac{\omega^2 r M_2(1 - \bar{v}_2 \rho)}{RT} \tag{9.51}$$

This describes the concentration gradient at equilibrium for a single solute under ideal solution conditions, and integration between the meniscus r_m' and any point 'r' in the cell gives

$$\ln\frac{c(r)}{c(r_m)} = \frac{\omega^2 M_2(1 - \bar{v}_2 \rho)(r^2 - r_m^2)}{2RT} \tag{9.52}$$

A graph of $\ln c(r)$ against r^2 can be constructed by measuring the concentrations at different points in the cell and M_2 can be calculated from the slope. The main experimental problem, however, is being able to calculate the concentration at each point in the cell which is not always easy. A more widely used method is to calculate the difference in concentration between that at the meniscus (cm) and the cell bottom (cb).

Rearranging and integrating equation (9.51) gives

$$\int_{r_m}^{r_b} \frac{dc}{dr} \cdot dr = \frac{M_2(1 - \bar{v}_2\rho)\omega^2}{RT} \int_{r_m}^{r_b} rc(r)\, dr$$

and the integral on the r.h.s. can be evaluated by considering mass conservation in a sector shaped cell, so that

$$c(\mathrm{b}) - c(\mathrm{m}) = \frac{M_2(1 - \bar{v}_2\rho)\omega^2}{RT} \frac{(r_b^2 - r_m^2)c_0}{2} \tag{9.53}$$

where c_0 is the initial concentration.

For a polydisperse polymer this gives a weight average molar mass M_w and the z average M_z can be calculated from the concentration gradients at the top and bottom

$$M_z = RT = \left\{ \frac{1}{r_b} \frac{dc}{dr_b} - \frac{1}{r_m} \frac{dc}{dr_m} \right\} \bigg/ (c(\mathrm{b}) - c(\mathrm{m}))(1 - \bar{v}_2\rho)\omega^2 \tag{9.54}$$

of the cell.

The main disadvantage of the method lies in the long periods of time required to reach equilibrium. Several variations exist which reduce this time scale such as studying the approach to equilibrium, using short columns, or meniscus depletion techniques can be employed but all are outside the scope of this text.

The value of M_w calculated from equation (9.53) is, of course, an apparent value relating to the initial concentration of the solution, and extrapolation to zero concentration is necessary.

9.13 Viscosity

When a polymer dissolves in a liquid, the interaction of the two components stimulates an increase in polymer dimensions over that in the unsolvated state. Because of the vast difference in size between solvent and solute, the frictional properties of the solvent in the mixture are drastically altered, and an increase in viscosity occurs which should reflect the size and shape of the dissolved solute, even in dilute solutions. This was first recognized in 1930 by Staudinger, who found that an empirical relation existed between the relative magnitude of the increase in viscosity and the molar mass of the polymer.

One of the simplest methods of examining this effect is by capillary viscometry. It has been shown that the ratio of the flow time of a polymer solution t to that of the pure solvent t_0 is effectively equal to the ratio of their viscosity (η/η_0) if the densities are equal. This latter approximation is reasonable for dilute solutions and provides a measure of the relative viscosity η_r.

$$\eta_r = (t/t_0) = (\eta/\eta_0) \tag{9.55}$$

As this has a limiting value of unity, a more useful quantity is the specific viscosity

$$\eta_{sp} = \eta_r - 1 = (t - t_0)/t_0 \tag{9.56}$$

Even in dilute solutions molecular interference is likely to occur and η_{sp} is extrapolated to zero concentration to obtain a measure of the influence of an isolated polymer coil. This is accomplished in either of two ways; η_{sp} can be expressed as a reduced quantity (η_{sp}/c) and extrapolated to $c = 0$ according to the relation

$$(\eta_{sp}/c) = [\eta] + k'[\eta]^2 c, \tag{9.57}$$

and the intercept is the limiting viscosity number $[\eta]$ which is a characteristic parameter for the polymer in a particular solvent, k' is a shape dependent factor called the Huggins constant and has values between 0.3 and 0.9 for randomly coiling vinyl polymers. The alternative extrapolation method uses the inherent viscosity as

$$(\log \eta_r)/c = [\eta] + k''[\eta]^2 c, \tag{9.58}$$

where k'' is another shape dependent factor. The dimensions of $[\eta]$ are the same as the reciprocal of the concentration.

When measuring $[\eta]$ solutions are filtered to remove spurious particles, then flow times for solvent and solutions are recorded in U-tube viscometers such as the "Cannon-Fenske" or the "Ubbelohde suspended level dilution" models. Dilution

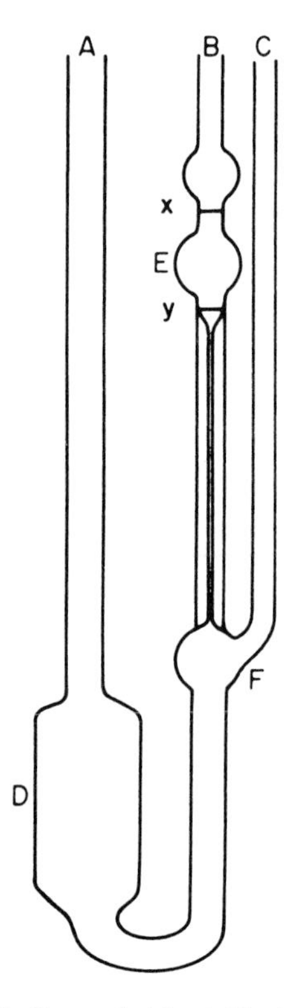

FIGURE 9.11. Suspended level dilution viscometer.

TABLE 9.3. A comparison of viscosity and sedimentation constants and exponents for several polymer + solvent systems from equations (9.59) and (9.48)

Polymer	Solvent	T/K	$10^2 K_v$/$cm^3\,g^{-1}$	v	$10^5 K_s$/s^{-1}	b
	Cyclohexene	298	1.63	0.68	3.85	0.42
Polystyrene	Chloroform	298	0.716	0.76	8.36	0.415
	Cyclohexane θ =	308	8.6	0.50	1.50	0.502
Poly(α-methylstyrene)	Cyclohexane θ =	310	7.8	0.50	1.86	0.50
	Toluene	310	1.0	0.72	4.02	0.43
Poly(vinyl acetate)	Butanone	298	4.2	0.62	9.8	0.38
Cellulose nitrate	Ethyl acetate	303	0.25	1.01	0.304	0.29
Cellulose	Cadoxen	298	250	0.75	19	0.40

viscometers are most convenient when a concentration series is to be measured. In these the concentration can be changed *in situ*, whereas fresh solution concentrations of exactly the same volume must be introduced for each measurement in the non dilution Cannon-Fenske.

In the Ubbelohde viscometer an aliquot of solution of known volume is pipetted into bulb D through A. The solution is then pumped into E, by applying a pressure down A with C closed off; the pressure is released and C is opened to allow the excess solution to drain back into D. This leaves the end of the capillary open or suspended. Solution then flows down the capillary and drains round the sides of the bulb back into D, but as no back pressure from the excess solution exists, the volume in D plays no part in determining the flow time t. This suspended level is the feature which allows dilution to be carried out in D without affecting t. Thus addition of a known amount of solvent to the solution in D, followed by mixing, gives the next concentration in the series. The flow time t, is the time taken for the solution meniscus to pass from x to y in bulb E.

For a given polymer + solvent system at a specified temperature, $[\eta]$ can be related to M through the Mark-Houwink equation

$$[\eta] = K_v M^v. \tag{9.59}$$

K_v and v can be established by calibrating with polymer fractions of known molar mass, and once this has been established for a system, $[\eta]$ alone will give M for an unknown fraction. This is normally achieved by plotting log $[\eta]$ against log M and interpolation is then quite straightforward. Values of v lie between 0.5 for a polymer dissolved in a theta-solvent to about 0.8 in very good solvents for linear randomly coiling vinyl polymers, and typical values for systems studied by viscosity and sedimentation are given in table 9.3. The exponents v and b are indicative of solvent quality. When the solvent is ideal, *i.e.* a theta-solvent, both v and b are 0.5, but as the solvent becomes thermodynamically better, and deviations from ideality larger, then v increases and b decreases.

VISCOSITY AVERAGE MOLECULAR WEIGHT

Polymer samples are normally polydisperse and it is of interest to examine the type of average molecular weight that might be expected from a measurement of $[\eta]$. As the specific viscosity will depend on the contributions from each of the polymer molecules in the sample we can write

$$\eta_{sp} = K \Sigma c_i M_i^v \tag{9.60}$$

If we now divide by the total concentration $c = \Sigma c_i$ and substitute for $c_i = N_i M_i / N_A V$ then

$$\frac{\eta_{sp}}{c} = \frac{K \Sigma N_i M_i}{N_A V} \frac{N_A V}{\Sigma N_i M_i} \cdot \Sigma M_i^v$$

or

$$[\eta] = \frac{\eta_{sp}}{c_{c \to 0}} = \frac{K \Sigma N_i M_i^{1+v}}{\Sigma N_i M_i} \tag{9.61}$$

Comparison with equation (9.59) shows that the viscosity average M_v is then

$$M_v = \left[\frac{\Sigma N_i M_i^{1+v}}{\Sigma N_i M_i} \right]^{1/v} \tag{9.62}$$

and that this lies between M_n and M_w in magnitude, but will be usually closer to M_w.

9.14 Gel permeation chromatography

The molar mass distribution (MMD) of a polymer sample has a significant influence on its properties and a knowledge of the shape of this distribution is fundamental to the full characterization of a polymer. The determination of the MMD by conventional fractionation techniques is time consuming, and a rapid, efficient and reliable method which can provide a measure of the MMD in a matter of hours has been developed. This is gel permeation chromatography (GPC). Known alternatively by its more descriptive name "size exclusion chromatography" (SEC), the method depends on the use of mechanically stable, highly crosslinked gels which have a distribution of different pore sizes and can, by means of a sieving action, effect separation of a polymer sample into fractions, dictated by their molecular volume.

The non-ionic gel stationary phase is commonly composed of crosslinked polystyrene or macroporous silica particles, which do not swell significantly in the carrier solvents. A range of pore sizes is fundamental to the success of this size fractionation procedure which depends on two processes. These are (a) separation by size exclusion alone, which is the more important feature, and (b) a dispersion process, controlled by molecular diffusion which may lead to an artificial broadening of the MMD.

Looking first at the mechanism of the separation process (a); in simple terms, the large molecules, which occupy the greatest effective volume in solution, are excluded from the smaller pore sizes in the gel and pass quickly through the larger channels between the gel particles. This results in their being eluted first from the column. As the molecular size of the polymer decreases there is an increasing probability that the molecules can diffuse into the smaller pores and channels in the gel which slows their time of passage through the column by providing a potentially longer path length before being eluted. By choosing a series of gel columns with an appropriate range of pore sizes, an effective size separation can be obtained.

The efficiency of the separation process is then a function of the dependence of the retention (or elution) volume V_R on the molar mass M, and a reliable relationship between the two parameters must be established. The value of V_R depends on the interstitial void volume V_0 and the accessible part of the pore volume in the gel,

$$V_R = V_0 + K_D V_i \tag{9.63}$$

where V_i is the total internal pore volume and K_D is the partition coefficient between V_i and the portion accessible to a given solute. Thus $K_D = 0$ for very large molecules ($V_R = V_0$), and rapid elution takes place, whereas $K = 1$ for very small molecules which can penetrate all the available pore volume. This is shown schematically in figure 9.12 and clearly the technique cannot discriminate amongst molecular sizes with $V_R \leqslant V_0$ or $V_R \geqslant V_0 + V_i$. For samples which fall within the appropriate range it has been suggested that a universal calibration curve can be constructed to relate V_R and M, by assuming that the hydrodynamic volume of a macromolecule is related to the product $[\eta].M$, where $[\eta]$ is the intrinsic viscosity of the polymer in the carrier solvent used, at the temperature of measurement. A universal calibration curve is then obtained by plotting $\log[\eta].M$ against V_R for a given carrier solvent and a fixed temperature. Experimental verification of this is shown in figure 9.13 for a variety of different polymers and can be utilized in the following way.

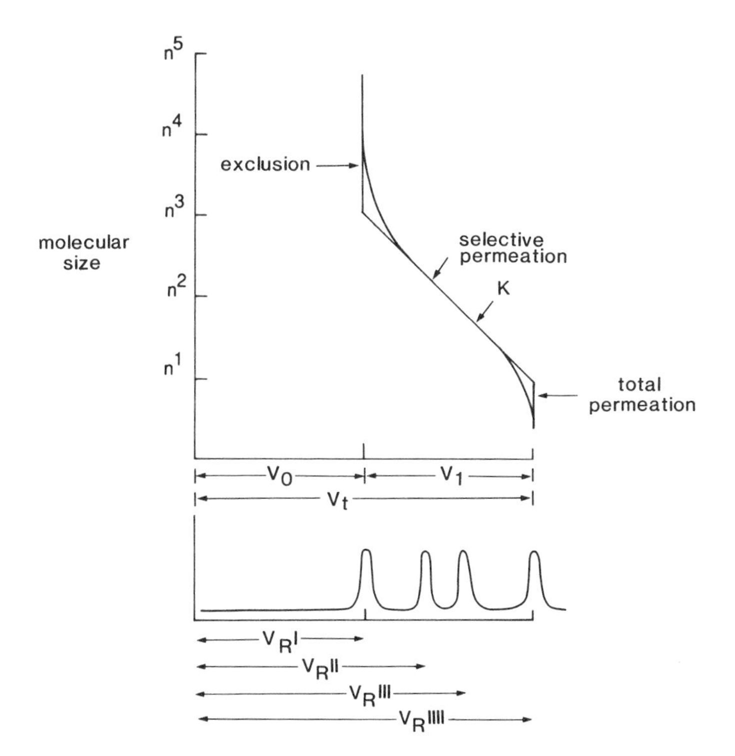

FIGURE 9.12. Elution curve showing schematically the range of elution volumes which are valid for a particular column. In this case molecules with molecular size $> n^3$ are totally excluded and eluted without discrimination, while those $< n^1$ tend to become absorbed or are partitioned if a mixed solvent is used and $K_D = 1$.

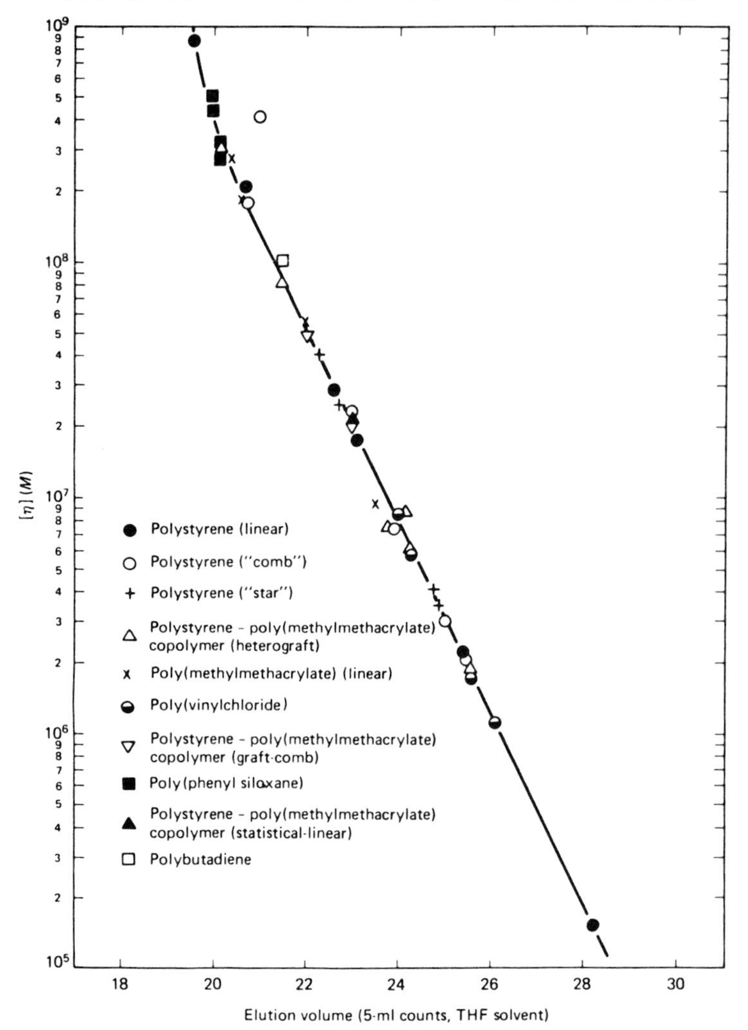

FIGURE 9.13. A universal calibration curve for several polymers in tetrahydrofuran. (Reproduced from data by Z. Grubisic, P. Rempp and H. Benoit, *Polymer Letters*, **5**, 753 (1967). © John Wiley and Sons Inc., N.Y.).

To obtain the MMD, the mass of the polymer being eluted must be measured. This can be achieved continuously using refractive index, UV or IR detectors, which will give a mass distribution as a function of V_R. It is still necessary to estimate the molar mass of each fraction before the MMD curve can be constructed. If the universal calibration curve is valid for the system then

$$\log [\eta]_s M_s = \log [\eta]_u M_u \tag{9.64}$$

where the subscripts s and u denote the standard calibration and the polymer under study, respectively. As the chains of the polymer under examination may swell in the carrier solvent to a different extent compared to an equal molar mass sample of the standard, the hydrodynamic volumes will not necessarily be equivalent. This can be compensated for by applying a correction based on the knowledge of the appropriate Mark-Houwink relations for each, the standard and the unknown, measured in the solvent used for elution (*i.e.* $[\eta]_s = K_s M^{\nu_s}$s and $[\eta]_u = K_u M^{\nu_u}$u). The molar mass M_u can then be obtained from

$$\log M_u = \frac{1}{1+\nu_u} \cdot \log\left[\frac{K_s}{K_u}\right] + \frac{1+\nu_s}{1+\nu_u} \cdot \log M_s \tag{9.65}$$

Thus a calibration curve constructed for standard samples of polystyrene can be used to determine M for other polymers if the Mark-Houwink relations are also known. This can be avoided if a "Viscotek", which is a combined differential refractometer and viscometer, is attached to the end of the column, as this measures both concentration and the η_{sp} for the fraction. By assuming $\eta_{sp}/c \approx [\eta]$ in dilute solutions the molar mass can be obtained from the Mark-Houwink relation. Alternatively, a low angle laser light scattering (LALLS) instrument can be attached in series with a concentration detector. This gives a direct measurement of M_w, using equation (9.28), if all the parameters in the equation are known.

When using SEC, care must be taken not to overload the columns with too large a polymer sample as this results in a non-linear response, characterized by losses in resolution and column efficiency. Also, although the band broadening referred to earlier can be minimized by using long, efficient columns, it may never be entirely eliminated. The MMD curves may then be broadened by this phenomenon and appropriate corrections must be applied. Unfortunately these are often difficult to calculate accurately, although it has been shown by Tung that the error introduced by broadening is negligible if $M_w/M_n > 2$ for the sample.

General Reading

G. Allen and J. C. Bevington, Eds, *Comprehensive Polymer Science*. Vol. 1. Pergamon Press (1989).

N. M. Bikales, *Characterization of Polymers*, Wiley Interscience (1971).

N. C. Billingham, *Molar Mass Measurements in Polymer Science*. Kogan Page (1977).

F. W. Billmeyer, *Textbook of Polymer Science*, 3rd Ed. John Wiley and Sons (1979).

M. Bohdanecky and J. Kovar, *Viscosity of Polymer Solutions*. Elsevier (1982).

T. J. Bowen, *An Introduction to Ultracentrifugation*. Wiley Interscience (1970).

B. Carroll, *Physical Methods in Macromolecular Chemistry*, Vol. 2. Marcel Dekker (1972).

H. Fujita, *Foundations of Ultracentrifugal Analysis*. John Wiley and Sons Ltd, N.Y. (1975).

M. B. Huglin, *Light Scattering from Polymer Solutions*. Academic Press (1972).

J. F. Johnson and R. F. Porter, *Analytical Gel Permeation Chromatography*. John Wiley and Sons (1968).

P. Kratochvil, *Classical Light Scattering from Polymer Solutions*. Elsevier (1987).

H. Morawetz, *Macromolecules in Solution*. 2nd Ed. John Wiley and Sons Ltd. (1975).

J. F. Rabek, *Experimental Methods in Polymer Chemistry*. John Wiley and Sons Ltd (1980).

W. W. Yau, J. J. Kirkland and D. D. Bly, *Modern Size-Exclusion Liquid Chromatography. Practice of Gel Permeation and Gel Filtration Chromatography*. John Wiley and Sons Inc. N.Y. (1979).

References

1. T. G. Fox, J. B. Kinsinger, H. F. Mason and E. M. Schuele, *Polymer*, **3**, 71 (1962).
2. Z. Grubisic, P. Rempp and H. Benoit, *Polymer Letters*, **5**, 753 (1967).

CHAPTER 10

Polymer Characterization—Chain Dimensions and Structures

As the size and shape of a polymer chain are of considerable interest to the polymer scientist it is useful to know how these factors can be assessed. Much of the information can be derived from studies of dilute solutions; an absolute measurement of polymer chain size can be obtained from light scattering, when the polymer is large compared with the wavelength of the incident light. Sometimes the absolute measurement cannot be used but the size can be deduced indirectly from viscosity measurements, which are related to the volume occupied by the chain in solution. Armed with this information we must now determine how meaningful it is and to do this a clearer understanding of the factors governing the shape of the polymer is required. We can confine ourselves to models of the random coil, as this is usually believed to be most appropriate for synthetic polymers; other models – rods, discs, spheres, spheroids – are also postulated, but need not concern us at this level.

10.1 Average chain dimensions

A polymer chain in dilute solution can be pictured as a coil, continuously changing its shape under the action of random thermal motions. This means, that at any time, the volume occupied by a chain in solution, could differ from that occupied by its neighbours, and these size differences are further accentuated by the fact that each sample will contain a variety of chain lengths. Taking these two points into consideration leads us to the conclusion that meaningful chain dimensions can only be values averaged over the many conformations assumed. Two such averages have been defined: (a) the average root mean square distance between the chain ends $\langle \bar{r}^2 \rangle^{1/2}$; and (b) the average root mean square radius of gyration $\langle \bar{S}^2 \rangle^{1/2}$ which is a measure of the average distance of a chain element from the centre of gravity of the coil. The angular brackets denote averaging due to chain polydispersity in the sample and the bar indicates averaging for the many conformational sizes available to chains of the same molar mass.

The two quantities are related, in the absence of excluded volume effects, for simple chains by

$$\langle \bar{r}^2 \rangle^{1/2} = \langle 6\bar{S}^2 \rangle^{1/2}, \tag{10.1}$$

but as the actual dimensions obtained can depend on the conditions of the measurement, other factors must also be considered.

10.2 Freely-jointed chain model

The initial attempts to arrive at a theoretical representation of the dimensions of a linear chain, treated the molecule as a number n of chain elements, joined by bonds of length l. By assuming the bonds act like universal joints, complete freedom of rotation about the chain bonds can be postulated. This model allows the chain to be pictured as in figure 10.1(a) which resembles the path of a diffusing gas molecule and as random flight statistics have proved useful in describing gases, a similar approach is used here. In two dimensions the diagram is more picturesquely called the "drunkard's walk" and $\boldsymbol{r}_{\mathrm{f}}$ is estimated by considering first the simplest case of two links. The end-to-end distance $\boldsymbol{r}_{\mathrm{f}}$ follows from the cosine law that

$$\mathrm{OB}^2 = \mathrm{OA}^2 + \mathrm{AB}^2 - 2(\mathrm{OA})(\mathrm{AB})\cos\theta \tag{10.2}$$

see figure 10.1(b), or

$$r_{\mathrm{f}}^2 = 2l^2 - 2l^2\cos\theta. \tag{10.3}$$

When the number of bonds, n is large, the angle θ will vary over all possible values so that the sum of all these terms will be zero, and as $\cos\theta = -\cos(\theta + \pi)$ equation (10.3) will reduce to

$$r_{\mathrm{f}}^2 = nl^2. \tag{10.4}$$

This shows that the distance between the chain ends, for this model is proportional to the square root of the number of bonds and so, is considerably shorter than a fully extended chain.

The result is the same if the molecule is thought to occupy three-dimensional space, but if it is centred on a co-ordinate system both positive and negative contributions occur with equal probability. To overcome this the dimension is expressed always as the square which eliminates negative signs. This model is, however, unrealistic. Polymer

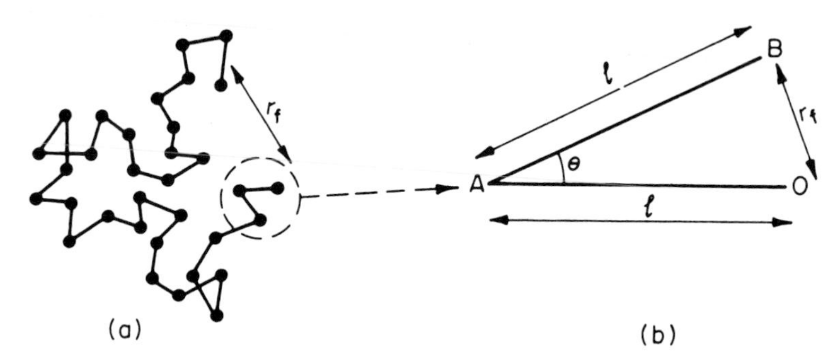

FIGURE 10.1. (a) Random walk chain of 32 steps, length l and (b) cosine law for two bonds.

chains occupy a volume in space, and the dimensions of any macromolecule are influenced by the bond angles and by interactions between the chain elements. These interactions can be classified into two groups: (i) *Short range* interactions which occur between neighbouring atoms or groups, and are usually forces of steric repulsion caused by the overlapping of electron clouds; (ii) *Long range* interactions which are comprised of attractive and repulsive forces between segments, widely separated in a chain, that occasionally approach one another during molecular flexing, and between segments and solvent molecules. These are often termed *excluded volume* effects.

10.3 Short range effects

The expansion of a covalently bonded polymer chain will be restricted by the valence angles between each chain atom. In general this angle is θ for a homoatomic chain and equation (10.4) can be modified to allow for these short range interactions.

$$\langle \bar{r}^2 \rangle_{\text{of}} = nl^2(1 - \cos\theta)/(1 + \cos\theta) \tag{10.5}$$

For the simplest case of an all carbon backbone chain such as polyethylene, $\theta \approx 109°$ and $\cos\theta = -\frac{1}{3}$ so that equation (10.5) reduces to

$$\langle \bar{r}^2 \rangle_{\text{of}} = 2nl^2. \tag{10.6}$$

This indicates that the polyethylene chain is twice as extended as the freely jointed chain model when short range interactions are considered.

10.4 Chain stiffness

As we have already seen in chapter 1 for butane and polyethylene, steric repulsions impose restrictions to bond rotation. This means that equation (10.5) has to be modified further and now becomes

$$\langle \bar{r}^2 \rangle_0 = nl^2 \frac{(1 - \cos\theta)}{(1 + \cos\theta)} \cdot \frac{(1 - \langle \cos\phi \rangle)}{(1 + \langle \cos\phi \rangle)} \tag{10.7}$$

where $\langle \cos\phi \rangle$ is the average cosine of the angle of rotation of the bonds in the backbone chain. The parameter $\langle \bar{r}^2 \rangle_0$ is the average mean square of the *unperturbed dimension*, which is a characteristic parameter for a given polymer chain.

The freely jointed dimensions are now more realistic when restricted by the factor ζ the skeletal factor – composed of the two terms

$$\zeta = \sigma^2(1 - \cos\theta)/(1 + \cos\theta), \tag{10.8}$$

where σ is known as the steric parameter and is $(1 - \langle \cos\phi \rangle)/(1 + \langle \cos\phi \rangle)$ for simple chains.

For more complex chains, containing rings or heteroatomic chains, *e.g.* polydienes, polyethers, polysaccharides, and proteins, an estimate of σ is obtained from

$$\sigma^2 = \langle \bar{r}^2 \rangle_0 / \langle \bar{r}^2 \rangle_{\text{of}}. \tag{10.9}$$

TABLE 10.1. Chain stiffness parameters and typical dimensions

Polymer	T/K	σ	$[\langle \bar{r}^2 \rangle_0/nl^2]$
Polypropylene(isotactic)	408	1.53	4.67
Polypropylene(atactic)	408	1.65	5.44
Natural rubber	293	1.67	4.70
Guttapercha	333	1.38	7.35
Polystyrene	308	2.23	10.00
Poly(methyl methacrylate)			
(Isotactic)	298	2.28	10.40
(Atactic)	298	2.01	8.10
(Syndiotactic)	308	1.94	7.50

Values of the unperturbed dimension can be obtained experimentally from dilute solution measurements made either directly in a theta-solvent (see section 9.9) or by using indirect measurements in non-ideal solvents and employing an extrapolation procedure. The geometry of each chain allows the calculation of $\langle \bar{r}^2 \rangle_{of}$ and results are expressed either as σ or as the characteristic ratio $\{\langle \bar{r}^2 \rangle_0/nl^2\}$. Both provide a measure of chain stiffness in dilute solution. The range of values normally found for σ is from about 1.5 to 2.5 as shown in table 10.1.

10.5 Treatment of dilute solution data

We can now examine some of the ways of calculating the polymer dimensions from experimental data.

THE SECOND VIRIAL COEFFICIENT

An investigation of the dilute solution behaviour of a polymer can provide useful information about the size and shape of the coil, the extent of polymer-solvent interaction and the molar mass. Deviations from ideality, as we have seen in section 9.7, are conveniently expressed in terms of virial expansions, and when solutions are sufficiently dilute, the results can be adequately described by the terms up to the second virial coefficient A_2 while neglecting higher terms. The value of A_2 is a measure of solvent-polymer compatibility, as the parameter reflects the tendency of a polymer segment to exclude its neighbours from the volume it occupies. Thus a large positive A_2 indicates a good solvent for the polymer while a low value (sometimes even negative) shows that the solvent is relatively poor. The virial coefficient can be related to the Flory dilute solution parameters by

$$A_2 = \psi_1(1 - \Theta/T)(\bar{v}_2^2/V_1)F(x) \tag{10.10}$$

where $F(x)$ is a molar mass dependent function of the excluded volume. The exact form of $F(x)$ can be defined explicitly by one of several theories, and while each leads to a slightly different form, all predict that $F(x)$ is unity when theta conditions are attained and the excluded volume effect vanishes. Equation (10.10) can be used to analyse data such as that in figure 8.7. Once Θ has been located, the entropy parameter ψ_1 can be calculated by replotting the data as $\psi_1 F(x)$ against T. Extrapolation to $T = \Theta$, where $F(x) = 1$, allows ψ_1 to be estimated for the system under theta conditions. This method of measuring Θ and ψ_1 is only accurate when the solvent is poor, and extrapolations are short.

The dependence of A_2 on M can often be predicted, for good solvents, by a simple equation

$$A_2 = kM^{-\gamma}, \tag{10.11}$$

where γ varies from 0.15 to 0.4, depending on the system and k is a constant.

EXPANSION FACTOR α

The value of A_2 will tell us whether or not the size of the polymer coil, which is dissolved in a particular solvent, will be perturbed or expanded over that of the unperturbed state, but the extent of this expansion is best estimated by calculating the expansion factor α.

If the temperature of a system, containing a polymer of finite M, drops much below Θ the number of polymer-polymer contacts increases until precipitation of the polymer occurs. Above this temperature, the chains are expanded, or perturbed, from the equilibrium size attained under pseudo-ideal conditions, by long range interactions. The extent of this coil expansion is determined by two long range effects. The first results from the physical exclusion of one polymer segment by another from a hypothetical lattice site which reduces the number of possible conformations available to the chain. This serves to lower the probability that tightly coiled conformations will be favoured. The second is observed in very good solvents, where the tendency is for polymer-solvent interactions to predominate, and leads to a preference for even more extended conformations. In a given solvent an equilibrium conformation is eventually achieved when the forces of expansion are balanced by forces of contraction in the molecule. The tendency to contract arises from the both the polymer-polymer interactions and the resistance to expansion of the chain into over extended and energetically less favoured conformations.

The extent of this coil perturbation by long range effects is measured by an expansion factor α, introduced by Flory. This relates the perturbed and unperturbed dimensions by

$$\langle \bar{S}^2 \rangle^{1/2} = \alpha \langle \bar{S}^2 \rangle_0^{1/2} \tag{10.12}$$

In good solvents (large, positive A_2) the coil is more extended than in poor solvents (low A_2) and α is correspondingly larger. Since α is solvent and temperature dependent a more characteristic dimension to measure for the polymer is $\langle \bar{S}^2 \rangle_0^{1/2}$, which can be calculated from light scattering in a theta-solvent, or indirectly as next described.

FLORY-FOX THEORY

The molecular dimensions of a polymer chain in any solvent can be calculated directly from light scattering measurements, using equation (9.36), if the coil is large enough to scatter light in an asymmetric manner, but when the chain is too short to be measured accurately in this way an alternative technique has to be used.

Flory and Fox suggested that as the viscosity of a polymer solution will depend on the volume occupied by the polymer chain, it should be feasible to relate coil size and $[\eta]$. They assumed that if the unperturbed polymer is approximated by a hydrodynamic sphere, then $[\eta]_\theta$, the limiting viscosity number in a theta solvent, could be related to the square root of the molar mass by

$$[\eta]_\theta = K_\theta M^{1/2}, \tag{10.13}$$

where

$$K_\theta = \Phi(\bar{r}_0^2/M)^{3/2}. \tag{10.14}$$

Equations (10.13) and (10.14) are actually derived for monodisperse samples, and when measurements are performed with heterodisperse polymers, the appropriate averages to use are M_n and $\langle \bar{r}^2 \rangle_{0n}$. The parameter Φ was originally considered to be a universal constant, but experimental work suggests that it is a function of the solvent, molar mass, and heterogeneity. Values can vary from an experimental one of 2.1×10^{23} to a theoretical limit of about 2.84×10^{23} when $[\eta]$ is expressed in $cm^3\,g^{-1}$. A most probable value of 2.5×10^{23} has been found to be acceptable for most flexible heterodisperse polymers in good solvents.

For non-ideal solvents equation (10.13) can be expanded to give

$$[\eta] = K_\theta M^{1/2} \alpha_\eta^3, \tag{10.15}$$

where $\alpha_\eta^3 = [\eta]/[\eta]_\theta$ is the linear expansion factor, pertaining to viscosity measurements, and is a measure of long range interactions. As the derivation is based on an unrealistic Gaussian distribution of segments in good solvents, it has been suggested that α_η is related to the more direct measurement of α in equation (10.12) by

$$\alpha_\eta^3 = \alpha^{2.43}. \tag{10.16}$$

Considerable experimental evidence exists to support this conclusion.

INDIRECT ESTIMATES OF $\langle \bar{r}^2 \rangle_0^{1/2}$

It is not always possible to find a suitable theta-solvent for a polymer and methods have been developed which allow unperturbed dimensions to be estimated in non-ideal (good) solvents.

Several methods of extrapolating data for $[\eta]$ have been suggested. The most useful of these was proposed by Stockmayer and Fixman, using the equation:

$$[\eta] M^{-1/2} = K_\theta + 0.51 \Phi B' M^{1/2}, \tag{10.17}$$

when Φ is assumed to adopt its limiting theoretical value, B' is related to the thermodynamic interaction parameter χ_1 by

$$B' = \bar{v}_2^2(1 - 2\chi_1)/V_1 N_A, \tag{10.18}$$

and examination of equation (10.10) shows that B' is also proportional to A_2. The unperturbed dimension can be estimated by plotting $[\eta]M^{-1/2}$ against $M^{1/2}$; K_θ is obtained from the intercept and $\langle \bar{r}^2 \rangle_0$ is calculated from equation (10.14).

A similar procedure has been proposed by Cowie and Bywater, in which the intrinsic frictional coefficient $[f]$ measured from sedimentation or diffusion experiments, will provide the same information using

$$[f]M^{-1/2} = K_f + 0.201 K_f^{-2} P_0^3 B' M^{1/2}, \ldots \qquad (10.19)$$

where

$$K_f = P_0[\langle \bar{r}_0^2 \rangle / M]^{1/2}$$

and P_0 is a "constant" with a limiting value of 5.2.

These extrapolation procedures all depend on the validity of the theoretical treatment and reliability must be judged in this light. Fortunately, it has been demonstrated that most non-polar polymers can be treated in this way and results agree well with direct measurements of $\langle \bar{r}^2 \rangle_0^{1/2}$. For more polar polymers, specific solvent effects become more pronounced and extrapolations have to be regarded with corresponding caution.

INFLUENCE OF TACTICITY ON CHAIN DIMENSIONS

Studies of the dilute solution behaviour of polymers with a specific stereostructure have revealed that the unperturbed dimensions may depend on the chain configuration. This can be seen from the data in table 10.1 where isotactic, syndiotactic, and atactic poly(methyl methacrylate) have different σ values. If the size of a polymer chain can be affected by its configuration, the microstructure must be well characterized before an accurate assessment of experimental data can be made. This can be achieved using n.m.r. and infrared techniques.

10.6 Nuclear magnetic resonance (n.m.r.)

High resolution n.m.r. has proved to be a particularly useful tool in the study of the microstructure of polymers in solution, where the extensive molecular motion reduces the effect of long range interactions and allows the short range effects to dominate. Interpretation of chain tacticity, based on the work of Bovey and Tiers, can be illustrated using poly(methyl methacrylate). The three possible steric configurations are shown in figure 10.2 where R is the group $—COOCH_3$.

For the purposes of n.m.r. measurements three consecutive monomer units in a chain are considered to define a configuration and called a triad. The term heterotactic is used now to define a triad which is neither isotatic nor syndiotactic. In the structures shown, the three equivalent protons of the α-methyl group absorb radiation at a single frequency, but this frequency will be different for each of the three kinds of triad, because the environment of the α-methyl groups in each is different. For poly(methyl methacrylate) samples, which were prepared under different conditions to give the three forms, resonances at $\tau = 8.78$, 8.95 and 9.09 were observed, which were assigned to the isotatic, heterotactic, and syndiotactic triads respectively. Thus in a sample with a mixture of configurations a triple peak will be observed and the area under each of these

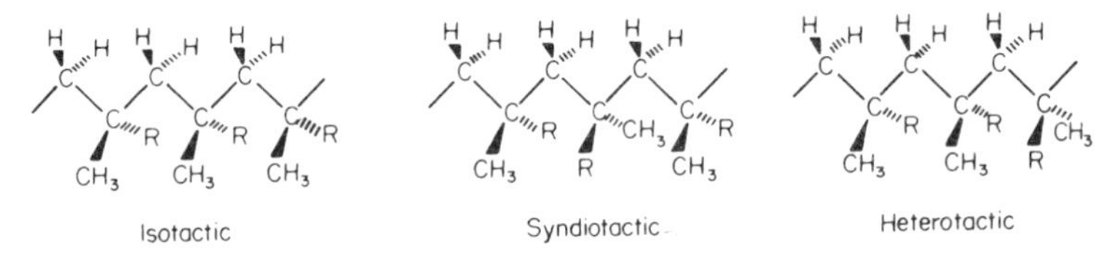

FIGURE 10.2. Stereoregular triads for poly(methyl methacrylate) where R = —$COOCH_3$.

peaks will correspond to the amount of each triad present in the polymer chain. This is illustrated in figure 10.3, where one sample is predominantly isotactic, but also contains smaller percentages of the heterotactic and syndiotactic configurations.

The analysis can be carried further. The fraction of each configuration, P_i, P_h, and P_s, measured from the respective peak areas, can be related to p_m the probability that a monomer adding on to the end of a growing chain will have the same configuration as the unit it is joining. This leads to the relations

$$P_i = p_m^2,\ P_s = (1 - p_m)^2, \qquad \text{and} \qquad P_h = 2p_m(1 - p_m).$$

Curves plotted according to this simple analysis are shown in figure 10.4 where they are compared with experimental data obtained for various tactic forms of poly(α-methyl styrene).

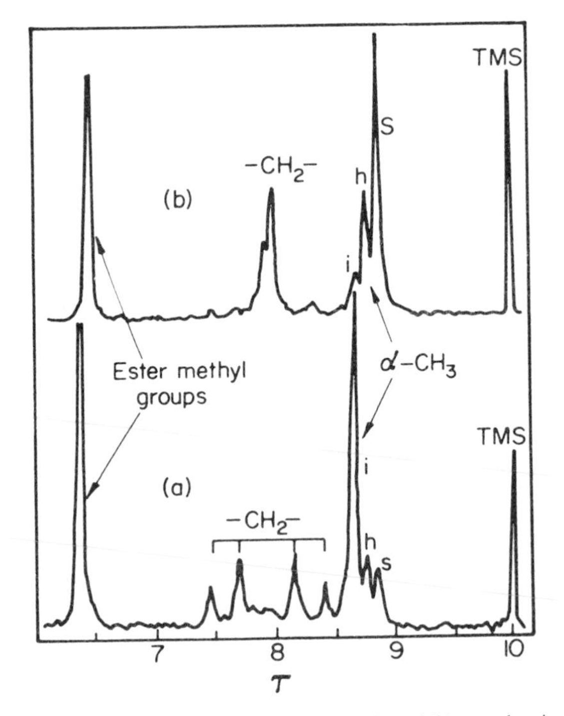

FIGURE 10.3. The n.m.r. spectra for (a) an isotactic sample and (b) a predominantly syndiotactic sample of poly(methyl methacrylate).

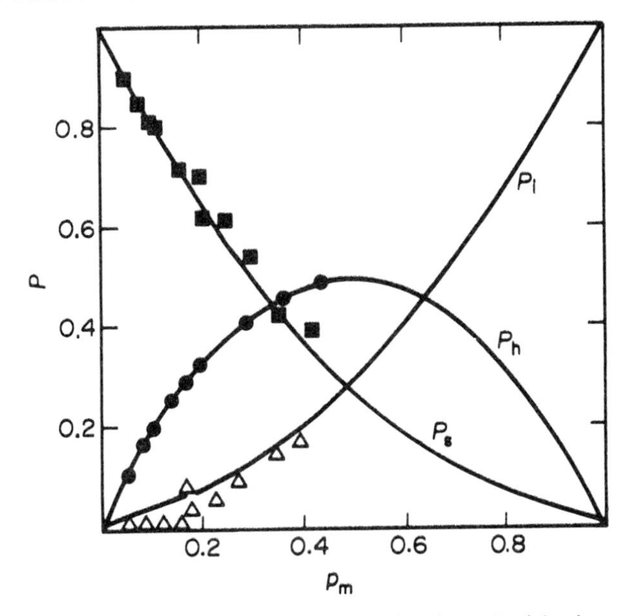

FIGURE 10.4. Theoretical curves for P as a function of p_m for each of the three configurations. Points represent experimental data for poly(α-methyl styrene) and illustrate the validity of the analysis. (From data by Brownstein, Bywater and Worsfold (1961).)

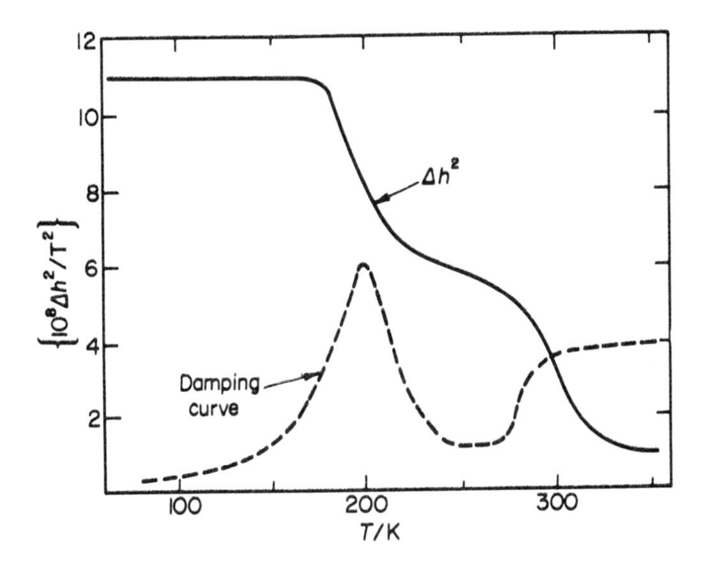

FIGURE 10.5. The n.m.r. line width Δh (the unit is the tesla T) as a function of temperature for poly(tetrafluoroethylene). The mechanical damping curve (-----) is included for comparison. (Adapted from Sauer and Woodward (1960).)

Differences in the microstructure of polydienes and copolymers can also be made using n.m.r. In the polydienes the difference between 1,2- and 1,4-addition can be distinguished on examination of the resonance peaks corresponding to terminal olefinic protons, found at $\tau = 4.9$ to 5.0, and non-terminal olefinic protons observed at $\tau = 4.6$ to 4.7.

Not only is the local field acting on the nucleus altered by environment, it is also sensitive to molecular motion, and it has been observed that as the molecular motion within a sample increases, the resonance lines become narrower. Determination of the width, or second moment, of an n.m.r. resonance line, then provides a sensitive measure of low frequency internal motions in solid polymers and can be used to study transitions and segmental rotations in the polymer sample. Line widths are also altered by the polymer crystallinity. Partially crystalline polymers present complex spectra as they are multi-phase materials, in which the molecular motions are more restricted in the crystalline phase than in the amorphous phase. However, attempts to estimate percentage crystallinity in a sample using n.m.r. have not been particularly successful. The method is illustrated in figure 10.5 for poly(tetrafluoroethylene) where glass and other transitions are readily detected. Below 200 K the chains are virtually immobile, but above 200 K the lines sharpen as $—CF_2—$ rotation begins. This is associated with the glass transition, but the way the line width increases in this region is governed by sample crystallinity.

10.7 Infrared spectroscopy

Infrared spectroscopy can be used to characterize long chain polymers because the infrared active groups, present along the chain, absorb as if each was a localized group in a simple molecule. Identification of polymer samples can be made by making use of the "finger-print" region, where it is least likely for one polymer to exhibit exactly the same spectrum as another. This region lies within the range 6.67 to 12.50 μm.

In addition to identification, the technique has been used to elucidate certain aspects of polymer microstructure, such as branching, crystallinity, tacticity, and *cis-trans* isomerism. The relative proportions of *cis*-1,4-, *trans*-1,4-, and 1,2-addition in polybutadienes can be ascertained by making use of the differences in absorption between (CH) out of plane bending vibrations, which depend on the type of substitution at the olefinic bond. Terminal and internal groups can also be distinguished, as an absorption band at about 11.0 μm is characteristic of a vinyl group and indicates 1,2-addition. The *cis*-1,4-addition is characterized by an absorption band at about 13.6 μm, whereas the *trans*-1,4 configuration exhibits a band at about 10.4 μm. An estimate of *cis-trans* isomerism can be made by measuring the absorbance A of each band, where $A = \log_{10}(I_0/I)$ and I_0 and I are the intensities of the incident and transmitted radiation respectively. This is calculated by locating a base line across the minima on either side of the absorption band and the vertical height to the top of the band from the base line is converted into a composition using the equation

$$P_{cis} = 3.65\,A_{cis}/(3.65\,A_{cis} + A_{trans}),$$

where P_{cis} is the fraction of *cis* configuration, A_{cis} is the absorbance at 13.6 μm, A_{trans} the absorbance at 10.4 μm, and if we assume that the 1,2 content is negligible. Polyisoprenes can also be analyzed in this way, only now the bands at 11.0 and 11.25 μm are used to estimate the 1,2- and 3,4-addition, while a band at 8.7 μm corresponds to the *trans*-1,4 linkage.

The infrared spectra of highly stereoregular polymers are distinguishable from those of their less regular counterparts, but many of the differences can be attributed to crystallinity rather than tacticity as such. The application of infrared to stereostructure determination in polymers is less reliable than n.m.r., but has achieved moderate success for poly(methyl methacrylate) and polypropylene. In poly(methyl methacrylate), a methyl deformation at 7.25 μm is unaffected by microstructure, and comparison of this with a band at 9.40 μm, which is present only in atactic or syndiotactic polymers allows an estimate of the syndiotacticity to be made from the ratio $\{A(9.40\ \mu\text{m})/A(7.25\ \mu\text{m})\}$. Similarly $\{A(6.75\ \mu\text{m})/A(7.25\ \mu\text{m})\}$ provides a measure of the isotactic content. An alternative method is to calculate the quantity J as an average of the two equations

$$J_1 = 179\{A(9.40\ \mu\text{m})/A(10.10\ \mu\text{m})\} + 27$$
$$J_2 = 81.4\ \{A(6.75\ \mu\text{m})/A(7.25\ \mu\text{m})\} - 43$$

where the absorption band at 10.10 μm is now used. If J lies between 100 and 115 a highly syndiotactic polymer is indicated, if between 25 and 30 the polymer is highly isotactic. For polypropylene, the characteristic band for the syndiotactic polymer appears at 11.53 μm, and the syndiotactic index I_s is $2A(11.53\ \mu\text{m})/\{A(2.32\ \mu\text{m}) + A(2.35\ \mu\text{m})\}$. Values of I_s about 0.8 indicate highly syndiotactic samples. Spectra can be measured in a number of ways; for soluble polymers a film can be cast, perhaps even on the NaCl plate to be used and examined directly. Measurements can also be made in solution, if the solvent absorption in any important region is low, or by a differential method.

10.8 X-ray diffraction

The extent of sample crystallinity can influence the behaviour of a polymer sample greatly. A particularly effective way of examining partially crystalline polymers is by X-ray diffraction. The crystallites present in a powdered or unoriented polymer sample diffract X-ray beams from parallel planes for incident angles θ which are determined by the Bragg equation

$$n\lambda = 2d \sin \theta, \tag{10.21}$$

where λ is the wavelength of the radiation, d is the distance between the parallel planes in the crystallites, and n is an integer. The reinforced waves reflected by all the small crystallites produce diffraction rings, or haloes, which are sharply defined for highly crystalline materials and become increasingly diffuse when the amorphous content is high.

If the polymer sample is oriented, by drawing a fibre, or by applying tension to a film, the crystallites tend to become aligned in the direction of the stress and the X-ray pattern is improved. In some samples of stereoregular or symmetrical polymers, the degree of three-dimensional ordering of the chains may be sufficiently high to allow a structural analysis of the polymer to be accomplished.

Sample crystallinity can be estimated from the X-ray patterns by plotting the density

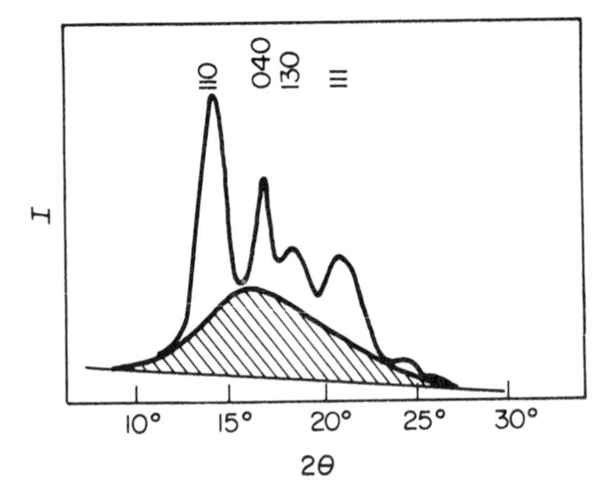

FIGURE 10.6. X-ray diffraction curves; the intensity I as a function of angle for totally amorphous polypropylene (shaded area), and for a sample with a 50 per cent crystalline content.

of the scattered beam against the angle of incidence. If this can be done for an amorphous sample and a corresponding sample which is highly crystalline, a relative measure of crystallinity for other samples of the same polymer can be obtained. In figure 10.6 the shaded portion is the amorphous polypropylene, while the maxima arise from the crystallites.

10.9 Thermal analysis

When a substance undergoes a physical or chemical change a corresponding change in enthalpy is observed. This forms the basis of the technique known as differential thermal analysis (DTA) in which the change is detected by measuring the enthalpy difference between the material under study and an inert standard.

The sample is placed in a heating block and warmed at a uniform rate. The sample temperature is then monitored by means of a thermocouple and compared with the temperature of an inert reference such as powdered alumina, or simply an empty sample pan, which is subjected to the same linear heating programme. As the temperature of the block is raised at a constant rate (5 to 20 K min^{-1}) the sample temperature T_s and that of the reference, T_r will keep pace until a change in the sample takes place. If the change is exothermic T_s will exceed T_r for a short period, but if it is endothermic T_s will temporarily lag behind T_r. This temperature difference ΔT is recorded and transmitted to a chart recorder where changes such as melting or crystallization are recorded as peaks. A third type of change can be detected. Since the heat capacities of sample and reference are different ΔT is never actually zero, and a change in heat capacity, such as that associated with a glass transition, will cause a shift in the base line. All three possibilities are shown in figure 10.7 for quenched terylene.

Other changes such as sample decomposition, crosslinking, and the existence of polymorphic forms can also be detected. As ΔT measured in DTA is a function of the thermal conductivity and bulk density of the sample, it is nonquantitative and relatively uninformative. To overcome these drawbacks an alternative procedure known as

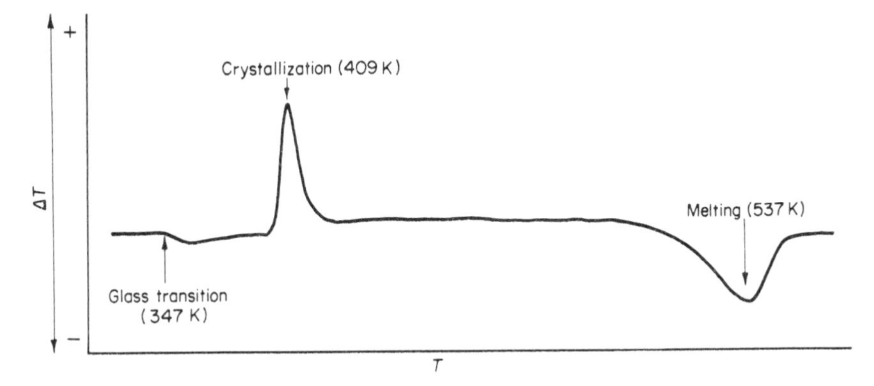

FIGURE 10.7. A DTA curve for quenched terylene showing the glass transition, melting endotherm, and a crystallization exotherm.

differential scanning calorimetry (DSC) is used. This technique retains the constant mean heat input but instead of measuring the temperature difference during a change a servo-system immediately increases the energy input to either sample or reference to maintain both at the same temperature. The thermograms obtained are similar to DTA, but actually represent the amount of electrical energy supplied to the system, not ΔT, and so the areas under the peaks will be proportional to the change in enthalpy which occurred. An actual reference sample can be dispensed with in practice and an empty sample pan used instead. Calibration of the instrument will allow the heat capacity of a sample to be calculated in a quantitative manner. This information is additional to that gained on crystallization, melting, glass transitions, and decompositions.

General Reading

G. Allen and J. C. Bevington, Eds, *Comprehensive Polymer Science*, Vols 1 and 2. Pergamon Press (1989).

F. J. Balta-Calleja and C. G. Vonk, *X-ray Scattering of Synthetic Polymers.* Elsevier Science Publishers (1989).

N. M. Bikales, *Characterization of Polymers.* Wiley-Interscience (1971).

F. A. Bovey, *Polymer Conformation and Configuration.* Academic Press (1969).

B. Carroll, *Physical Methods in Macromolecular Chemistry*, Vol. 2. Marcel Dekker (1972).

C. D. Craver, Ed., *Polymer Characterization: Spectroscopic Chromatographic and Physical Instrumental Methods.* Advances in Chemistry, Series 203 (193).

A. Elliott, *Infrared Spectra and Structure of Organic Long Chain Polymers.* Edward Arnold (1969).

P. J. Flory, *Principles of Polymer Chemistry*, Chapters 10 and 14. Cornell Univ. Press, Ithaca, N.Y. (1953).

P. J. Flory, *Statistical Mechanics of Chain Molecules.* Interscience Publishers Inc. (1969).

W. C. Forsman, Ed., *Polymers in Solution. Theoretical Considerations and New Methods of Characterization.* Plenum Press (1986).

P. G. de Gennes *Scaling Concepts in Polymer Physics.* Cornell University Press (1979).
J. F. Johnson and R. F. Porter, *Analytical Calorimetry.* Plenum Press (1968).
B. Ke, *Newer Methods of Polymer Characterization.* Interscience Publishers Inc. (1964).
J. L. Koenig, *Chemical Microstructure of Polymer Chains.* John Wiley and Sons Ltd (1980).
J. Mitchell, Ed., *Applied Polymer Analysis and Characterization: Recent Developments in Techniques, Instrumentation and Problem Solving.* Hanser Publishers (1987).
P. C. Painter, M. M. Coleman and J. Koenig, *The Theory of Vibrational Spectroscopy and its Application to Polymeric Materials.* John Wiley and Sons Ltd (1982).
J. C. Rabek, *Experimental Methods in Polymer Chemistry.* John Wiley and Sons Ltd (1980).
A. V. Tobolsky and H. Mark, *Polymer Science and Materials*, Chapter 3. Wiley-Interscience (1971).
E. A. Turi, *Thermal Characterization of Polymeric Materials.* Academic Press (1981).

References

1. S. Brownstein, S. Bywater and D. J. Worsfold, *Makromol. Chem.*, **58**, 127 (1961).
2. J. A. Sauer and A. E. Woodward, *Reviews in Modern Physics*, **32**, 88 (1960).

CHAPTER 11

The Crystalline State

11.1 Introduction

When polymers are irradiated by a beam of X-rays, scattering produces diffuse haloes on the photographic plate for some polymers, while for others a series of sharply defined rings superimposed on a diffuse background is recorded. The former are characteristic of amorphous polymers, and illustrate that a limited amount of short range order exists in most polymeric solids. The latter patterns are indicative of considerable three-dimensional order and are typical of polycrystalline samples containing a large number of unoriented crystallites associated with amorphous regions. The rings are observed to sharpen into arcs, or discrete spots, if the polymer is drawn or stretched, a process which orients the axes of the crystallites in one direction.

The occurrence of significant crystallinity in a polymer sample is of considerable consequence to a materials scientist. The properties of the sample – the density, optical clarity, modulus, and general mechanical response – all change dramatically when crystallites are present and the polymer is no longer subject to the rules of linear visco-elasticity, which apply to amorphous polymers as outlined in Chapter 13. However, a polymer sample is rarely completely crystalline and the properties also depend on the amount of crystalline order.

It is important than to examine crystallinity in polymers and determine the factors which control the extent of crystallinity.

11.2 Mechanism of crystallization

A polymer in very dilute solution can be effectively regarded as an isolated chain whose shape is governed by short and long range inter- and intra-molecular interactions. In the aggregated state this is no longer true, the behaviour of the chain is now influenced largely by the proximity of the neighbouring chains and the secondary valence forces which act between them. These factors determine the orientation of chains relative to each other in the undiluted state, and this is essentially an interplay between the entropy and internal energy of the system which is expressed in the usual thermo-dynamic form

$$G = (U + pV) - TS.$$

In the melt, polymers normally attain a state of maximum entropy consistent with a stable state of minimum free energy. Crystallization is a process involving the orderly arrangement of chains and is consequently associated with a large negative entropy of activation. If a favourable free energy change is to be obtained for crystallite formation, the entropy term has to be offset by a large negative energy contribution.

The alignment of polymer chains at specific distances from one another to form crystalline nuclei will be assisted when intermolecular forces are strong. The greater this interaction between chains the more favourable will be the energy parameter and this provides some indication of the type of chain which might be expected to crystallize from the melt, *viz.*

(1) Symmetrical chains which allow the regular close packing required for crystallite formation.

(2) Chains possessing groups which encourage strong intermolecular attraction thereby stabilizing the alignment.

In addition to the thermodynamic requirements, kinetic factors relating to the flexibility and mobility of a chain in the melt must also be considered. Thus polyisobutylene $+CH_2C(CH_3)_2+_n$ might be expected to crystallize because the chain is symmetrical, but it will only do so if maintained at an optimum temperature for several months. This is presumably a result of the flexibility of the chain which allows extensive convolution thereby impeding stabilization of the required long range alignment.

The creation of a three-dimensional ordered phase from a disordered state is a two stage process. Just above its melting temperature a polymer behaves like a highly viscous liquid in which the chains are all tangled up with their neighbours. Each chain pervades a given volume in the sample, but as the temperature decreases the volume available to the molecule also decreases. This in turn restricts the number of disordered conformational states available to the chain due to the constraining influence of intramolecular interactions among chains in juxtaposition. As a result there is an increasing tendency for the polymer to assume an ordered conformation in which the chain bonds are in the rotational states of lowest energy. However, various other factors will tend to oppose crystallization; chain entanglements will hinder the diffusion of chains into suitable orientations and if the temperature is above the melting temperature, thermal motions will be sufficient to disrupt the potential nuclei before significant growth can take place. This restricts crystallization to a range of temperatures between T_g and T_m.

The first step in crystallite formation is the creation of a stable nucleus brought about by the ordering of chains in a parallel array, stimulated by intramolecular forces, followed by the stabilization of long range order by the secondary valence forces which aid the packing of molecules into a three-dimensional ordered structure.

The second stage is the growth of the crystalline region, the size of which is governed by the rate of addition of other chains to the nucleus. As this growth is counteracted by thermal redispersion of the chains at the crystal-melt interface, the temperature must be low enough to ensure that this disordering process is minimal.

11.3 Temperature and growth rate

Measurable rates of crystallization occur between $(T_m - 10\,K)$ and $(T_g + 30\,K)$, a range in which the thermal motion of the polymer chains is conducive to the formation of stable ordered regions. The growth rate of crystalline areas passes through a

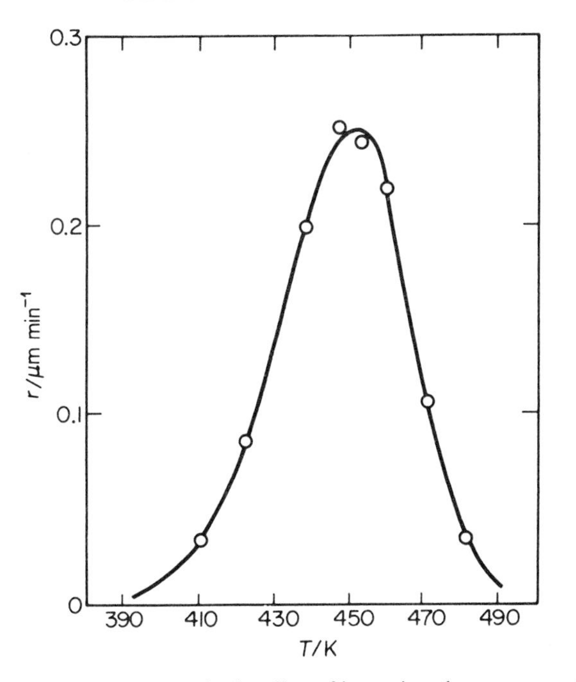

FIGURE 11.1. Radial growth rate r of spherulites of isotactic polystyrene as a function of the crystallization temperature.

maximum in this range as illustrated in figure 11.1 for isotactic polystyrene. Close to T_m the segmental motion is too great to allow many stable nuclei to form, while near T_g the melt is so viscous that molecular motion is extremely slow.

As the temperature drops from T_m, the melt viscosity, which is a function of the molar mass, increases and the diffusion rate decreases, thereby giving the chains greater opportunity to rearrange themselves to form a nucleus. This means that there will exist an optimum temperature of crystallization, which depends largely on the interval T_m to T_g, but also on the molar mass of the sample.

The melt usually has to be supercooled by about 5 to 20 K before a significant number of nuclei appear which possess the critical dimensions required for stability and further growth. If a nucleating agent is added to the system, crystallization can be induced at higher temperatures. This is known as heterogeneous nucleation and only affects the crystallization rate, not the spherulitic growth rate, at a given temperature.

11.4 Melting

The melting of a perfectly crystalline substance is an equilibrium process characterized by a marked volume change and a well-defined melting temperature. Polymers are never perfectly crystalline, but contain disordered regions and crystallites of varying size. The process is normally incomplete because crystallization takes place when the polymer is a viscous liquid. In this state, the chains are highly entangled, and as sufficient time must be allowed for the chains to diffuse into the three-dimensional order required for crystallite formation, the crystalline perfection of the sample is affected by the thermal history. Thus, rapid cooling from the melt usually prevents the development of significant crystallinity. The result is that melting takes place over a range of temperatures, and this range is a useful indication of sample crystallinity.

Effect of crystallite size on melting. The range of temperature, which covers the melting of a polymer, is indicative of the size and perfection of the crystallites in the sample. This is illustrated in a study of the melting of natural rubber samples, which has shown that the melting range is a function of the temperature of crystallization. At low crystallization temperatures the nucleation density in the rubber melt is high, segmental diffusion rates are low, and small imperfect crystalline regions are formed. Thus broad melting ranges are measured for samples crystallized at these lower temperatures, and these become narrower as the crystallization temperature increases.

This suggests that careful annealing at the appropriate temperature could produce samples with a high degree of crystallinity. These samples might then exhibit almost perfect first order phase changes at the melting temperature. A close approximation to these conditions has been attained by Mandelkern, who annealed a linear polyethylene for 40 days. The improvement in the crystalline organization is obvious from examination of the resulting fusion curves in figure 11.2, where the variation of specific volume with temperature for this sample is compared with that for a branched

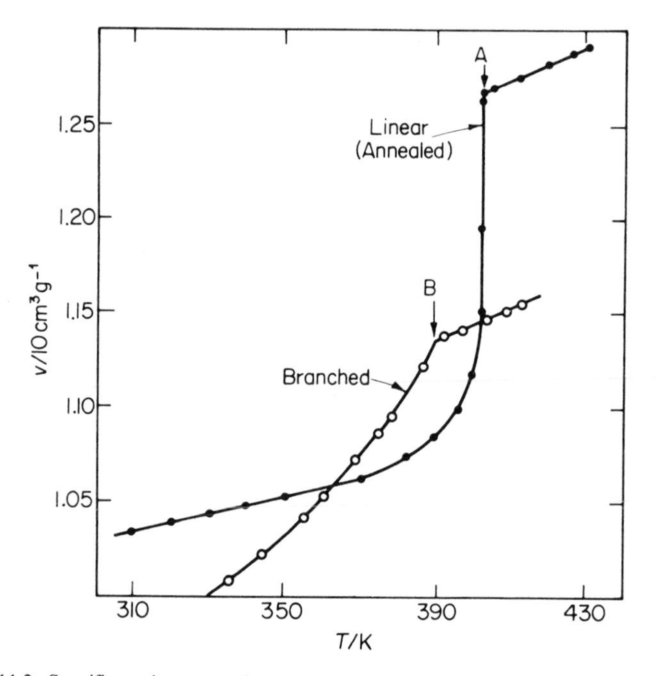

FIGURE 11.2. Specific volume v plotted against temperature T for a sample of linear polyethylene annealed for 40 days and a branched sample. Points A and B are the respective melting temperatures. (From data by Mandelkern.)

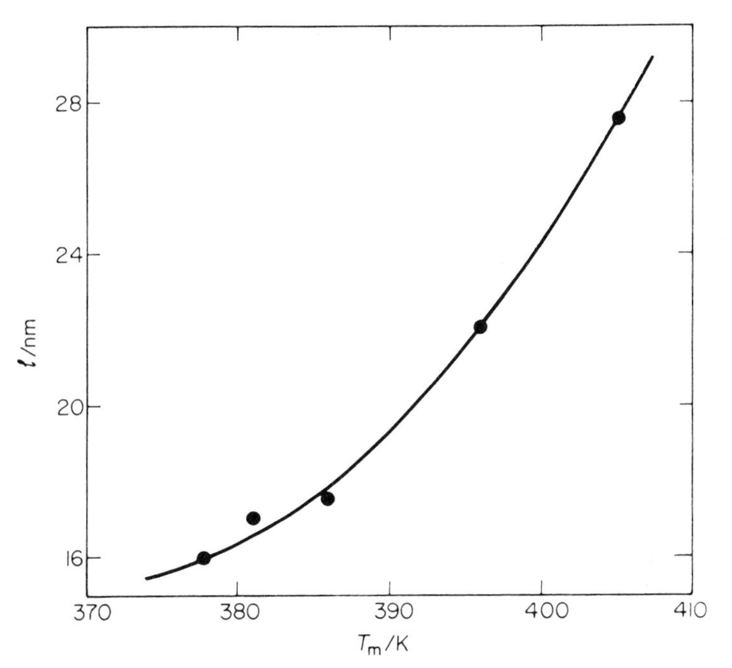

FIGURE 11.3. Dependence of T_m on the length l of the crystallite.

polyethylene of low crystallinity. The effect of branching is to decrease the percentage crystallinity, broaden the melting range, and reduce the average melting temperature. The points A and B in the diagram represent the temperatures at which the largest crystallites disappear and are regarded as the respective melting temperatures T_m for the samples.

The effect of crystal size on T_m is shown more clearly in figure 11.3. The small crystals melt about 30 K lower than the large ones due to the greater contribution from the interfacial free energy in the smaller crystallites, *i.e.* there is an excess of free energy associated with the disordered chains emerging from the ends of ordered crystallites and this is relatively greater for the small crystallites, resulting in lower melting temperatures.

11.5 Thermodynamic parameters

Even with carefully annealed specimens, it is thought that the equilibrium melting temperature of the completely crystalline polymer T_m° is never actually attained. The temperature T_m° is related to the change in enthalpy ΔH_u and the entropy change ΔS_u, for the first order melting transition of pure crystalline polymer to pure amorphous melt, by

$$T_m^\circ = \Delta H_u / \Delta S_u. \tag{11.1}$$

The enthalpy change can be estimated by adding varying quantities of a diluent to the

TABLE 11.1. Thermodynamic parameters derived from melting; the quantities refer to unit amount of basic unit shown

	T_m	ΔH_u	ΔS_u	
Polymer	K	J mol^{-1}	J K^{-1} mol^{-1}	*Basic Unit*
polyethylene	410	3970	9.70	$-(CH_2)-$
poly(tetrafluoroethylene)	645	2860	4.76	$-(CF_2)-$
cis-1,4-polyisoprene	301	4400	14.60	$-(CH_2-C(CH_3)=CH-CH_2)-$
trans-1,4-polyisoprene	347	12700	36.90	(same as above)
polypropylene	447	10880	24.40	$-(CH_2-CH(CH_3))-$
poly(decamethylene terephthalate)	411	46000	114.00	$-[(CH_2)_{10}-O-C(=O)-C_6H_4-C(=O)-O]-$

polymer, which serves to depress the observed melting temperature, and measuring T_m for each polymer + diluent mixture. The results are then plotted according to the Flory equation

$$(1/\phi_1)(1/T_m - 1/T_m^\circ) = (R/\Delta H_u)(V_u/V_1)(1 - BV_1\phi_1/RT_m), \qquad (11.2)$$

where (V_u/V_1) is the ratio of the molar volume of the repeating unit in the chain to that of the diluent, and ϕ_1 is the volume fraction of the diluent. The factor (BV_1/RT_m) is equivalent to the Flory interaction parameter χ_1, indicating that equation (11.2) is dependent on the polymer-diluent interaction. For practical purposes T_m° is taken to be the melting temperature of the undiluted polymer irrespective of the crystalline content. Typical values obtained in this way are shown in table 11.1

In many cases the entropy change is the most important influence on the magnitude of the melting temperature of a polymer. A large part of this entropy is due to the additional freedom which allows the chain conformational changes to occur in the melt, after the restrictions of the crystalline lattice. In the crystalline phase the chain bonds are in their lowest energy state. If the energy difference between the rotational states $\Delta\varepsilon$ is low, the population of the higher energy states will increase in the melt and considerable flexing of the chain is achieved. The contribution of ΔS_u is then high. When $\Delta\varepsilon$ is large, the tendency to populate the high energy states is not too great, consequently the chain is less flexible and ΔS_u is lower. Two polymers which exist in the all *trans* state in the crystal are polyethylene and poly(tetrafluoroethylene). For polyethylene $\Delta\varepsilon$ is about 3.0 kJ mol^{-1}, but it is as high as 18.0 kJ mol^{-1} for poly(tetrafluoroethylene). Hence the polyethylene chain is much more flexible in the melt and gains considerably more entropy on melting, so that T_m is correspondingly lower.

11.6 Crystalline arrangement of polymers

The formation of stable crystalline regions in a polymer requires that, (i) an economical close packed arrangement of the chains can be achieved in three dimensions, and that (ii) a favourable change in internal energy is obtained during this process. This imposes restrictions on the type of chain which can be crystallized with ease and, as mentioned earlier, one would expect symmetrical linear chains such as polyesters, polyamides, and polyethylene to crystallize most readily.

FACTORS AFFECTING CRYSTALLINITY AND T_m

These can be dealt with under the general headings, symmetry, intermolecular bonding, tacticity, branching and molar mass.

Symmetry. The symmetry of the chain shape influences both T_m and the ability to form crystallites. Polyethylene and poly(tetrafluoroethylene) are both sufficiently symmetrical to be considered as smooth stiff cylindrical rods. In the crystal these rods tend to roll over each other and change position when thermally agitated. This motion within the crystal lattice, called *premelting*, increases the entropy of the crystal and effectively stabilizes it. Consequently, more thermal energy is required before the crystal becomes unstable, and T_m is raised. Flat or irregularly shaped polymers, with bends and bumps in the chain, cannot move in this way without disrupting the crystal lattice, and so have lower T_m values. This is only one aspect.

For crystallite formation in a polymer, easy close-packing of the chains in a regular three-dimensional fashion is required. Again linear symmetrical molecules are best. Polyethylene, poly(tetrafluoroethylene) and other chains with more complex backbones containing $+O+$, $+COO+$, and $+CONH+$ groups all possess a suitable symmetry for crystallite formation and usually assume extended zig-zag conformations when aligned in the lattice.

Chains containing irregular units, which detract from the linear geometry, reduce the ability of a polymer to crystallize. Thus *cis*-double bonds (I), *o*- and *m*-phenylene groups (II), or *cis*-oriented puckered rings (III), all encourage bending and twisting in the

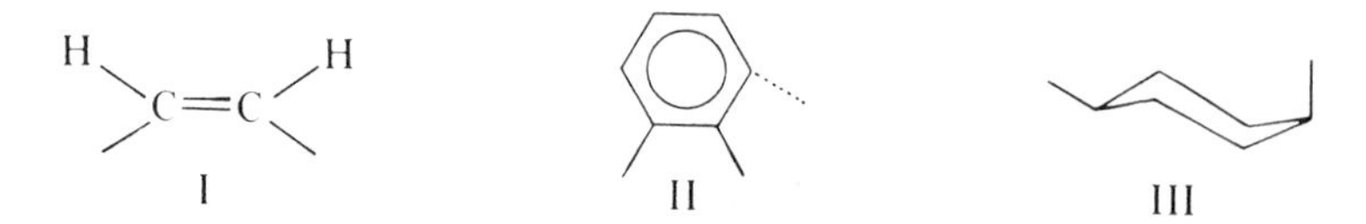

chains and make regular close-packing very difficult. If, however, the phenylene rings are *para*-oriented, the chains retain their axial symmetry and can crystallize more readily. Similarly, incorporation of a *trans*-double bond maintains the chain symmetry. This is highlighted when comparing the amorphous elastomeric *cis*-polyisoprene with the highly crystalline *trans*-polyisoprene which has no virtue as an elastomer, or *cis*-poly(1,3-butadiene) $T_m = 262$ K, with *trans*-poly(1,3-butadiene), $T_m = 421$ K.

Intermolecular bonding. In polyethylene crystallites, the close packing achieved by the chains allows the van der Waals forces to act co-operatively and provide additional stability to the crystallite. Any interaction between chains in the crystal lattice will help to hold the structure together more firmly and raise the melting temperature. Polymers containing polar groups, *e.g.* Cl, CN, or OH, can be held rigid, and aligned, in a polymer matrix by the strong dipole-dipole interactions between the substituents, but the effect is most obvious in the symmetrical polyamides. These polymers can form intermolecular hydrogen bonds which greatly enhance crystallite stability. This is illustrated in figure 11.4 for nylon-6,6, where the extended zig-zag conformation is ideally suited to allow regular intermolecular hydrogen bonding. The increased stability is reflected in T_m, which for nylon-6,6 is 540 K compared with 410 K for polyethylene.

The structures of related polyamides do not always lead to this neat arrangement of intermolecular bond formation; for example the geometry of an extended nylon-7,7 chain allows the formation of only every second possible hydrogen bond when the chains are aligned and fully extended. However, the process is so favourable energetically, that sufficient deformation of the chain takes place to enable formation of all possible hydrogen bonds. The added stability that this imparts to the crystallite far outweighs the limited loss of energy caused by chain flexing.

Secondary bonds can therefore lead to a stimulation of the crystallization process in the appropriate polymers.

Tacticity. Chain symmetry and flexibility both affect the crystallinity of a polymer sample. If a chain possesses large pendant groups, these will increase the rigidity but

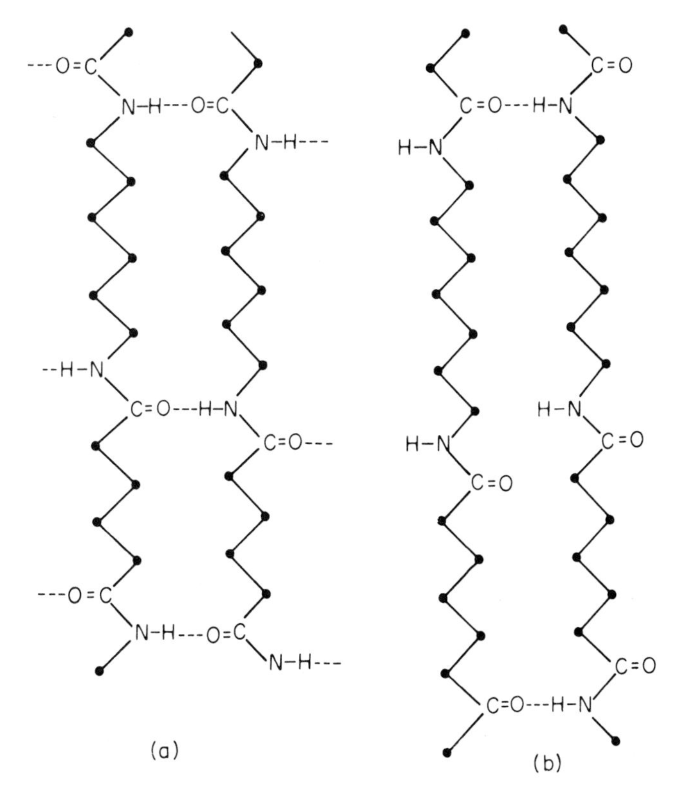

FIGURE 11.4. Extended zig-zag structures for (a) nylon-6,6 and (b) nylon-7,7 showing the allowed hydrogen bonding.

also increase the difficulty of close packing to form a crystalline array. This latter problem can be overcome if the groups are arranged in a regular fashion along the chain. Isotactic polymers tend to form helices to accommodate the substituents in the most stable steric positions; these helices are regular forms capable of regular alignment. Thus atactic polystyrene is amorphous but isotactic polystyrene is semi-crystalline ($T_m = 513$ K).

Syndiotactic polymers are also sufficiently regular to crystallize, but not necessarily as a helix, rather in glide planes.

Branching in the side group tends to stiffen the chain and raise T_m, as shown in the series poly(but-1-ene), $T_m = 399$ K; poly(3-methyl but-1-ene), $T_m = 418$ K; poly(3,3′-dimethyl but-1-ene), $T_m > 593$ K. If the side group is flexible and non-polar, T_m is lowered.

Branching and molar mass. If the chain is substantially branched, the packing efficiency deteriorates and the crystalline content is lowered. Polyethylene provides a good example of this (figure 11.2) where extensive branching lowers the density and T_m of the polymer.

Molar mass can also alter T_m. Chain ends are relatively free to move and if the number of chain ends is increased by reducing the molar mass, then T_m is lowered because of the decrease in energy required to stimulate chain motion and melting. For example, polypropylene, with $M = 2000\,\text{g mol}^{-1}$, has $T_m = 387\,\text{K}$, whereas a sample with $M = 30\,000\,\text{g mol}^{-1}$, has $T_m = 443\,\text{K}$.

11.7 Morphology and kinetics

Having once established that certain polymeric materials are capable of crystallizing, fundamental studies are directed along two main channels of interest centred on (a) the mode and kinetics of crystallization, and (b) the morphology of the sample on completion of the process.

Although the morphology depends largely on the crystallizing conditions, we shall consider the macro- and microscopic structure first before dealing with the kinetics of formation.

11.8 Morphology

A number of distinct morphological units have been identified during the crystallization of polymers from the melt, which have helped to clarify the mechanism. We shall now discuss the ordered forms which have been identified.

Crystallites. In an X-ray pattern produced by a semicrystalline polymer, the discrete maxima observed arise from the scattering by small regions of three-dimensional order, which are called crystallites. They are formed in the melt by diffusion of molecules, or sections of molecules, into close packed ordered arrays; these then crystallize. The sizes of these crystallites are small relative to the length of a fully extended polymer chain, but they are also found to be independent of the molar mass and rarely exceed 1 to 100 nm. As a result, various portions of one chain may become incorporated in more than one crystallite during growth, thereby imposing a strain on the polymer which retards the process of crystallite formation. This will also introduce imperfections in the crystallites which continue growing until the strains imposed by the surrounding crystallites eventually stop further enlargement. Thus a matrix of ordered regions with disordered interfacial areas is formed, but, unlike materials with small molar masses, the ordered and disordered regions are not discrete entities and cannot be separated by differential solution techniques unless the solvent causes selective degradation of the primary bonds in the amorphous regions.

Crystallites of cellulose have been isolated from wood pulp in this way by treatment with acid to hydrolyse and remove the amorphous regions. Typical dimensions of the remaining crystallites were 46 nm long by 7.3 nm wide corresponding to bundles of about 100 to 150 chains in each crystallite.

The first attempts to explain the crystalline structure of a polymer sample produced a model called the fringe-micelle structure. The chain was envisaged as meandering throughout the system, entering and leaving several ordered regions along its length. The whole structure was thus made up of crystalline regions imbedded randomly in a continuous amorphous matrix. This model has now been virtually discarded in the light of more recent research which has revealed features incompatible with this picture.

Single crystals. When a polymer is crystallized from the melt, imperfect polycrystalline aggregations are formed in association with a substantial amorphous content. This is a consequence of chain entanglement and the high viscosity of the melt combining to hinder the diffusion of chains into the ordered arrays necessary for crystallite formation.

If these restrictions to free movement are reduced and a polymer is allowed to crystallize from a dilute solution, it is possible to obtain well-defined single crystals. By working with solutions in which the amount of polymer is considerably less than 0.1 per cent the chance of a chain being incorporated in more than one crystal is greatly reduced, thereby increasing the possibility of isolated single crystals being formed.

These crystals are usually very small, but they have been detected for a range of polymers including polyesters, polyamides, polyethylene, cellulose acetate, and poly(4-methyl pentene-1). Although small, these single crystals can be studied using an electron microscope. This reveals that they are made up of thin lamellae, often lozenge shaped, sometimes oval, about 10 to 20 nm thick, depending on the temperature of crystallization. The most surprising feature of these lamellae is that while the molecular chains may be as long as 1000 nm, the direction of the chain axis is across the thickness of the platelet. This means that the chain must be folded many times like a concertina to be accommodated in the crystal.

For a polymer such as polyethylene, the fold in the chain is completed using only 3 or 4 monomer units with bonds in the *gauche* conformation. The extended portions in between have about 40 monomers units all in the *trans* conformation.

The crystals, thus formed, have a hollow pyramid shape, because of the requirement that the chain folding must involve a staggering of the chains if the most efficient packing is to be achieved. There is also a remarkable constancy of lamellar thickness, but this increases as the temperature increases. While opinions vary between kinetic and thermodynamic reasons for this constancy of fold distance, it is suggested that the fold structure allows the maximum amount of crystallization of the molecule at a length which produces a free energy minimum in the crystal. One suggestion is that the folding maintains the appropriate kinetic unit of the chain at any given temperature; as this would be expected to lengthen with increasing temperature, it would account for the observed thickening of the lamellae.

Hedrites. If the concentration of the polymer solution is increased a crystalline polyhedral structure emerges composed of lamellae joined together along a common plane. These have also been detected growing from a melt which suggests that lamellar growth can take place in the melt and may be a sub-unit of the spherulite.

Crystallization from the melt. Whereas crystallization from dilute solutions may result in the formation of single polymer crystals, this perfection is not achieved when dealing with polymers cooled from the melt. The basic characteristic feature is still the lamellar-like crystallite with amorphous surfaces or interfaces, but the way these are formed may be different based on the careful investigation of melt crystallized polymers using neutron scattering techniques. The two models that have been proposed to describe the fine structure of these lamellae and their surface characteristics in semicrystalline polymers, differ mainly in the way the chains are thought to enter and leave the ordered lamellae regions. These are:

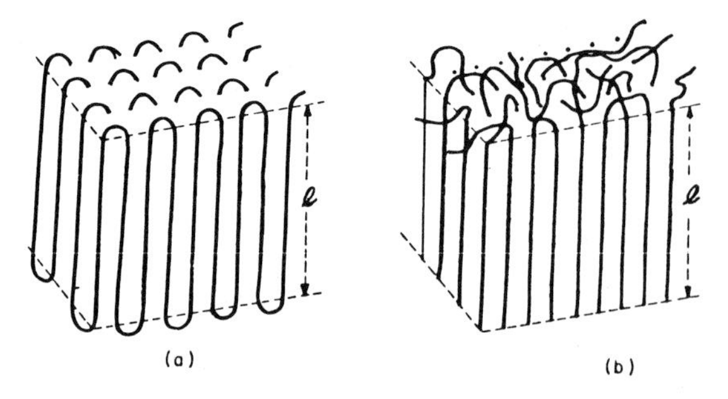

FIGURE 11.5. Schematics of possible chain morphology in a single polymer crystal. (a) regular folding with adjacent re-entry of chains; (b) "switchboard" model with random re-entry of chains.

(a) the *regular folded array* with adjacent re-entry of the chains, but with some loose folding and emergent chain ends or cilia that contribute to the disordered surface, or
(b) the *switchboard model*, where there is some folding of the chains but re-entry is now quite random.

Both are represented schematically in figure 11.5 but the exact nature of the structure has been the subject of considerable controversy. While the morphology of the single crystals grown from dilute solutions may be more regular and resemble the first model, for polymers that are crystallized from the melt (and this is by far the more important procedure technically) the mass of evidence tends to favour a form of the switchboard model.

Measurements of the densities of several semicrystalline polymers points to the fact that a significant fraction of the chain units are in a non crystalline environment. This is not consistent with the regular folded form of the crystallites where the amorphous part is associated only with loose folding of the chains and cilia. Even more persuasive are small angle neutron scattering studies. These have demonstrated that the radii of gyration of several semicrystalline polymers remain essentially unchanged on moving from the melt phase to the semicrystalline phase (table 11.2).

TABLE 11.2. Comparison between the radii of gyration expressed as $(\langle S^2\rangle/M_w)^{1/2}$ in the melt and in the semicrystalline state

Polymer	$M_w \cdot 10^{-3}$	$\left[\dfrac{\langle S^2\rangle}{M_w}\right]^{1/2}$	(Å(g/Mol)]$^{1/2}$
		crystallized	*melt*
Polyethylene, quenched	140	0.46 ± 0.05 0.45	0.46 ± 0.05 0.45
Polyethylene oxide, crystallized by slowly cooling	150	0.52	0.46
Isotactic polypropylene, isothermally crystallized	340	0.37	0.34

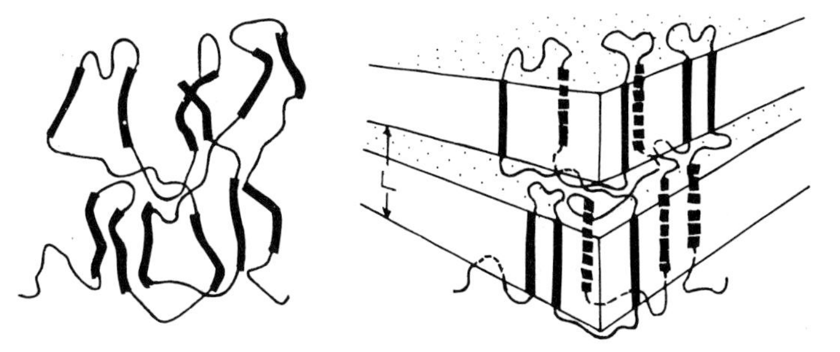

FIGURE 11.6. Solidification model illustrating crystallization from the melt. (Adapted from M. Dettenmaier, E. W. Fischer and M. Stamm (1980) with permission from Dr Dietrich Steinkopff Verlag, Darmstadt.)

This means that there is no significant reordering of the chain conformation when crystallization takes place after cooling from the melt, which would be required if a regularly folded chain structure was to be constructed in the lamellae. To explain these observations Fischer has proposed the *solidification model* in which crystallization is believed to take place by the straightening of sections of the polymer coil followed by alignment of these sequences in regular arrays forming the lamellar structure. This precludes the need for the extensive, long range, diffusion of the chain through a highly viscous medium that would be necessary if a regular chain folded structure was to be constructed. The process is shown schematically in figure 11.6 and the resulting structure is a variation of the switchboard model. This hypothesis can account for the fact that on cooling, rapid crystal growth is seen to occur which is inconsistent with the need for long range diffusion if the regularly folding lamellae were forming. The solidification model shows that the chains can be incorporated into the basic lamellar form with the minimum amount of movement and that there will be extensive meandering of chains between the lamellae forming the interfacial amorphous regions.

Spherulites. Examination of thin sections of semicrystalline polymers reveals that the crystallites themselves are not arranged randomly, but form regular birefringent structures with circular symmetry. These structures, which exhibit a characteristic Maltese cross optical extinction pattern, are called spherulites. While spherulites are characteristic of crystalline polymers, they have also been observed to form in low molar mass compounds which are crystallized from highly viscous media.

Each spherulite grows radially from a nucleus formed either by the density fluctuations which result in the initial chain ordering process or from an impurity in the system. As the structure is not a single crystal, the sizes found vary from somewhat greater than a crystallite to diameters of a few millimetres. The number, size, and fine structure depend on the temperature of crystallization, which determines the critical size of the nucleating centre. This means that large fibrous structures form near T_m, whereas greater numbers of small spherulites grow at lower temperatures. When the nucleation density is high, the spherical symmetry tends to be lost as the spherulite edges impinge on their neighbours to form a mass such as shown in figure 11.7.

A study of the fine structure of a spherulite shows that it is built up of fibrous

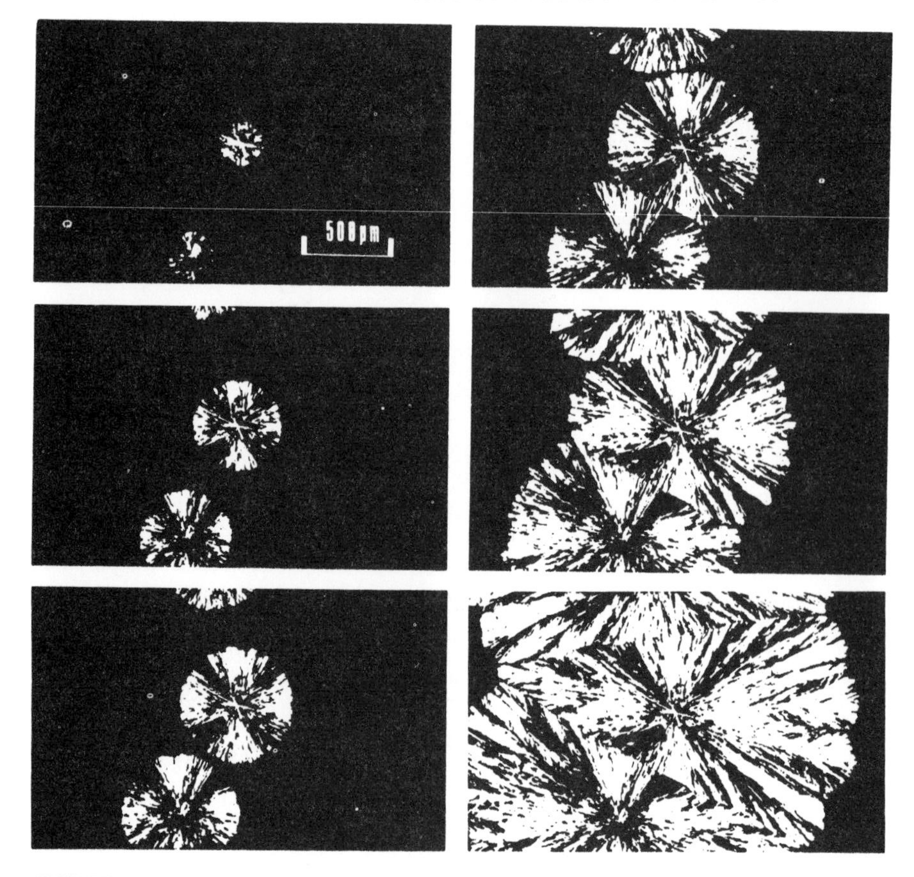

FIGURE 11.7. Sequence of photographs taken, under polarized light, over a period of about 1 min, showing the growth of poly(ethylene oxide) spherulites from the melt. From top left to bottom right; initially discrete spherulites with spherical symmetry are observed but the growing fronts eventually impinge on one another to form an irregular matrix. (Photographer: R. B. Stewart.)

sub-units, growth takes place by the formation of fibrils which spread outwards from the nucleus in bundles, into the surrounding amorphous phase. As this fibrillar growth advances, branching takes place, and at some intermediate stage in the development, the spherulite often resembles a sheaf of grain. This forms as the fibrils fan out and begin to create the spherical outline. Although the fibrils are arranged radially, the molecular chains lie at right angles to the fibril axis. This has led to the suggestion that the fine structure is created from a series of lamellar crystals winding helically along the spherulite radius. Growth proceeds from a small crystal nucleus which develops into a fibril. Low branching and twisting then produces bundles of diverging and spreading fibrils which eventually fill out into the characteristic spherical structure. In between the branches of the fibrils are amorphous areas and these, along with the amorphous interfacial regions between the lamellae, make up the disordered content of the semi-crystalline polymer (figure 11.8).

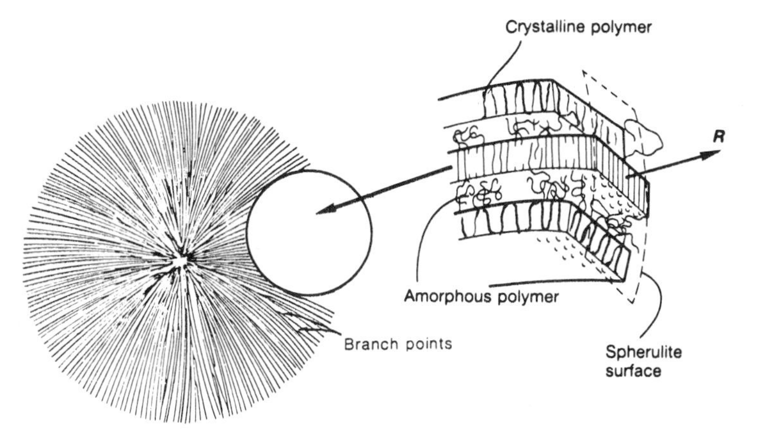

FIGURE 11.8. Fully-developed spherulite grown from the melt, comprising chain folded lamellae (magnified section) and branching points which help to inpart a spherical shape to the structure. Most rapid growth occurs in the direction of the spherulite radius R. (Adapted from McCrum *et al.* (1988) with permission from Oxford University Press.)

Spherulites are classified as positive when the refractive index of the polymer chain is greater across the chain than along the axis, and negative when the greater refractive index is in the axial direction. They also show various other features such as zig-zag patterns, concentric rings, and dendritic structures.

11.9 Kinetics of crystallization

The crystalline content of a polymer has a profound effect on its properties and it is important to know how the rate of crystallization will vary with the temperature, especially during the processing and manufacturing of polymeric articles. The chemical structure of the polymer is also an important feature in the crystallization; for example, polyethylene crystallizes readily and cannot be quenched rapidly enough to give a largely amorphous sample whereas this is readily accomplished for isotactic polystyrene. However, this aspect will be discussed more fully later.

Isothermal crystallization. Two main factors influence the rate of crystallization at any given temperature: (i) the rate of nucleation; and (ii) the subsequent rate of growth of these nuclei to macroscopic dimensions.

The kinetic treatment of crystallization from the melt is based on the radial growth of a front through space and can be likened to someone scattering a handful of gravel onto the surface of a pond. Each stone is a nucleus which, when it strikes the surface, generates expanding circles (similar to spherulites in two dimensions). These grow unimpeded for a while but the leading edges eventually collide with others and growth rates are altered. When a similar picture is adopted for the crystallization of a polymer certain basic assumptions are made first.

The formation of ordered growth centres by the alignment of chains from the melt is called *spontaneous nucleation.* When the temperature of crystallization is close to the melting temperature, nucleation is sporadic and only a few large spherulites will grow. At lower temperatures, nucleation is rapid and a large number of small spherulites are formed. The growth of the spherulites may occur in one, two, or three dimensions and the rate of radial growth is taken to be linear at any temperature. Finally the density ρ_c of the crystalline phase is considered to be uniform throughout but different from that of the melt ρ_L. A kinetic treatment has been developed taking account of these points.

The Avrami equation. The kinetic approach relies on the establishment of a relation between the density of the crystalline and melt phases and the time. This provides a measure of the overall crystallization rate. It is assumed that the spherulites grow from nuclei whose relative positions in the melt remain unaltered, and the analysis allows for the eventual impingement of the growing discs on one another. The final relation describing the process is known as the Avrami equation expressed as

$$w_L/w_0 = \exp(-kt^n), \tag{11.3}$$

where k is the rate constant, w_0 and w_L are the masses of the melt at zero time and that left after time t. The exponent n is the Avrami exponent and is an integer which can provide information on the geometric form of the growth.

Sporadic nucleation is assumed to be a first-order mechanism and if we consider that a two-dimensional disc is formed, then $n = 2 + 1 = 3$. Rapid nucleation is a zeroth-order process in which the growth centres are formed at the same time, and for each growth unit listed in table 11.3, the corresponding values of the exponent would be $(n - 1)$. Thus the Avrami exponent is the sum of the order of the rate process and the number of dimensions the morphological unit possesses.

Dilatometry. As crystallization involves the close packing of chains in regular three-dimensional structures, the economical use of space is accompanied by an increase in density. Thus the rate of crystallization can be followed by recording the density changes which are readily detected in a dilatometer. This is achieved by placing the polymer in a dilatometer with a confining liquid, such as mercury, so that any volume change can be recorded as a movement of the liquid meniscus in a capillary. A typical design is shown in figure 11.9.

The polymer is introduced into the dilatometer between the point A and the capillary. The apparatus is then pumped out and sealed under vacuum at the point A. Sufficient mercury is then added to enclose the polymer and extend into the capillary, after which the tube is sealed at B, and placed in a thermostat at a temperature somewhat higher than the melting temperature of the polymer. When

TABLE 11.3. Relation between the Avrami exponent and the morphological unit formed for sporadic nucleation

Growth unit	*Nucleation*	*Avrami exponent n*
Fibril	sporadic	2
Disc	sporadic	3
Spherulite	sporadic	4
Sheaf	sporadic	6

the sample is completely molten the dilatometer is transferred to a second thermostat set at the temperature selected for crystallization to take place and allowed to equilibrate. The initial period of temperature adjustment to the second temperature may make the initial height h_0 rather difficult to locate, but usually a plot such as shown in figure 10.9(b) is recorded. If secondary crystallization takes place the final portion of the curve may tail away making h_∞ more difficulty to measure.

The mass fraction of the uncrystallized polymer (w_L/w_0) can be related to the volume changes and to the heights measured in the dilatometer by

$$w_L/w_0 = (V_t - V_\infty)/(V_0 - V_\infty) = (h_t - h_\infty)/(h_0 - h_\infty) = \exp(-kt^n), \tag{11.4}$$

where h_t, h_0, and h_∞ are the heights at time t, the beginning, and the end of the process respectively, with V_t, V_0 and V_∞ the corresponding volumes. The slope of a plot of $-\ln\{(h_t - h_\infty)/(h_0 - h_\infty)\}$ against t allows evaluation of the Avrami exponent n while k can be calculated from the intercept.

Deviations from Avrami equation. The Avrami equation can describe some but not all systems investigated. The crystallization isotherms of poly(ethylene terephthalate) can be fitted by equation (11.3) using $n = 4$ above 473 K and $n = 2$ at 383 K. The equation should be used with caution, however, as noninteger values have been reported and the geometric shape of the morphological unit is not always that predicted by the value of n calculated from the experimental data.

Secondary crystallization. Deviations from the Avrami treatment may also be observed towards the end of the crystallization process and values of h_∞ are often

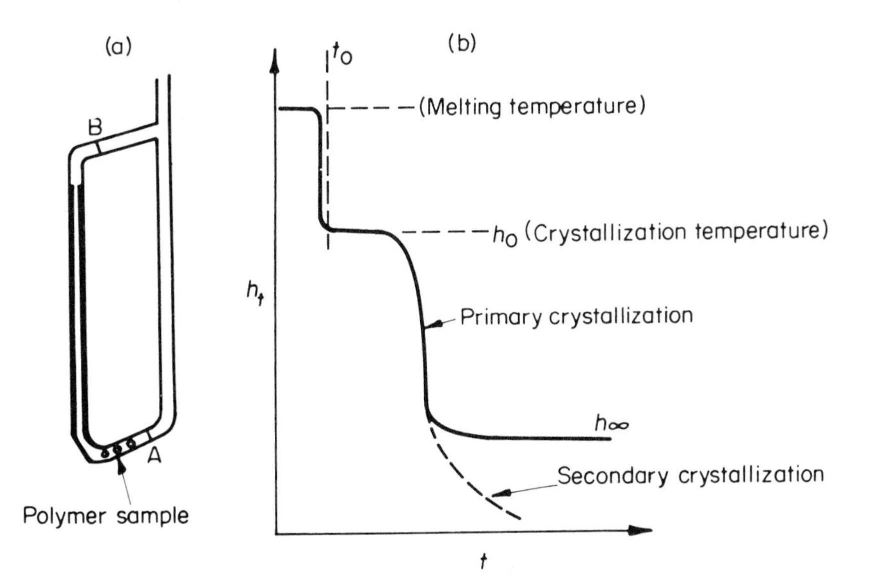

FIGURE 11.9. (a) Typical dilatometer for following crystallization kinetics and (b) general shape of a plot of dilatometer height h_t as a function of time t obtained in crystallization experiments.

difficult to determine accurately, as shown in the curve derived from dilatometric data. The tailing of the curve is a result of a secondary crystallization process which is a slower reorganization of the crystalline regions to produce more perfectly formed crystallites.

General reading

G. Allen and J. C. Bevington, Eds, *Comprehensive Polymer Science*. Vol. 2. Pergamon Press (1989).

D. C. Bassett, *Principles of Polymer Morphology*. Cambridge University Press (1981).

P. Geil, *Polymer Single Crystals*. Interscience Publishers Inc. (1963).

M. Gordon, *High Polymers*. Iliffe Books (1963).

I. H. Hall, Ed., *Structure of Crystalline Polymers*. Elsevier (1984).

L. Mandelkern, *Crystallization of Polymers*. McGraw-Hill (1964).

N. B. McCrum, C. P. Buckley and C. B. Bucknall, *Principles of Polymer Engineering*. Oxford University Press (1988).

D. M. Sadler, in *Static and Dynamic Properties of the Polymeric Solid State*. Eds, R. A. Pethrick and R. W. Richards. D. Reidel Publishing Co. (1982).

A. Sharples, *Introduction to Polymer Crystallization*. Edward Arnold (1966).

A. V. Tobolsky and H. Mark, *Polymer Science and Materials*, Chapter 8. Wiley-Interscience (1971).

References

1. M. Dettenmaier, E. W. Fischer and M. Stamm, *Colloid and Polymer Science*, **228**, 343 (1980).
2. L. Mandelkern, *Rubber Chem. Technol.*, **32**, 1392 (1949).
3. L. Marker, R. Early and S. L. Aggarwal, *J. Polym. Sci.*, **38**, 369 (1959).

CHAPTER 12

The Amorphous State

12.1 Molecular motion

A linear polymer chain can be treated as a "one-dimensional co-operative system" in which the rotation of a chain segment is restricted or aided by the neighbouring segments. For long chains, co-operative motion cannot be expected to extend along the entire length, and the polymer tends to act as if it were composed of a series of interconnected, but independent, kinetic units. Any significant movement of such a chain is generated by rotation about the single bonds connecting the atoms in the chain, and depends on the ease of interchange of any element from one rotational state to another. The height of the potential energy barrier ΔE (c.f. figure 1.3) will determine the rapidity of conformational change at any temperature, and when the temperature of the polymer increases, the additional thermal energy allows ΔE to be overcome more often. This encourages increasing molecular motion until eventually the polymer behaves like a viscous liquid (assuming that no thermal degradation takes place).

In the amorphous state the distribution of polymer chains in the matrix is completely random, with none of the strictures imposed by the ordering encountered in the crystallites of partially crystalline polymers. This allows the onset of molecular motion in amorphous polymers to take place at temperatures below the melting temperature of such crystallites. Consequently, as the molecular motion in an amorphous polymer increases, the sample passes from a glass, through a rubber-like state, until finally it becomes molten. These transitions lead to changes in the physical properties and material application of a polymer, and it is important to examine physical changes wrought in an amorphous polymer as a result of variations in the molecular motion.

12.2 The five regions of viscoelastic behaviour

The physical nature of an amorphous polymer is related to the extent of the molecular motion in the sample, which in turn is governed by the chain flexibility and the temperature of the system. Examination of the mechanical behaviour shows that there are five distinguishable states in which a linear amorphous polymer can exist and these are readily displayed if a parameter such as the elastic modulus is measured over a range of temperatures.

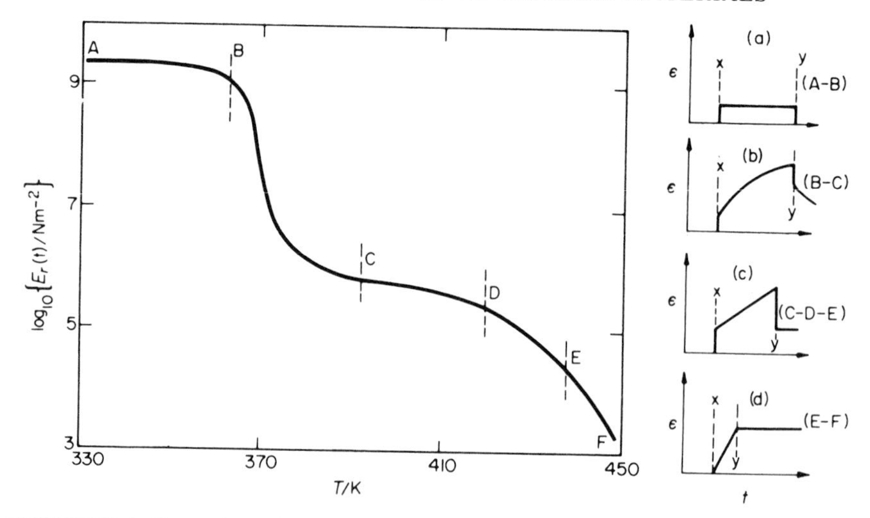

FIGURE 12.1. Five regions of viscoelasticity, illustrated using a polystyrene sample. Also shown are the strain–time curves for stress applied at x and removed at y: (a) glassy region; (b) leathery state; (c) rubbery state; and (d) viscous state.

The general behaviour of a polymer can be typified by results obtained for an amorphous atactic polystyrene sample. The relaxation modulus E_r was measured at a standard time interval of 10 s and $\log_{10}E_r$ is shown as a function of temperature in figure 12.1. Five distinct regions can be identified on this curve.

(i) *The glassy state.* This is section A to B lying below 363 K and it is characterized by a modulus between $10^{9.5}$ and $10^9\,\text{N m}^{-2}$. Here co-operative molecular motion along the chain is frozen, causing the material to respond like an elastic solid to a stress, and the strain–time curve is of the form shown in figure 12.1(a).

(ii) *Leathery or retarded highly elastic state.* This is the transition region B to C where the modulus drops sharply from about 10^9 to about $10^{5.7}\,\text{N m}^{-2}$ over the temperature range 363 to 393 K. The glass transition temperature T_g is located in this area and the rapid change in modulus reflects the constant increase in molecular motion as the temperature rises from T_g to about ($T_g + 30$ K). Just above T_g the movement of the chain segments is still rather slow, imparting what can best be described as leathery properties to the material. The strain–time curve is that shown in figure 12.1(b).

(iii) *The rubbery state.* At approximately 30 K above the glass transition the modulus curve begins to flatten out into the plateau region C to D in the modulus interval $10^{5.7}$ to $10^{5.4}\,\text{N m}^{-2}$ and extends up to about 420 K.

(iv) *Rubbery flow.* After the rubbery plateau the modulus again decreases from $10^{5.4}$ to $10^{4.5}\,\text{N m}^{-2}$ in the section D to E. The effect of applied stress to a polymer in states (iii) and (iv) is shown in figure 12.1(c) where there is instantaneous elastic response followed by a region of flow.

(v) *Viscous state.* Above a temperature of 450 K, in the section E to F, there is little evidence of any elastic recovery in the polymer and all the characteristics of a viscous liquid become evident (figure 12.1(d)). Here there is a steady decrease of the modulus from $10^{4.5}$ N m^{-2} as the temperature increases.

The overall shape of the curve shown in figure 12.1 is typical for linear amorphous polymers in general, although the temperatures quoted are specific to polystyrene and will differ for other polymers. Variations in shape are found for different molar masses and when the sample is crosslinked or partly crystalline. The value of the modulus provides a good indication of the state of the polymer and can be obtained from the curve.

12.3 The viscous region

Before considering the flow in polymer melts, the viscous behaviour of simple liquids will be examined.

The application of a force to a simple liquid of low molar mass is relieved by the flow of molecules past one another into new positions in the system. A liquid, forced to flow in this way by a shearing force σ, experiences a viscous resistance expressed by

$$\eta = \sigma(\mathrm{d}v/\mathrm{d}x)^{-1}, \tag{12.1}$$

where v is the velocity of flow along a tube of radius x, so that $(\mathrm{d}v/\mathrm{d}x)$ is the velocity gradient or shear rate $\dot{\gamma}$, and η is the viscosity coefficient of the liquid. A liquid is said to exhibit Newtonian flow if η is independent of $\dot{\gamma}$ but substances which show deviations from this flow pattern, with either decreasing or increasing $(\sigma/\dot{\gamma})$ ratios, are termed non-Newtonian. (See figure 12.2.) Most polymers fall into this latter category, with η decreasing as the shear rate increases.

The temperature dependence of η can normally be expressed in the form

$$\eta = A \exp(\Delta E_{\mathrm{D}}/RT), \tag{12.2}$$

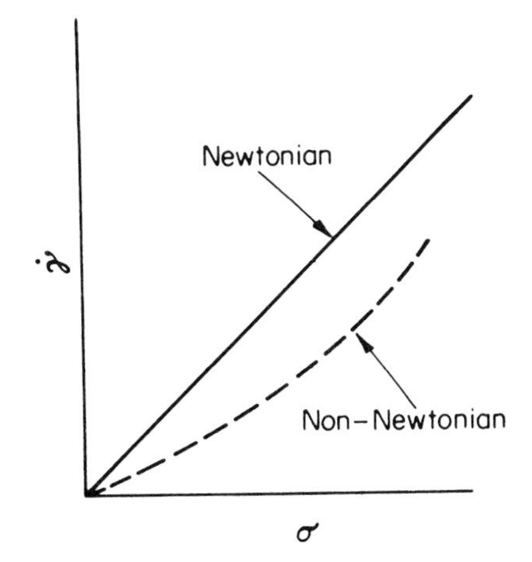

FIGURE 12.2. Newtonian and non-Newtonian flow curves.

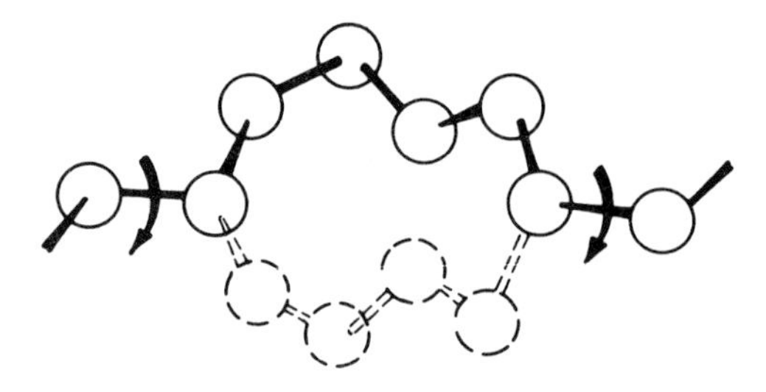

FIGURE 12.3. Crankshaft motion in a polymer chain.

where A is a constant and ΔE_D represents the activation energy required to create a hole big enough for a molecule to translate or "jump" into during flow. In liquids with larger or irregularly shaped molecules, the deformation is slower as the molecules restrict the easy translation of one past the other. This results in a high value of η.

12.4 Kinetic units in polymer chains

Resistance to flow in polymer systems is even greater, because now the molecules are covalently bonded into long chains which are coiled and entangled and translational motion must, of necessity, be a co-operative process. It would be unreasonable to expect easy co-operative motion along the entire polymer chain, but as there is normally some degree of flexibility in the chain, local segmental motion can take place more readily. The polymer can then be considered as a series of kinetic units; each of these moves in an independent manner and involves the co-operative movement of a number of consecutive chain atoms.

Crankshaft motion. If we now consider an arbitrary kinetic unit which involves the movement of six atoms by rotation about two chain bonds, the movement can be visualized as shown diagrammatically in figure 12.3. The amorphous or molten polymer is a conglomeration of badly packed interlacing chains and the extra empty space caused by this random molecular arrangement is called the *free volume* which essentially consists of all the holes in the matrix. When sufficient thermal energy is present in the system the vibrations can cause a segment to jump into a hole by co-operative bond rotation and a series of such jumps will enable the complete polymer chain eventually to change its position. Heating will cause a polymer sample to expand thereby creating more room for movement of each kinetic unit and the application of a stress in a particular direction will encourage flow by segmental motion in the direction of the stress. The segmental transposition involving six carbon atoms is called crankshaft motion and is believed to require an activation energy of about $25\,\mathrm{kJ\,mol^{-1}}$.

12.5 Effect of chain length

Although it is thought that translation of a polymer chain proceeds by means of a series of segmental jumps involving short kinetic units, which may each consist of between 15 and 30 chain atoms, the complete movement of a chain cannot remain unaffected by the surrounding chains. As stated previously, considerable entanglement exists in the melt and any motion will be retarded by other chains.

According to Bueche the polymer molecule may drag along several others during flow and the energy dissipation is then a combination of the friction between the chain plus those which are entangled and the neighbouring chains as they slip past each other. It would seem reasonable to assume from this, that the length of the chains in the sample must play a significant role in determining the resistance to flow and the effect of chain length on $\log \eta$, measured at low shear rates to ensure Newtonian flow, is illustrated in figure 12.4. The plot comprises two linear portions meeting at a critical chain length Z_c. Above Z_c the relation describing the flow behaviour is

$$\log \eta = 3.4 \log Z + \log K_2, \tag{12.3}$$

and η is proportional to the 3.4 power of Z. Below Z_c, η is directly proportional to Z and the expression becomes

$$\log \eta = \log Z + \log K_1, \tag{12.4}$$

where K_1 and K_2 are temperature dependent constants.

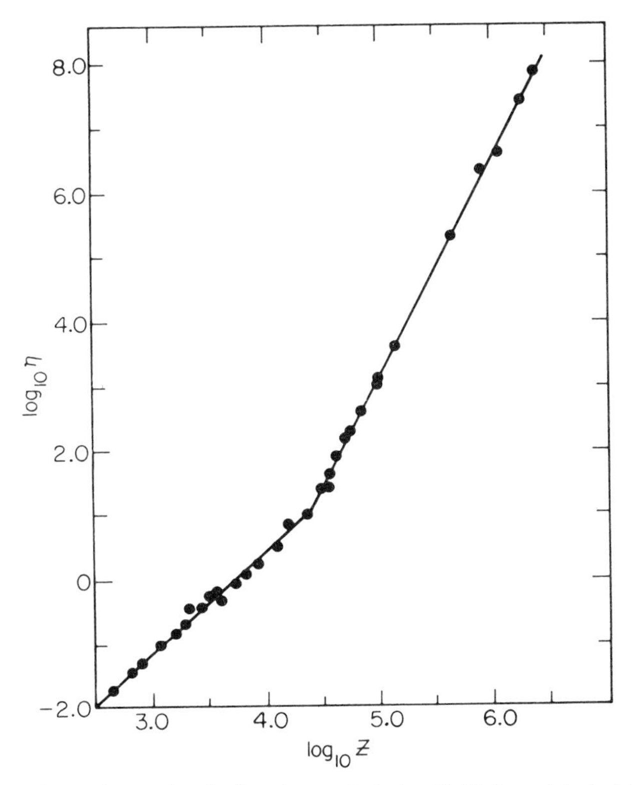

FIGURE 12.4. Dependence of melt viscosity on chain length Z, for polyisobutylene fractions measured at low shear rates and at 490 K. (Data by Fox and Flory, 1951.)

The critical chain length Z_c is interpreted as representing the dividing point between chains which are too short to provide a significant contribution to η from entanglement effects and those large enough to cause retardation of flow by intertwining with their neighbours. If Z is defined as the number of atoms in the backbone chain of a polymer then typical values for Z_c are 610 for polyisobutylene, 730 for polystyrene, and 208 for poly(methyl methacrylate). In general Z_c is lower for polar polymers than for non-polar polymers.

12.6 The reptation model

The theory proposed by Bueche tends to suggest a very clear-cut distinction between the movement of chains of length less than Z_c and the relative immobility of the entangled chains with lengths greater than Z_c. As independent chain mobility cannot be discounted for these longer chains after the onset of entanglement, a modified model is required to account for the ability of long chains to translate and diffuse through the polymer matrix, *i.e.* the entanglement network must be considered as being transient. Such a concept is embodied in the "reptation" model proposed by de Gennes. In this approach the chain is assumed to be contained in a hypothetical tube which is placed initially in a three dimensional network formed from the other entangled chains. Although for simplicity, these network "knots" are regarded as a set of fixed obstacles round which the isolated chain under consideration must wriggle during translation, in practice the network "knots" would also be in motion. The contours of the tube are then defined by the position of the entanglement points in the network.

Two types of chain motion can be envisaged, a conformational change taking place within the confines of the tube, and more importantly, reptation. The latter is imagined to be a snake-like movement that translates the chain through the tube and allows it to escape at the tube ends. Mechanistically it can be regarded as the movement of a kink in the chain along its length (see figure 12.5) until this reaches the end of the

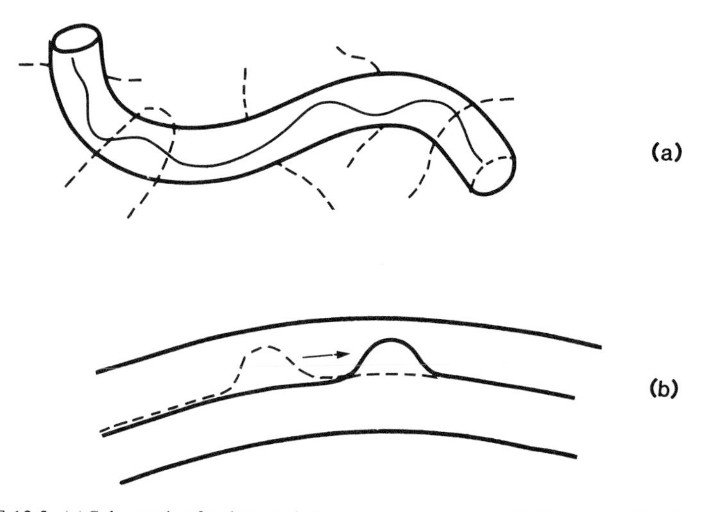

FIGURE 12.5. (a) Schematic of polymer chain restrained in a hypothetical tube. (b) Movement of a "kink" along the chain.

chain and leaves it. Motion of this kind translates the chain through the tube, like a snake moving through grass, and successive defects moving the chain in this way will eventually carry it completely out of the hypothetical tube.

The motion can be characterized by a reptation time, or more accurately by a relaxation time, τ, that is a measure of the time required for a chain to escape completely from its tube. If the tube is defined as having the same length as the unperturbed chain, nl_0, where l_0 is the bond length under θ conditions (corrected for short range interactions), then the time required for the chain to reptate out of the tube is proportional to the square of the distance travelled, *i.e.*

$$\tau = \frac{(nl_0)^2}{2D_t} \tag{12.5}$$

Here D_t is the diffusion constant within the tube, and is distinguished from translation outside the tube which will be slower and more difficult. This can be expressed as the frictional coefficient for the chain, again within the tube confines ($D_t = kT/f_t$). However, because the reptation is assumed to occur by migration of a segmental kink along the chain, the force needed to do this is applied one segment at a time and so it is more appropriate to use the frictional factor per segment ζ. Thus

$$\tau = (nl_0)^2(n\zeta/2kT) \tag{12.6a}$$

or

$$\tau = \left(\frac{l_0^2\zeta}{2kT}\right)n^3 = \tau_0 n^3 \tag{12.6b}$$

Equation (12.6b) shows that the relaxation time is proportional to the cube of the chain length. This is the fundamental result of the reptation model. The cube dependence is not a precise match with the 3.4 exponent obtained from viscosity measurements of long chains, but it is acceptable, particularly as the model gives a satisfactory picture of how a polymer chain can overcome the restraining influence of entanglements and move within the matrix.

Typically τ_0 is of the order of 10^{-10} seconds for $n = 1$ and so the relaxation time τ for a polymer chain with $n = 10^4$ would be about 100 seconds.

Reptation theory has been developed further by Doi and Edwards and is being applied to both viscoelastic and solution behaviour. It has been shown that for a chain moving in the melt, over time-scales that greatly exceed the lifetime of the tube τ, a reptation self-diffusion coefficient D_{rept} can be measured which is inversely proportional to n^2, *i.e.* the diffusion law is

$$D_{\text{rept}} \propto 1/n^2 \tag{12.7}$$

This law holds for the "welding" of polymers at an interface which can be explained by reptation. When two blocks of the same polymer are brought together and held at a temperature just above the T_g for a time t, interdiffusion of the chains takes place from each block across the interface (see figure 12.6) thereby joining the blocks together. The strength of the junction formed will depend on t which should be

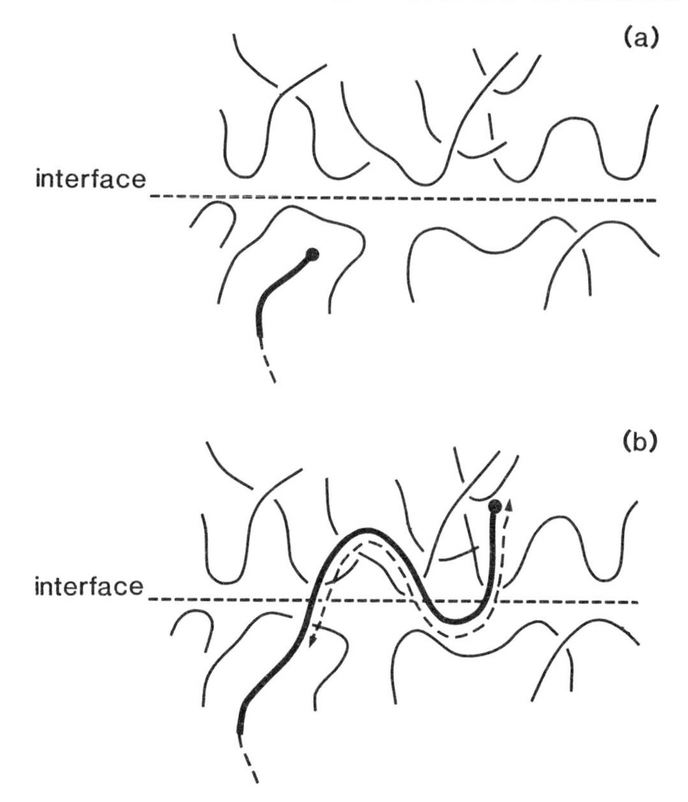

FIGURE 12.6. Schematic of chain movement across an interface: (a) first contact and (b) after having been in contact for some time, with a chain from one surface having *reptated* across into the interface of the adjoining polymer block.

smaller than the reptation time τ, *i.e.* the mixed layer ought to be smaller than the size of the coil if an interfacial link is to be formed.

The situation changes if the blocks are composed of two different polymers which, as a pair, can form a miscible blend. Although welding can again take place, the diffusion law (equation 12.7) is now altered. It has been found that if a block of poly(vinylchloride) is brought in contact with a block of polycaprolactone, at temperatures above T_g, then D_{rept} is higher than expected and is proportional to $(1/n)$. This has been interpreted as being a consequence of the negative enthalpy of mixing in the system which acts as an additional driving force for the chains on either side of the boundary to cross into the other matrix. This driving force will be proportional to the number of monomers in a chain, hence the change in the diffusion law.

Reptation theory can also be applied to polymer dissolution processes.

12.7 Temperature dependence of η

When a polymer is transformed into a melt without degradation and is stable at even higher temperatures, η is observed to decrease rapidly as the temperature increases. If it is still stable at temperatures in excess of 100 K above T_g, the temperature dependence has an exponential form

$$\eta = B \exp(\Delta H/RT), \tag{12.8}$$

where according to the Eyring rate theory ΔH is the activation enthalpy of viscous flow and is a more representative parameter than the energy. Values of ΔH vary slowly over a range from 20 to 120 kJ mol^{-1}. When the temperature is lowered towards T_g, ΔH changes dramatically and a simple equation such as (12.8) is no longer valid. The increase in ΔH, observed with temperature lowering, can be equated with a rapid loss of free volume as T_g is approached. Hence ΔH becomes dependent on the availability of a suitable hole for a segment to move into, rather than being representative of the potential energy barrier to rotation. This approach suggests that the jump frequency decreases when there is an increasing co-operative motion among the chains needed to produce holes.

12.8 Rubbery state

With a decrease in temperature, the flow of a polymer melt becomes increasingly sluggish as the chain motion becomes too slow to effect complete untangling of the polymer coils. The viscosity increases rapidly to a value of about 10^{12} Pa s as T_g is approached, but on passing from the melt to the glass a region of rubbery flow and elasticity is traversed. In this state the polymer exhibits several unique properties which are dealt with in chapter 14 and only a brief description of the chain behaviour in this region is given here.

Long range elasticity. The rubber-like region, which lies above T_g, appears when the rotation about the segment links is free enough to enable the chains to assume any of the immense number of equi-energetic conformations available, without significant chain untangling taking place. The majority of these shapes will be compact coils because the possibility of their occurrence is much greater than for the more extended forms.

When a polymer, which is not too crystalline and has a reasonably high molar mass ($> 20\,000$ g mol^{-1}), is in this elastic state it will elongate quite readily in the direction of an applied stress, *e.g.* natural rubber will stretch easily when pulled. If the stress is applied for a short time, then removed, the sample snaps back to its original length suggesting that some "memory" of its initial unstretched condition is retained. The ability of an elastomer to regain its former size, when extensions of up to 400 per cent have been experienced, is associated with the long chain character of the material. This retractive action of linear uncrosslinked polymers can be observed if the time interval between extension and release is short, but if the stress is maintained for some time, then a relaxation process takes place allowing the tension to decay eventually to zero.

This can be explained quite simply. The molecules are initially in highly coiled shapes but application of a force causes rotation about the chain bonds resulting in an elongation of the molecules in the direction of the stress. This produces a distribution of chain conformations which differs significantly from the most probable distribution, and as this is an unstable state the chains will rapidly recoil when the

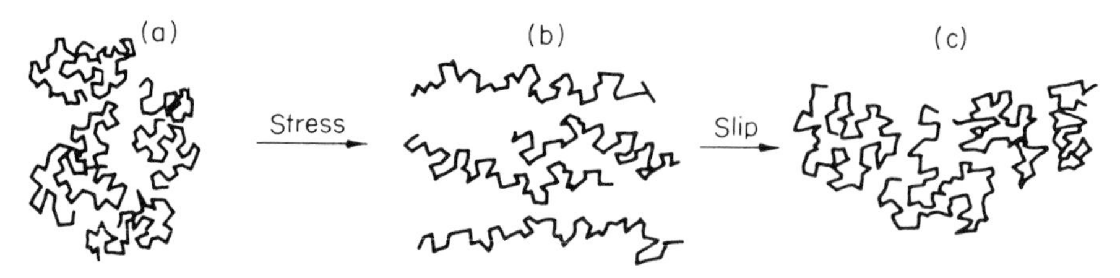

FIGURE 12.7. Schematic representations of an elastomer (a) under no stress, (b) chain alignment under an applied stress, and (c) stress relief produced by chains slipping past one another into new positions in the sample and recoiling.

stress is released in an attempt to regain their original shape distribution. For short periods of stress in an amorphous elastomer, the entanglement and intertwining of chains with their neighbours acts as a physical restraint to excessive chain movement and the elastomer regains its original length when the stress is removed. If however, the stress is maintained for a sufficient time, there is a general tendency for chains to unravel and slip past one another into new positions where the segments can relax and regain a stable coiled form. The resultant flow relieves the tension and produces the observed stress decay. The process is shown schematically in figure 12.7. When the molar mass is too low to produce sufficient entanglement, the material will flow more readily and behave like a viscous liquid. Similarly, as the temperature increases further and further above the glass transition, the enhanced segmental movement facilitates stress decay because of the greater ease of chain disentanglement.

12.9 Glass transition region

When the polymer is at a temperature below its glass temperature, chain motion is frozen. The polymer then behaves like a stiff spring storing all the available energy in stretching as potential energy, when work is performed on it. If sufficient thermal energy is supplied to the system to allow the chain segments to move co-operatively, a transition from the glass to the rubber-like state begins to take place. Motion is still restricted at this stage, but as the temperature increases further a larger number of chains begin to move with greater freedom. In mechanical terms the transition can be likened to the transformation of a stiff spring to a weak spring. As weak springs can only stored a fraction of the potential energy that a strong spring can hold, the remainder is lost as heat and if the change from a strong to a weak spring takes place over a period of time, equivalent to the observation time, then the energy loss is detected as mechanical damping. Finally when molecular motion increases to a sufficiently high level, all the chains behave like weak springs the whole time. This means that the modulus is much lower, but so too is the damping, which passes through a maximum in the vicinity of T_g. The maximum appears because the polymer is passing from the low-damping glassy state, through the high-damping transition region, to the lower-damping rubber-like state.

Treloar has described a very apt demonstration of the transition. A thin rubber rod is wound round a cylinder to create the shape of a spring and then frozen in this shape using liquid nitrogen. The cylinder (possibly of paper) is then removed leaving the rubber spring. The rubber is now in the glassy state and it acts like a stiff metal spring by regaining its shape rapidly after an extension. As the temperature is raised a gradual loss in the elastic recovery is observed after each applied stress, until a stage is reached when there is no recovery and the rubber remains in the deformed shape. With a further increase in temperature the rod straightens under its own weight and eventually regains its rubber-like elasticity at slightly higher temperatures.

THE GLASS TRANSITION TEMPERATURE, T_g

The transition from the glass to the rubber-like state is an important feature of polymer behaviour, marking as it does a region where dramatic changes in the physical properties, such as hardness and elasticity, are observed. The changes are completely reversible, however, and the transition from a glass to a rubber is a function of molecular motion, not polymer structure. In the rubber-like state or in the melt the chains are in relatively rapid motion, but as the temperature is lowered the movement becomes progressively slower until eventually the available thermal energy is insufficient to overcome the rotational energy barriers in the chain. At this temperature, which is known as the glass transition temperature T_g, the chains become locked in whichever conformation they possessed when T_g was reached. Below T_g the polymer is in the glassy state and is, in effect, a frozen liquid with a completely random structure.

Although the glass-rubber transition itself does not depend on polymer structure, the temperature at which T_g is observed depends largely on the chemical nature of the polymer chain and for most common synthetic polymers lies between 170 and 500 K. It is quite obvious that T_g is an important characteristic property of any polymer as it has an important bearing on the potential application of a polymer. Thus for a polymer with a flexible chain, such as polyisoprene, the thermal energy available at about 300 K is sufficient to cause the chain to change shape many thousands of times in a second. This polymer has $T_g = 200$ K. On the other hand, virtually no motion can be detected in atactic poly(methyl methacrylate) at 300 K, but at 450 K, the chains are in rapid motion. In this case $T_g = 378$ K. This means that at 300 K polyisoprene is likely to exhibit rubber-like behaviour and be useful as an elastomer, whereas poly(methyl methacrylate) will be a glassy material. If the operating temperature was lowered to 100 K, both polymers would be glasses.

EXPERIMENTAL DEMONSTRATION OF T_g

The glass transition is not specific to long chain polymers. Any substance, which can be cooled to a sufficient degree below its melting temperature without crystallizing, will form a glass. The phenomenon can be conveniently demonstrated using glucose penta-acetate (GPA). A crystalline sample of GPA is melted, then chilled rapidly in ice-water to form a brittle amorphous mass. By working the hard material between one's fingers, the transition from glass to rubber will be felt as the sample warms up. A little perseverence, with further rubbing and pulling, will eventually result in the recrystallization of the rubbery phase, which then crumbles to a powder.

DETECTION OF T_g

The transition from a glass to a rubber-like state is accompanied by marked changes in the specific volume, the modulus, the heat capacity, the refractive index, and other physical properties of the polymer. The glass transition is not a first-order transition, in the thermodynamic sense, as no discontinuities are observed when the entropy or volume of the polymer are measured as a function of temperature. If the first derivative of the property-temperature curve is measured, a change in the vicinity of T_g is found; for this reason it is sometimes called a second-order transition. Thus while the change in a physical property can be used to locate T_g, the transition bears many of the characteristics of a relaxation process and the precise value of T_g can depend on the method used and the rate of the measurement.

Techniques for locating T_g can be divided into two categories, dynamic and static. In the static methods, changes in the temperature dependence of an intensive property, such as density or heat capacity are followed and measurements are carried out slowly, to allow the sample to equilibrate and relax at each observation temperature. In dynamic mechanical methods a rapid change in modulus is indicative of the glass transition, but now the transition region is dependent on the frequency of the applied force. If we assume that, in the transition region, the restrictions to motion still present in the sample, allow only a few segments to move in some time interval, say 10 s, then considerably fewer will have moved if the observation time is less than 10 s. This means that the location of the transition region and T_g will depend on the experimental approach used, and T_g is found to increase 5 to 7 K for every tenfold increase in the frequency of the measuring techniques. This time dependence of segmental motion corresponds to the strong-weak transformation of a hypothetical spring and results in the high damping which imparts the lifeless leathery consistency to the polymer in this region. The temperature of maximum damping is usually associated with T_g, and at low frequencies the value assigned to T_g is within a few kelvins of that obtained from the static methods. As the static methods lead to more consistent values some of these can be described.

Measurement of T_g from V–T curves. One of the most frequently used methods of locating T_g is to follow the change in the volume of the polymer as a function of the temperature. The polymer sample is placed in the bulb of a dilatometer, degassed, and a confining liquid such as mercury added. If the bulb is attached to a capillary the change in polymer volume can be traced by noting the overall change in volume registered by the movement of the mercury level in the capillary. A variation of this method makes use of a density gradient column. A small sample of polymer suspended in this column provides a direct measure of the polymer density which can be measured easily as the temperature is varied.

Typical specific-volume-temperature curves are shown in figure 12.8 for poly(vinyl acetate). These consist of two linear portions whose slopes differ and closer inspection reveals that over a narrow range of temperature of between 2 and 5 K the slope changes continuously. To locate T_g, the linear portions are extrapolated and intersect at the point which is taken to be the characteristic transition temperature of the material. Each point on the curve is normally recorded after allowing the polymer time to equilibrate at the chosen temperature and as the rate of measurement affects the magnitude of T_g quite noticeably the equilibration time should be several hours at least. The effect of the measuring rate on T_g was demonstrated by Kovacs who recorded the volume of a polymer at each temperature, over a range including the transition, using two rates of cooling. If the sample was cooled rapidly (0.02 h) to each temperature the value of T_g derived from the resulting curve was some 8 K higher than that measured from results obtained using a slow cooling rate (100 h).

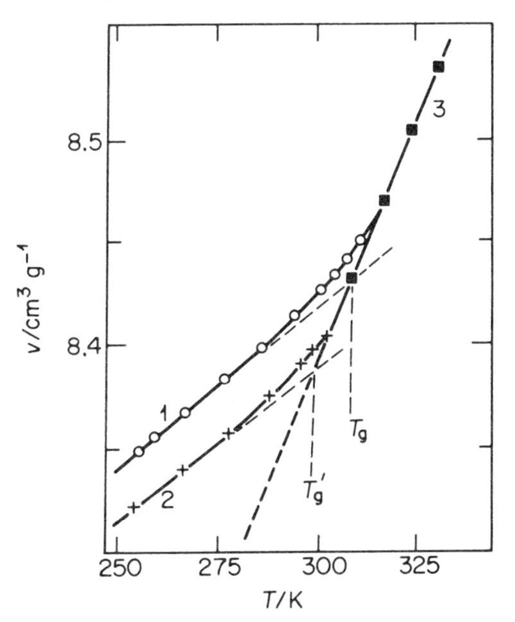

FIGURE 12.8. Specific volume v plotted against temperature for poly(vinyl acetate) measured after rapid cooling from above the T_g; 1, measured 0.02 hour after cooling; 2, measured 100 hours after cooling; T_g and T'_g are the glass transition temperatures measured for the different equilibration times. (After Kovacs, 1958.)

Refractive index measurements. The change in refractive index of the polymer with temperature has been used by several workers to establish T_g. A linear decrease in refractive index is observed as the temperature increases, and as the transition is passed, the rate of decrease becomes greater; T_g is again taken as the intersection of the linear extrapolation.

Heat capacity and other methods. The glass transition temperature can be detected calorimetrically by following the change in heat capacity with change in temperature. The curve for atactic polypropylene is shown in figure 12.9 where the abrupt increase in c_p at about 260 K, corresponds to the glass transition.

Among other reported techniques the most useful include differential thermal analysis, dielectric loss measurements, X- and β-ray absorption, and gas permeability studies. All indicate the existence of the phenomenon which we call the glass transition.

12.10 Factors affecting T_g

We have seen that the magnitude of T_g varies over a wide temperature range for different polymers. As T_g depends largely on the amount of thermal energy required to keep the polymer chains moving, a number of factors which affect rotation about chain links, will also influence T_g. These include (1) chain flexibility, (2) molecular structure (steric effects), (3) molar mass (see section 12.12), (4) branching and crosslinking.

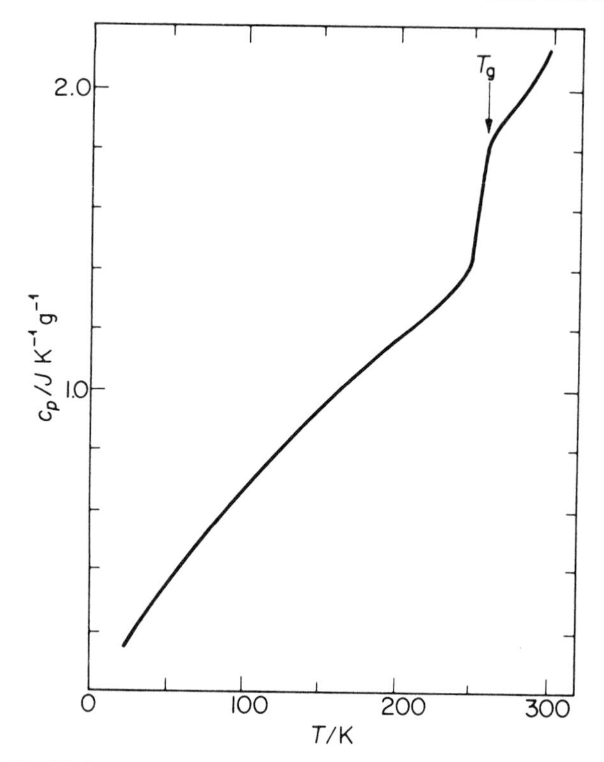

FIGURE 12.9. Specific heat capacity c_p plotted against temperature for atactic polypropylene showing the glass transition in the region of 260 K. (O'Reilly and Karasz, 1966.)

Chain flexibility. The flexibility of the chain is undoubtedly the most important factor influencing T_g. It is a measure of the ability of a chain to rotate about the constituent chain bonds, hence a flexibile chain has a low T_g whereas a rigid chain has a high T_g.

For symmetrical polymers, the chemical nature of the chain backbone is all important. Flexibility is obtained when the chains are made up of bond sequences which are able to rotate easily, and polymers containing $-\!\!(CH_2-CH_2)\!\!-$, $-\!\!(CH_2-O-CH_2)\!\!-$, or $-\!\!(Si-O-Si)\!\!-$ links will have correspondingly low values of T_g. The value of T_g is raised markedly by inserting groups which stiffen the chain by impeding rotation, so that more thermal energy is required to set the chain in motion. The *p*-phenylene ring is particularly effective in this respect, but when carried to extremes, produces a highly intractable, rigid structure, poly(*p*-phenylene)

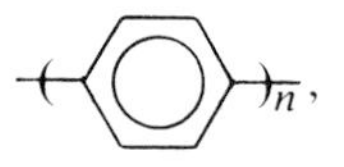

with no softening point. The basic structure can be modified by introducing flexible groups in the chain and some examples are given in table 12.1.

TABLE 12.1. Influence of bond flexibility on T_g

Polymers	*Repeat unit*	T_g/K
poly(dimethylsiloxane)	$-\!\!\left[Si(CH_3)_2-O-Si(CH_3)_2\right]\!\!-$	150
polyethylene	$-\!\!\left[CH_2-CH_2\right]\!\!-$	180
cis-polybutadiene	$-\!\!\left[CH_2-CH=CH-CH_2\right]\!\!-$	188
poly(oxyethylene)	$-\!\!\left[CH_2-CH_2-O\right]\!\!-$	206
Poly(phenylene oxide)	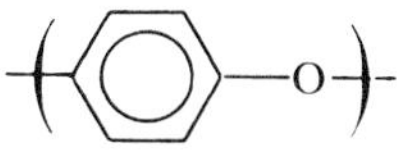 $-\!\!\left[C_6H_4-O\right]\!\!-$	356
Poly(arylene sulphone)	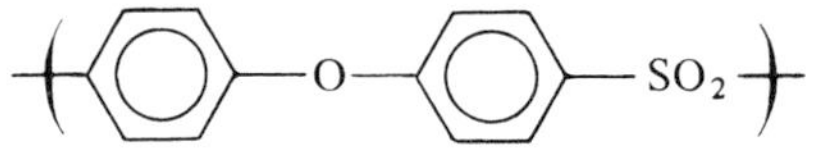 $-\!\!\left[C_6H_4-O-C_6H_4-SO_2\right]\!\!-$	523
poly(*p*-xylylene)	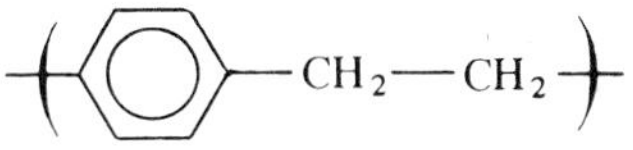 $-\!\!\left[C_6H_4-CH_2-CH_2\right]\!\!-$	about 553

Steric effects. When the polymer chains are unsymmetrical, with repeat units of the type $-\!\!\left[CH_2CHX\right]\!\!-$, an additional restriction to rotation is imposed by steric effects. These arise when bulky pendant groups hinder the rotation about the backbone and cause T_g to increase. The effect is accentuated by increasing the size of the side group and there is some evidence of a correlation between T_g and the molar volume V_X of the pendant group. It can be seen in table 12.2, that T_g increases with increasing V_X in the progressive series, polyethylene, polypropylene, polystyrene, and poly(vinyl naphthalene). Superimposed on this group size factor are the effects of polarity and the intrinsic flexibility of the pendant group itself. An increase in the lateral forces in the bulk state will hinder molecular motion and increase T_g. Thus polar groups tend to encourage a higher T_g than non-polar groups of similar size, as seen when comparing polypropylene, poly(vinyl chloride) and polyacrylonitrile. The influence of side chain flexibility is evident on examination of the polyacrylate series from methyl through butyl, and also in the polypropylene to poly(hex-1-ene) series.

A further increase in steric hindrance is imposed by substituting an α-methyl group, which restricts rotation even further and leads to higher T_g. For the pair polystyrene-poly(α-methyl styrene), the increase in T_g is 70 K, while the difference between poly(methyl methacrylate) and poly(methyl acrylate) is 100 K.

These steric factors all affect the chain flexibility and are simply additional contributions to the main chain effects.

Configurational effects. Cis-trans isomerism in polydienes and tacticity variations in certain α-methyl substituted polymers alter chain flexibility and affect T_g. Some examples are shown in table 12.3. It is interesting to note that when no α-methyl group is present in a polymer, tacticity has little influence on T_g.

TABLE 12.2. Glass transition temperatures for atactic polymers of the general type $+CH_2{\cdot}CXY+_n$

Polymer	T_g/K	V_X/cm^3 mol^{-1}†	*Group* X
Type $+CH_2CHX+_n$			
polyethylene	188	3.7	$-H$
polypropylene	253	25.9	$-CH_3$
poly(but-1-ene)	249	48.1	$-C_2H_5$
poly(pent-1-ene)	233	70.3	$-C_3H_7$
poly(hex-1-ene)	223	92.5	$-C_4H_9$
poly(4-methyl pent-1-ene)	302	92.5	$-CH_2-CH(CH_3)_2$
poly(vinyl alcohol)	358	11.1	$-OH$
poly(vinyl chloride)	354	22.1	$-Cl$
polyacrylonitrile	378	30.0	$-CN$
poly(vinyl acetate)	301	60.1	$-O{\cdot}C(=O)-CH_3$
poly(methyl acrylate)	279	60.1	$-C(=O)-O-CH_3$
poly(ethyl acrylate)	249	82.3	$-COOC_2H_5$
poly(propyl acrylate)	225	104.5	$-COOC_3H_7$
poly(butyl acrylate)	218	126.7	$-COOC_4H_9$
polystyrene	373	92.3	
poly(α-vinylnaphthalene)	408	143.9	
poly(vinyl biphenyl)	418	184.0	
Type $+CH_2C(CH_3)X+_n$	T_g/K	$V(X+Y)$ cm^3 mol^{-1}†	
poly(methyl methacrylate)	378	86.0	
poly(ethyl methacrylate)	338	108.2	
poly(propyl methacrylate)	308	130.4	
polymethacrylonitrile	393	55.9	
poly(α-methylstyrene)	445	118.2	

†Calculated using LeBas volume equivalents. (See Glasstone "Textbook of Physical Chemistry" Macmillan, 1951, Chapter 8.)

Effect of crosslinks on T_g. When crosslinks are introduced into a polymer, the density of the sample is increased proportionally. As the density increases, the molecular motion in the sample is restricted and T_g rises. For a high crosslink density the transition is broad and ill-defined, but at lower values, T_g is found to increase linearly with the number of crosslinks.

TABLE 12.3. The effect of microstructure on T_g

Polymer	*Stereostructure*	T_g/K
poly(methyl methacrylate)	isotactic	318
	atactic	378
	syndiotactic	388
polybutadiene	*cis*	165
	trans	255
polyisoprene	*cis*	200
	trans	220

12.11 Theoretical treatments

Before embarking on a rather brief description of the theoretical interpretations of the glass transition a word of caution should be given. In the foregoing sections several features of the results point to the fact that, in the vicinity of T_g, rate effects are closely associated with changes in certain thermodynamic properties. This has engendered two schools of thought on the origins of this phenomenon, together with variations on each theme. The elementary level of this text precludes detailed critical discussion of the relative merits of any particular treatment, and to avoid prejudicing the issue with personal comment the main ideas of each are outlined together with a more recent and possibly unifying approach to complete the picture.

THE FREE VOLUME THEORY

The free volume concept has been touched on in previous sections but it is instructive now to consider this idea more closely and to draw together the various points alluded to earlier. The free volume, V_f, is defined as the unoccupied space in a sample, arising from the inefficient packing of disordered chains in the amorphous regions of a polymer sample. The presence of these empty spaces can be inferred from the fact that when a polystyrene glass is dissolved in benzene there is a contraction in the total volume. This and similar observations indicate that the polymer can occupy less volume when surrounded by benzene molecules and that there must have been unused space in the glassy matrix to allow this increase in packing efficiency to occur.

On that basis, the observed specific volume of a sample, V, will be composed of the volume actually occupied by the polymer molecules, V_0, and the free volume in the system, *i.e.*

$$V = V_0 + V_f \tag{12.9}$$

Each term will, of course, be temperature dependent. The free volume is a measure of the space available for the polymer to undergo rotation and translation, and when the polymer is in the liquid or rubber-like states the amount of free volume will increase with temperature as the molecular motion increases. If the temperature is decreased, this free volume will contract and eventually reach a critical value when there is insufficient free space to allow large scale segmental motion to take place. The temperature at which this critical value is reached is the glass transition temperature. Below T_g the free volume will remain essentially constant as the temperature decreases further since the chains have now been immobilized and frozen in position. In contrast, the occupied volume will alter because of the changing amplitude of thermal vibrations in the chains and, to the first approximation, will be a linear function of temperature irrespective of whether the polymer is in the liquid or glassy state.

The glass transition can then be visualized as the onset of co-ordinated segmental motion made possible by an increase of the holes in the polymer matrix to a size sufficient to allow this type of motion to occur. This is manifest as a change in the specific volume due solely to an increase in the free volume and is shown schematically as the cross hatched area in figure 12.10, where the broken line indicates the temperature dependence of V_0.

The precise definition of the average amount of free volume present in a totally amorphous polymer remains unclear, but it must also depend to some extent on the thermal history of the sample. A number of suggestions have been made.

Simha and Boyer observed that a general empirical relationship exists between the T_g and the difference in expansion coefficients of the liquid and glass states. From the examination of a wide range of polymers they concluded that

$$(\alpha_l - \alpha_g)T_g = K_1 \tag{12.10}$$

where K_1 is a constant with a value of 0.113. This implies that the free volume fraction is the same for all polymers, *i.e.* 11.3 per cent of the total volume in the glassy state. The

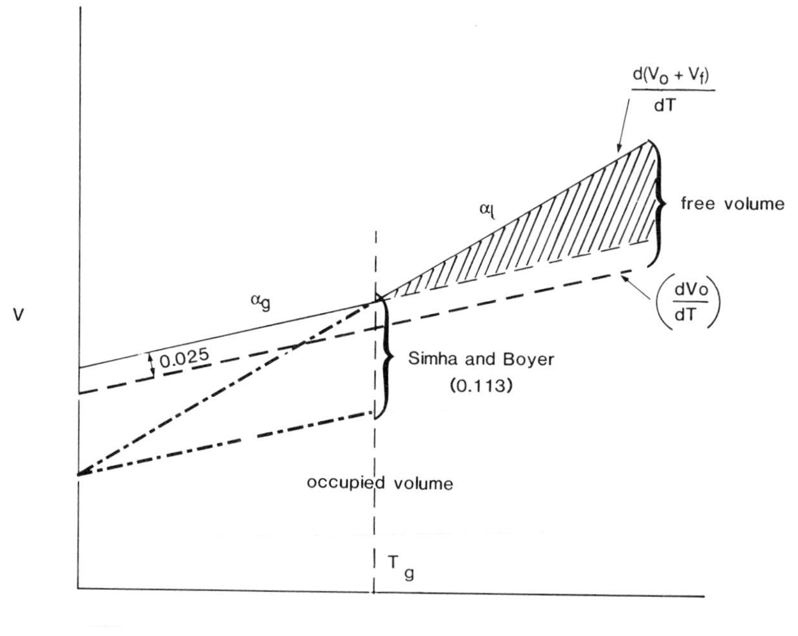

FIGURE 12.10. Schematic representation of the free volume as defined by Flory and Fox (0.025) and by Simha and Boyer (0.113).

definition of the S-B free volume can be seen from figure 12.10 to be

$$V_f = V - V_{0l}(1 + \alpha_g T) \tag{12.11}$$

where V_{0l} is the hypothetical liquid volume at absolute zero. This definition is perhaps too rigid and discounts differing chain flexibilities, so a more accurate representation is thought to be given by

$$\Delta\alpha \cdot T_g = 0.07 + 10^{-4} T_g \tag{12.12}$$

The values obtained are still much higher than the estimates from the Williams, Landel, Ferry (WLF) equation. This is an empirical equation but it can also be derived from free volume considerations by starting with a description of the viscosity of the system. In section 12.7 the Arrhenius equation was used to describe the temperature dependence of viscous flow, but an empirical equation proposed by Doolittle gives a much better description of viscous flow, and has a similar form

$$\ln \eta = \ln A + B\left\{\frac{V - V_f}{V_f}\right\} \tag{12.13}$$

where A and B are constants and $(V - V_f) = V_0$. On a molecular level, the ratio (V_0/V_f) is then a measure of the average volume of the polymer relative to that of the holes. Thus when $V_0 > V_f$, *i.e.* the polymer chain is larger than the average hole size, the viscosity will be correspondingly high, whereas when $V_0 < V_f$, the viscosity will be low.

We can now introduce a free volume fraction f

$$f = \left[\frac{V_f}{V_0 + V_f}\right] \approx \left[\frac{V_f}{V_0}\right] \tag{12.14}$$

and substitute in equation (12.13)

$$\ln \eta = \ln A + B/f \tag{12.15}$$

Next, a comparison can be made between the viscosity of a polymer melt at a temperature $T(\eta_T)$, and that at a reference temperature such as $T_g(\eta_g)$ and so

$$\ln\left(\frac{\eta_T}{\eta_g}\right) = B\left[\frac{1}{f_T} - \frac{1}{f_g}\right] \tag{12.16}$$

Here f_T and f_g are the fractional free volumes at T and T_g respectively. From figure 12.10 it can be seen that V_f is assumed to remain constant during expansion of the polymer in the glassy state but that above T_g there is a steady increase with rising temperature. If α_f is the expansion coefficient of the free volume above T_g, then the temperature dependence of f_T can be written

$$f_T = f_g + \alpha_f(T - T_g) \tag{12.17}$$

Substitution of equation (12.17) in equation (12.16) gives

$$\ln\left(\frac{\eta_T}{\eta_g}\right) = B\left[\frac{1}{f_g + \alpha_f(T - T_g)} - \frac{1}{f_g}\right] \tag{12.18a}$$

$$= B\left[\frac{f_g - \{f_g + \alpha_f(T - T_g)\}}{f_g\{f_g + \alpha_f(T - T_g)\}}\right] \tag{12.18b}$$

$$= -\frac{B\alpha_f(T - T_g)}{f_g\{f_g + \alpha_f(T - T_g)\}} \tag{12.18c}$$

Rearranging and dividing by α_f

$$\ln\left(\frac{\eta_T}{\eta_g}\right) = -\frac{(B/f_g)(T - T_g)}{(f_g/\alpha_f) + (T - T_g)} \tag{12.19}$$

Equation (12.19) is one form of the WLF equation, but as viscosity is a time dependent quantity and is proportional to the flow time, t, and density, ρ, then

$$\left(\frac{\eta_T}{\eta_g}\right) = \left(\frac{\rho_T}{\rho_g}\cdot\frac{t_T}{t_g}\right) \tag{12.20}$$

and

$$\log_{10}\left(\frac{t_T}{t_g}\right) \approx \frac{-(B/2.303 f_g)(T - T_g)}{(f_g/\alpha_f) + (T - T_g)} \tag{12.21}$$

where the small differences in density have been neglected. This can be compared with the form of the WLF equation

$$\log_{10} a_T = \frac{-C_1(T - T_g)}{C_2 + (T - T_g)} \tag{12.22}$$

where a_T is the reduced variables shift factor, C_1 and C_2 are constants that can be evaluated from experimental data, and are found to be $C_1 = 17.44$ and $C_2 = 51.6$ when T_g is the reference temperature. A more general description can be used

$$\log_{10} a_T = \frac{-8.86(T - T_s)}{101.6 + (T - T_s)} \tag{12.23}$$

where T_s is an arbitrary reference temperature usually located ≈ 50 K above T_g. C_1 and C_2 now have different values, and the shift factor is expressed as a ratio of relaxation times, τ, at T and T_s

$$a_T = \tau(T)/\tau(T_s) \tag{12.24}$$

As we shall see in chapter 13, the relaxation time is a function of the viscosity and modulus (G) of the polymer and, according to the Maxwell model, $\tau = (\eta/G)$. The modulus will be much less temperature dependent than the viscosity so we can write $a_T = (\eta_T/\eta_s)$, which demonstrates the equivalence of the empirical equation (12.22) with that derived from the free volume theory, equations (12.19) and 12.21).

The WLF equation can be used to describe the temperature dependence of dynamic mechanical, and dielectric relaxation behaviour of polymers near the glass transition where the response is no longer described by an Arrhenius relation. This will be dealt with in chapter 13.

The equations (12.19) and (12.22) can be used to evaluate f_g as we see that $(B/2.303 f_g) = 17.44$ and $(f_g/\alpha_f) = 51.6$. On the basis of viscosity data, B can be assigned a value of unity, leading to $f_g = 0.025$ and $\alpha_f = 4.8 \times 10^{-4}\,K^{-1}$. If α_f is assumed to be equivalent to $(\alpha_l - \alpha_g)$, this value compares well with the average value of $\Delta\alpha = 3.6 \times 10^{-4}\,K^{-1}$ determined for 18 polymers covering a wide range of T_g. The free volume fraction of 2.5 per cent is low compared with the S-B estimation but is comparable to that derived from the Gibbs-Di Marzio theory. Other values of 8 per cent from the "hole" theory of Hirai and Eyring, and 12 per cent calculated by Miller from heats of vaporization and liquid compressibilities, illustrate the uncertainty surrounding the magnitude of this free volume parameter. The free volume theory deals with the need for space to be available before co-operative motion, characteristic of the glass transition, can be initiated, but it tells us little about the molecular motion itself. Other approaches have chosen to base their description of the glass transition on a thermodynamic analysis.

GIBBS-DI MARZIO THERMODYNAMIC THEORY

Comments on the thermodynamic theories will be restricted to the proposals of Gibbs and Di Marzio (G-D) who, while acknowledging that kinetic effects are inevitably encountered when measuring T_g, consider the fundamental transition to be a true equilibrium. The data reported by Kovacs in section 12.9 imply that the observed T_g would decrease further if a sufficiently long time for measurement was allowed. This aspect is considered in the G-D theory by defining a new transition temperature T_2 at which the configurational entropy of the system is zero. This temperature can be considered in effect to be the limiting value T_g would reach in a hypothetical experiment taking an infinitely long time. On this basis the experimentally detectable T_g is a time dependent relaxation process and the observed value is a function of the time scale of the measuring technique. The theoretical derivation is based on a lattice treatment. The configurational entropy is found by calculating the number of ways that n_x linear chains each x segments long can be placed on a diamond lattice, for which the coordination number $z = 4$, together with n_0 holes. The restrictions imposed on the placing of a chain on the lattice are embodied in the hindered rotation which is expressed as the "flex energy" $\Delta\epsilon$, and ϵ_h which is the energy of formation of a hole. The flex energy is the energy difference between the potential energy minimum of the located bond and the potential minima of the remaining $(z-2)$ possible orientations which may be used on the lattice. Thus for polyethylene the *trans* position is considered most stable and the *gauche* positions are the flexed ones with $\Delta\epsilon$ the energy difference between the ground and flexed states. This of course varies with the nature of the polymer. The quantity ϵ_h is a measure of the cohesive energy. The configurational entropy S_{conf} is derived from the partition function describing the location of holes and polymer molecules.

As the temperature drops towards T_2 the number of available configurational states in the system decreases until at the temperature T_2 the system possesses only one degree of freedom. This leads to

$$\frac{S_{conf}(T_2)}{n_x k T_2} = 0 = \phi\left(\frac{\epsilon_h}{kT_2}\right) + \lambda\left(\frac{\Delta\epsilon}{kT_2}\right) + \frac{1}{x}\ln\left[\{(z-2)x+2\}\frac{(z-1)}{2}\right] \quad (12.25)$$

where $\phi(\epsilon_h/kT) = \ln(\epsilon_h/S_0)^{z/2-1} + f_0/f_x \ln(f_0/S_0)$
and

$$\lambda\left(\frac{\Delta\epsilon}{kT}\right) = \frac{x-3}{x}\ln\left\{1+(z-2)\exp(-\Delta\epsilon/kT) + (\Delta\epsilon/kT)\right.$$

$$\left.\times\left[\frac{(z-2)\exp(-\Delta\epsilon/kT)}{1+(z-2)\exp(-\Delta\epsilon/kT)}\right]\right\}$$

The fractions of unoccupied and occupied sites are f_0 and f_x respectively while S_0 is a function of f_0, f_x, and z. The main weaknesses of this theory are (a) that a chain of zero stiffness would have a T_g of 0 K and (b) that the T_g would be essentially independent of any intermolecular interactions In spite of these limitations, various aspects of the behaviour of copolymers, plasticized polymers, and the chain length dependence of T_g, can be predicted in a reasonably satisfactory manner. The temperature T_2 is not of course an experimentally measurable quantity but is calculated to lie approximately 50 K below the experimental T_g and can be related to T_g on this basis.

ADAM-GIBBS THEORY

While the kinetic approach embodied in the WLF equation and the equilibrium treatment of the G-D theory have both been successful in their way, the one-sided aspect of each probably masks the fact that they are not entirely incompatible with one another. An attempt to reunite both channels of thought has been made by Adam and Gibbs who have outlined a molecular kinetic theory.

In this they relate the temperature dependence of the relaxation process to the temperature dependence of the size of a region, which is defined as a volume large enough to allow co-operative rearrangement to take place without affecting a neighbouring region. This "co-operatively rearranging region" is large enough to allow a transition to a new conformation, hence is determined by the chain conformation and by definition will equal the sample size at T_2 where only one conformation is available to each molecule. Evaluation of the temperature dependence of the size of such a region leads to an expression for the co-operative transition probability, $W(T)$, which is simply the reciprocal of the relaxation time.

The polymer sample is described as an ensemble of co-operative regions, or subsystems, each containing Z monomeric segments. The transition probability of such a co-operative region is then calculated as a function of its size, to be

$$W(T) = A\exp(-Z\Delta\mu/kT) \quad (12.26)$$

where $\Delta\mu$ is the activation energy for a co-operative rearrangement per monomer segment. A lower limit to Z is then defined; this is the smallest size Z^* capable of having two configurations available to it, with a critical configurational entropy S_c^* which, from the definition, can be approximated by $k \ln 2$. Thus

$$Z^* = N_A S_c^*/S_c \tag{12.27}$$

where S_c is the macroscopic configurational entropy for the ensemble. Substitution gives

$$W(T) = A \exp(-\Delta\mu S_c^*/kTS_c) \tag{12.28}$$

and expressing this in the WLF form

$$-\log_{10} a_T = \log[W(T_s)/W(T)] \tag{12.29a}$$

$$= 2.303 \left\{ \frac{\Delta\mu S_c^*}{k} \left(\frac{1}{T_s S_c(T_s)} - \frac{1}{T S_c(T)} \right) \right\} \tag{12.29b}$$

The following approximations for S_c can now be used

$$S_c(T) - S_c(T_s) = \Delta C_p(\ln[T/T_s]) \tag{12.30}$$

and, remembering that $S_c(T_2) = 0$, then

$$S_c(T_s) = \Delta C_p(\ln[T_s/T_2]) \tag{12.31}$$

Substitution in equation (12.29b) gives a WLF equation

$$-\log_{10} a_T = C_1(T - T_s)/C_2 + (T - T_s)$$

where

$$C_1 = \left\{ \frac{2.303\, \Delta\mu\, S_c^*}{k\, \Delta C_p T_s \ln(T_s/T_2)} \right\} \tag{12.32a}$$

$$C_2 = \frac{T_s \ln(T_s/T_2)}{1 + \ln(T_s/T_2)} \tag{12.32b}$$

Results plotted according to the WLF equation could be predicted also from the molecular kinetic equation and show that the two approaches are compatible. The Adam–Gibbs equations also lead to a value of $(T_g - T_2) = 55$ K, so the theory appears to resolve most of the differences between the kinetic and thermodynamic interpretations of the glass transition.

These theories point to the fundamental importance of T_2 as a true second-order transition temperature and to the experimental T_g as the temperature governed by the time scale of the measuring technique. The latter value has great practical significance, however, and is a parameter which is essential to the understanding of the physical behaviour of a polymer.

12.12 Dependence of T_g on molar mass

The value of T_g depends on the way in which it is measured but it is also found to be a function of the polymer chain length. At high molar masses the glass temperature is essentially constant when measured by any given method, but decreases as the molar mass of the sample is lowered. In terms of the simple free volume concept each chain end requires more free volume in which to move about than a segment in the chain interior. With increasing thermal energy the chain ends will be able to rotate more readily than the rest of the chain and the more chain ends a sample has the greater the contribution to the free volume when these begin moving, consequently the glass transition temperature is lowered. Bueche expressed this as

$$T_g(\infty) = T_g + K/M = T_g + (2\rho N_A \theta/\alpha_f M_n), \tag{12.33}$$

where $T_g(\infty)$ is the glass temperature of a polymer with a very large molar mass, θ is the free volume contribution of one chain end and is 2θ for a linear polymer, ρ is the polymer density, N_A is Avogradro's constant, and α_f is the free volume expansivity defined as

$$\alpha_f = (\alpha_L - \alpha_G). \tag{12.34}$$

The linear expression in equation (12.33) has been widely used and describes the

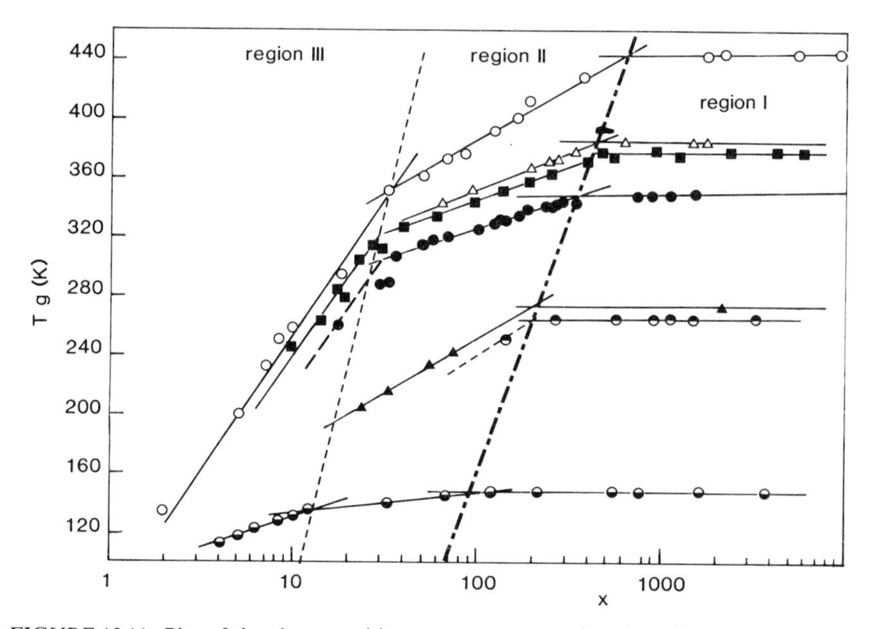

FIGURE 12.11. Plot of the glass transition temperature as a function of log x, where x is the number of chain atoms or bonds in the backbone. Data for (–○–) poly(α-methyl styrene); (–△–) poly(methyl methacrylate) (–■–) polystyrene; (–●–) poly(vinylchloride); (–▲–) isotactic polypropylene; (–◓–) atactic polypropylene; (–◒–) poly(dimethylsiloxane). (Reproduced from J. M. G. Cowie, *Europ. Polym. J.* **11**, 297 (1975) with permission of Pergamon Press PLC.)

behaviour of many polymer systems over a reasonable range of molar mass (>5000). For short chains the relationship is no longer valid and it has been shown that if T_g is plotted against log x, where x is the number of atoms or bonds in the polymer backbone, then three distinct regions can be identified for a number of common amorphous polymers (figure 12.11). Region I denotes the range of chain lengths at which T_g reaches its asymptotic value $T_g(\infty)$, and the critical value x_c at which this occurs increases as the chain becomes more rigid. Thus x_c is approximately 90 for a flexible polymer poly(dimethyl siloxane) but nearer 600 for the more rigid poly(α-methyl styrene).

The relationship between $T_g(\infty)$ and x_c is then

$$T_g(\infty) = 372.6 \log x_c - 595 \tag{12.35}$$

In region II, T_g is dependent on the molar mass and can be described by equation (12.35), but, on entering region III where the decrease in T_g accelerates, this is no longer true. The latter region incorporates the material that is oligomeric, and the line separating II and III represents the oligomer-polymer transition where the chains begin to become long enough to be considered capable of adopting a gaussian coil conformation.

12.13 The glassy state

When a linear amorphous polymer is in the glassy state, the material is rigid and brittle because the flow units of the chain are co-operatively immobile and effectively frozen in position. The polymer sample is also optically transparent, as the chains are distributed in a random fashion and present no definite boundaries or discontinuities from which light can be reflected. An amorphous polymer in this state has been likened to a plate of frozen spaghetti. If a small stress is applied to a polymer glass, it exhibits a rapid elastic response resulting from purely local, bond angle, deformation. Consequently, although the modulus is high, specimen deformation is limited to about 1 per cent, due to the lack of glide planes in the disordered mass. This means that the sample has no way of dissipating a large applied stress, other than by bond rupture, and so a polymer glass is prone to brittle fracture.

12.14 Relaxation processes in the glassy state

Polymers do not form perfectly elastic solids, as a limited amount of bond rotation can occur in the glass which allows slight plastic deformation; this makes them somewhat tougher than an inorganic glass.

There is now ample evidence to support the suggestion that relaxation processes can be active in polymer glasses at temperature well below T_g. While the co-operative, long-range, chain motion which is released on passing from the glass to the rubber-like state is not possible at $T < T_g$, other relaxations can take place. Many of these processes can be identified as secondary loss peaks in dynamic mechanical, or dielectric measurements, as will be seen in chapter 13. They often find their origin in the movement of groups that are attached pendant to the main chain, but relaxation of limited sections of the main chain can also be identified.

The molecular mechanisms for a number of these sub-glass transition relaxations have now been established, and by way of illustration some examples of group motions that have been found to be active in a series of poly(alkyl methacrylate)s will be described.

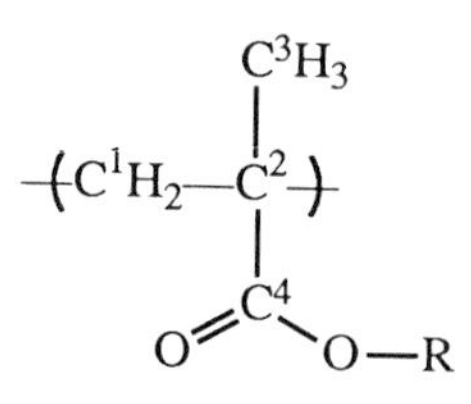

For the methyl, ethyl, and propyl derivatives a broad, mechanically active, damping peak is observed at 280 K (1 Hz), which is below the T_g of each polymer. Rotation of the oxycarbonyl group about (C^2-C^4) has been identified as the cause. However, if the group R is an alkyl or cycloalkyl unit then these can relax at even lower temperatures. Thus when R is a methyl unit, rotation is possible in the glass at temperatures below 100 K, and the α-methyl unit will also be capable of rotation at low temperatures. When R is larger, $-(CH_2)_n-CH_3$, another relaxation process is seen at around 120 K, which is common to all of these polymers with $n = 3$ to 11. This is believed to involve the relaxation of a four atom unit (-O-C-C-C-) or (-C-C-C-C-) which has been variously described by mechanisms including, and similar to, the Schatzki or Boyer crankshaft motions shown schematically in figures 12.3 and 12.12. The latter mechanisms have also been used to account for limited segmental relaxations in the backbone of all carbon chain, single strand, polymers. Even larger units can relax, and Heijboer has demonstrated that if R is a cyclohexyl ring, a relaxation at 180 K (1 Hz) can be located in the glass. This can be attributed to an intramolecular chair-chair transition in the ring.

As these relaxations require energy, and are associated with a characteristic activation energy, it has been suggested that they may improve the impact resistance of some materials. This point still requires confirmation as a general phenomenon, but

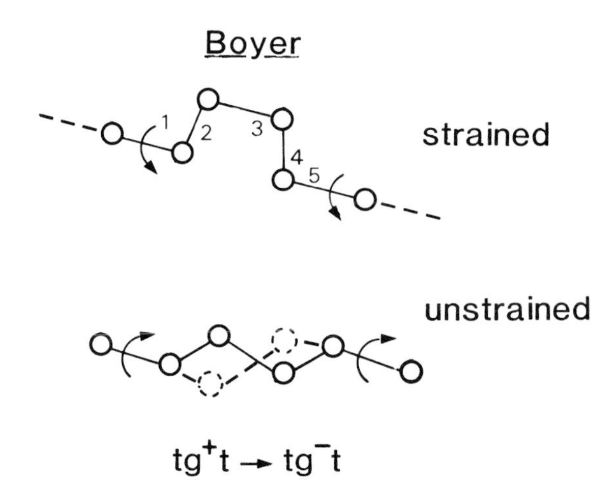

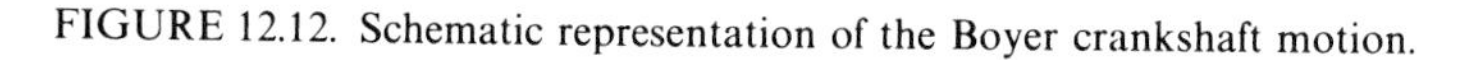

FIGURE 12.12. Schematic representation of the Boyer crankshaft motion.

there is little doubt that polymer molecules are not totally frozen or immobile when in the glassy state and that small sub units in the chain can remain mechanically and dielectrically active below T_g.

General Reading

G. Allen and J. C. Bevington, Eds, *Comprehensive Polymer Science*, Vol. 2. Pergamon Press (1989).
F. Bueche, *Physical Properties of Polymers.* Interscience Publishers Inc. (1962).
M. Doi and S. F. Edwards, *The Theory of Polymer Dynamics.* Oxford University Press (1986).
J. D. Ferry, *Viscoelastic Properties of Polymers*, 3rd edition. John Wiley and Sons Ltd (1979).
P. G. de Gennes, *Scaling Concepts in Polymer Physics.* Cornell University Press (1979).
M. Gordon, *High Polymers.* Iliffe (1963).
J. E. Mark, A. Eisenberg, W. W. Graessley, L. Mandelkern and J. L. Koenig, *Physical Properties of Polymers.* American Chemical Society (1984).
P. Meares, *Polymers: Structures and Bulk Properties*, Chapter 10. Van Nostrand (1965).
R. A. Pethrick and R. W. Richards, Eds, *Static and Dynamic Properties of the Polymeric Solid State.* D. Reidel Publishing Co. (1982).
L. H. Sperling, *Introduction to Physical Polymer Science.* John Wiley and Sons Ltd (1986).
A. V. Tobolsky and H. Mark, *Polymer Science and Materials*, Chapter 6. Wiley-Interscience (1971).
I. M. Ward, *Mechanical Properties of Solid Polymers*, 2nd edition. John Wiley and Sons Ltd (1983).

References

1. R. F. Boyer, *Rubber Chem. Technol.*, **36**, 1303 (1963).
2. R. B. Beevers and E. F. T. White, *Trans. Farad. Soc.*, **56**, 744 (1960).
3. J. M. G. Cowie, *Europ. Polym. J.*, **11**, 297 (1975).
4. T. G. Fox and P. J. Flory, *J. Phys. Chem.*, **55**, 221 (1951).
5. P. G. de Gennes, *Physics Today*, **33** (1983).
6. A. J. Kovacs, *J. Polym. Sci.*, **30**, 131 (1958).
7. J. M. O'Reilly and F. E. Karasz, *J. Polym. Sci.*, **C, No. 14**, 49 (1966).

CHAPTER 13

Mechanical Properties

13.1 Viscoelastic state

The fabrication of an article from a polymeric material in the bulk state, whether it be the moulding of a thermosetting plastic or the spinning of a fibre from the melt, involves deformation of the material by applied forces. Afterwards, the finished article is inevitably subjected to stresses, hence it is important to be aware of the mechanical and rheological properties of each material and understand the basic principles underlying their response to such forces.

In classical terms the mechanical properties of elastic solids can be described by Hooke's law, which states that an applied stress is proportional to the resultant strain, but is independent of the rate of strain. For liquids the corresponding statement is known as Newton's law, with the stress now independent of the strain, but proportional to the rate of strain. Both are limiting laws, valid only for small strains or rates of strain, and while it is essential that conditions involving large stresses, leading to eventual mechanical failure, be studied, it is also important to examine the response to small mechanical stresses. Both laws can prove useful under these circumstances.

In many cases, a material may exhibit the characteristics of both a liquid and a solid and neither of the limiting laws will adequately describe its behaviour. The system is then said to be in a *viscoelastic state.* A particularly good illustration of a viscoelastic material is provided by a silicone polymer known as "bouncing putty". If a sample is rolled into the shape of a sphere it can be bounced like a rubber ball, *i.e.* the rapid application and removal of a stress causes the material to behave like an elastic body. If on the other hand, a stress is applied slowly over a longer period the material flows like a viscous liquid so that the spherical shape is soon lost if left to stand for some time. Pitch behaves in a similar, if less spectacular, manner.

Before examining the viscoelastic behaviour of amorphous polymeric substances in more detail, some of the fundamental terms used will be defined.

13.2 Mechanical properties

Homogeneous, isotropic, elastic materials possess the simplest mechanical properties and three elementary types of elastic deformation can be observed when such a body is subjected to (i) simple tension, (ii) simple shear, and (iii) uniform compression.

Simple tension. Consider a parallelepiped of length x_0 and cross-sectional area $A_0 =$

y_0z_0. If this is subjected to a balanced pair of tensile forces F, its length changes by an increment dx so that $x_0 + dx = x$. When dx is small, Hooke's law is obeyed, and the tensile *stress* σ is proportional to the tensile *strain* ϵ. The constant of proportionality is known as the *modulus*, and for elastic solids

$$\sigma = E\epsilon, \tag{13.1}$$

where E is Young's modulus

The stress σ is a measure of the force per unit area (F/A), and the strain or elongation is defined as the extension per unit length, *i.e.* $\epsilon = (dx/x_0)$. It should be pointed out, however, that other definitions of strain will be met with in the literature, most notably, $\epsilon = \ln(x/x_0)$ is often called the *true strain*, while an expression arising from the kinetic theory of elasticity has the form

$$\epsilon = (1/3)\{(x/x_0) - (x_0/x)^2\}.$$

Of course, the extension dx will be accompanied by lateral contractions dy and dz, but although normally negative and equal, they can usually be assumed to be zero.

For an isotropic body, the change in length per unit length is related to the change in width per unit of length, such that

$$\nu_P = (dy/y_0)/(dx/x_0) \tag{13.2}$$

where ν_P is known as Poisson's ratio and varies from 0.5, when no volume change occurs, to about 0.2.

Simple shear. In simple shear the shape change is not accompanied by any change in volume. If the base of the body, shown shaded in the diagram, figure 13.1(b) is firmly

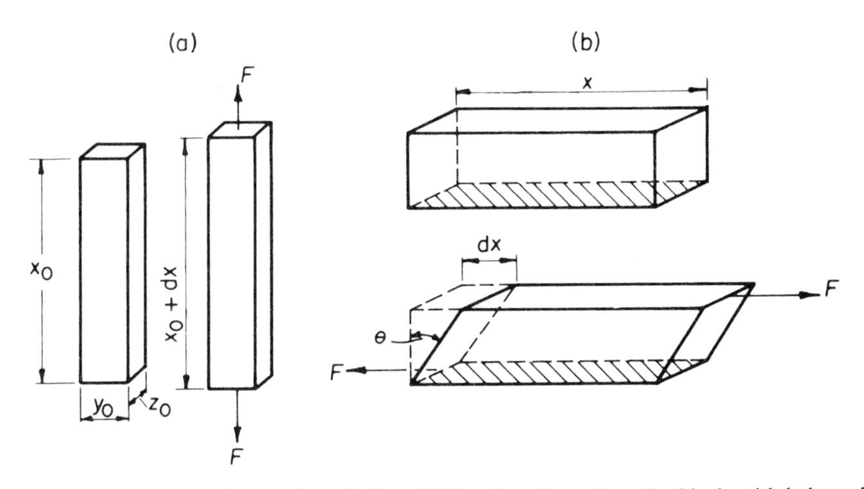

FIGURE 13.1. (a) Tensile stressing of a bar. (b) Shearing of a rectangular block, with balanced pair of forces F.

fixed, a transverse force F applied to the opposite face is sufficient to cause a deformation dx through an angle θ. The shear modulus G is then given by the quotient of the shearing force per unit area and the shear per unit distance between shearing surfaces; and so

$$G = \sigma_s/\epsilon_s = (F/yz)/(\mathrm{d}x/y) = F/A \tan\theta.$$

For very small shearing strains $\tan\theta \approx \theta$ and

$$G = F/A\theta. \tag{13.3}$$

Both E and G depend on the shape of the specimen and it is usually necessary to define the shape carefully for any measurement.

Uniform compression. When a hydrostatic pressure $-p$ is applied to a body of volume V_0, causing a change in volume ΔV, a bulk modulus B can be defined as

$$B = -p/(\Delta V/V_0). \tag{13.4}$$

The quantity B is often expressed in terms of the compressibility which is the reciprocal of the bulk modulus. Similarly E^{-1} and G^{-1} are known as the tensile and shear compliances and given the symbols D and J respectively.

13.3 Interrelation of moduli

The relations given above pertain to isotropic bodies and for non-isotropic bodies the equations are considerably more complex. Polymeric materials are normally either amorphous, or partially crystalline with randomly oriented crystallites embedded in a disordered matrix. However, any symmetry possessed by an individual crystallite can be disregarded and the body as a whole is treated as being isotropic.

The various moduli can be related to each other in a simple manner, because an isotropic body is considered to possess only two independent elastic constants and so

$$E = 3B(1 - 2\nu_P) = 2(1 + \nu_P)G. \tag{13.5}$$

This indicates, that for an incompressible elastic solid, *i.e.* one having a Poisson ratio of 0.5, Young's modulus is three times larger than the shear modulus. These moduli have dimensions of pressure and typical values for several polymeric and non-polymeric materials can be compared at ambient temperatures in table 13.1.

TABLE 13.1. Comparison of various moduli for some common materials

Material	E/GN m^{-2}	ν_P	G/GN m^{-2}
Steel	220	0.28	85.9
Copper	120	0.35	44.4
Glass	60	0.23	24.4
Granite	30	0.30	15.5
Polystyrene	34	0.33	1.28
Nylon-6,6	20	—	—
Polyethylene	24	0.38	0.087
Natural Rubber	0.02	0.49	0.00067

The response of polymers to mechanical stresses can vary widely, and depends on the particular state the polymer is in at any given temperature.

13.4 Mechanical models describing viscoelasticity

A perfectly elastic material obeying Hooke's law behaves like a perfect spring. The stress-strain diagram is shown in figure 13.2(a), and can be represented in mechanical terms by the model of a weightless *spring* whose modulus of extension represents the modulus of the material.

The application of a shear stress to a viscous liquid on the other hand, is relieved by viscous flow, and for small values of σ_s can be described by Newton's law

$$\sigma_s = \eta \mathrm{d}\epsilon_s/\mathrm{d}t, \tag{13.6}$$

where η is the coefficient of viscosity and $(\mathrm{d}\epsilon_s/\mathrm{d}t)$ is the rate of shear sometimes denoted by $\dot{\gamma}$. As stress is now independent of the strain the form of the diagram changes and can be represented by a *dashpot* which is a loose fitting piston in a cylinder containing a liquid of viscosity η. (Figure 13.2(b).)

Comparison of the two models shows that the spring represents a system storing energy which is recoverable, whereas the dashpot represents the dissipation of energy in the form of heat by a viscous material subjected to a deforming force. The dashpot is used to denote the retarded nature of the response of a material to any applied stress.

Because of their chain-like structure, polymers are not perfectly elastic bodies and deformation is accompanied by a complex series of long and short range co-operative molecular rearrangements. Consequently, the mechanical behaviour is dominated by viscoelastic phenomena, in contrast to materials such as metal and glass where atomic adjustments under stress are more localized and limited.

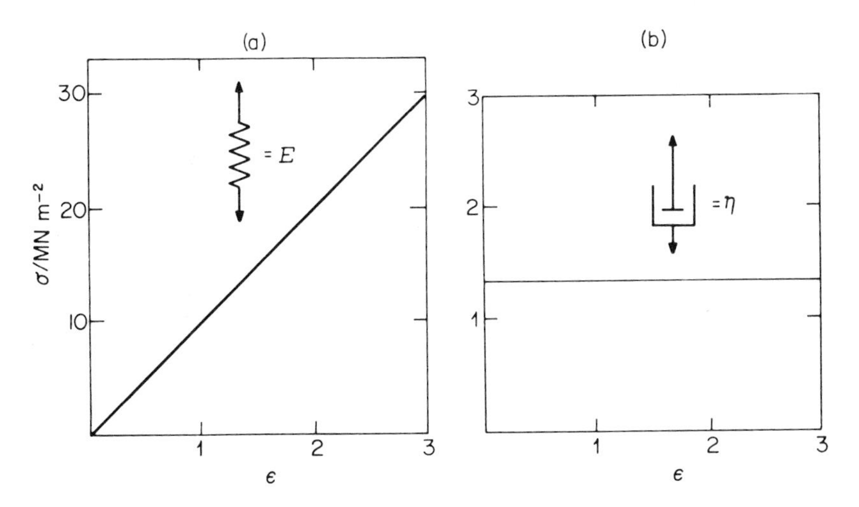

FIGURE 13.2. Stress-strain ($\sigma - \epsilon$) behaviour of (a) spring of modulus E and (b) a dashpot of viscosity η.

The Maxwell model. One of the first attempts to explain the mechanical behaviour of materials such as pitch and tar was made by James Clark Maxwell. He argued that when a material can undergo viscous flow and also respond elastically to a stress it should be described by a combination of both the Newton and Hooke laws. This assumes that both contributions to the strain are additive so that $\epsilon = \epsilon_{elast} + \epsilon_{visc}$. Expressing this as the differential equation leads to the equation of motion of a Maxwell unit

$$d\epsilon/dt = (1/G)(d\sigma/dt) + \sigma/\eta \tag{13.7}$$

Under conditions of constant shear strain ($d\epsilon/dt = 0$) the relation becomes

$$d\sigma/dt + G\sigma/\eta = 0, \tag{13.8}$$

and if the boundary condition is assumed that $\sigma = \sigma_0$ at zero time, the solution to this equation is

$$\sigma = \sigma_0 \exp(-tG/\eta), \tag{13.9}$$

where σ_0 is the initial stress immediately after stretching the polymer. This shows that when a Maxwell element is held at a fixed shear strain, the shearing stress will relax exponentially with time. At a time $t = (\eta/G)$ the stress is reduced to $1/e$ times the original value and this characteristic time is known as the *relaxation time* τ.

The equations can be generalized for both shear and tension and G can be replaced by E. The mechanical analogue for the Maxwell unit can be represented by a combination of a spring and a dashpot arranged in series so that the stress is the same on both elements. This means that the total strain is the sum of the strains on each element as expressed by equation (13.7). A typical stress-strain curve predicted by the Maxwell model, is shown in figure 13.3(a). Under conditions of constant stress, a

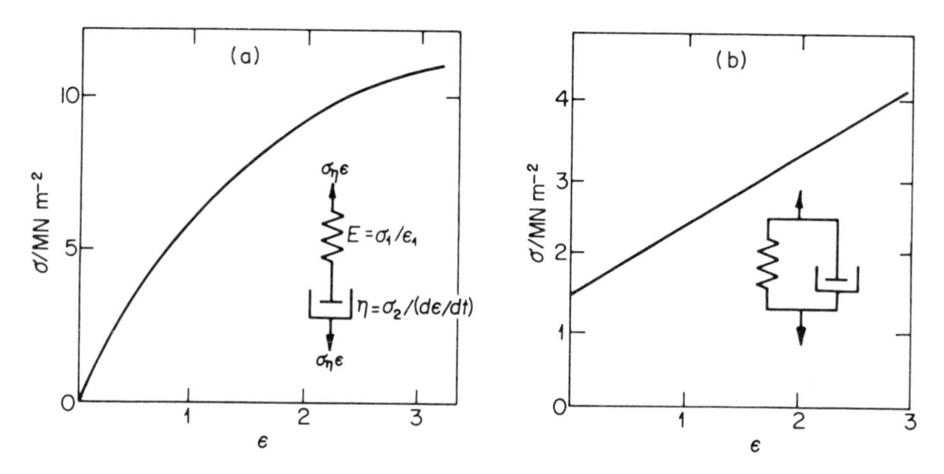

FIGURE 13.3. Stress-strain ($\sigma - \epsilon$) behaviour of two simple mechanical models, (a) the Maxwell model and (b) the Voigt-Kelvin model.

Maxwell body shows instantaneous elastic deformation first, followed by a viscous flow.

Voigt–Kelvin model. A second simple mechanical model can be constructed from the ideal elements by placing a spring and dashpot in parallel. This is known as a Voigt–Kelvin model. Any applied stress is now shared between the elements and each is subjected to the same deformation. The corresponding expression for strain is

$$\epsilon(t) = \sigma_0 J\{1 - \exp(t/\tau_R)\}. \tag{13.10}$$

Here $\tau_R = (\eta/G)$ is known as the *retardation time* and is a measure of the time delay in the strain after imposition of the stress. For high values of the viscosity, the retardation time is long and this represents the length of time the model takes to attain $(1 - 1/e)$ or 0.632 of the equilibrium elongation.

Such models are much too simple to describe the complex viscoelastic behaviour of a polymer, nor do they provide any real insight into the molecular mechanism of the process, but in certain instances they can prove useful in assisting the understanding of the viscoelastic process.

13.5 Linear viscoelastic behaviour of amorphous polymers

A polymer can possess a wide range of material properties and of these the hardness, deformability, toughness, and ultimate strength, are amongst the most significant. Certain features, such as high rigidity (modulus) and impact strength, combined with low creep characteristics are desirable in a polymer if eventually it is to be subjected to loading. Unfortunately, these are conflicting properties, as a polymer with a high modulus and low creep response does not absorb energy by deforming easily, hence has poor impact strength. This means a compromise must be sought depending on the use to which the polymer will be put, and this requires a knowledge of the mechanical response in detail.

The early work on viscoelasticity was performed on silk, rubber, and glass, and it was concluded that these materials exhibited a "delayed elasticity" manifest in the observation, that the imposition of a stress resulted in an instantaneous strain which continued to increase more slowly, with time. It is this delay between cause and effect that is fundamental to the observed viscoelastic response and the three major examples of this hysteresis effect are (1) *Creep*, where there is a delayed strain response after the rapid application of a stress, (2) *Stress-relaxation* (section 13.7) in which the material is quickly subjected to a strain and a subsequent decay of stress is observed, and (3) *Dynamic response* (section 13.9) of a body to the imposition of a steady sinusoidal stress. This produces a strain oscillating with the same frequency as, but out of phase with, the stress. For maximum usefulness, these measurements must be carried out over a wide range of temperature.

CREEP

To be of any practical use, an object made from a polymeric material must be able to retain its shape when subjected to even small tensions or compressions over long periods of time. This dimensional stability is an important consideration in choosing a polymer to use in the manufacture of an item. No one wants a plastic telephone receiver which sags after sitting in its cradle for several weeks, or a car tyre that develops a flat spot if parked in one position for too long, or clothes made from synthetic fibres which become baggy and deformed after short periods of wear. Creep tests provide a measure of this tendency to deform and are relatively easy to carry out.

Creep can be defined as a progressive increase in strain, observed over an extended time period, in a polymer subjected to a constant stress. Measurements are carried out on a sample clamped in a thermostat. A constant load is firmly fixed to one end and the elongation is followed by measuring the relative movement of two fiducial marks, made initially on the polymer, as a function of time. To avoid excessive changes in the sample cross section, elongations are limited to a few per cent and are followed over approximately three decades of time.

The initial, almost instantaneous, elongation produced by the application of the tensile stress is inversely proportional to the rigidity or modulus of the material, *i.e.* an elastomer with a low modulus stretches considerably more than a material in the glassy state with a high modulus. The initial deformation corresponds to portion OA of the curve (figure 13.4), increment a. This rapid response is followed by a region of creep, A to B, initially fast but eventually slowing down to a constant rate represented by the section B to C. When the stress is removed the instantaneous elastic response OA is completely recovered and the curve drops from C to D, *i.e.* the distance $a' = a$. There follows a slower recovery in the region D to E which is never complete, falling short of the initial state by an increment $c' = c$. This is a measure of the viscous flow experienced by the sample and is a completely non-recoverable response. If the tensile load is enlarged, both the elongation and the creep rate increase, so results are usually reported in terms of the *creep compliance* $J(t)$, defined as the ratio of the relative elongation γ at

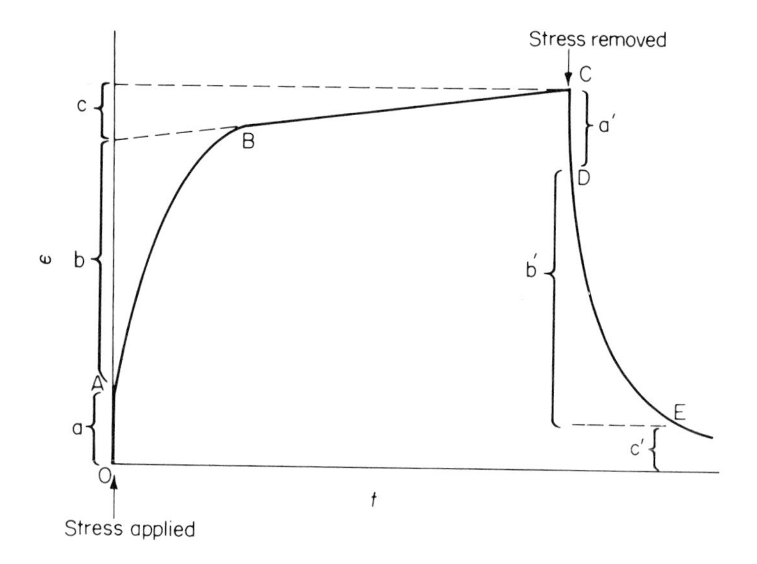

FIGURE 13.4. Schematic representation of a creep curve: a, initial elastic response; b, region of creep; c, irrecoverable viscous flow. This curve can be represented by the four element model shown in figure 13.5.

time t to the stress so that

$$J(t) = yE/\sigma \tag{13.11}$$

At low loads $J(t)$ is independent of the load.

This idealized picture of creep behaviour in a polymer has its mechanical equivalent constructed from the springs and dashpots described earlier. The changes a and a' correspond to the elastic response of the polymer and so we can begin with a Hookean spring. The Voigt–Kelvin model is embodied in equation (13.11) and this reproduces the changes b and b'. The final changes c and c' represent viscous flow and can be represented by a dashpot so that the whole model is a four element model – figure 13.5.

The behaviour can be explained in the following series of steps. In diagram (i) the system is at rest. The stress σ is applied to spring E_1 and dashpot η_3; it is also shared by E_2 and η_2 but in a manner which varies with time. In diagram (ii), representing zero time, the spring E_1 extends by an amount $\sigma/E_1(=a)$. This is followed by a decreasing rate of creep with a progressively increasing amount of stress being carried by E_2 until eventually none is carried by η_2 and E_2 is fully extended – diagram (iii). Such behaviour is described by

$$\epsilon(t) = (\sigma_0/E_2)\{1 - \exp(-t/\tau_R)\} \tag{13.12}$$

where the retardation time τ_R provides a measure of the time required for E_2 and η_2 to reach 0.632 of their total deformation. A considerably longer time is required for complete deformation to occur. When spring E_2 is fully extended the creep attains a constant rate corresponding to movement in the dashpot η_3. Viscous flow continues and the dashpot η_3 is deformed until the stress is removed. At that time, E_1 retracts quickly along section a' and a period of recovery ensues (b'). During this time spring E_2 forces the dashpot plunger in η_2 back to its original position. As no force acts on η_3 it remains in the extended state, and corresponds to the non-recoverable viscous flow; region $c' = \sigma t/\eta_3$. The system is then as shown in diagram (v). In practice, a substance possesses a large number of retardation times which can be expressed as a distribution

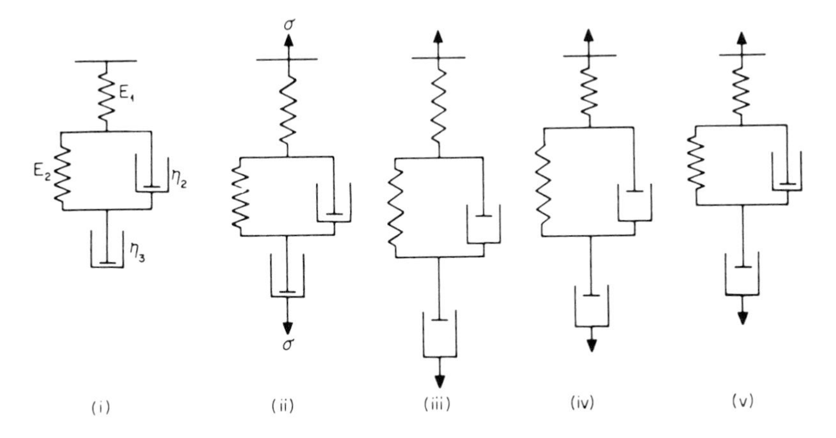

FIGURE 13.5. Use of mechanical models to describe the creep behaviour of a polymeric material.

function $L_1(\tau)$ where

$$L_1(\tau) = \mathrm{d}\{J(t) - (t/\eta)\}/\mathrm{d}\ln t. \tag{13.13}$$

To the first approximation, this is estimated from a plot of creep compliance against $\log_e t$, and (t/η) is the contribution from viscous flow.

STRESS-STRAIN MEASUREMENTS

The data derived from stress-strain measurements on thermoplastics are important from a practical viewpoint, providing as they do, information on the modulus, the brittleness, and the ultimate and yield strengths of the polymer. By subjecting the specimen to a tensile force applied at a uniform rate and measuring the resulting deformation, a curve of the type shown in figure 13.6 can be constructed.

The shape of such a curve is dependent on the rate of testing, consequently, this must be specified if a meaningful comparison of data is to be made. The initial portion of the curve is linear and the tensile modulus E is obtained from its slope. The point L represents the stress beyond which a brittle material will fracture, and the area under the curve to this point is proportional to the energy required for brittle fracture. If the material is tough no fracture occurs, and the curve then passes through a maximum or inflection point Y, known as the yield point. Beyond this, the ultimate elongation is eventually reached and the polymer breaks at B. The area under this part of the curve is the energy required for tough fracture to take place.

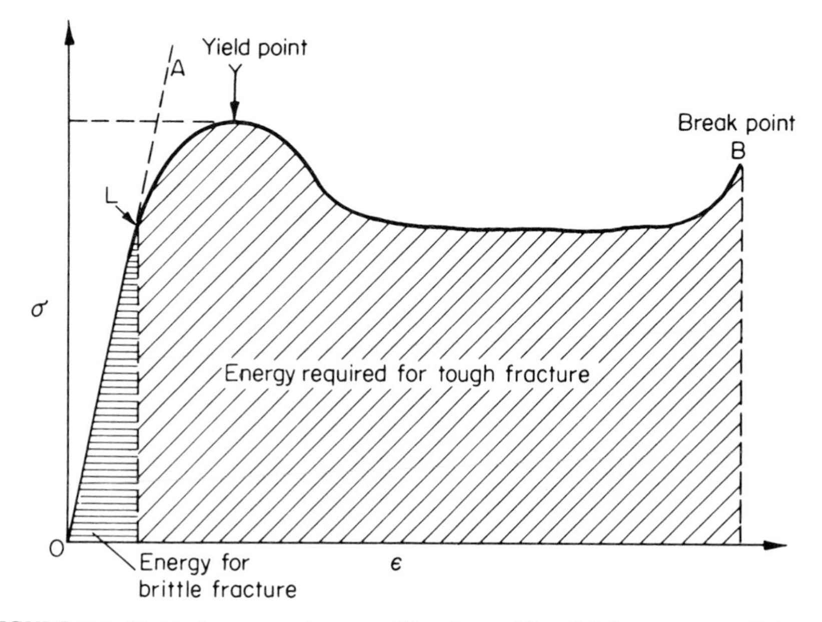

FIGURE 13.6. Idealized stress-strain curve. The slope of line OA is a measure of the true modulus.

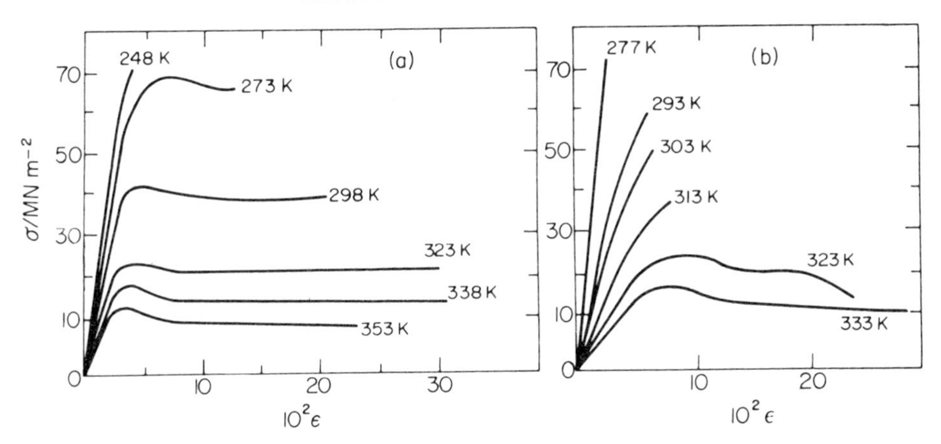

FIGURE 13.7. Influence of temperature on the stress-strain response of (a) cellulose acetate and (b) poly(methyl methacrylate). (From data by Carswell and Nason.)

EFFECT OF TEMPERATURE ON STRESS-STRAIN RESPONSE

Polymers such as polystyrene and poly(methyl methacrylate) with a high E at ambient temperatures fall into the category of hard brittle materials which break before point Y is reached. Hard tough polymers can be typified by cellulose acetate and several curves measured at different temperatures are shown in figure 13.7(a). Stress-strain curves for poly(methyl methacrylate) are also shown for comparison (figure 13.7(b)).

It can be seen that the effect of temperature on the characteristic shape of the curve is significant. As the temperature increases both the rigidity and the yield strength decrease while the elongation generally increases. For cellulose acetate there is a transformation from a hard brittle state below 273 K to a softer but tougher type of polymer at temperatures above 273 K. For poly(methyl methacrylate) the hard brittle characteristics are retained to a much higher temperature, but it eventually reaches a soft tough state at about 320 K. Thus if the requirements of high rigidity and toughness are to be met, the temperature is important. Cellulose acetate meets these requirements if used at 298 K more satisfactorily than when used at 350 K where the modulus is smaller and the ability to absorb energy, represented by the area under the curve, is also lower.

13.6 Boltzmann superposition principle

If a Hookean spring is subjected to a series of incremental stresses at various times, the resulting extensions will be independent of the loading or past history of the spring. A Newtonian dashpot also behaves in a predictable manner. For viscoelastic materials the response to mechanical testing is time dependent, but the behaviour at any time can be predicted by applying a superposition principle proposed by Boltzmann. This can be illustrated by a creep test using a simple Voigt–Kelvin model with a single retardation time τ_R, placed initially under a stress σ_0 at time t_0. If after times $t_1, t_2, t_3, \ldots$ the system is subjected to additional stresses $\sigma_1, \sigma_2, \sigma_3, \ldots$ then the principle states that the creep response of the system can be predicted simply by summing the individual responses from each stress increment. Thus if the stress alters continually, the summation can be

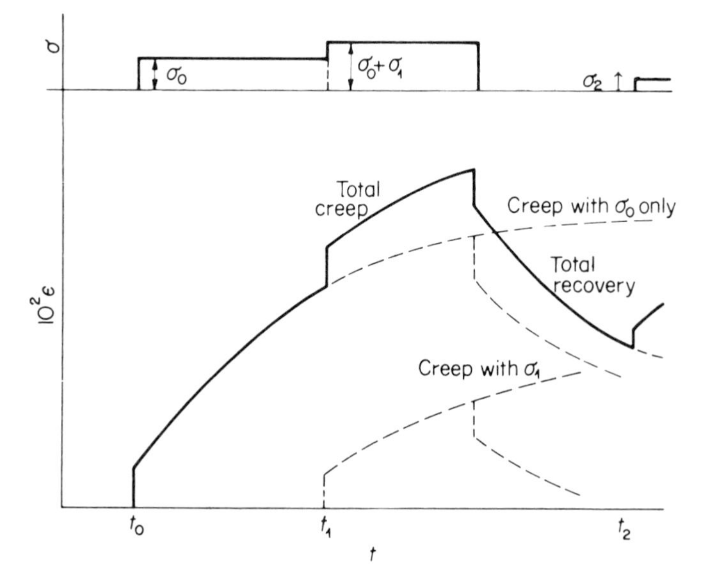

FIGURE 13.8. Application of the Boltzmann superposition principle to a creep experiment.

replaced by an integral, and σ_n by a continually varying function, so that at time t^* when the stress $\sigma(t^*)$ existed, the strain is given by

$$\epsilon(t^*) = \int_0^{t^*} \frac{d\sigma(t^*)}{dt^*} \phi(t^* - t_n)\,dt. \tag{13.14}$$

The principle has been applied successfully to the tensile creep of amorphous and rubber-like polymers, but it is not too successful if appreciable crystallinity exists in the sample. Graphical representation of the principle is shown in figure 13.8.

13.7 Stress-relaxation

Stress-relaxation experiments involve the measurement of the force required to maintain the deformation produced initially by an applied stress as a function of time. Stress-relaxation tests are not performed as often as creep tests because many investigators believe they are less readily understood. The latter point is debatable and it may only be that the practical aspects of creep measurements are simpler. As will be shown later, all the mechanical parameters are in theory interchangeable, and so all such measurements will contribute to the understanding of viscoelastic theory. While stress-relaxation measurements are useful in a general study of polymeric behaviour, they are particularly useful in the evaluation of antioxidants in polymers, especially elastomers, because measurements on such systems are relatively easy to perform and are sensitive to bond rupture in the network.

Experimental stress-relaxation technique. In a stress-relaxation experiment, the sample under study is deformed by a rapidly applied stress. As the stress is normally observed to reach a maximum as soon as the material deforms and then decreases thereafter, it is necessary to alter this continually in order to maintain a constant deformation or measure the stress that would be required to accomplish this operation.

The apparatus used varies in complexity with the physical nature of the sample, being simplest for an elastomer and becoming more sophisticated when the polymer is more rigid. One type of experimental set up is shown in figure 13.9. The sample is fixed in position by means of clamps, one being attached to a spring beam above and the other to an adjustable rod R below. A stress is applied to the sample by rapidly pulling rod R downwards and clamping it in position. This causes the beam to bend and the displacement is measured by means of a strain gauge or a differential transformer. The beam deflection is then fed to a recorder and a trace of stress against time is obtained.

The results are expressed as a relaxation modulus $E_r(t)$ which is a function of the time of observation. Typical data for polyisobutylene are shown in section 13.14, figure 13.21, where the logarithm of the relaxation modulus $\log E_r(t)$ is plotted against $\log t$. From the curves it can be seen that there is a rapid change in $\log E_r(t)$ over a narrow range of temperature corresponding to the glass transition.

Again a simple model with a single relaxation time is too crude, and the stress-relaxation modulus $E_r(t)$ is better represented by

$$E_r(t) = \int_0^\infty H(\tau)\exp(-t/\tau)\,d(\ln\tau), \tag{13.15}$$

where $H(\tau)$ is the distribution function of relaxation times. This is suitable for a linear polymer but requires the additional term E_∞ if the material is crosslinked.

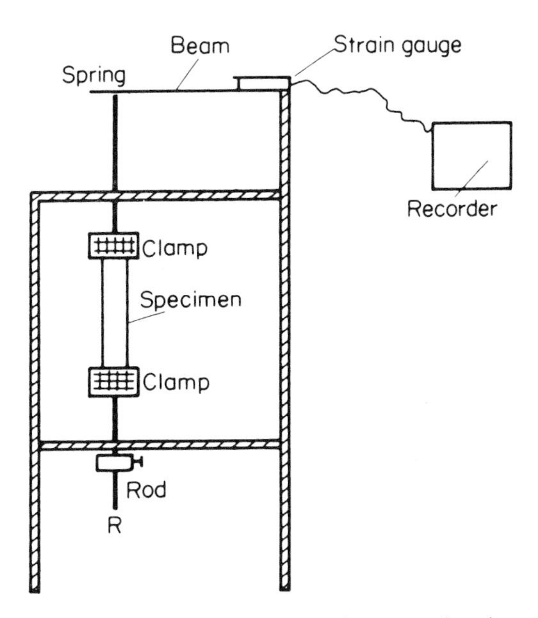

FIGURE 13.9. Simple apparatus to measure the stress-relaxation of a polymer.

13.8 Dynamic mechanical and dielectric thermal analysis

Non-destructive testing methods are particularly useful for assessing the physical properties of polymeric materials when an understanding of the performance at a molecular level is important. The foregoing techniques for measuring mechanical properties are transient or non periodic methods and typically cover time intervals of up to 10^6 s. For information relating to short times, two approaches that have been widely used are dynamic mechanical thermal analysis (DMTA) and dielectric thermal analysis (DETA). These are both particular kinds of relaxation spectroscopy in which the sample is perturbed by a sinusoidal force (either mechanical or electrical) and the response of the material is measured over a range of temperatures and at different frequencies of the applied force. From an analysis of the material response it is possible to derive information about the molecular motions in the sample, and how these can affect the modulus, damping characteristics and structural transitions. Both techniques can be used to probe molecular motions in liquid or solid polymers, but when dielectric spectroscopy is used the relaxation or transition must involve movement of a dipole or a charge displacement if it is to be detected. Thus while both DMTA and DETA can provide similar information about a sample, they can also be used in a complementary fashion, particularly when trying to identify the molecular mechanism of a particular process and in ascertaining whether or not the group is polar.

13.9 Dynamic mechanical thermal analysis (DMTA)

In DMTA a small sinusoidal stress is imparted to the sample in the form of a torque, push-pull, or a flexing mode, of angular frequency ω. If the polymer is treated as a classical damped harmonic oscillator, both the elastic modulus and the damping characteristics can be obtained. Elastic materials convert mechanical work into potential energy which is recoverable; for example an ideal spring, if deformed by a stress, stores the energy and uses it to recover its original shape after removal of the stress. No energy is converted into heat during the cycle and so no damping is experienced.

Liquids on the other hand flow if subjected to a stress; they do not store the energy but dissipate it almost entirely as heat and thus possess high damping characteristics. Viscoelastic polymers exhibit both elastic and damping behaviour. Hence if a sinusoidal stress is applied to a linear viscoelastic material, the resulting stress will also be sinusoidal, but will be out of phase when there is energy dissipation or damping in the polymer.

Harmonic motion of a Maxwell element. The application of a sinusoidal stress to a Maxwell element produces a strain with the same frequency as, but out of phase with, the stress. This can be represented schematically in figure 13.10 where δ is the phase angle between the stress and the strain. The resulting strain can be described in the terms of its angular frequency ω and the maximum amplitude ϵ_0 using complex notation, by

$$\epsilon^* = \epsilon_0 \exp(i\omega t), \tag{13.16}$$

where $\omega = 2\pi v$, the frequency is v and $i = -1^{1/2}$. The relation between the alternating

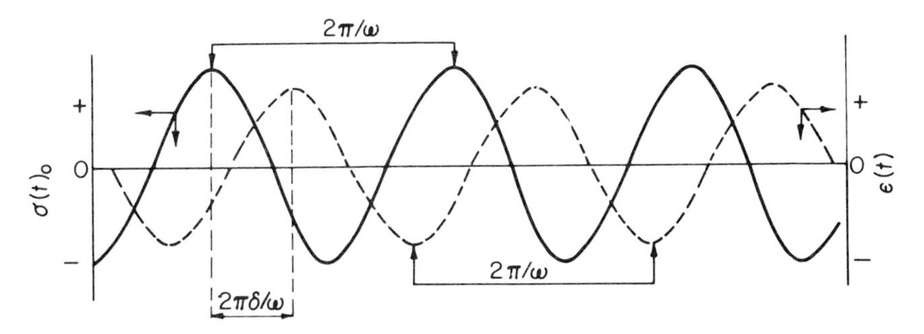

FIGURE 13.10. Harmonic oscillation of a Maxwell model, with the solid line representing the stress and the broken line the strain curve. This shows the lag in the response to an applied stress.

stress and strain is written as

$$\sigma^* = \epsilon^* E^*(\omega), \tag{13.17}$$

where $E^*(\omega)$ is the frequency dependent complex dynamic modulus defined by

$$E^*(\omega) = E'(\omega) + iE''(\omega). \tag{13.18}$$

This shows that $E^*(\omega)$ is composed of two frequency dependent components; $E'(\omega)$ is the real part in phase with the strain called the *storage modulus*, and $E''(\omega)$ is the *loss modulus* defined as the ratio of the component 90° out of phase with the stress to the stress itself. Hence $E'(\omega)$ measures the amount of stored energy and $E''(\omega)$, sometimes called the imaginary part, is actually a real quantity measuring the amount of energy dissipated by the material.

The response is often expressed as a complex dynamic compliance

$$J^*(\omega) = J'(\omega) - iJ''(\omega), \tag{13.19}$$

especially if a generalized Voigt model is used. For a Maxwell model

$$\sigma^*/\epsilon^* = E'\omega^2\tau^2/(1 + \omega^2\tau^2) + iE''\omega\tau/(1 + \omega^2\tau^2). \tag{13.20}$$

In more realistic terms, there is a distribution of relaxation times and a continuous distribution function can be derived, if required.

The damping in the system or the energy loss per cycle can be measured from the "loss tangent" tan δ. This is a measure of the internal friction and is related to the complex moduli by

$$\tan\delta = 1/\omega\tau = E''(\omega)/E'(\omega) = J''(\omega)/J'(\omega). \tag{13.21}$$

The onset of molecular motion in a polymer sample is reflected in the behaviour of E' and E''. A schematic diagram (figure 13.11) of the variation of E' and E'' as a function of ω, assuming only a single value for τ in the model, shows that a maximum in the loss angle is observed where $\omega = 1/\tau$. This represents a transition point such as T_g, T_m, or

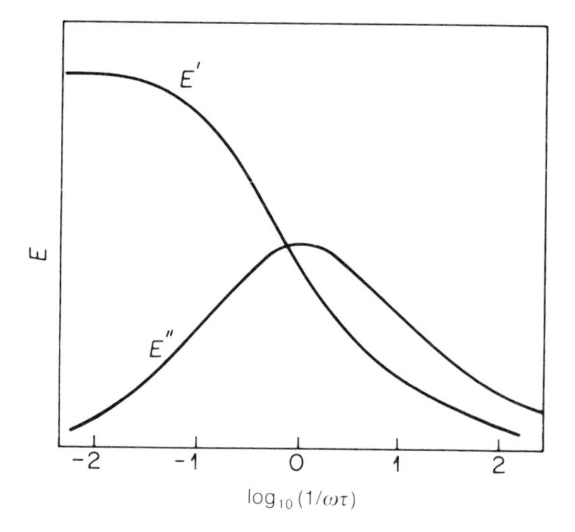

FIGURE 13.11. Behaviour of E' and E'' as a function of the angular frequency for a system with a single relaxation time.

some other region where significant molecular motion occurs in the sample. The maximum is characteristic of the dynamic method as the creep and relaxation techniques merely show a change in the modulus level.

13.10 Experimental methods

There are three main experimental approaches for measuring the dynamic mechanical properties of a sample, (a) free vibration, (b) forced vibration – resonance, (c) forced vibration – non-resonance. The mechanical response is usually determined at low frequencies and over as wide a temperature range as possible and examples of each are described in the following section.

TORSIONAL PENDULUM – FREE VIBRATION

A study of the mechanical damping and shear modulus under free vibration can be made using a torsional pendulum. The specimen is firmly fixed at one end and the other end is clamped to a disc, with a large moment of inertia, which can move freely. As the polymer sample should not be under a tensile stress, the suspension wire supporting the disc is passed over a pulley and the weight of the disc and sample are counterbalanced by loading the end. If the disc is subjected to an angular displacement and then released, the sample will twist backwards and forwards about the vertical axis. The oscillations stimulated in the sample are picked up by an arm attached to the rigidly fixed end held in torsion bars, and transmitted to a recorder by a linear variable differential transformer. The sample movements are traced as a series of oscillations whose frequency is a function of the physical state of the sample. The period of oscillation P is taken as the distance between adjacent maxima or minima and the amplitude A is a measure of the height from one minimum to the preceding maximum. The exponential decay of the amplitude along the axis provides an indication of the mechanical

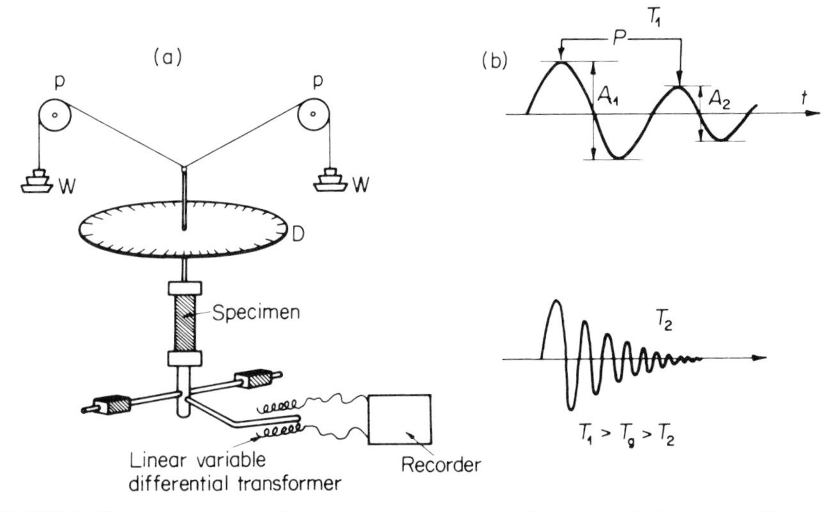

FIGURE 13.12. (a) Schematic diagram of a torsional pendulum in which the weight of the disc D is counterbalanced by weights W suspended over pulleys p. (b) Typical curves from a sample at T_2 below and at T_1 above its glass transition temperature.

damping. At a temperature $T_1 > T_g$ the sample absorbs most of the energy and damping is high whereas at a much lower temperature $T_2 < T_g$ the material tends to store the energy and mechanical damping is considerably lower.

A quantitative measure of the damping is provided by the logarithmic decrement Δ defined as the logarithmic decrease in amplitude per cycle. It is calculated from the ratio of amplitudes of any two successive oscillations using the relation

$$\Delta = \ln(A_1/A_2) = \ln(A_2/A_3) = \cdots\cdots = \ln(A_n/A_{n+1}). \tag{13.22}$$

The shear modulus can also be derived from the data, being inversely proportional to the square of the period $G = KI/P^2$, where K is a factor depending on the shape and the size of the sample and I is the polar moment of inertia.

The method can cover the complete range of moduli encountered in polymeric systems but is confined to a relatively narrow frequency range of 0.01 to 10 Hz.

VIBRATING REED – RESONANCE

For resonance forced vibration measurements a sample in the form of a thin strip is clamped firmly at one end leaving the other end free. The clamped end of the system is then vibrated laterally at a given frequency ν and the amplitude of the vibration induced at the free end of the sample is recorded. A range of frequencies wide enough to ensure that it encompasses the resonant frequency of the sample ν_r is then examined. The resonant frequency is detected as the maximum of a graph of amplitude against frequency. The results provide information on the elastic modulus E since it is related to the square of the resonance frequency by

$$E = cL^4\rho{\nu_r}^2/D^2 \tag{13.23}$$

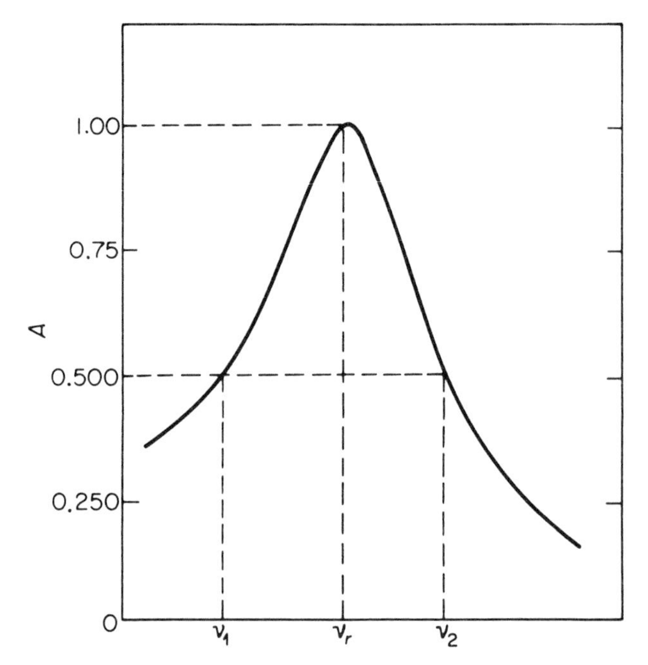

FIGURE 13.13. Typical curve obtained from a vibrating reed apparatus.

where c is a numerical constant, L is the free length of the sample, D is its thickness, and ρ is the sample density.

If the amplitudes are expressed as ratios of the amplitude to the maximum amplitude, then damping is measured from the half-width h of the curve, *i.e.*

$$h = (\nu_2 - \nu_1)/\nu_r. \tag{13.24}$$

This technique is not as useful as the torsional pendulum but covers the higher frequency range 10 to 10^3 Hz.

FORCED VIBRATION – NON-RESONANCE

Several types of instrument can be used for this type of test, and these are usually limited to measurements on rigid polymers or rubbers. One such instrument is shown in the block diagram 13.14. The sample C is attached firmly at each end to a strain gauge; one of these is a force transducer measuring the applied sinusoidal force and the other records the sample deformation. A sinusoidal tensile stress of a given frequency can be generated in the vibrator A and if the electrical vectors from the force and displacement are represented by $\bar{\alpha}_1$ and $\bar{\alpha}_2$ then by satisfying the condition $|\bar{\alpha}_1| = |\bar{\alpha}_2| = 1$ the tangent of the phase angle δ between the stress and the strain may be calculated from

$$\bar{\alpha}_1 - \bar{\alpha}_2 = 2\sin(\delta/2) \approx \tan\delta. \tag{13.25}$$

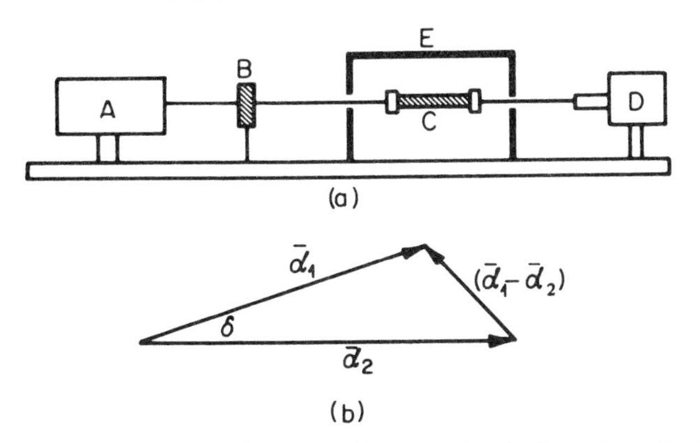

FIGURE 13.14. (a) Block diagram of apparatus for measuring the dynamic mechanical response using a non-resonance technique, (b) Vector diagram showing relation between α and δ.

This operation of adjustment followed by subtraction of the electrical vectors is performed directly in the recording circuit.

The complex elastic modulus E^* is given by

$$E^* = FL/\Delta LA, \tag{13.26}$$

where F is the amplitude of the tensile force, A is the sample cross-sectional area, L is the sample length, and ΔL is the amplitude of elongation. Tensile storage and loss moduli E' and E'' follow from $E' = E^* \cos \delta$ and $E'' = E^* \sin \delta$.

A second version now widely used for these measurements is the Polymer Laboratories DMTA instrument and a schematic diagram of the working head is shown in figure 13.15. Several damping arrangements are available for the sample so that measurements may be made in the bending, shear, or tensile, modes.

In the bending mode the sample, in the form of a small bar, is clamped firmly at both

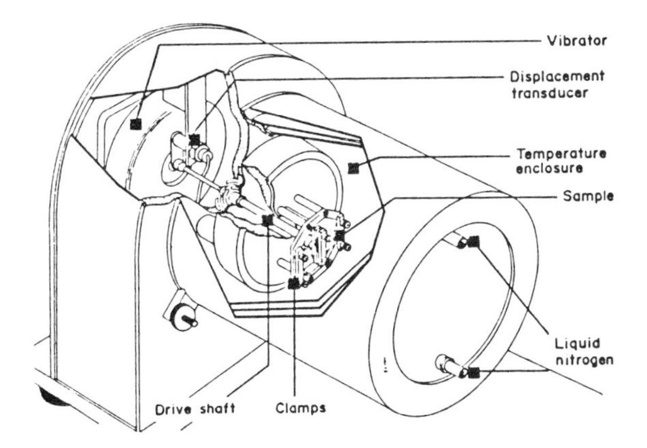

FIGURE 13.15. A schematic diagram of the measuring head for a DMTA.

ends and the central point is vibrated by means a ceramic drive shaft. This can be driven at frequencies selected from the range 0.01 to 200 Hz. The applied stress is proportional to the a.c. current fed to the drive shaft and the strain is detected using a transducer that measures the displacement of the drive clamp. Temperature can be controlled over the range 120 to 770 K, either isothermally or more normally by ramping up and down at various fixed rates.

13.11 Correlation of mechanical damping terms

The several practical methods described express the damping and moduli in slightly different forms but these can all be interrelated quite simply. In general, one can select a dissipation factor or loss tangent derived from the ratio (G''/G') or (E''/E') to represent the energy conversion per cycle. This leads to the equivalent forms

$$G''/G' = \Delta/\pi = (1/\pi n)\ln(A_1/A_n), \tag{13.27}$$

$$E''/E' = (1/\sqrt{3})(v_2 - v_1)/v_r, \tag{13.28}$$

and

$$E''/E' = \tan\delta. \tag{13.29}$$

To a first approximation it is also possible to write

$$(E''/E') = (G''/G')$$

thereby allowing use of the data from either type of measurement to characterize the sample. It should also be noted that if complex moduli are used the corresponding complex compliances are given by $(G''/G') = (J''/J')$. Moduli can also be related to the viscosity, $G' = \omega\eta''$ and $G'' = \omega\eta'$ where η is known as the dynamic viscosity.

The approximations $G \approx G'$ and $E \approx E'$ can be made when damping is low, and the absolute value for the modulus $|G|$ or $|E|$ can be related to the complex components by $|E| = \{(E')^2 + (E'')^2\}^{1/2}$. A similar expression holds for $|G|$.

13.12 Dielectric thermal analysis (DETA)

Dry polymers are very poor conductors of electricity and can be regarded as insulators. Application of an electric field to a polymer can lead to polarization of the sample, which is a surface effect, but if the polymer contains groups that can act as permanent dipoles then the applied field will cause them to align in the direction of the field. When the electric field is released, the dipoles can relax back into a random orientation, but, due to the frictional resistance experienced by the groups in the bulk polymer this will not be instantaneous. The process of disordering can be characterized by a relaxation time, but may not be easily measured. It is more convenient to apply a sinusoidally varying voltage to the sample and to study the dipole polarization under steady state conditions.

In DETA a small alternating electric field is applied to the sample and the electric charge displacement Q is measured by following the current i ($= \mathrm{d}Q/\mathrm{d}t$). The complex dielectric permittivity ε^* can be measured from the change in amplitude and, if the phase lag between the applied voltage and the outcoming current is determined (see figure 13.16), then ε^* can be resolved into the two components ε', the storage (dielectric

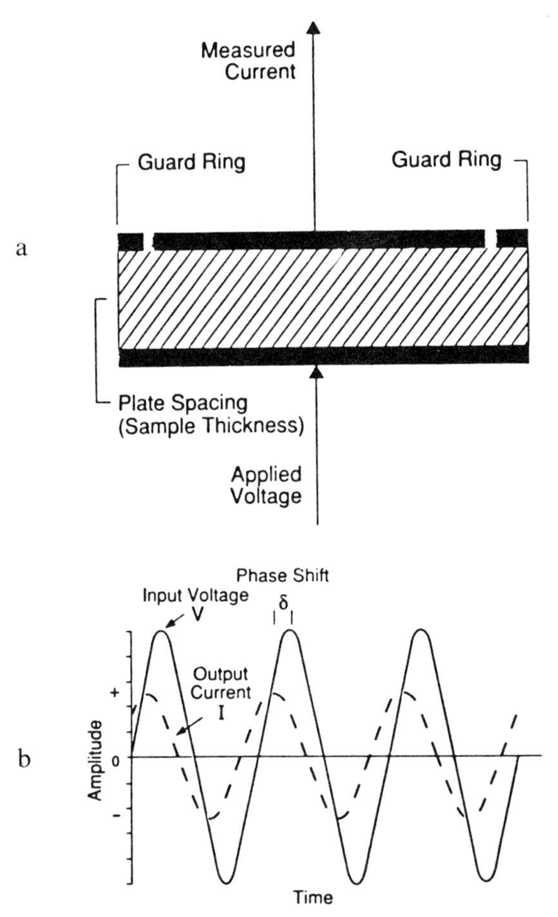

FIGURE 13.16. A schematic diagram of (a) a measuring cell and (b) the behaviour of input voltage and output current in DETA. (Adapted from T. Grentzer and J. Leckenby. *International Laboratory*, **19**(6), 34–38, July/August (1989). Copyright 1989 by International Scientific Communications Inc.)

constant) and ε'', the loss (dielectric loss). The frequencies used in the measurements must now be in the range where orientational polarization of the dipoles in the polymer is active. These frequencies are much higher than used normally for DMTA and typically lie in the range 20 Hz to 100 kHz.

While the main variable is temperature, the factors ε' and ε'' can be studied as a function of the angular frequency ω, and in the frequency region where there is a relaxation, ε' decreases as shown in figure 13.17. The magnitude of this decrease $(\varepsilon_0 - \varepsilon_\infty)$ is a measure of the strength of the molecular dipole involved in the relaxation, where ε_0 is the static dielectric constant related to the actual dipole moment of the polymer and ε_∞ is the dielectric constant measured at high frequencies. When the dielectric loss factor is measured at a characteristic frequency ω_{max}, and a given temperature, it passes through a maximum when a relaxation occurs, and the dipole

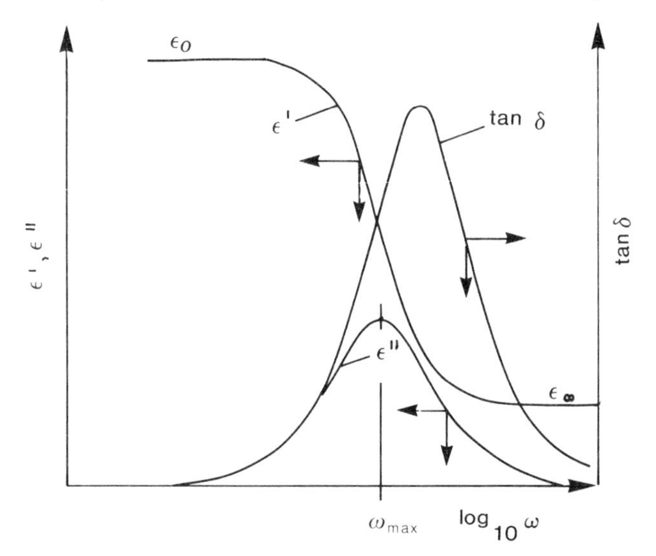

FIGURE 13.17. A schematic representation of the behaviour of the storage and loss factors ε' and ε'' as a function of the log of angular frequency ω. Also shown is the dielectric loss tangent as a function of ω.

relaxation time $\tau = 1/\omega_{max}$, can be obtained. At frequencies above ω_{max}, the dipoles cannot move fast enough to follow the alternating field so both ε' and ε'' are low. When the frequency is lower than ω_{max} the permanent dipoles can follow the field quite closely and so ε' is high because the dipoles align easily with each change in polarity; ε'' on the other hand is low again because now the voltage and the current are approximately 90° out of phase.

Dielectric relaxation processes can be described formally by the following relations:

$$\varepsilon' = \varepsilon_0 + \frac{(\varepsilon_0 - \varepsilon_\infty)}{(1 + \omega^2\tau^2)} \tag{13.30}$$

and

$$\varepsilon'' = \omega\tau + \frac{(\varepsilon_0 - \varepsilon_\infty)}{(1 + \omega^2\tau^2)} \tag{13.31}$$

A useful way of examining the data is to measure the ratio of the two factors, which gives the dielectric loss tangent

$$\tan\delta_D = (\varepsilon''/\varepsilon') \tag{13.32}$$

Dipolar groups in a polymer coil may not all be able to relax at the same speed because of the variable steric restrictions they may experience, imposed by their environment. This can be caused by the disordered packing of chains in the amorphous glassy phase, and a random distribution of the available free volume, or perhaps even by the random coil structure of the chain itself causing local environmental changes. The result is that a distribution of relaxation times is to be expected for a given process and this results in a broadening of the dielectric loss peak. Thus, the more mobile a dipolar group, the easier it is for it to follow the electric field up to higher frequencies, whereas the less mobile groups can only orient at lower frequencies.

13.13 Comparison between DMTA and DETA

Data from mechanical and dielectric measurements can be related, certainly in a qualitative, if not always in a quantitative way. Formally, the dielectric constant (ε') can be regarded as the equivalent of the mechanical compliance (J'), rather than the modulus, and this highlights the fact that mechanical techniques measure the ability of the system to resist movement, whereas the dielectric approach is a measurement of the ability of the system to move, given that the groups involved must also be dipolar. Interestingly, the dielectric loss (ε'') appears to match the loss modulus (E'' or G'') more closely than the loss compliance when data are compared for the same system.

Both techniques respond in a similar fashion to a change in the frequency of the measurement. When the frequency is increased, the transitions and relaxations that are observed in a sample appear at higher temperatures. This is illustrated from work on poly(ethylene terephthalate) where the loss peak representing the glass transition has been measured by both DMTA and DETA at several frequencies between 0.01 Hz and 100 kHz (figure 13.18). The maximum of this loss peak (T_{max}) is seen to move from a

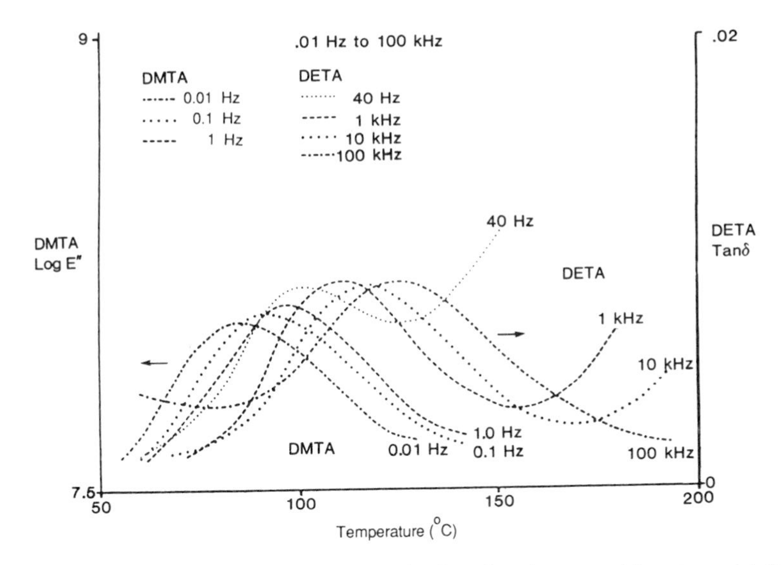

FIGURE 13.18. Comparison of the loss peaks for the glass transition in poly(ethylene terephthalate) measured at different frequencies from both dynamic mechanical and dielectric techniques. This shows the shift of the maximum temperature of the relaxation (T_{max}) to higher temperatures as the frequency of the measurement increases. (Reproduced from R. E. Wetton, M. R. Morton and A. M. Rowe, *International Laboratory*, March (1986) (figure 7, p. 80) with permission from International Laboratory).

temperature of about 360 K (0.01 Hz) to about 400 K (100 kHz), which is an increase of 40 K over a frequency change of seven orders of magnitude. This is close to the rule of thumb that the temperature for the maximum of a loss peak (T_{max}) (or a relaxation process) will change by approximately 7 K for each decade of change in frequency.

This type of measurement can be used to estimate the activation energy (ΔE^*) for a transition or relaxation process, if the frequency ν, at T_{max} is expressed as a function of reciprocal temperature according to the relation

$$\Delta E^* = \mathrm{d}(\log \nu)/\mathrm{d}(1/T_{max}) \tag{13.33}$$

Data plotted using equation (13.33) for the β-relaxation process in a series of poly(alkylmethacrylate)s are shown in figure 13.19. Both techniques have been used and separately give good straight lines with the same slope, but the fact that the lines do not overlap precisely indicates that the measurements may not be exactly equivalent.

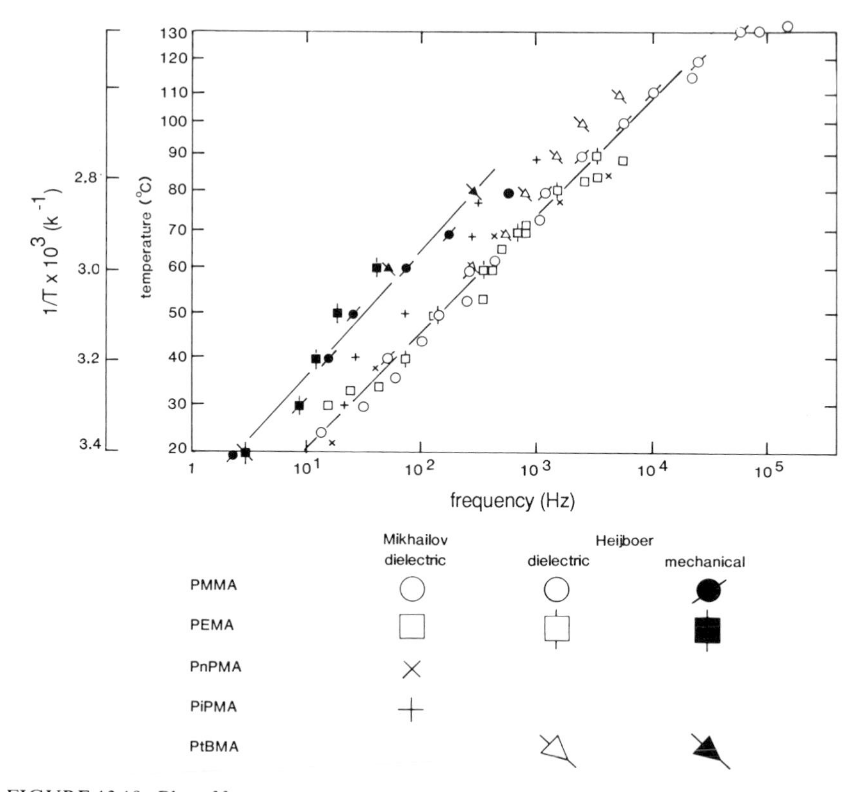

FIGURE 13.19. Plot of frequency against reciprocal temperature of the T_{max} for the β-relaxation process in poly(methyl methacrylate) PMMA; poly(ethyl methacrylate) PEMA; poly(n-propyl methacrylate) PnPMA; poly(iso-propyl methacrylate) PiPMA; and poly(t-butyl methacrylate) PtBMA. The data are measured by both dielectric (open symbols) and dynamic mechanical (filled symbols) methods, but do not fit on a common line. (Reproduced from N. G. McCrum, B. E. Read and G. Williams, *Anelastic and Dielectric Effects in Polymeric Solids*. © John Wiley and Sons Inc., N.Y. (1976)).

The results from DMTA and DETA can be used in a complementary manner to distinguish between relaxations involving polar and non polar units relaxing in the polymeric system. Thus if the response of poly(ethylene terephthalate) to both DMTA and DETA is examined, two major loss peaks can be identified in each, as seen in figure 13.20. The high temperature loss peak (α-peak) can be assigned to the glass transition and this can be confirmed by d.s.c. measurements. There is a second loss (β-relaxation) which appears at lower temperatures and suggests that there is a relaxation process active in the glassy state. It is not immediately obvious which group is responsible for this process, but it is active both mechanically and dielectrically. Examination of the polymer structures suggests that the relaxation in the glass could involve libration of the phenyl ring, motion of the oxycarbonyl unit or rearrangement of the (-O-C-C-O-) unit. From the spectra it can be seen that the intensity of the β-peak relative to the α-peak is much stronger in the dielectric response compared with

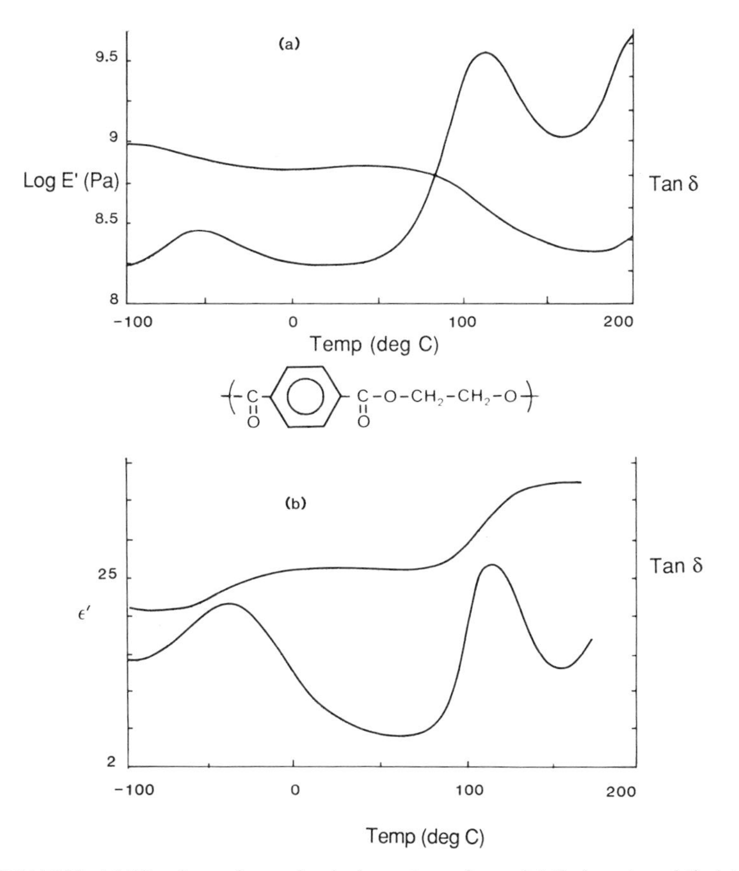

FIGURE 13.20. (a) The dynamic mechanical spectrum for poly(ethylene terephthalate) PET showing the storage modulus and tan δ as a function of temperature. (b) The dielectric (storage and tan δ) behaviour for PET in comparison with the mechanical response. (Reproduced from R. E. Wetton, M. R. Morton and A. M. Rowe, *International Laboratory*, March (1986) (Figure 13(a) and (b), p. 60 with permission from International Laboratory).

the mechanical measurements. This suggests that the group undergoing relaxation is associated with a dipole moment and thus rules out the phenyl ring libration as a likely process. This does not give irrefutable evidence of the participation of the oxycarbonyl unit but it does point in this direction.

13.14 Time-temperature superposition principle

A curve of the logarithm of the modulus against time and temperature is shown in figure 13.21. This provides a particularly useful description of the behaviour of a polymer and allows one to estimate, among other things, either the relaxation or retardation spectrum.

The practical time scale for most stress-relaxation measurements ranges from 10^1 to 10^6 s but a wider range of temperature is desirable. Such a range can be covered relatively easily by making use of the observation, first made by Leaderman, that for viscoelastic materials time is equivalent to temperature. A composite isothermal curve covering the required extensive time scale can then be constructed from data collected at different temperatures. This is accomplished by translation of the small curves along the log t axis until they are all superimposed to form a large composite curve. The technique can be illustrated using data for polyisobutylene at several temperatures. An arbitrary temperature T_0 is first chosen to serve as a reference which in the present case

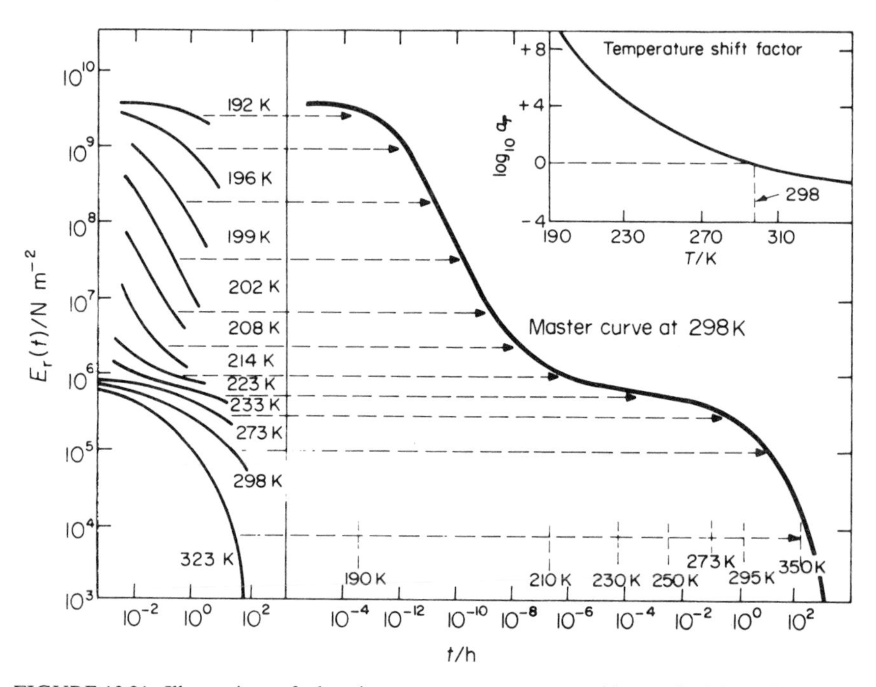

FIGURE 13.21. Illustration of the time-temperature superposition principle using stress-relaxation data for polyisobutylene. Curves are shifted along the axis by an amount represented by a_T as shown in the insert. The reference temperature in this instance is 298 K. (Adapted from Castiff and Tobolsky).

is 298 K. As values of the relaxation modulus $E_r(t)$ have been measured at widely differing temperatures, they must be corrected for changes in the sample density with temperature to give a reduced modulus, where ρ and ρ_0 are the polymer densities at T and T_0 respectively. This correction is small and can often be neglected.

$$[E_r(t)]_{red} = (T_0\rho_0/T_\rho)E_r(t). \tag{13.34}$$

Each curve of reduced modulus is shifted with respect to the curve at T_0 until all fit together forming one master curve. The curve obtained at each temperature is shifted by an amount

$$(\log t - \log t_0) = \log(t/t_0) = \log a_T \tag{13.35}$$

The parameter a_T is the shift factor and is positive if the movement of the curve is to the left of the reference and negative for a move to the right. The shift factor is a function of temperature only and decreases with increasing temperature, it is, of course, unity at T_0. The superposition principle can also be applied to creep data. Curves exhibiting the creep behaviour of polymers at different temperatures can be compared by plotting $J(t)T$ against $\log t$. This reduces all the curves at various temperatures to the same shape but displaced along the $\log t$ axis. Superposition to form a master curve is readily achieved by movement along the $\log t$ axis, where the shift factor a_T has the same characteristics as for the relaxation data. This shift factor has also been defined as the ratio of relaxation or retardation times at the temperatures T and T_0, *i.e.*

$$a_T = \tau/\tau_0 = (\eta/\eta_0)(T_0\rho_0/T\rho), \tag{13.36}$$

and is related to the viscosities. If the viscosities obey the Arrhenius equation, then by neglecting the correction factor, we can express a_T in an exponential form as

$$a_T = \exp b(1/T - 1/T_0) \tag{13.37}$$

or

$$\log_{10} a_T = -b(T - T_0)/2.303\,TT_0, \tag{13.38}$$

where b is a constant.

This equation is very similar in form to the WLF equation,

$$\log_{10} a_T = -a_1(T - T_0)/(a_2 + T - T_0) \tag{13.39}$$

For polyisobutylene, the shift factor a_T can be predicted if $T_0 = (T_g + 45\,\text{K})$ is used with $a_1 = 8.86$ and $a_2 = 101.6$ K. As outlined in chapter 12, the reference temperature is often chosen to be T_g with $a_1 = 17.44$ and $a_2 = 51.6$ K, from which a_T can be calculated for various amorphous polymers.

The superposition principle can be used to predict the creep and relaxation behaviour at any temperature if some results are already available, with the proviso that the most reliable predictions can be made for interpolated temperatures rather than long extrapolations.

The principle can also be applied to dielectric data which can be shifted either along the temperature or the frequency axis. An example of the latter type of shift is shown in

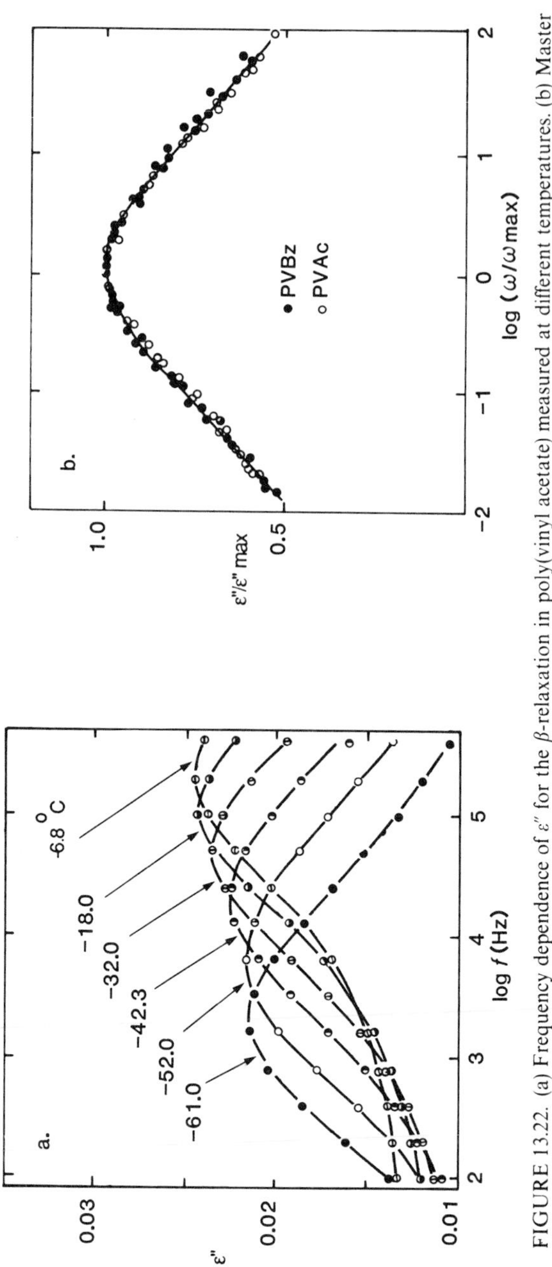

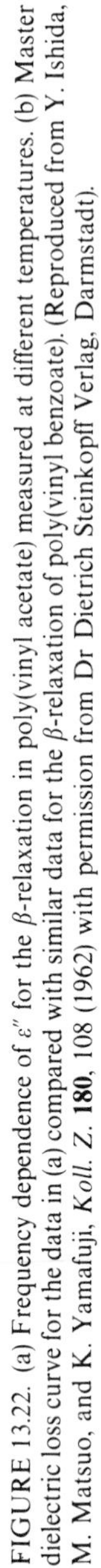

FIGURE 13.22. (a) Frequency dependence of ε'' for the β-relaxation in poly(vinyl acetate) measured at different temperatures. (b) Master dielectric loss curve for the data in (a) compared with similar data for the β-relaxation of poly(vinyl benzoate). (Reproduced from Y. Ishida, M. Matsuo, and K. Yamafuji, *Koll. Z.* **180**, 108 (1962) with permission from Dr Dietrich Steinkopff Verlag, Darmstadt).

figure 13.22, where instead of time dependence measurements the frequency dependence of the β-relaxation in poly(vinyl acetate) has been studied at fixed temperatures in the range 212 to 266 K. A master curve can be constructed for this relaxation region by plotting $(\varepsilon''/\varepsilon''_{max})$ against $\log_{10}(\omega/\omega_{max})$, where the 'max' subscript refers to the peak maximum at each experimental temperature.

13.15 A molecular theory for viscoelasticity

So far the interpretation of viscoelastic behaviour has been largely phenomenological, relying on the application of mechanical models to aid the elucidation of the observed phenomena. These are, at best, no more than useful physical aids to illustrate the mechanical response and suffer from the disadvantage that a given process may be described in this way using more than one arrangement of springs and dashpots. In an attempt to gain a deeper understanding on a molecular level, Rouse, Zimm, Bueche, and others have attempted to formulate a theory of polymer viscoelasticity based on a chain model consisting of a series of sub-units. Each sub-unit is assumed to behave like an entropy spring and is expected to be large enough to realize a Gaussian distribution of segments (*i.e.* > 50 carbon atoms). This approach, although still somewhat restrictive has led to reasonable predictions of relaxation and retardation spectra.

One starts with a single isolated chain and the assumption that it exhibits both viscous and elastic behaviour. If the chain is left undisturbed it will also adopt the most notable conformation or segmental distribution, so that, with the exception of high frequencies, the observed elasticity is predominantly entropic. Thus the application of a stress to the molecule will cause distortion, by altering the equilibrium conformation to a less probable one, resulting in a decrease in the entropy and a corresponding increase in the free energy of the system. When the stress is removed the chain segments will diffuse back to their unstressed positions even though the whole molecule may have changed its spatial position in the meantime. If on the other hand, the stress is maintained, strain relief is sought by converting the excess free energy into heat, thereby stimulating the thermal motion of the segments back to their original positions. *Stress-relaxation* is then said to have occurred. For a chain molecule composed of a large number of segments, movement of the complete molecule depends on the co-operative movement of all the segments, and as stress-relaxation depends on the number of ways the molecule can regain its most probable conformation, each possible co-ordinated movement is treated as a mode of motion with a characteristic relaxation time. For simplicity we can represent the polymer as in figure 13.23.

The first mode $p = 1$ represents translation of the molecule as a whole and has the longest relaxation time τ_1 because the maximum number of co-ordinated segmental movements are involved. The second mode $p = 2$ corresponds to the movement of the chain ends in opposite directions; for $p = 3$, both chain ends move in the same

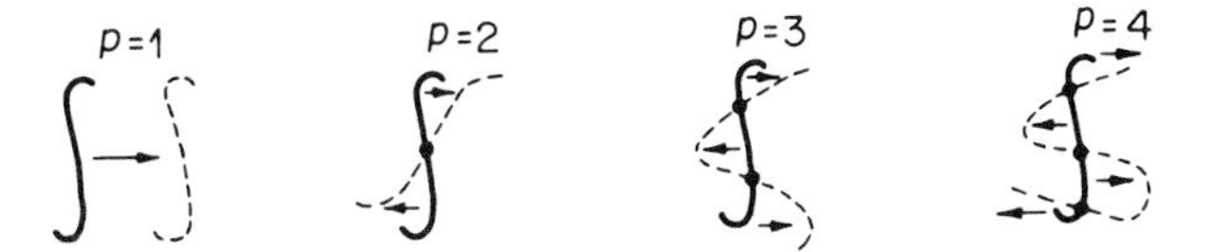

FIGURE 13.23. First four normal modes of movement of a flexible polymer molecule.

direction, but the centre moves in the opposite direction. Higher modes 4, 5... m follow involving a progressively decreasing degree of co-operation for each succeeding mode and correspondingly lower relaxation times τ_p. This means that a single polymer chain possesses a wide distribution of relaxation times. Using this concept, Rouse considered a molecule in dilute solution under sinusoidal shear and derived the relations

$$\eta' = (G''/\omega) = (nkT/\omega) \sum_{p=1}^{m} \omega^2\tau_p^2/(1 + \omega^2\tau_p^2), \tag{13.40}$$

$$\eta' = (G''/\omega) = (nkT/\omega) \sum_{p=1}^{m} \omega^2\tau_p^2/(1 + \omega^2\tau_p^2), \tag{13.41}$$

$$\tau_p = 6(\eta - \eta_s)/(\pi^2 p^2 nkT), \tag{13.42}$$

where η and η_s are the viscosities of the solution and the solvent respectively, n is the number of molecules per unit volume, k is the Boltzmann constant, and ω is the angular frequency of the applied stress which is zero for steady flow. These equations are strictly applicable only to dilute solutions of non-draining monodisperse coils, but can be extended to undiluted polymers above their glass temperature if suitably modified. This becomes necessary when chain entanglements begin to have a significant effect on the relaxation times. The undiluted system is represented as a collection of polymer segments dissolved in a liquid matrix composed of other polymer segments and η_s can be replaced by a monomeric frictional coefficient ζ_0. This provides a measure of the viscous resistance experienced by a chain and is characteristic of a given polymer at a particular temperature. The continuous relaxation and retardation spectra calculated from the Rouse theory are

$$H(\tau) = (\rho N_A/2\pi M)(r_0^2 NkT\zeta_0/6\tau)^{1/2}, \tag{13.43}$$

and

$$L(\tau) = (2M/\pi\rho N_A)(6\tau_R/r_0^2 NkT\zeta_0)^{1/2}, \tag{13.44}$$

where r_0^2 is the unperturbed mean square end-to-end distance of a chain of molar mass M and density ρ containing N monomer units. The equations predict linearity in the plots log $H(\tau)$ and log $L(\tau_R)$ against log τ with slopes of $-\frac{1}{2}$ and $+\frac{1}{2}$ respectively. Comparison with experimental results for poly(methyl acrylate) shows validity only for longer values of the relaxation and retardation times.

The Rouse model only pertains to the region covering intermediate τ values. The reason for this lies in the response of a polymer to an alternating stress. At low frequencies Brownian motion can relieve the deformation caused by the stress before the next cycle takes place, but as the frequency increases the conformational change begins to lag behind the stress and energy is not only dissipated but stored as well. Finally at very high frequencies only enough time exists for bond deformation to occur. As it was stipulated that each segment be long enough to obey Gaussian statistics, short relaxation times may not allow a segment sufficient time to rearrange and regain this distribution. Thus the contribution from short segments to the distribution functions tends to be lost and deviations from the theoretical represent departure from ideal Gaussian behaviour.

This approach to viscoelastic theory is reasonably successful in the low modulus regions but it requires considerable modification if the high modulus and rubbery plateau regions are to be described.

General Reading

J. J. Aklonis and W. J. MacKnight, *Introduction to Polymer Viscoelasticity.* John Wiley and Sons Ltd (1983).
G. Allen and J. C. Bevington, Eds, *Comprehensive Polymer Science.* Vol. 2. Pergamon Press (1989).
R. T. Bailey, A. M. North, and R. A. Pethrick, *Molecular Motion in High Polymers.* Oxford University Press (1981).
F. Bueche, *Physical Properties of Polymers.* Interscience Publishers Inc. (1962).
M. Doi and S. F. Edwards, *The Theory of Polymer Dynamics.* Oxford University Press (1986).
J. D. Ferry, *Viscoelastic Properties of Polymers.* John Wiley and Sons (1979).
P. Hedvig, *Dielectric Spectroscopy of Polymers.* Adam Hilger Ltd (1977).
J. E. Mark, A. Eisenberg, W. W. Graessley, L. Mandelkern and J. L. Koenig, *Physical Properties of Polymers.* American Chemical Society (1984).
N. G. McCrum, B. E. Read and G. Williams, *Anelastic and Dielectric Effects in Polymeric Solids.* John Wiley and Sons Ltd (1967).
P. Meares, *Polymers: Structure and Bulk Properties,* Chapters 9 and 11. Van Nostrand (1965).
L. E. Nielsen, *Mechanical Properties of Polymers.* Reinhold Publishing Corp. (1962).
L. E. Nielsen, *Mechanical Properties of Polymers and Composites.* Vols 1 and 2. Marcel Dekker Inc. (1974).
L. H. Sperling, *Introduction to Physical Polymer Science.* John Wiley and Sons Ltd. (1986).
A. V. Tobolsky, *Properties and Structure of Polymers.* Interscience Publishers Inc. (1960).

CHAPTER 14

The Elastomeric State

14.1 General introduction

Most materials, when stressed, exhibit a limited elastic region where the material regains its original dimensions if the stress is removed. As the resulting strain is related to the extent of movement of atoms from their equilibrium conditions, substances such as metals and glass have elastic limits rarely exceeding 1 per cent because atomic adjustments are localized. For long chain polymers, under certain conditions, the situation is different; the extensive covalent bonding between the atoms to form chains allows considerable deformation, which is accompanied by long and short range co-operative molecular rearrangement arising from the rotation about chain bonds.

One of the first materials found to exhibit a sizeable elastic region was a natural substance obtained from the tree *Hevea brasiliensis*, now known to us as *cis*-polyisoprene, or more commonly referred to as rubber. A large number of polymers, with rubber-like characteristics at ambient temperatures, are now available and it is preferable to call the general group of polymers, elastomers. Elastomers possess several significant characteristics:

(1) The materials are above their glass temperature;
(2) They possess the ability to stretch and retract rapidly;
(3) They have high modulus and strength when stretched;
(4) The polymers have a low or negligible crystalline content;
(5) The molar mass is large enough for network formation or they must be readily crosslinked.

The most important factor is of course the T_g as this determines the range of temperatures where elastomeric behaviour is important and defines its lower limiting temperature. Hence polymers with T_g below ambient may be useful elastomers, if they are essentially amorphous, whereas this is unlikely if T_g is in excess of 400 K. Environmental temperature is, of course, an important factor, and on a different planet, much colder than our own, latex rubber could prove to be a good glassy material, while on a hot planet, perspex could well be adapted for use as an elastomer.

One of the features of an elastomer is its ability to deform elastically by elongating up to several hundred per cent, but as we have seen in chapter 12 chain slippage occurs under prolonged tension and the sample deforms. This flow under stress can be greatly reduced by introducing crosslinks between the chains. In effect, these crosslinks act as anchors or permanent entanglements and prevent the chains slipping past each other.

The process of crosslinking is known generally as vulcanization and the resultant polymer is a network of interlinked molecules now capable of maintaining an equilibrium tension. This is most important as it changes the properties of an elastomer to a marked degree and extends the usefulness of the polymer as a material.

NATURAL RUBBER (NR)

Naturally occurring rubber is a linear polymer of isoprene units linked 1,4 and because

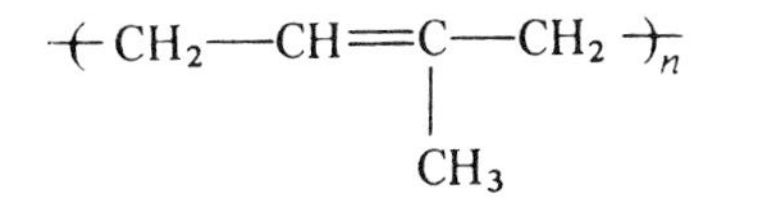

the chain is unsaturated, two forms are found. Natural rubber is the *cis* form; it has low crystallinity, $T_g = 200$ K and $T_m = 301$ K, whereas the *trans* form, called gutta percha

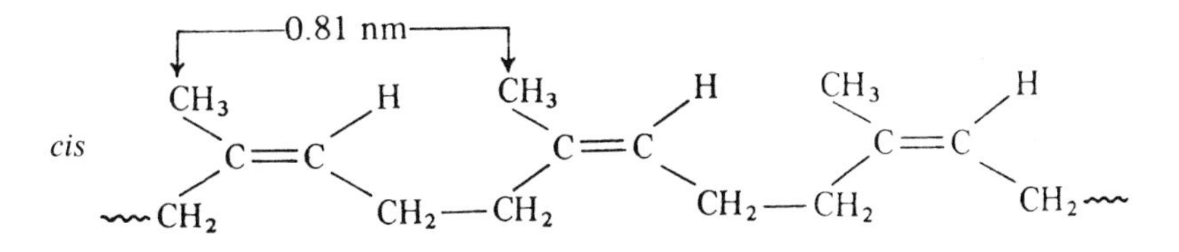

or balata, is of medium crystallinity with $T_g = 200$ K and $T_m = 347$ K.

The remarkable effect of *cis* and *trans* isomerism on the properties is well illustrated in the case of the polyisoprenes. The more extended all *trans* form of balata allows the polymer to develop a greater degree of crystallinity and order. This is reflected in its hard tough consistency which makes it suitable, in the vulcanized state, for golfball covers. X-ray diffraction reveals two forms of gutta percha: an α-form with a molecular repeat distance of 0.88 nm, slightly larger than a *cis* unit

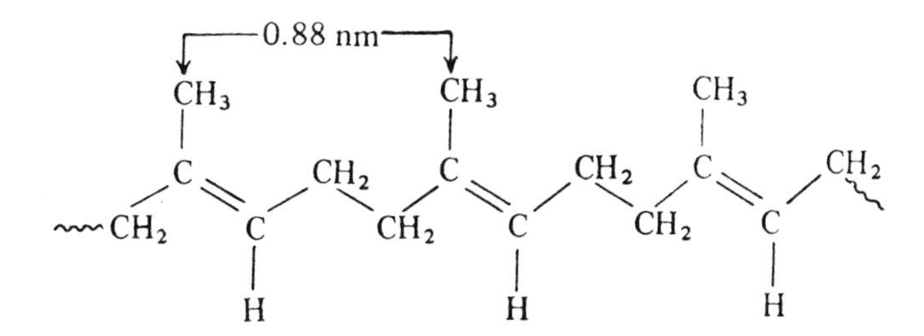

and a more compact β-form with a repeat distance of 0.47 nm.

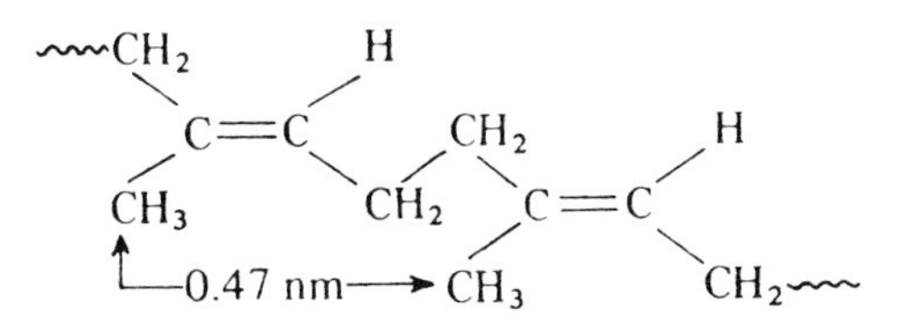

The chains of the *cis*-polyisoprene are more easily rotated than their *trans* counterparts and as a result the molecules prefer to coil up into a compact conformation. The viscoelastic behaviour and long range elasticity arise from this random arrangement of long, freely moving chains, and recoverable deformations of up to 1000 per cent can be observed. In the raw state natural rubber is a tacky substance rather difficult to handle, has poor abrasion resistance, and is sensitive to oxidative degradation. It can be used for making crêpe soles and adhesives but it is immensely improved if vulcanized, a process which cross-links the rubber. Vulcanization enhances resistance to degradation and increases both the tensile strength and elasticity.

14.2 Experimental vulcanization

The process of vulcanization was discovered independently by Goodyear (1839) in the U.S.A. and Hancock (1843) in the U.K. Both found that when natural rubber was heated with sulphur, the undesirable properties of surface tackiness and creep under stress could be eliminated. The chemical reaction involves the formation of interchain links, composed of two, three, or four sulphur atoms, between sites of unsaturation on adjacent chains. It has been found that about three parts sulphur per hundred parts rubber produces a useful elastomer, capable of reversible extensions of up to 700 per cent. Increasing the sulphur, up to 30 parts per hundred, alters the material drastically and produces a hard, highly crosslinked substance called Ebonite. The actual mechanism of the cross-linking reaction is still in some doubt but it is thought to proceed via an ionic route.

The quantitative introduction of crosslinks using sulphur is difficult to achieve. An alternative method involves heating the polymer with either dicumyl peroxide or ditertiary butyl peroxide and this is applicable to both polydienes and elastomers with no unsaturated sites in the chain, *e.g.* the ethylene-propylene copolymers or the polysiloxanes. The peroxide radical abstracts hydrogen from the polymer chain and creates a radical site on the interior of the chain. Two such sites interact to form the crosslink. It is claimed that one peroxide molecule produces one crosslink, but side reactions may impair the attainment of such a precise ratio. The major disadvantage of this technique is its commercial inefficiency; better results can be obtained by synthesizing precursors which are more suitable for crosslinking, such as in ethylene-propylene terpolymers. Recently room temperature vulcanization techniques have been developed for silicone elastomers. These are based on linear polydimethylsiloxane chains terminated by hydroxyl groups. Curing can be achieved either by adding a crosslinking agent and a metallic salt catalyst such as tri- or tetra-alkoxysilane with stannous octoate or by incorporating in the mixture a crosslinking agent sensitive to atmospheric water which initiates vulcanization.

A crosslinking technique, recently developed, makes use of the reactive nitrene intermediates formed from compounds of the type $N_3COO(CH_2)_nOOCN_3$. If these are heated in the presence of linear polymers such as polyethylene or polypropylene, nitrogen is lost and the resulting dinitrene reacts with the polymer chains to crosslink them. This reaction is particularly useful when no unsaturated sites exist in the chain. The crosslinking also serves to improve the resilience of polyethylene and other related polymers.

14.3 Properties of elastomers

Elastomers exhibit several other unusual properties which can be attributed to their chain-like structure. It has been found that (a) as the temperature of an elastomer increases, so too does the elastic modulus, (b) an elastomer becomes warm when stretched, and (c) the expansivity is positive for an unstretched sample but negative for a sample under tension. As these properties are so different from those observed for other materials they are worth examining in detail.

In simple mechanistic terms, the elastic modulus is simply a measure of the resistance to the uncoiling of randomly oriented chains in an elastomer sample under stress. Application of a stress eventually tends to untangle the chains and align them in the direction of the stress, but an increase in temperature will increase the thermal motion of the chains and make it harder to induce orientation. This leads to a higher elastic modulus. Under a constant force some chain orientation will take place, but an increase in temperature will stimulate a reversion to a randomly coiled conformation and the elastomer will contract.

While this is a satisfactory picture to describe properties (a) and (c), it is possible to derive the more rigorous thermodynamic explanation which follows.

14.4 Thermodynamic aspects of rubber-like elasticity

As early as 1806 John Gough made two interesting discoveries when studying natural rubber. He found that (1) the temperature of rubber changed when a rapid change in sample length was induced, and (2) a rubber sample under constant tension changed length as the temperature changed, *i.e.* observations (b) and (c) above.

The first point is readily demonstrated with a rubber band. If the centre of the band is placed lightly touching the lips and then extended rapidly by pulling on both ends, a sensation of warmth can be felt as the temperature of the rubber rises.

In thermodynamic terms process (1) is analogous to the change in temperature undergone by a gas subjected to a rapid volume change and can be treated formally in a like manner. An ideal gas can only store energy in the form of kinetic energy and when work is performed on a gas during compression, the energy appears as kinetic energy or heat causing the temperature to rise. Extension of an elastomer results in the evolution of heat for similar reasons.

This effect can be examined further by studying the *reversible adiabatic extension* of an elastomer. Although this experiment is more easily carried out under conditions of constant pressure rather than constant volume, it is best to derive the relevant equations for constant volume. For this reason we consider first the Helmholtz function for the system

$$A = U - TS. \tag{14.1}$$

If the applied force is f, and l° and l are the lengths of the sample in the unextended and extended states then differentiation of equation (14.1) with respect to l, at constant temperature gives

$$(\partial A/\partial l)_T = (\partial U/\partial l)_T - T(\partial S/\partial l)_T. \tag{14.2}$$

The work done by the system during a reversible extension of the sample by an amount dl against the restoring force f, is given by d$A = -f$ dl and so

$$f = (\partial U/\partial l)_T - T(\partial S/\partial l)_T = f_U + f_S. \tag{14.3}$$

The force f is seen to be composed of two contributions; the energy f_U and the entropy f_S. For an ideal elastomer the contribution of f_U to the total force is negligible because there is no energy change during extension and

$$f = -T(\partial S/\partial l)_T. \tag{14.4}$$

This is the expression for an entropy spring and shows that the strain in a stretched elastomer is caused by a reduction in conformational entropy (see section 8.1) of the chains under stress. If the interdependence of length and temperature is now examined, the quantity of interest is $(\partial T/\partial l)$ and

$$(\partial T/\partial l)_{S,p} = -(\partial T/\partial S)_{l,p}(\partial S/\partial l)_{T,p}. \tag{14.5}$$

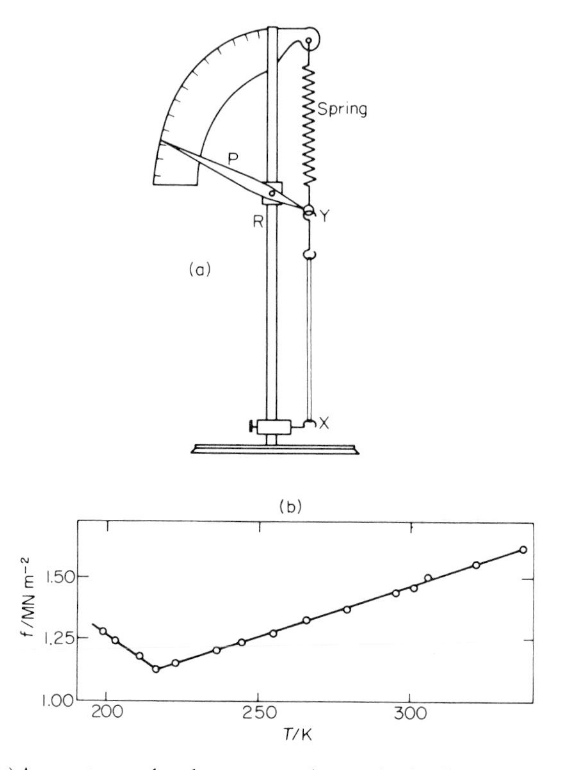

FIGURE 14.1. (a) Apparatus used to demonstrate thermoelastic effects in elastomers. (b) Tensile force f – temperature plot for rubber. The minimum occurs in the region of the glass transition temperature. (From data of Meyer and Ferri, 1935.)

Each factor can now be evaluated separately as follows

$$(\partial T/\partial S)_{l,p} = (\partial T/\partial H)_{l,p}(\partial H/\partial S)_{l,p} = T/C_{p,l}, \tag{14.6}$$

while starting from the Maxwell equation $(\partial A/\partial T)_V = -S$ and differentiating both sides with respect to l, leads to

$$(\partial/\partial l)(\partial A/\partial T) = (\partial/\partial T)(\partial A/\partial l) = (\partial f/\partial T)_l = -(\partial S/\partial l)_T. \tag{14.7}$$

It follows then from equations (14.5), (14.6), and (14.7) that

$$(\partial T/\partial l)_{S,p} = (T/C_{p,l})(\partial f/\partial T)_{l,p}. \tag{14.8}$$

This shows that for a rapid reversible adiabatic extension when $(\mathrm{d}f/\mathrm{d}T)_{l,p}$ is positive, the temperature of the elastomer increases. The equation also tells us that the elastomer will contract, not expand on heating.

Demonstration of the Gough effect. This thermoelastic effect can be conveniently demonstrated using an apparatus like that shown in figure 14.1(a). A spring with one end free and the other attached to a frame is hooked onto a pointer P, pivoted on the support rod R. A rubber band is stretched between a fixed hook X and the other end Y of the spring-pointer arrangement. If the band is encased in a glass tube it can be heated with a bunsen producing a contraction which is indicated by the movement of the pointer. Results showing the temperature dependent behaviour of an elastomer held under constant tension, are plotted in figure 14.1(b). Between 210 and 330 K the tensile force increases as T increases indicating a contraction of the elastomer, but the behaviour is reversed below 210 K as the material passes through the glass transition and the polymer reacts normally like a glassy solid.

14.5 Non-ideal elastomers

The behaviour of most elastomers under stress is far from ideal and a significant contribution from f_U is found. This can be expressed in several ways and following on from equations (14.3) and (14.7):

$$f_s = T(\partial f/\partial T)_{V,l}, \tag{14.9}$$

so that

$$(f_U/f) = (1 - f_S/f) = 1 - (T/f)(\partial f/\partial T)_{V,l}, \tag{14.10}$$

or

$$(f_U/f) = -\{\partial \ln(f/T)/\partial \ln T\}_{V,l}. \tag{14.11}$$

The experimental determination of the quantity of $(\partial f/\partial T)_{V,l}$ at constant volume is extremely difficult to perform and it is much more convenient to work at constant pressure. Approximate relations between the quantities have been suggested by Flory who proposed that equation (14.11) be modified by an additional term

$$f_U/f = -\{\partial \ln(f/T)/\partial \ln T\}_{p,l} - \alpha T/\lambda^3 - 1) \tag{14.12}$$

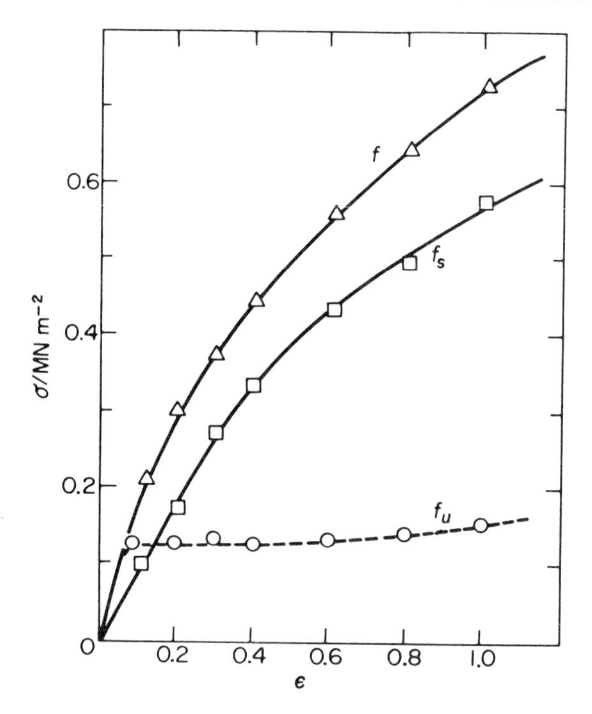

FIGURE 14.2. Stress-strain ($\sigma-\varepsilon$) curves derived from data in figure 14.3, showing the magnitude of the contributions from f_U and f_S. (Adapted from Beevers *Experiments in Fibre Physics*.)

where λ is the extension ratio (l/l^o) and α is the expansivity. The modifying factor is the difference between the values of $(\partial f/\partial T)$ measured at constant volume and constant pressure.

Thermoelastic data showing the relative contributions of f_U and f_S appear as the curves in figure 14.2 and it can be seen that at low extensions f_U remains small but begins to increase as the elongation rises.

Experimental determination of f_U *and* f_S. The curves in figure 14.2 can be derived using a stress-relaxation balance. (c.f. chapter 13.) A rubber band is stretched between two hooks and the force required to maintain a constant deformation is measured at various temperatures. If a number of extensions are examined, a family of curves (figure 14.3) can be constructed starting with the highest temperature at each strain. As the temperature is reduced incrementally, the load must be adjusted accordingly. Here the stress is taken as the ratio of the load to the unstrained cross-sectional area of the elastomer (best measured simply with an accurate ruler), while the strain is just the extension of the sample. Values of f_S and f_U at each strain are calculated from the slope and the intercept at 0 K respectively. In the diagram $f_S = -(T\partial S/\partial l)$ has been calculated for $T = 298$ K.

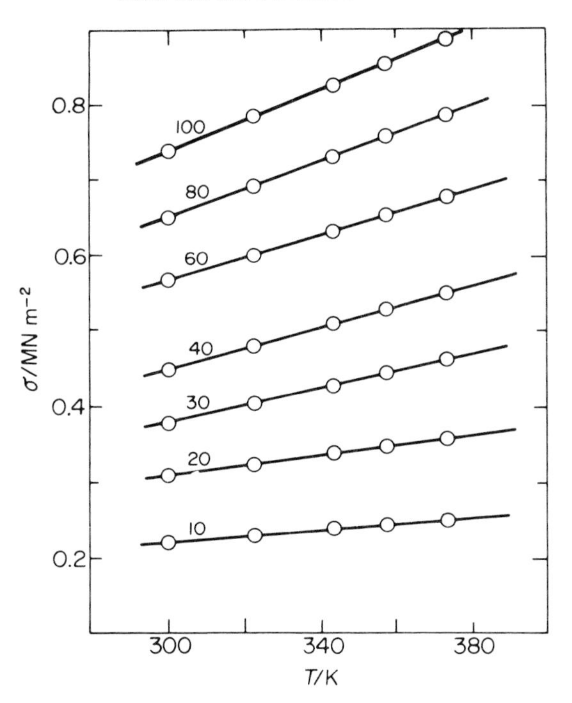

FIGURE 14.3. Thermoelastic behaviour of a rubber sample. Stress-temperature (σ–T) curves for a series of extension values. The percentage strain is shown against each curve. (Adapted from Beevers.)

14.6 Distribution function for polymer conformation

The retractive force in a rubbery material is a direct result of the chain in the extended form wanting to regain its most probable, highly coiled conformation. Thus it is of considerable interest to calculate, in addition to the average dimensions of the polymer chain, the distribution of all the possible shapes available to the molecules experiencing thermal vibrations.

This can be accomplished by first considering a chain in three-dimensional space with one end located at the origin of a set of Cartesian co-ordinates (see figure 14.4). The probability that the other end will be found in a volume element (dx, dy, dz) at the point (x, y, z) is given by $p(x, y, z)\mathrm{d}x, \mathrm{d}y, \mathrm{d}z$, where $p(x, y, z)$ measures the number of possible conformations the chain can adopt in the range $(x + \mathrm{d}x)$, $(y + \mathrm{d}y)$, and $(z + \mathrm{d}z)$. This is known as the *probability density* and can be expressed in terms of the parameter $\beta = (3/2nl^2)^{1/2}$ as

$$(\mathrm{d}x, \mathrm{d}y, \mathrm{d}z)p(x, y, z) = (\beta^2/\pi^{3/2})\exp\{-\beta^2(x^2 + y^2 + z^2)\}\,\mathrm{d}x.\mathrm{d}y.\mathrm{d}z \qquad (14.13)$$

It now remains to calculate the probability $p(r)$ that the chain is located in a spherical shell of thickness dr and distance r from the origin. This is given by

$$p(r)\mathrm{d}r = 4\pi r^2\mathrm{d}r\{(\beta^3/\pi^{3/2})\exp(-\beta^2 r^2)\}, \qquad (14.14)$$

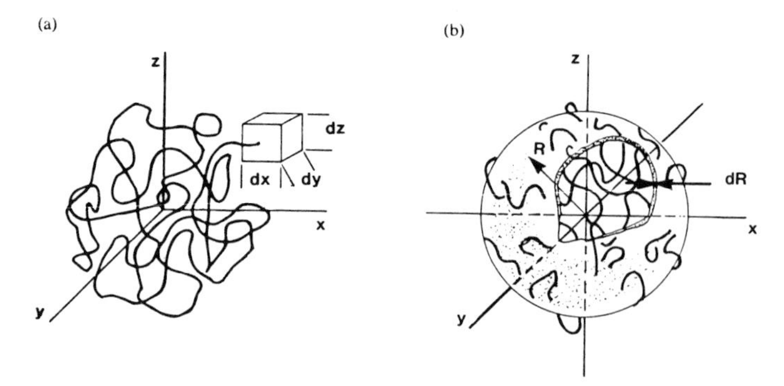

FIGURE 14.4. (a) Schematic diagram of a flexible polymer coil with one end contained in a small volume element (dx, dy, dz) and the other located at the origin of the co-ordinates. (b) Hypothetical sphere containing the coil.

where $4\pi r^2\,\mathrm{d}r$ is the volume of the shell. This function has the form shown in figure 14.5 where the maximum corresponds to the most probable distance between chain ends and is given by

$$r^2 = \int_0^\infty r^2 p(r)\mathrm{d}r = 3/2\beta^2 = l^2 n. \tag{14.15}$$

It has already been stated that while a highly coiled conformation is the most probable for an elastomer the energy of an ideal elastomer in the extended form is the same as that in the coiled state. The elastic retractive force is entropic and not energetic in origin and depends on the fact that the number of possible ways a polymer coil can exist in a highly compact form is overwhelmingly greater than the number of available arrangements of the chain segments in an extended ordered form. This means that the

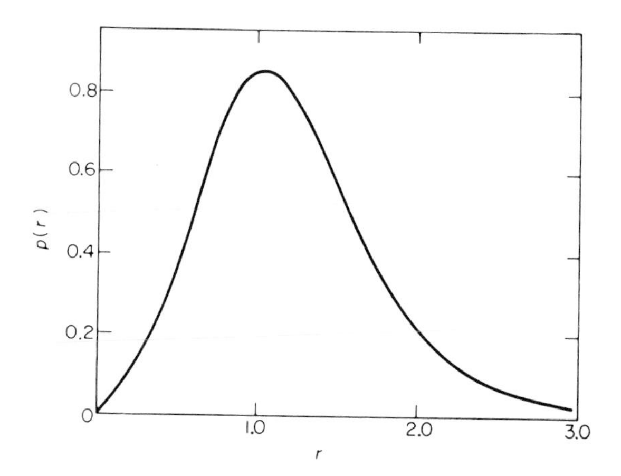

FIGURE 14.5. Distribution function $p(r)$ for end-to-end distances r calculated from equation (14.14). The maximum occurs at $r = 1/\beta$.

probability of finding a chain in a coiled state is high and as probability and entropy are related by the Boltzmann equation,

$$S = k \ln p, \tag{14.16}$$

then the most likely state for the chain is one of maximum entropy. Substitution in equation (14.14) leads to

$$S = C - k\beta^2 r^2, \tag{14.17}$$

where C is a constant. This relation provides a measure of the entropy of an ideal flexible chain whose ends are held a distance r apart.

While an elastomer is considered to be solely an entropy spring, it is never perfect and small changes in the internal energy are observed when it is under stress.

14.7 Statistical approach

Having examined the thermodynamic approach we can now outline briefly the stress-strain behaviour of an elastomer in terms of the chain conformations.

Consider a lightly crosslinked network with the junction points sufficiently well spaced to ensure that the freedom of movement of each chain section is unrestricted. If the length of a chain between two crosslinking points is assumed to be r, the probability distribution calculated above can be applied to the network structure. The entropy of a single chain, as described by equation (14.17) can be used to calculate S for a chain in the network, but if the stress-strain relations are required, then S for chains in both the deformed and undeformed states must be calculated, and for the complete network an integration over all the chains in the sample should be performed.

When a unit cube of elastomer is stretched, the resulting entropy change is

$$\Delta S = -\tfrac{1}{2} Nk(\lambda_1^2 + \lambda_2^2 + \lambda_3^2 - 3), \tag{14.18}$$

where N is the number of individual chain segments between successive crosslinks per unit volume, and λ_1, λ_2, and λ_3 are the principal extension ratios. For an ideal elastomer there is no change in the internal energy and the work of deformation w is derived entirely from $w = -T\Delta S$ so that

$$w = \tfrac{1}{2} NkT(\lambda_1^2 + \lambda_2^2 + \lambda_3^2 - 3). \tag{14.19}$$

EXPERIMENTAL STRESS-STRAIN RESULTS

The treatment of mechanical deformation in elastomers is simplified when it is realized that the Poisson ratio (table 13.1) is almost 0.5. This means that the volume of an elastomer remains constant when deformed and if one also assumes that it is essentially incompressible ($\lambda_1\lambda_2\lambda_3 = 1$) the stress-strain relations can be derived for simple extension and compression using the stored energy function w.

Simple extension. The required conditions are $\lambda = \lambda_1$ and $\lambda_2 = \lambda_3 = \lambda^{-1/2}$. Elimination of λ_3 and substitution of λ leads to

$$w = \tfrac{1}{2}NkT(\lambda^2 + 2/\lambda - 3). \tag{14.20}$$

As the force f is simply $= (dw/d\lambda)$, differentiation provides a relation between f and λ

$$f = G(\lambda - \lambda^{-2}), \tag{14.21}$$

where the modulus factor is $G = NkT$ and shows that G will depend on the number of crosslinks in the sample. If we neglect the fact that the chains in a network possess free ends and assume that all network chains end at two cross-linking points then

$$G = RT\rho M_s^{-1}, \tag{14.22}$$

where R is the gas constant, ρ the polymer density, and M_s is the number average molar mass of a chain section between two junction points in the network. The product (ρM_s^{-1}) is then a measure of the crosslink density of a sample. This also shows that the modulus will increase with temperature. The statistically derived equation (14.21) was tested first by Treloar for both extension and compression of a rubber vulcanizate. A value of $G = 0.392\ \mathrm{MN\,m^{-2}}$ was chosen to fit the data at low extensions where the application of Gaussian statistics to the chain might be expected to be valid. The experimental data for extension, figure 14.6 drops below the theoretical curve at $\lambda \approx 2$ then sweeps up sharply when λ exceeds 6. This rapid increase was originally attributed to the crystallization of the polymer under tension but it is now believed to reflect the departure of the network from the assumed Gaussian distribution.

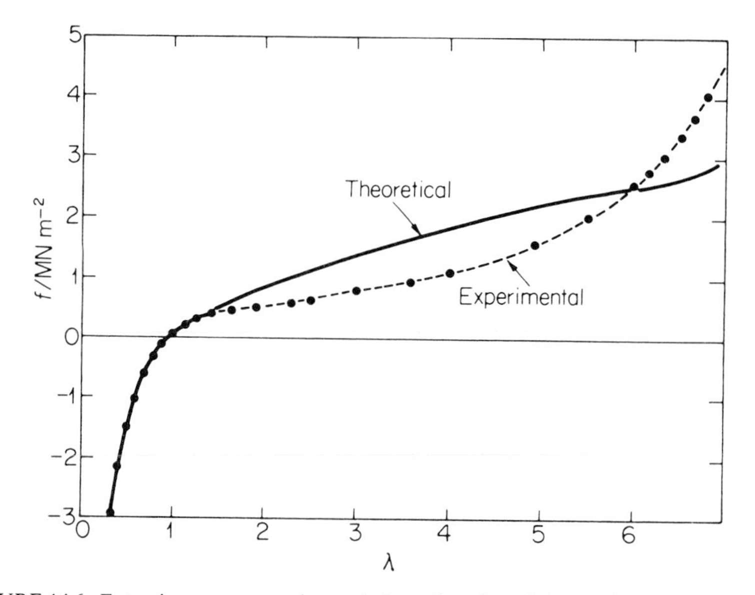

FIGURE 14.6. Extension or compression ratio λ as a function of the tensile or compressive force f for a rubber vulcanizate. Theoretical curve is derived from equation (14.21) using $G = 0.392\ \mathrm{MN\,m^{-2}}$. (From data by Treloar, 1944.)

Simple compression. The agreement between experiment and theory for the equivalent of simple compression is somewhat better. The experimental determination of compression is rather difficult to carry out, but a two-dimensional extension with $\lambda_2 = \lambda_3$ serves to provide the same information. This is achieved by clamping a circular sheet of rubber around the circumference and then inflating it to provide a stress. Correlation between the results and a curve derived from equation (14.21), using the same G factor as before, is good.

The statistical approach, using a Gaussian distribution, thus appears to predict the stress-strain response except at moderately high elongations.

Pure shear. The corresponding equation to describe the behaviour of an elastomer under shear is $f = G(\lambda - \lambda^{-3})$ but again agreement is reasonable only at low λ.

Large elastic deformation. At high extensions, departure from the Gaussian chain approximation becomes significant, and has led to the development of a more general, but semi-empirical, theory based on experimental observations. This is expressed in the Mooney, Rivlin, and Saunders, MRS equation

$$\tfrac{1}{2}f(\lambda - \lambda^{-2})^{-1} = C_1 + C_2\lambda^{-1}, \tag{14.23}$$

where C_1 and C_2 are constants. Unfortunately this simple form is still only capable of predicting data over a range of low to moderate extension, but not for samples under compression.

14.8 Swelling of elastomeric networks

A crosslinked elastomer cannot dissolve in a solvent. Dispersion is resisted because the crosslinks restrict the movement and complete separation of the chains, but the elastomer does swell when the solvent molecules diffuse into the network and cause the chains to expand. This expansion is counteracted by the tendency for the chains to coil up and eventually an equilibrium degree of swelling is established which depends on the solvent and the crosslink density, *i.e.* the higher the crosslink density the lower the swelling.

The behaviour is predicted in the Flory-Huggins treatment of swelling which leads to a relation between the degree of swelling Q, for a particular solvent and the shear modulus G of the unswollen rubber

$$G = RTA/V_1Q^{5/3}, \tag{14.24}$$

where A is a constant. The validity of this expression was tested by Flory as shown in figure 14.7, where the expected slope of $-5/3$ was obtained. The theory also predicts that the equilibrium swelling of an elastomer increases when under a tensile stress.

The statistical theory will also describe the response under stress of elastomers swollen by solvents and in general it is found that the greater the degree of swelling the better the agreement between theory and experiment. The modifications necessary for the treatment of swollen networks are relatively straight forward and the stored energy function becomes

$$w = \tfrac{1}{2}NkT\phi_{\text{re}}^{1/3}(\lambda_1'^2 + \lambda_2'^2 + \lambda_3'^2 - 3), \tag{14.25}$$

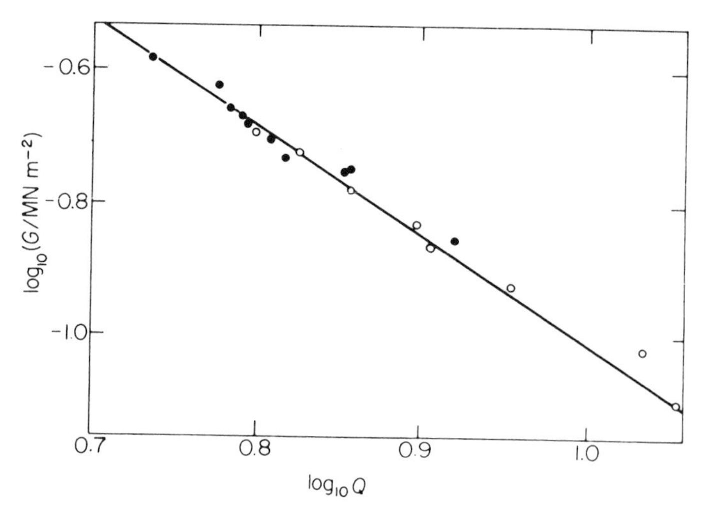

FIGURE 14.7. The swelling of butyl vulcanizates by cyclohexane plotted as a function of the elastic modulus, using equation (14.23). (Flory, 1946.)

where ϕ_{re} is the volume fraction of the elastomer and the prime represents the swollen unstrained state of the network. Similarly for simple extension $\phi_{re}^{1/3}$ is inserted in the right-hand side of equation (14.21).

14.9 Network defects

The number average molar mass M_s of a chain section between two junction points in the network is an important factor controlling elastomeric behaviour; when M_s is small the network is rigid and exhibits limited swelling, but when M_s is large the network is more elastic and swells rapidly when in contact with a compatible liquid. Values of M_s can be estimated from the extent of swelling of a network, which is considered to be ideal but rarely is, and interpretation of the data is complicated by the presence of network imperfections. A real elastomer is never composed of chains linked solely at tetra-functional junction points, but will inevitably contain defects such as (a) loose chain ends, (b) intramolecular chain loops, and (c) entangled chain loops.

Both M_s and the interaction parameter χ_1 can be calculated from measurements of network swelling when the elastomer is brought in contact with a solvent. We can assume that the total free energy change will then be composed of

$$\Delta G = \Delta G^{el} + \Delta G^{M} \tag{14.26}$$

where ΔG^M, the free energy change on mixing the elastomer chains with solvent is given by the Flory-Huggins equation (8.32), in which $N_2 = 1$ for a crosslinked network, and ΔG^{el} is the change in the elastic free energy. The latter can be estimated from the simple statistical theory, but we now have a choice. The Flory *affine network model* assumes that the junction points are embedded in the network and hence affine deformation means that any movement of the network chains is in proportion to the change in the macroscopic dimensions of the sample, whereas the James and Guth *phantom network model* assumes that the junction points can fluctuate independently of the macroscopic deformation.

The general form for ΔG^{el} derived is essentially the same for both theories and is given by equation (14.19) if N is replaced by a front factor F whose value depends on the model used. Experimental work has shown that the behaviour of a swollen network is best described by the phantom network model and further equations are derived on that basis.

For isotropic swelling $\lambda_1 = \lambda_2 = \lambda_3 = (\phi_r/\phi_{re})^{1/3}$ where ϕ_r is the volume fraction of the dry elastomer and usually has a value of unity in the unswollen state, and ϕ_{re} is the volume fraction of the elastomer at the maximum degree of swelling (equilibrium) when the network is in contact with excess solvent. Also for a phantom network,

$$\Delta G^{el} = \tfrac{3}{2} kT\gamma \left[\frac{\phi_r^{2/3}}{\phi_{re}} - 1 \right] \tag{14.27}$$

where

$$\gamma = \frac{V_0 \rho N_A}{M_s}(1 - 2/f) \tag{14.28}$$

Here f is the functionality of a junction in the network, ρ is the network density, and V_0 is the volume of solvent plus polymer. Equation (14.27) can be differentiated with respect to N_1 and combined with equation (8.33) (the corresponding mixing term) to given at equilibrium

$$\ln(1 - \phi_{re}) + \chi_1 \phi_{re}^2 + \phi_{re} + B(\phi_{re}/\phi_r)^{1/3} = 0 \tag{14.29}$$

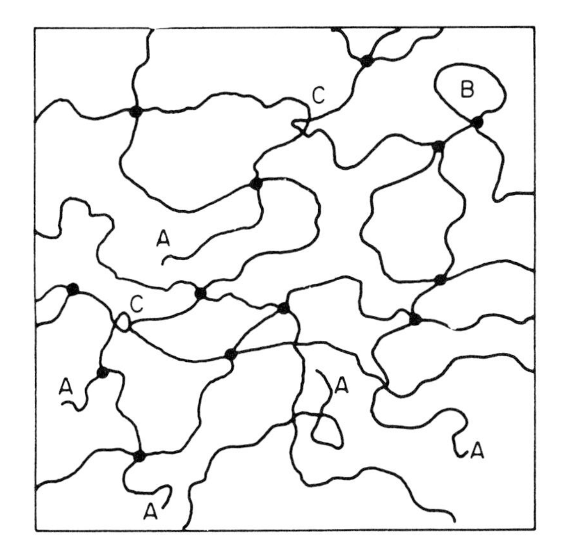

FIGURE 14.8. Diagram showing defects in an elastomeric network; A, loose chain ends; B, intramolecular chain loops; C, entangled chain loops.

where $B = (V_1/RT)(\gamma kt V_0)$ and V_1 is the molar volume of the solvent. From this it follows that

$$M_s = \frac{-\rho(1 - 2/f)V_1\phi_r^{2/3}\phi_{re}^{1/3}}{\ln(1 - \phi_{re}) + \chi_1\phi_{re}^2 + \phi_{re}} \quad (14.30)$$

or alternatively

$$\chi_1 = \frac{-\ln(1 - \phi_{re}) + \phi_{re} - B\phi_r^{2/3}\phi_{re}^{1/3}}{\phi_{re}^2} \quad (14.31)$$

14.10 Resilience of elastomers

When an elastomer, in the form of a ball, is dropped from a given height onto a hard surface, the extent of the rebound provides an indication of the resilience of the elastomer. A set of elastomer balls, manufactured by Polysar Corporation in Canada, provide an excellent demonstration of this phenomenon and have been used to measure the rebound height of several elastomers, recorded schematically in figure 14.9. If h_0 is the original height and h is the recovery height, then the rebound resilience is defined as (h/h_0) and the relative energy loss per half cycle is $(1 - h/h_0)$.

It should be remembered that an elastomer exhibiting good elastic properties under slow deformation may not possess good resilience; the relative recoveries of natural and butyl rubber provide a good example of this fact. As resilience is the ability of an elastomer to store and return energy when subjected to a rapid deformation, it can be shown that temperature also plays an important part in determining resilience. If the

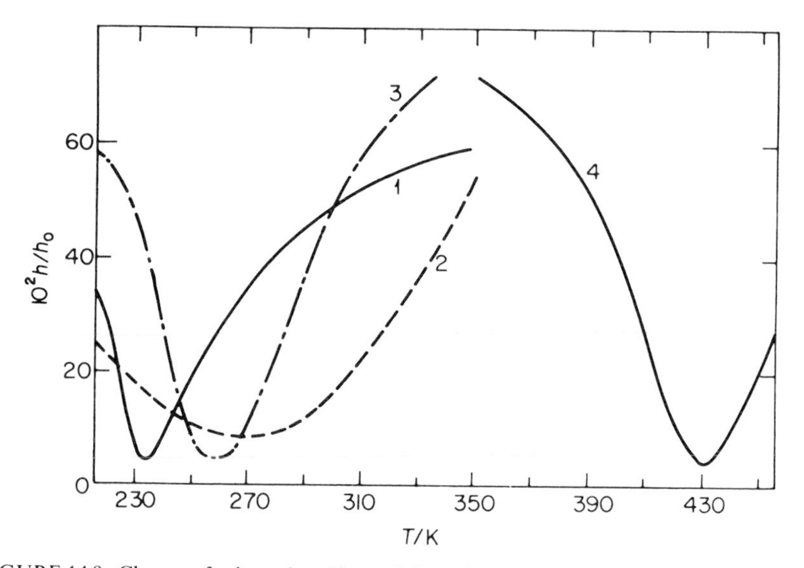

FIGURE 14.9. Change of rebound resilience (h/h_0) with temperature T for: 1, natural rubber; 2, butyl rubber; 3, neoprene; 4, poly(methyl methacrylate). (After Mullins (1947) and Gordon (1957).)

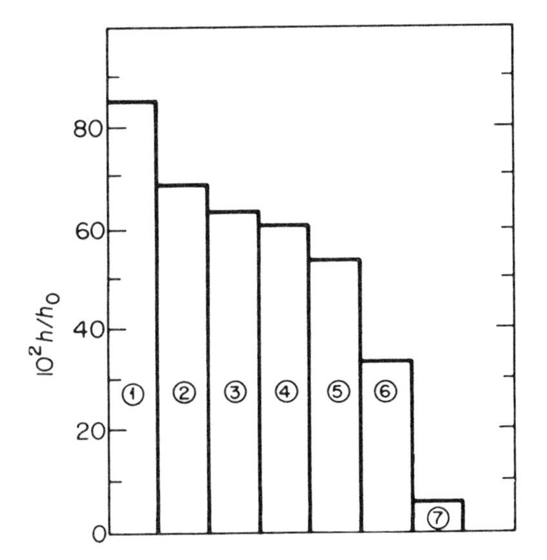

FIGURE 14.10. Percentage rebound recovery measured at 298 K, with balls made from: 1, *cis*-polybutadiene; (2), synthetic *cis*-polyisoprene; (3), natural *cis*-polyisoprene; (4), ethylene-propylene copolymer; (5), SBR; (6), *trans*-polyisoprene; (7), butyl rubber.

butyl and natural rubber balls are now heated to about 373 K, both will rebound to about the same extent. The importance of the two variables, time and temperature, is illustrated in a plot of the rebound resilience against temperature for three elastomers and a recognized plastic, figure 14.10. The sharply defined minima are characteristic of such curves and the broad butyl curve in anomalous. The minimum for curve 1 is closely related to the loss of long range elasticity at the glass temperature $T_g = 218$ K although it actually occurs at the higher temperature of 238 K. A similar situation is found for neoprene and poly(methyl methacrylate). One can conclude from this that resilience is closely related to the molecular structure and the intermolecular forces affecting the ability of the chain to rotate.

When the chains are deformed during a bounce, a stress is applied and then rapidly removed. The time required for the chains to regain their original positions is measured by the relaxation time τ, defined in section 13.4. Thus relaxation times are a measure of the ability of the chains to rotate. At room temperature the butyl rubber with the bulky methyl groups will not rotate as readily as the *cis*-polyisoprene so that when the deformation of chains in the sample of butyl rubber occurs, the chains do not return to their equilibrium positions as rapidly as the natural rubber, *i.e.* τ is longer.

The elastomer showing superior rebound potential at room temperature is *cis*-polybutadiene. This sample is non-crystalline and has no pendant groups to impede free segmental rotation, so the relaxation time is correspondingly shorter than other elastomers. The response of the butyl sample improves as the temperature increases because additional thermal energy is available to enhance chain rotation and decrease the relaxation time correspondingly. This leads to an improved resilience and the rebound potential now matches the natural rubber whose τ is not so sensitive to temperature change in this range.

General Reading

G. Allen and J. C. Bevington, Eds, *Comprehensive Polymer Science*. Vol. 2. Pergamon Press (1989).

R. B. Beevers, *Experiments in Fibre Physics*. Butterworths (1970).

D. C. Blackley, *Synthetic Rubbers: Their Chemistry and Technology*. Elsevier (1983).

F. Bueche, *Physical Properties of Polymers*, Chapter 1. Interscience Publishers Inc. (1962).

P. J. Flory, *Principle of Polymer Chemistry*, Chapter 11. Cornell Univ. Press, Ithaca, N.Y. (1953).

J. E. Mark, A. Eisenberg, W. W. Graessley, L. Mandelkern and J. L. Koenig, *Physical Properties of Polymers*. American Chemical Society (1984).

L. Koenig, *Physical Properties of Polymers*. American Chemical Society (1984).

J. E. Mark and B. Erman, *Rubberlike Elasticity–A Molecular Primer*. John Wiley and Sons (1988).

L. H. Sperling, *Introduction to Physical Polymer Science*. John Wiley and Sons Ltd (1986).

A. V. Tobolsky and H. Mark, *Polymer Science and Materials*, Chapter 9. Wiley-Interscience (1971).

L. R. G. Treloar, *Physics of Rubber Elasticity*. Clarendon Press (1958).

References

1. P. J. Flory, *Ind. Eng. Chem.*, **38**, 417 (1946).
2. K. H. Meyer and C. Ferri, *Helv. Chim. Acta*, **18**, 570 (1935).
3. L. Mullins, *I.R.I. Trans.*, **22**, 235 (1947).
4. L. R. G. Treloar, *Trans. Farad. Soc.*, **40**, 59 (1944).

CHAPTER 15

Structure–Property Relations

15.1 General considerations

The increasing use of synthetic polymers by industrialists and engineers to replace or supplement more traditional materials, such as wood, metals, ceramics, and natural fibres, has stimulated the search for even more versatile polymeric structures covering a wide range of properties. For such a quest to be efficient, a fundamental knowledge of structure–property relations is required.

The problem can be examined initially on two broad planes:

(a) *The chemical level.* This deals with information on the fine structure, namely what type of monomer constitutes the chain and whether more than one type of monomer is used (copolymer), *i.e.* the parameters which relate ultimately to the three-dimensional aggregated structure, and influence the extent of sample crystallinity and the physical properties.
(b) *The architectural aspects.* These are concerned with the chain as a whole, and now we are required to ask such questions as: is the polymer linear, branched, or crosslinked; what distribution of chain lengths exist; what is the chain conformation and rigidity?

Having considered these general points, one must then establish the suitability of a polymer for a particular purpose. This depends on whether it is glass-like, rubber-like, or fibre forming, and the characteristics depend primarily on chain flexibility, chain symmetry, intermolecular attractions, and of course environmental conditions. Excluding the environment, these parameters, in turn, are reflected in the more tangible factors, T_m, T_g, modulus, and crystallinity, which, being easier to assess, are commonly used to characterize the polymer and ascertain its potential use.

As the relative values of both T_m and T_g play such an important part in determining the ultimate behaviour of a polymer, we can begin an examination of structure and properties by finding out how a polymer scientist can attempt to control these parameters.

15.2 Control of T_m and T_g

We have already seen, in earlier chapters, how chain symmetry, flexibility, and tacticity, can influence the individual values of both T_m and T_g. Thus a highly flexible chain has a low T_g, which increases as the rigidity of the chain becomes greater. Similarly, strong intermolecular forces tend to raise T_g and also increase crystallinity. Steric factors play an important role. A high T_g is obtained when large pendant groups attached to the chain restrict its internal rotation, and bulky pendant groups tend to impede crystallization, except when arranged regularly in isotactic or syndiotactic chains.

Chain flexibility is undoubtedly the controlling factor in determining T_g, but it also has a strong influence on T_m. Hence we must consider these parameters together from now on and determine how to effect control of both.

CHAIN STIFFNESS

It is important to be able to regulate the degree of chain stiffness, as rigid chains are preferred for fibre formation, while flexible chains make better elastomers. The flexibility of a polymer depends on the ease with which the backbone chain bonds can rotate. Highly flexible chains will be able to rotate easily into the various available conformations, while the internal rotations of bonds in a stiff chain are hindered and impeded.

Variations in chain stiffness can be brought about by incorporating different groups in linear chains and the results can be appraised by following the changes in T_m and T_g in a series of different polymers. The effects can be assessed more easily if an arbitrary reference is chosen and the simplest synthetic organic polymer, polyethylene with $T_m \approx 400$ K and $T_g \approx 188$ K, is suitable for this purpose.

One can begin by considering a general structure $-[(CH_2)_m-X]_n-$, where m and X vary. The effect on T_m of incorporating different links in the carbon chain is illustrated in table 15.1

The chain flexibility is increased by groups such as $-(O)-$, $-(CO.O)-$, and $-(OCO.O)-$, and as the length of the $-(CH_2)-$ section grows. This is shown by a lowering of T_m relative to polyethylene. Insertion of the polar $-(SO_2)-$ and $-(CONH)-$ groups raises T_m, because the intermolecular bonding now assists in stabilizing the extended forms in the crystallites.

Chain stiffness is also greatly increased when a ring is incorporated in the chain, as this restricts the rotation in the backbone and reduces the number of conformations a polymer can adopt. This is an important aspect, as fibre properties are enhanced by stiffening the chain, and the effect of aromatic rings on T_g and T_m is shown in table 15.2.

The *p*-phenylene group in structure 3 causes a big increase in T_m and this can be modified by introducing a flexible group as in terylene, structure 4. This shows that a judicious combination of units can lead to a wide variety of chain flexibilities and

TABLE 15.1. Influence of various links on T_m when incorporated in an all-carbon chain

			T_m/K				
Polymer group	*Repeat unit*	*m*	2	3	4	5	6
Polyethylene	$-[(CH_2)_m]-$		400	–	–	–	–
Polyester	$-[(CH_2)_mCO\cdot O]-$		395	335	329	335	325
Polycarbonate	$-[(CH_2)_m-O\cdot CO\cdot O]-$		312	320	330	318	320
Polyether	$-[(CH_2)_m\cdot CH_2-O]-$		308	333	–	–	–
Polyamide	$-[(CH_2)_m-CO\cdot NH]-$		598	538	532	496	506
Polysulphone	$-[(CH_2)_mCH_2\cdot SO_2]-$		573	544	516	493	–

TABLE 15.2. Effect of aromatic rings on chain stiffness, as shown by the values of T_m and T_g

Structure	T_g/K	T_m/K
1. $+CH_2—CH_2+_n$	188	400
2. $+CH_2—CH_2—O+_n$	206	339
3. $+CH_2—C_6H_4—CH_2+_n$	–	about 653
4. $[+(CH_2)_2—O·CO—C_6H_4—CO\ O]_n$	342	538
5. $+NH(CH_2)_6NHCO·(CH_2)_4CO+_n$	320	538
6. $[NH—C_6H_4—NHCO(CH_2)_4—CO]_n$	–	613
7. $[NH—C_6H_4—NHCO—C_6H_4—CO]_n$ (meta)	546	about 635 (Decomposition)
8. $[NH—C_6H_4—NHCO—C_6H_4—CO]_n$ (para)	–	about 773

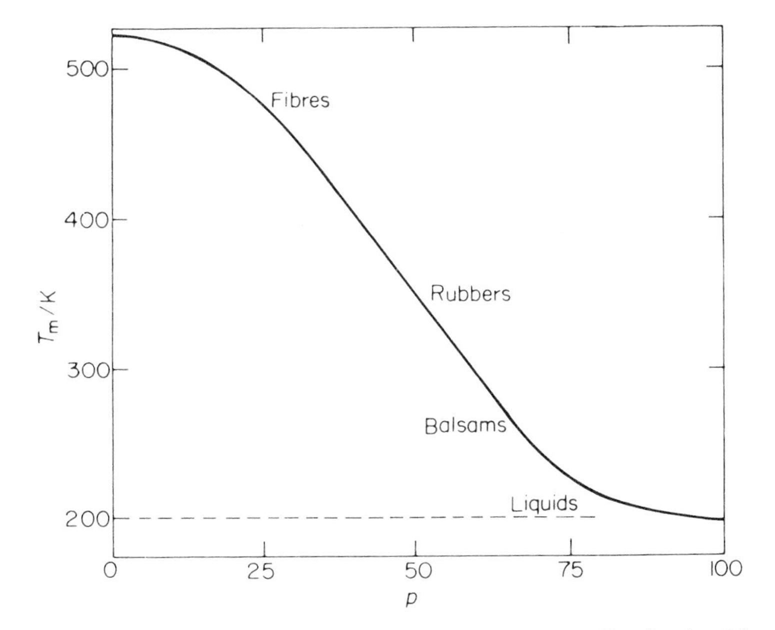

FIGURE 15.1. Change in the properties and melting temperature T_m of nylon-6,6 as the hydrogen-bonding capacity is reduced by changing the percentage p of amide substitution. (Adapted from R. Hill, *Fibres from Synthetic Polymers*.)

physical properties. The effect of the aromatic ring is again obvious in structures 5 through 8. The influence of chain symmetry is also noticeable when comparing the decomposition temperature of the symmetrical chain 8 with the unsymmetrical chain 7. The latter is also an inefficient close-packing structure, as sufficient disorder exists for the glass transition to appear.

INTERMOLECULAR BONDING

An increase in the lattice energy of a crystallite is obtained when the three-dimensional order is stabilized by intermolecular bonding. In the polyamide and polyurethane series, the additional cohesive energy of the hydrogen bond (about $24\,kJ\,mol^{-1}$) strengthens the crystalline regions and raises T_m. The effect is strongest when regular, evenly spaced, groups exist in the chain, as with nylon-6,6. The importance of secondary bonding in the polyamide series is illustrated quite dramatically when the crucial hydrogen atom of the amide group is replaced by a methylol group. The loss of the hydrogen-bonding capability impairs the tendency for regular chain alignment to take place and the character of the polyamide changes dramatically. With little substitution, they are suitable fibres, but, as the hydrogen is replaced, they change and become more elastomeric, then eventually like balsams, and finally liquids.

An alternative method of reducing the hydrogen-bonding potential, and so T_m, in the polyamides, is to increase the length of the $\text{+}CH_2\text{+}_n$ sequence between each bonding site. This leads to nylons with a variety of properties, *e.g.* nylon-12 has properties intermediate between those of nylon-6 and polyethylene. The effect is shown in table 15.3.

15.3 Relation between T_m and T_g

Most of the factors discussed so far influence T_g and T_m in much the same way, but in spite of this, the fact that T_m is a first-order thermodynamic transition, whereas T_g is not, precludes the possibility of a simple relation between them.

There is, however, a crude correlation, represented in figure 15.2, where a broad band covers most of the results for linear homopolymers, and the ratio (T_g/T_m) lies between 0.5 and 0.8 for about 80 per cent of these.

Obviously then, a synthetic chemist attempting to control the T_m and T_g of a simple chain structure by varying flexibility, symmetry, tacticity, *etc*., is limited to structures

TABLE 15.3. Melting temperatures T_m of linear aliphatic polyamides

Monadic nylon	T_m/K	*Dyadic nylon*	T_m/K
4	533	4,6	581
6	496	5,6	496
7	506	6,6	538
8	473	4,10	509
9	482	5,10	459
10	461	6,10	495
11	463	6,12	482
12	452		

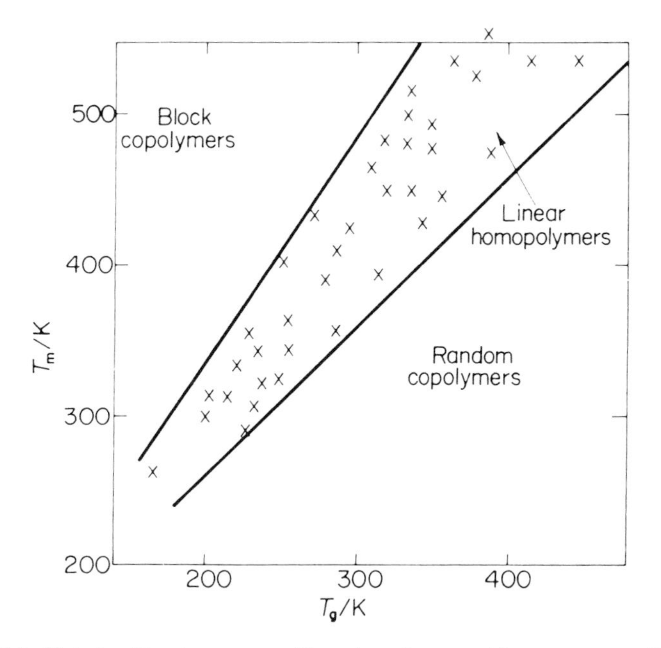

FIGURE 15.2. Plot of melting temperature T_m against glass transition temperature T_g for linear homopolymers with (T_g/T_m) lying in the range 0.5 to 0.8.

with either a high T_m and T_g or a low T_m and T_g. In effect neither T_m nor T_g can be controlled separately to any great degree.

To exercise this additional control another method of chain modification must be sought and this leads to the use of copolymers.

15.4 Random copolymers

Axial symmetry in a chain is a major factor in determining the ability of a chain to form crystallites, and one method of altering the crystalline content is to incorporate some structural irregularity in the chain. The controlled inclusion of linear symmetrical homopolymer chains $\text{-(}A\text{)-}_n$ in a crystal lattice can be achieved by copolymerizing A with varying quantities of monomer B, whose purpose is to destroy the regularity of the structure.

This leads to a gradual decrease in T_m as shown schematically in figure 15.3. The broken line represents the possibility that, in the middle composition range, the decrease in regularity is so great that the material is amorphous. This situation is sometimes obtained when a terpolymer is prepared.

A practical application is found in the polyamides. An improvement in the elastic qualities of the polyamide fibre is obtained if the modulus is reduced and, as the factors which affect the melting temperature affect the modulus, this can be achieved by starting from nylon-6,6 or nylon-6,10 and forming (66/610) copolymers. The random inclusion of the two types of unit in the chain disturbs both the symmetry and the regular spacing of the hydrogen-bonding sites, resulting in a drop in T_m.

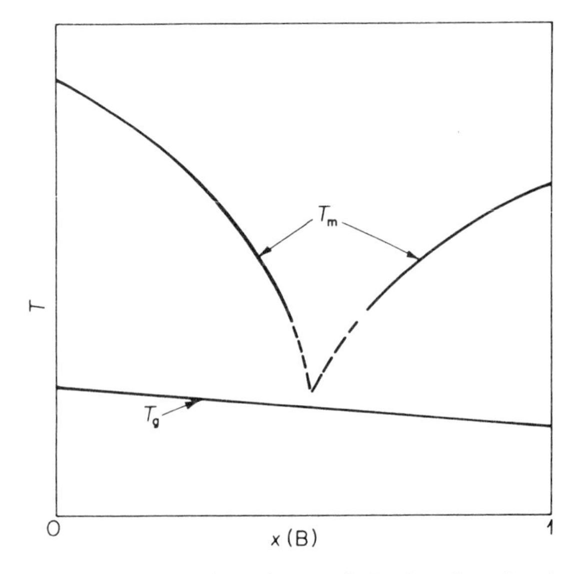

FIGURE 15.3. Schematic representation of T_m and T_g plotted as functions of copolymer composition shown as mole fraction x(B) of B. The broken lines represent the possibility that structural irregularities are so great that no crystallization of the copolymer can occur.

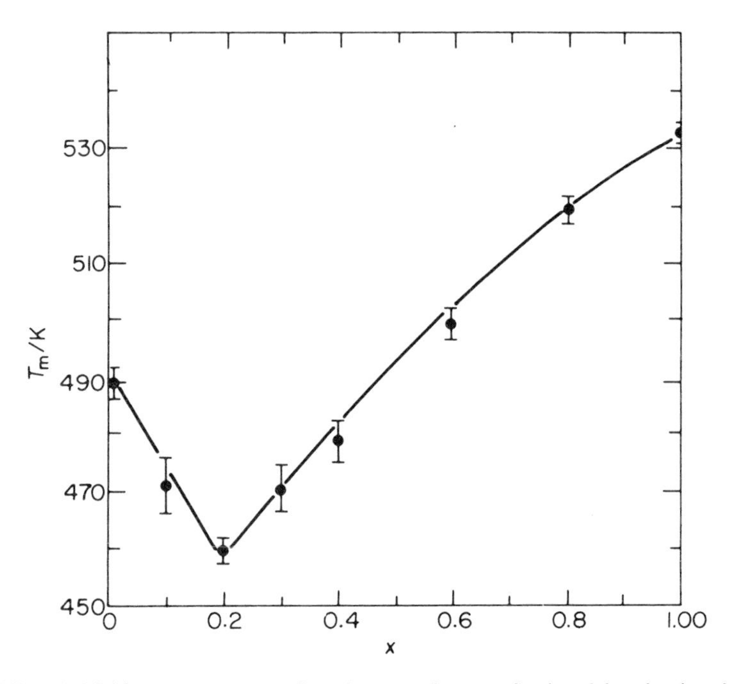

FIGURE 15.4. Melting temperatures of random copolymers of nylon-6,6 and nylon-6,10 as a function of the mole fraction of adipamide in the copolymer. (From data by Cowie and Mudie.)

The glass transition T_g is not affected in the same way as T_m, because T_g is more a function of the differences in chain flexibility, than the packing efficiency. This means that the response of T_g to a change in copolymer composition is quite different and this affords a means of controlling the magnitudes of T_m and T_g independently, a most important feature not readily achieved by other methods.

15.5 Dependence of T_m and T_g on copolymer composition

A quantitative expression for the depression of the melting temperature can be derived, thermodynamically, in terms of the composition and enthalpy of fusion ΔH_u of polymer A by

$$1/T_m^{AB} - 1/T_m^{A} = -(R/\Delta H_u)\ln x_A, \tag{15.1}$$

where T_m^A and T_m^{AB} are the melting temperature of pure polymer A and the copolymer AB respectively, and x_A is the mole fraction of A in the copolymer.

The simple linear relation between T_g and x shown in figure 15.3 is found only for a few copolymers composed of compatible monomer pairs, such as styrene copolymerized with either methyl acrylate or butadiene. A simple ideal mixing rule can be applied to these systems, but when the comonomer properties differ markedly, the linear dependence is lost, and a non-linear equation has to be developed.

One simple relation, which usefully describes the behaviour of many vinyl monomer pairs is

$$1/T_g^{AB} = w_A/T_g^A + w_B/T_g^B, \tag{15.2}$$

where w_A and w_B are the mass fractions of monomers A and B. For a system where the conditions $T_B^A < T_g^{AB} < T_g^B$ hold, the free volume concept can be used to formulate a relation between T_g and w, and Gordon and Taylor have proposed

$$(T_g^{AB} - T_g^A)w_A + K(T_g^{AB} - T_g^B)w_B = 0. \tag{15.3}$$

This expression assumes that the free volume contribution from a monomer is the same as both homo- and copolymers. For a given pair of monomers the constant K is calculated from the corresponding expansivities of the homopolymers.

$$K = (\alpha_1^B - \alpha_g^B)/(\alpha_1^A - \alpha_g^A). \tag{15.4}$$

A similar relation has been proposed by Gibbs and Di Marzio,

$$(T_g^{AB} - T_g^A)n_o^A + (T_g^{AB} - T_g^B)n_o^B = 0, \tag{15.5}$$

where now the fraction of rotatable bonds n_o is introduced in place of the composition term.

15.6 Block copolymers

Random or statistical copolymers can be prepared if one wishes to narrow the gap between T_m and T_g in a sample, and so cover one property region not readily satisfied by homopolymers.

If a wider interval between T_m and T_g is required, a different class of co-polymer – the block copolymer – must be investigated. These are usually {AB} or {ABA} block sequences. By synthesizing sequences, which are long enough to crystallize independently, the combination of a high melting block A with a low melting block B will provide a material with a high T_m from A and a low T_g from B. A slight depression of T_m, arising from the presence of block B, is sometimes encountered, but this is rarely large. Combinations of this type allow the scientist to cover the remaining area of property combinations shown in figure 15.2.

The change of T_g in block copolymers is rather variable and certain pairs of monomers will form a block copolymer possessing two glass transitions. Interesting changes in the mechanical properties can be obtained when SBR block copolymers are synthesized using a lithium catalyst. A material is produced which behaves as though crosslinked at ambient temperatures. This is due to the presence of the two glass transitions associated with each block; the butadiene block has one at 210 K and the styrene block has one at 373 K. Above 373 K plastic flow is observed, but between 210 and 373 K the glassy polystyrene blocks act as crosslinks for the elastomeric polybutadiene and the copolymer exhibits high resilience and low creep characteristics.

The arrangement of the blocks is important; high tensile strength materials, with elastomeric properties similar to a filler reinforced vulcanizate, are obtained only when the copolymer contains two or more polystyrene (S) blocks per molecule. Thus copolymers with the structure {S.B.} or {B.S.B.}, where B is a polybutadiene block, are as brittle as polystyrene, but {S.B.S} and {S.B.S.B} copolymers are much tougher. At ambient temperatures these behave like conventional crosslinked rubbers but they have the additional advantage that their thermal behaviour is reproducible.

The property enhancement of these block copolymers is usually explained in terms of the "*domain concept*". The glassy polystyrene blocks tend to aggregate in domains (see figure 15.5) which act as both crosslinking points and filler particles. The glassy regions serve to anchor the central elastomeric polydiene blocks securely at both ends and act as effective cross-linking points, thereby precluding the necessity to vulcanize the material.

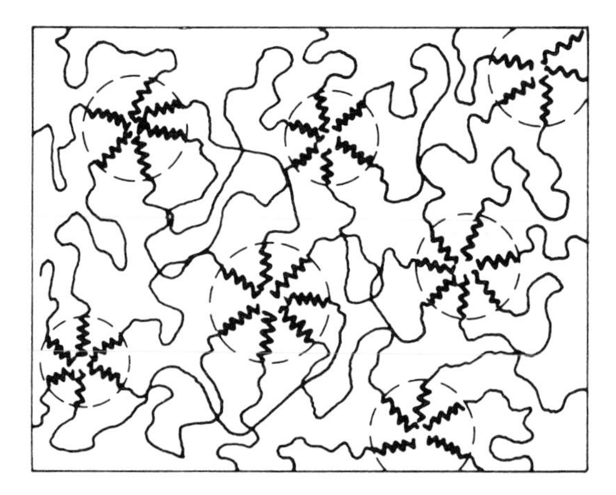

FIGURE 15.5. Schematic representation of elastoplastic sandwich block copolymers, showing areas of aggregation of the glassy {A} blocks, joined by the amorphous rubber-like chains of {B}.

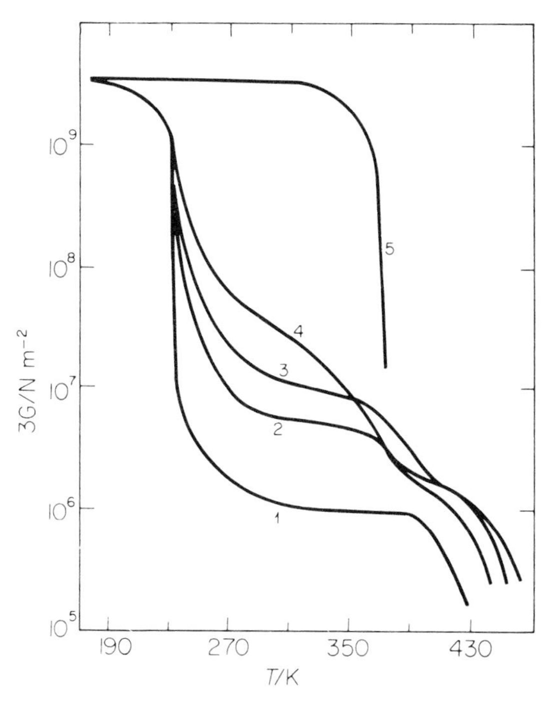

FIGURE 15.6. Modulus-temperature behaviour of polyester-polystyrene block copolymers: 1, polyester; 2, polyester containing 20 per cent of polystyrene; 3, containing 45 per cent polystyrene; 4, containing 60 per cent polystyrene; 5, pure polystyrene.

One unexpected application arises from the observation that the presence of more than 10 per cent block copolymer in natural rubber prevents bacterial growth on the polymer surface. Thus, incorporation of the copolymer in butchers' chopping blocks can lead to more hygienic conditions in meat handling.

The synthesis of {ABA} blocks, from a glassy thermoplastic A and an elastomeric B, produces other "elastoplastics" with attractive properties. Polyester chains can be extended with diisocyanate, which is then treated with cumene hydroperoxide to leave a peroxide group at both ends of the chain. By heating this in the presence of styrene, a vinyl polymerization is initiated and an {ABA} block created. The modulus-temperature curves show how the mechanical properties can be modified in this way (figure 15.6).

These block copolymers are known as *thermoplastic elastomers*.

15.7 Plasticizers

A polymer sample can be made more pliable by lowering its T_g, and this can be achieved by incorporating quantities of high boiling, low molar mass, compounds in the material. These are called plasticizers and must be compatible with the polymer. The extent to which T_g is depressed depends on the amount of plasticizer present and can be predicted from the relation

$$1/T_g^M = w/T_g + w_1/T_g^1, \tag{15.6}$$

where T_g^M and T_g^1 correspond to the mixture and the liquid, respectively, while w and w_1 are the mass fractions of the polymer and plasticizer in the system.

The action of the plasticizer is one of a lubricant, where the small molecules ease the movement of the polymer chains by pushing them further apart. As this lowers both T_g and the modulus, their main use is to increase the flexibility of a polymer for use in tubing and films.

Poly(vinyl chloride), whose T_g is 354 K, usually contains 30 to 40 mass per cent of plasticizers, such as dioctyl or dinonyl phthalate, to increase its toughness and flexibility at ambient temperatures. This depresses T_g to about 270 K and makes the polymer suitable for plastic raincoats, curtains, and "leather-cloth". The low volatility of the plasticizer ensures that it is not lost by evaporation, a mistake made in the early post-war years, which led eventually to a brittle product and considerable customer disaffection. In the rubber industry, plasticizers are usually called oil extenders.

In fibre technology, water absorption is an important factor governing the mechanical response, because the water tends to act as a plasticizer. Thus, as the moisture content increases, the modulus drops, but there is a corresponding improvement in the impact strength. In fibres, such as nylon-6,6, water acts as a plasticizer to depress T_g below room temperature. Thus, when nylon shirts are washed and hung up to drip-dry, the polymer is above the T_g, and this helps creases to straighten out, thereby giving the clothing an "ironed" appearance.

15.8 Crystallinity and mechanical response

The mechanical properties are dependent on both the chemical and physical nature of the polymer and the environment in which it is used. For amorphous polymers, the principles of linear viscoelasticity apply, but these are no longer valid for a semicrystalline polymer.

The mechanical response of a polymer is profoundly influenced by the degree of crystallinity in the sample.

The importance of both crystallinity and molar mass is illustrated by the range of properties displayed by polyethylene. This is shown schematically in figure 15.7 and provides some indication of the effect of these variables.

The interpretation of the mechanical behaviour is further complicated by the presence of glide planes and dislocations, which lead to plastic deformation, but these also serve to provide the materials scientist with a wider variety of property combinations, and can prove useful.

The major effect of the crystallite in a sample is to act as a crosslink in the polymer matrix. This makes the polymer behave as though it was a crosslinked network, but, as the crystallite anchoring points are thermally labile, they disintegrate as the temperature approaches the melting temperature, and the material undergoes a progressive change in structure until beyond T_m, when it is molten. Thus crystallinity has been aptly described by Bawn as a form of "thermoreversible crosslinking".

The restraining influence of the crystallite alters the mechanical behaviour by raising the relaxation time τ and changing the distribution of relaxation and retardation times

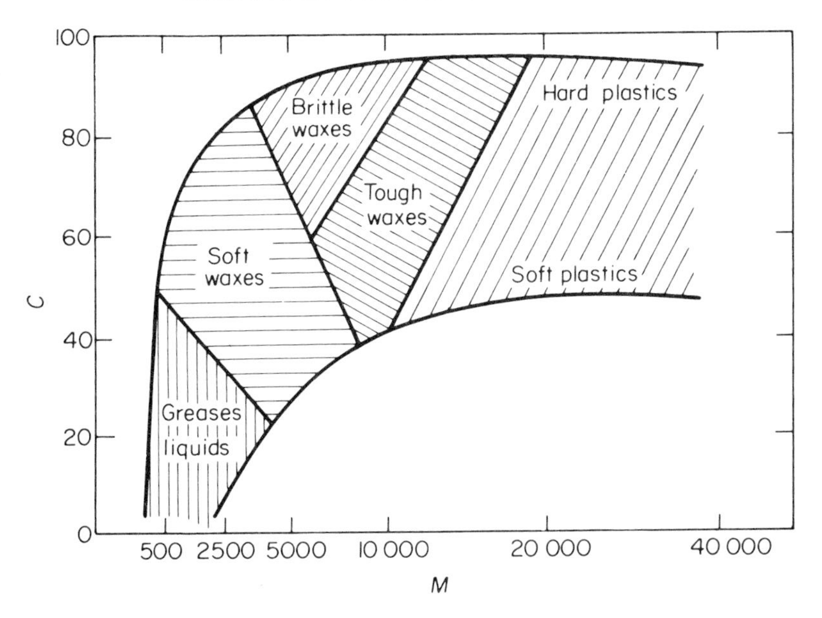

FIGURE 15.7. Influence of crystallinity and chain length on the physical properties of polyethylene. (After Richards, *J. Appl. Chem.*, 1951.) The percentage crystallinity c is plotted against molar mass M.

in the sample. Consequently, there is an effective loss of short τ, causing both the modulus and yield point to increase. The creep behaviour is also curtailed and stress-relaxation takes place over much longer periods. Semi-crystalline polymers are also observed to maintain a relatively higher modulus over a wider temperature range than an amorphous sample.

These points can be illustrated by comparing the elastic relaxation modulus $E_r(t)$ for crystalline (isotactic), amorphous, and chemically crosslinked (atactic), polystyrene samples; see figure 15.8. Crystallinity has little effect below T_g, but as the molecular motion increases above T_g, the modulus of the amorphous polymer drops more sharply. The value of $E_r(t)$ remains high for the crystalline polymer throughout this range until the rapid decrease at the melting temperature is recorded. The crosslinked sample maintains its modulus level at this temperature as the crosslinks are not thermally labile and do not melt.

Rapid quenching of the isotactic polymer destroys the crystallinity and produces behaviour identical to the atactic material. The spherulite size also affects the response; slow cooling from the melt promotes the formation of large spherulites and produces a polymer with a lower impact strength than one cooled rapidly from the melt, whose spherulites are much smaller and more numerous. This effect can be seen as a shift in the damping maxima.

In practical terms, the use of poly(vinyl chloride) in the manufacture of plastic raincoats provides a good illustration of the effect of crystallite crosslinking. The polymer is plasticized, until T_g is below ambient, to make the material flexible, and one might expect that if the coat was hung on a hook (*i.e.* subjected to a tensile load), it

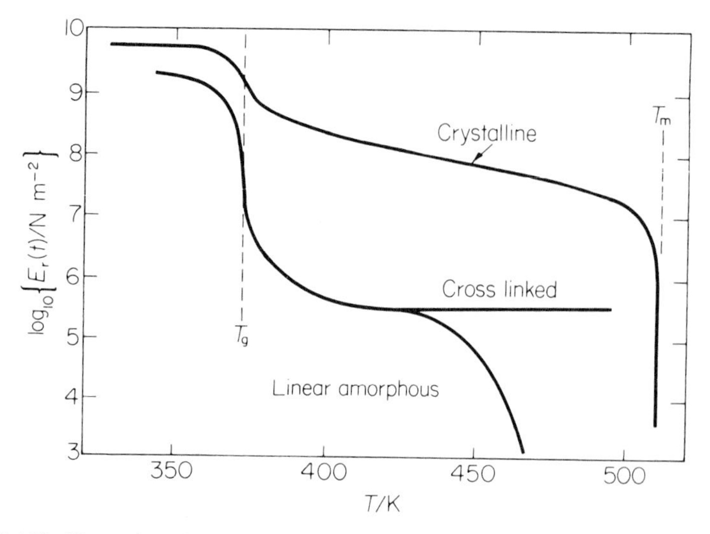

FIGURE 15.8. Illustration of the variation in the modulus–temperature curves for three types of polystyrene.

would eventually flow onto the floor after prolonged tension. This is not so; the material behaves as though it was a chemically crosslinked elastomer, because it contains a sufficient number of crystallites to act as restraining points and prevent flow.

Similarly, the glass transition of polyethylene is well below ambient temperature, and if the polymer was amorphous, it is likely that it would be a viscous liquid, at room temperature. It is, in fact, a tough, leathery or semi-rigid plastic, because it is highly crystalline and the crystallite crosslinks impart a high modulus and increased strength to the polymer between 188 and 409 K, a very useful temperature range.

The main points can now be restated briefly:

(1) Crystallinity only affects the mechanical response in the temperature range T_g to T_m, and below T_g the effect on the modulus is small.
(2) The modulus of a semi-crystalline polymer is directly proportional to the degree of crystallinity, and remains independent of temperature if the amount of crystalline order remains unchanged.

15.9 Application to fibres, elastomers and plastics

We have seen how various parameters can be altered and combined to produce a material with given responses, but the lines of demarkation dividing polymers into the three major areas of application – fibres, elastomers, and plastics – are by no means well defined. It is most important then to establish criteria which determine that a polymer is a superior fibre, an excellent elastomer, or a particularly suitable plastic before over-indulging in the interesting stages of molecular design and engineering.

15.10 Fibres

Superficially a fibre is a polymer with a very high length to diameter ratio (at least 100:1), but most polymers, capable of being melted or dissolved, can be drawn into

TABLE 15.4. Values of T_m and T_g of some typical fibres

Polymer	*Structure*	T_g/K	T_m/K
Poly(ethylene terephthalate)	$-[(CH_2)_2O\cdot OC-C_6H_4-CO\cdot O]-$	343	538
nylon-6,6	$-[NH(CH_2)_6NHCO(CH_2)_4CO]-$	333	538
polyacrylonitrile	$-[CH_2-CH(CN)]-$	378	590
(isotactic) polypropylene	$-[CH_2-CH(CH_3)]-$	268	435

filaments. They may, however, have no technical advantages if they cannot meet the requirements of a good fibre; these are high tensile strength, pliability, and resistance to abrasion. In addition, to be useful for clothing, it is preferable that the polymer has $T_m > 470$ K, to allow ironing without damage, but lower than 570 K, to enable spinning from the melt. Also T_g must not be so high that ironing is ineffective. Some typical fibres with useful temperature ranges are shown in table 15.4. These all have T_g lower than 380 K but above room temperature, so that in a cloth the fibres will soften when ironed at about 420 K. This will remove creases or allow pleats to be made which will be retained on cooling. Subsequent washing is normally carried out at temperatures too low to resoften the polymer significantly and so destroy the pleats. This "permanent" crease is a desirable feature of some clothing.

The main distinguishing feature of a fibre is that it is an oriented polymer, and as such is anisotropic, being much stronger along the fibre axis than across it. Thus, the most important technical requirement for fibre formation is the ability to draw or orient the chains in the direction of the fibre axis, and *retain* this after removal of the drawing force. Clearly then, factors which aid this retention of orientation are prime requirements for a good fibre, and these will include all structural features contributing to intermolecular binding.

This means that a polymer should be symmetrical and unbranched to encourage a high degree of crystallinity; it should preferably have a high cohesive energy; and it should have an average length of about 100 nm fully extended. These properties can be conveniently examined under two main headings – the chemical requirements and the mechanical response – and the important factors to consider are: (i) melting and glass transition temperatures; (ii) modulus; (iii) elasticity; (iv) tensile strength, and (v) moisture absorption and dyeability.

CHEMICAL REQUIREMENTS

If the polymer chains are quite short, they are not entangled to any great extent in the solid and are relatively free to move, hence they cannot add to the fibre strength. As the chain length increases (and so the intertwining), the fibre strength improves and the optimum range of molar mass for a good fibre is 10 000 to 50 000 g mol^{-1}. It has been found that fibre properties deteriorate outside these limits. However, as we shall see later, chain entanglement can detract from the tensile strength and modulus, therefore alignment of long chains is an important and desirable feature.

We have already mentioned the importance of T_g and T_m and know that these can be affected by chain symmetry, stiffness, and intermolecular bonding. The tensile strength of a fibre is observed to increase with crystallinity, consequently this is a desirable quality and linear chains will be preferred for fibre formation. As the shape and symmetry of a linear chain governs its ability to crystallize, chains containing irregular units, which detract from the linear geometry, should be avoided in fibre forming polymers. This is obvious when comparing terylene I, which is an excellent fibre, with its isomer II, prepared using *o*-phthalic acid.

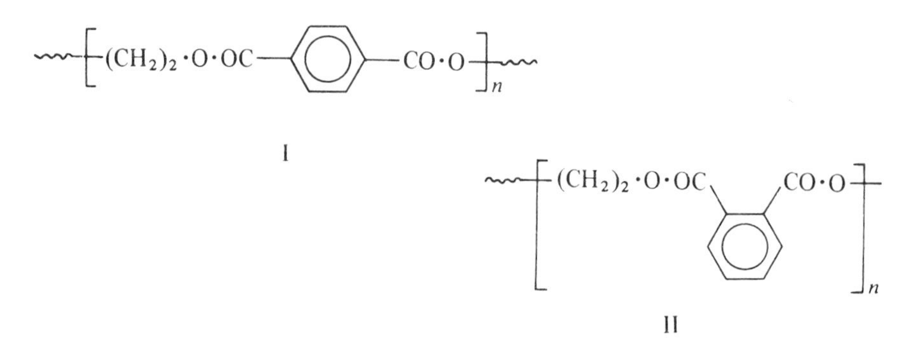

Structure II has lost its regularity, is less crystalline, has a lower T_g, and makes a much poorer fibre. This is also true in the polyamide series, where the regular polymer III has $T_m = 643$ K and $T_g = 453$ K.

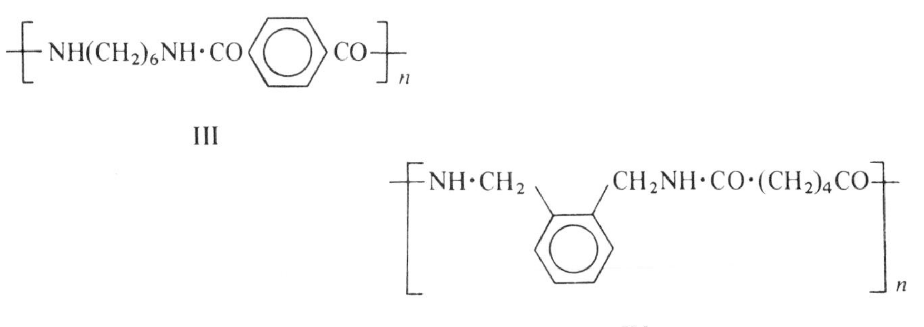

but the irregular form IV has $T_m = 516$ K and $T_g = 363$ K.

Stereoregular polymers also have symmetrical structures and the helices of isotactic polymers can be close packed to produce highly crystalline material. Isotactic polypropylene is crystalline and an important fibre forming polymer, whereas the atactic form has virtually no crystalline content and has little value as a fibre; indeed it is considerably more elastomeric in nature.

Although crystallinity and stereoregularity are important factors in fibre formation, atactic amorphous polymers can also prove useful, if there are intermolecular forces present. Dipolar interactions between side groups such as $\leftarrow$CN) (energy of interaction about 36 kJ mol^{-1}) are significantly stronger than hydrogen bonds or van der Waals forces and serve to improve the molecular alignment immensely. This interaction stabilizes orientation during fibre manufacture and enhances the fibre forming potential of polymers such as polyacrylonitrile and poly(vinyl chloride), both essentially amorphous and atactic. This point highlights the fact that molecular alignment is the most important factor in fibre formation, not crystallinity, which is only one method of obtaining a stable orientation of chains.

The importance of hydrogen bonding has already been described and will not be dealt with further.

Linear polyesters. Many of the general points discussed can be illustrated conveniently by referring to the numerous linear polyesters which have been prepared. These are grouped together in table 15.5.

A comparison of structure 1 with 2(i) and 2(ii) indicates a drop in T_m caused by the increase in chain flexibility arising from the ethylene and ethylene dioxy groups inserted between the phenylene rings. The change is even more dramatic on comparing 1 with 6 when two phenylene rings are used instead of the $\leftarrow CH_2 \rightarrow_4$ sequence and the difference in T_m is 205 K.

The influence of symmetry is seen in the terephthalic 3, and isophthalic 4 series. The unsymmetrical ring placement in 4(i) and 4(ii) lowers T_m by 25 K and 77 K respectively, compared with their counterparts 3(i) and 3(ii).

Bulky side groups interfere with the close packing capabilities of a chain, as evidenced by the effect on T_m of the methyl groups in 3(iii), 3(iv), and 5 compared with 3(i) and 3(ii). The additional asymmetry in 3(iv) actually prevents crystallization occurring.

The added stability of secondary bonding in the crystallite is reflected in the increase in T_m in the series 2(i) to 2(iii), as we move from simply van der Waals forces to dipolar and hydrogen-bond interactions. The hydrogen bonding is also sufficient to raise T_m of 2(iii) above that of 3(i), in spite of the extra flexible sequence present in the 2(iii) chain.

These points cover most of the chemical requirements and we can now look at the mechanical properties.

MECHANICAL REQUIREMENTS FOR FIBRES

Fibres are subject to a multitude of mechanical deformations; stretching, abrasion, bending, twisting, shearing, and now the properties of interest are: (i) *tenacity*, which is the stress at the breaking point of the material; (ii) *toughness*, defined as the total energy input to the breaking point; (iii) *initial modulus*, the measure of resistance to stretching (portion A–B of the stress-strain curve, figure 15.9); and (iv) the extent of *permanent set*.

In technological terminology, the textile industry recognizes the following qualities as suitable: (a) tenacity: 1 to 10 g denier^{-1} (about 5 g denier^{-1} optimum for clothing), (b) modulus of elasticity: 20 to 200 g denier^{-1}, and (c) extensibility: 2 to 50 per cent. The denier is the mass in grams of 9000 m of yarn.

As the mechanical response of a fibre can be controlled to some extent in the spinning process, this will be discussed briefly.

Spinning techniques. The process of converting a bulk polymer sample into a thread or yarn is known as *spinning*, and several methods can be used depending on the nature of the sample.

TABLE 15.5. Values of T_m and T_g for linear polyesters

	Structure	*Group R*	T_m/K	T_g/K
1.	$-[OC-C_6H_4-C_6H_4-CO\cdot O(CH_2)_2O]_n-$		528	—
2.	$-[OC-C_6H_4-R-C_6H_4-CO\cdot O(CH_2)_2O]_n-$	(i) $-(CH_2)_4-$	443	—
		(ii) $-O-(CH_2)_2-O-$	513	—
		(iii) $-NH-(CH_2)_2-NH-$	546	—
3.	$-[OC-C_6H_4-CO\cdot O\cdot R\cdot O]_n-$	(i) $-(CH_2)_2-$	538	342
		(ii) $-(CH_2)_4$	503	353
		(iii) $-CH_2-C(CH_3)_2-CH_2-$	413	—
		(iv) $-CH_2-CH(CH_3)-$	non-crystalline	341
4.	$[OC-C_6H_4(meta)-CO\cdot O\cdot R\cdot O]_n-$	(i) $-(CH_2)_2-$	513	324
		(ii) $-(CH_2)_4-$	426	—
5.	$-[OC-C_6H_3(CH_3)-CO\cdot O(CH_2)_2\cdot O]_n-$		343	—
6.	$-[OC\cdot(CH_2)_4CO\cdot O\cdot(CH_2)_2\cdot O]_n-$		323	—

Melt spinning is used when polymers are readily melted without degradation and the molten polymer is forced through a spinnaret comprising of 50 to 1000 fine holes. On emerging from the holes, the threads solidify, often in an amorphous glassy state, and are wound into a yarn. Orientation and crystallinity are important requirements in fibres and the yarn is subjected to a drawing procedure which orients the chains and strengthens the fibre. This technique is applied to polyesters, polyamides, and polyolefins.

Wet and dry spinning. Acrylic polymers cannot be melt spun because they are thermally labile, and spinning is carried out using concentrated solutions of the polymer. The solvent is removed by evaporation, after extrusion, leaving an amorphous filament, which is then said to have been *dry* spun. When the solution filaments are extruded into a vat of a non-solvent, the polymer precipitates in the form of a thread, and is then a *wet* spun fibre.

Drawing, orientation, and crystallinity. A fibre, in its amorphous state, can be strengthened by *drawing*, a process which extends its length by several times the original, and in doing so aligns the chains in the sample. The process is irreversible and corresponds to the section C-D of the stress-strain curve in figure 15.9, where deformation up to the yield point C is elastic, but beyond this irreversible plastic deformation occurs.

At C the polymer suddenly thins down or "necks" at one point, and subsequent drawing increases the length of the reduced region at the expense of the undrawn region, until the process is complete. Further extension causes rupture at D, the breaking point.

The effect of molecular order is far more important in fibre production than any other area of polymer application and drawing ability is a fundamental requirement in good fibre-forming materials. It is worth pointing out again that crystallinity and orientation are not necessarily synonymous terms and that there is a difference between orientation of crystallites and orientation of chains in the amorphous regions of a polymer. It is the amorphous part of a fibre which will distort and elongate under stress, and these are the areas which must be oriented to improve the intermolecular attraction, if the fibre modulus is to be enhanced. Drawing only improves a highly

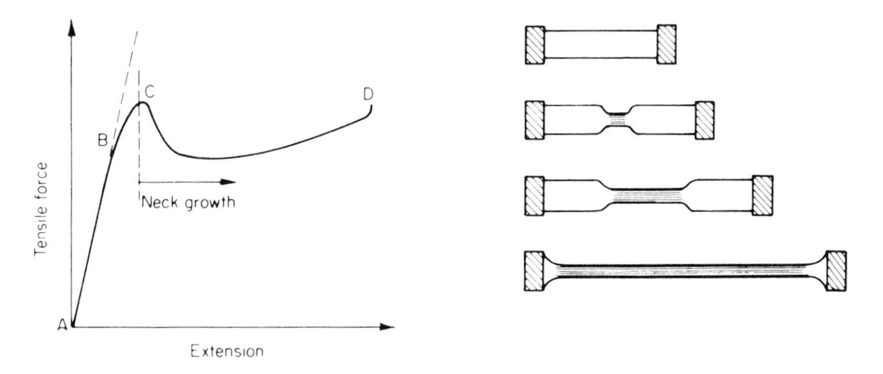

FIGURE 15.9. Successive stages in the drawing of polymer, showing the necking down and subsequent neck growth resulting in increased chain alignment.

crystalline fibre to a small extent by orienting the crystallites, but the amorphous fibre is improved immensely.

Drawing affects the mechanical properties of a fibre in several ways. It makes the fibre tough and tenacious, it can increase the modulus and the density, and can alter T_g by orienting the chains in the amorphous regions. For example, the $-\!\!(O-(CH_2)_2-O)\!\!-$ group in poly(ethylene terephthalate), which has a *gauche* conformation in the amorphous phase, is "drawn" into the *trans* conformation. This improves the sample crystalline and T_g also rises 10 to 15 K.

The tenacity and physical characteristics of the fibre can also be controlled by the extent of the draw. Limited orientation, produced at low draw stresses, leads to a medium tenacity nylon yarn with low tensile strength, low modulus, and high extensibility, which are all properties associated with a flexible soft material suitable for clothing. Higher draw rates yield high tenacity, high strength yarns, more suited to tyre cord production. Thus some fibre properties are subject to the art of the spinner.

Both tenacity and modulus can be controlled by crystallinity in a fibre. Low pressure polyethylene is highly crystalline and has a fibre tenacity of about 6 g denier^{-1}, but the high pressure, highly branched, and consequently less crystalline polyethylene has a fibre tenacity of only 1.2 g denier^{-1}. We have already seen that conversion of the amide group to a non-hydrogen-bonding group such as a methylol group $-CON(CH_2OH)-$ with formaldehyde, curtails the intermolecular bonding in polyamides. This also increases hydrophilicity and makes the polymer increasingly water soluble, but the hydrophobic characteristics can be restored by methylating the group to $-CON(CH_2OCH_3)-$. At low degrees of substitution the modulus is reduced and a more elastic fibre is obtained. As the substitution increases, the crystallinity is completely destroyed, and the fibre forming capacity disappears.

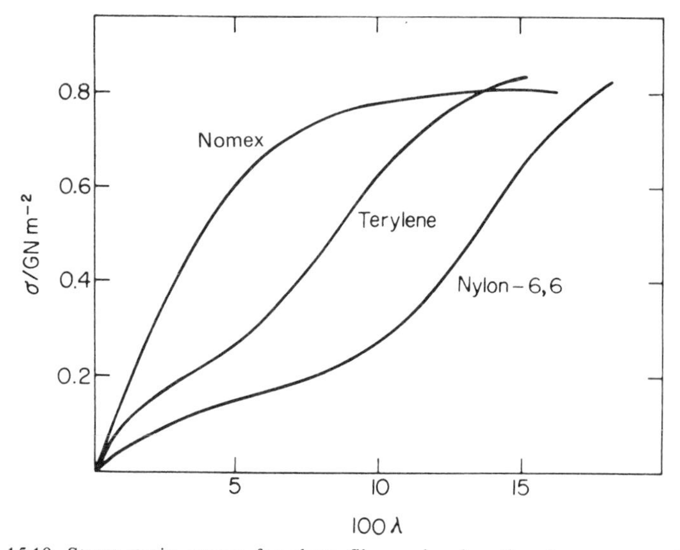

FIGURE 15.10. Stress-strain curves for three fibres, showing the changes wrought by the introduction of one (terylene), then two (Nomex) phenylene groups into the repeat unit of the chain, the stress σ is plotted against 100λ the percentage elongation.

Modulus and chain stiffness. Fibre modulus can be regulated by orientation and crystallinity, but a third parameter, chain stiffness, is available for modification, if additional control is required.

The effect of chain stiffness on the initial modulus is seen in figure 15.10. The increase in chain rigidity on moving from nylon-6.6 $\left[NH(CH_2)_6NHCO(CH_2)_4CO \right]_n$ to terylene

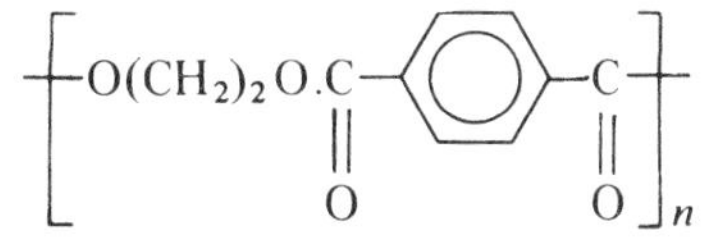

to poly(*m*-phenylene isophthalimide), Nomex,

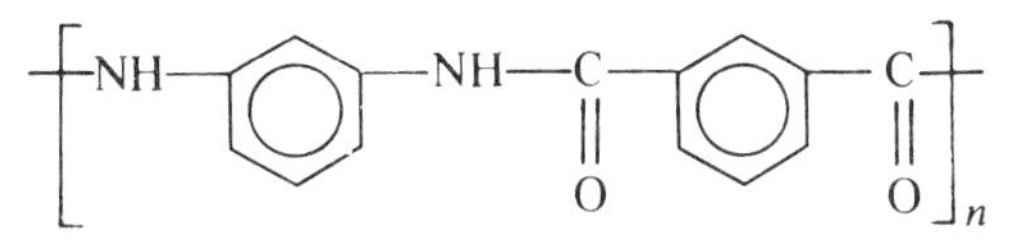

is manifest in an increase in the initial modulus, with the inclusion of aromatic rings in the chain. This is discussed more fully in section 15.11.

Other factors. The moisture regain of a fibre is important when comfort is being considered. In hot weather the ability to absorb perspiration makes clothing more comfortable, and polar polymers are best adapted for this purpose. High moisture retention also decreases the resistivity of the fibre and reduces the tendency to build up static charges which attract dirt and increase the discomfort. Most synthetic fibres have poor moisture regain characteristics and have to be modified in some way to improve this defect. Grafting of poly(ethylene oxide) or acrylic acid on to nylons improves the moisture uptake immensely without affecting the mechanical properties.

Dyeing is also a problem and chemical modification is often necessary. Sites are provided for dyeing by substituting a number of $-SO_3H$ groups in the phenylene rings of the terylene chain or by copolymerizing acrylonitrile with small quantities of vinyl sulphonic acid. These modifications also improve the moisture uptake.

When selecting a fibre for clothing, one should avoid material with a high value of "permanent set". This is a measure of the amount of irreversible flow (C-D in figure 15.9) left in the polymer and is reflected in an increase in fibre length after being subjected to a stress. Obviously in clothing, where the amount of knee or elbow bending is great, a high permanent set value will result in gross fibre deformation and "baggy trousers" or "kneed" stockings. This is partially or totally offset by drawing, but high draw ratios may make the fibre hard. Hence, while poor creep recovery in an article results in loss of shape, its capacity to absorb energy may deteriorate, and if overcompensated for, may result in actual material failure.

It is not uncommon to be forced into a compromise when faced with a choice between two incompatible properties.

15.11 Aromatic polyamides

Synthetic fibre forming polymers with the qualities of stiffness and heat resistance, to make them competitive substitutes for steel wire or glass fibres, have been much sought after for use as reinforcing materials in composites, or in the production of ropes, cables, hoses, and coated fabrics. One of the most successful groups developed for this purpose

TABLE 15.6. Commercially important aramid of fibre forming polymers

Polymer		*Producer*	*Trade name*
I [—NH—C_6H_4(m)—NH—CO—C_6H_4(m)—CO—]	MPD-1	DuPont	Nomex
		DuPont	Nomex II
		Teijin	Conex
		USSR	Phenylon
		DyPont	HT-4
		Monsanto } Firesafe }	Durette
II [—NH—C_6H_4—CO—]	PPB	DuPont	Fibre B
		USSR	Terlon
III [—NH—C_6H_4—NH—CO—C_6H_4—CO—]	PPD-T	DuPont	Kevlar
		Enka	Arenka
		USSR	Vniivlon
		Akzo	Twaron
IV [—NH—C_6H_4—NH—CO—C_6H_4—CO— / —NH—C_6H_4—O—C_6H_4(m)—NH—CO—C_6H_4—CO—]		Teijin	HM-50
V [—NH—C_6H_4—SO_2—C_6H_4—NH—CO—C_6H_4(m)—CO—]		USSR	Sulfon 1
VI [—NH—C_6H_4—SO_2—C_6H_4—NH—CO—C_6H_4—CO—]		USSR	Sulfon T
[(NH—benzazole(X)—C_6H_4—NH)—CO—C_6H_4—CO—]; X = —O—, —S—, —NH—; also copolymers with *p*-phenylenediamine		USSR	SVM
[—NH—CO—phthalimide(C=O, N, C=O)—N—C_6H_4—X—C_6H_4—]		Rhone-Poulenc	Kermel ($X = CH_2$ or O)

TABLE 15.6. Commercially important aramid of fibre forming polymers (*continued*)

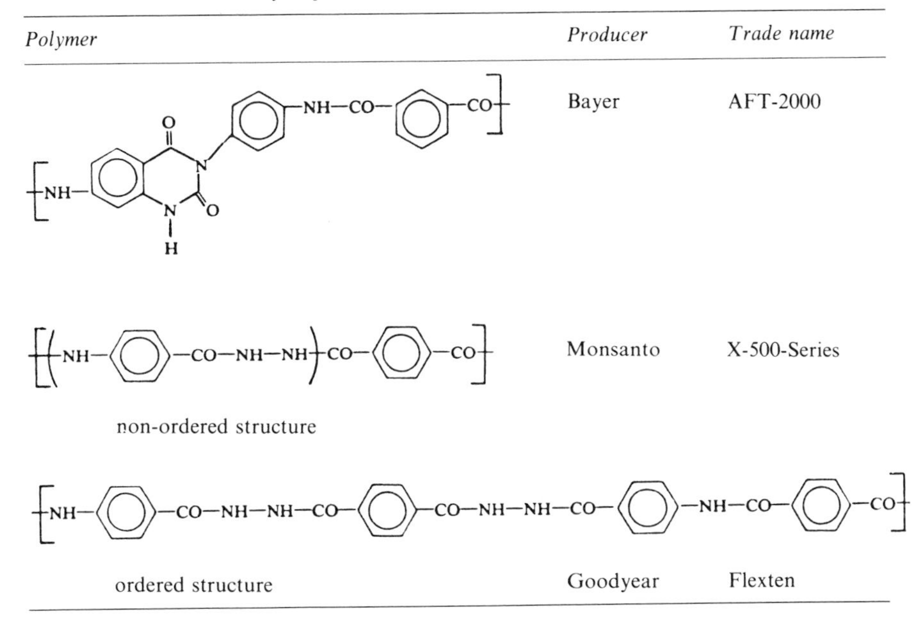

Polymer	*Producer*	*Trade name*
	Bayer	AFT-2000
non-ordered structure	Monsanto	X-500-Series
ordered structure	Goodyear	Flexten

is that of the aromatic polyamides or aramids, defined as fibre forming substances comprising long chain polyamides with > 85 per cent of the amide groups attached directly to aromatic rings.

The structures of several of the commercially significant aramids are shown in table 15.6, whilst the lyotropic liquid crystalline behaviour of the first three members is described in section 16.4. Of these, the most important is poly(*p*-phenylene terephthalamide) PPD-T, which exhibits superior properties and can serve as a model for the others. The polymer is prepared by the condensation polymerization of *p*-phenylene diamine and terephthaloyl chloride in a solvent comprising N-methyl pyrrolidone (NMP) and $CaCl_2$. This solvent combination keeps the growing polymer in solution longer, thereby increasing the molar mass of the product to levels which make it suitable for fibre formation. This is achieved by reducing the intermolecular hydrogen bonding (the salt is a competitive hydrogen bonding agent) while NMP acts as an acid acceptor and good solvator of the polymer chain. High speed stirring is also necessary. The final product is relatively insoluble but can be redissolved in 98 per cent sulphuric acid, and solutions of PPD-T, with greater than 6–7 per cent solids, form anisotropic solutions characteristic of lyotropic liquid crystalline polymers c.f. Section 16.4.

When a flexible, non aromatic, polyamide such a nylon-6,6 is dissolved in a solvent, the chains behave like random coils. As the polymer concentration increases these become entangled. Subsequent spinning and drawing produces a fibre in which the chains retain this entangled structure and are only partially extended, thereby reducing their potential modulus and tenacity. In solutions of the rigid, rod-like, aramids, random coil structures do not form and instead the rigid chains pack in quasi-parallel bundles when the polymer concentration in solution is increased. When spinning from these solutions, the shearing forces orient these bundles in the direction of the applied force, and the resulting fibres are composed of highly oriented, fully extended chains which can also crystallize easily. Thus a product with a high modulus is obtained.

Fibres must be prepared from solutions of PPD-T in sulphuric acid, and conventional spinning techniques have proved ineffective. Two innovations have helped to solve this problem. First it was found that, on heating with H_2SO_4, PPD-T formed a stable complex that melts around 343 K and has a (1:10) PPD-T:H_2SO_4 composition. By using this complex, much higher concentrations can be used in the spinning process than were previously possible. Secondly, a new method called dry-jet-wet-spinning was developed in which an air gap is left between the spinnaret opening and the cold water quench bath. This design allows the chains time to orient in the solution, after emerging from the spinnaret hole, and before final quenching into the fibre form. Spinning PPD-T in this way produces fibres with remarkable properties.

This material, developed by DuPont, has the trade name Kevlar and when formed as described above gives a Kevlar 29 grade. An improved version, Kevlar 49, can be produced by hot drawing the fibre in an inert atmosphere at temperatures above 520 K, whilst a third grade, Kevlar "Hp", has recently been produced with intermediate properties.

Comparison of the stress-strain curves for these aramids and other fibre forming materials is shown in figure 15.11 (a) and (b), and demonstrates the superior properties of PPD-T. It is believed that the reason for the high strength ($\approx$ 2.6 GPa) and modulus (60 to 120 GPa) lies in the three dimensional order coming from both longitudinal and radial orientation. The PPD-T fibre can then be pictured as in figure 15.12 with ordered sheets of hydrogen bonded, extended chains, radiating from the fibre core.

The aramids are very resistant to heat and only begin to decompose and char at temperatures in excess of 670 K. Kevlar is often combined with carbon fibres and embedded in epoxy resins to form hybrid composites that have the ability to withstand catastrophic impact and so find use in aircraft body and wing structures. Other applications include rope and cable manufacture where the high strength per unit weight is a distinct advantage. Compared with steel, Kevlar has a much higher breaking strength, but is six times lighter, thus providing a tremendous weight saving. This makes for ease of handling and it can be used as cables for mooring lines, for off-shore drilling platforms, for parachute lines, fishing lines, mountaineering ropes and pulley ropes. Other uses include cord for reinforcing car tyres, and in protective clothing and body armour.

15.12 Polyethylene

The influence of chain branching on the properties of a polymer is usefully illustrated with reference to polyethylene. Chemically, this is one of the simplest of the synthetic polymers with the repeating unit $\text{-}(CH_2\text{—}CH_2)_n\text{-}$, but it can be prepared in several different ways and these determine the extent of chain branching in the product.

Polyethylene now tends to be marketed in three general grades, high density polyethylene (HDPE), linear low density polyethylene (LLDPE or 1-LDPE) and low density polyethylene (LDPE). The essential structural differences for each are shown on page 344.

HDPE, prepared using organometallic catalysts, is a structurally regular chain material with very few small branch points (less than 7 per 1000 carbon atoms). Because of this regularity, the polymer chains can pack efficiently resulting in a highly

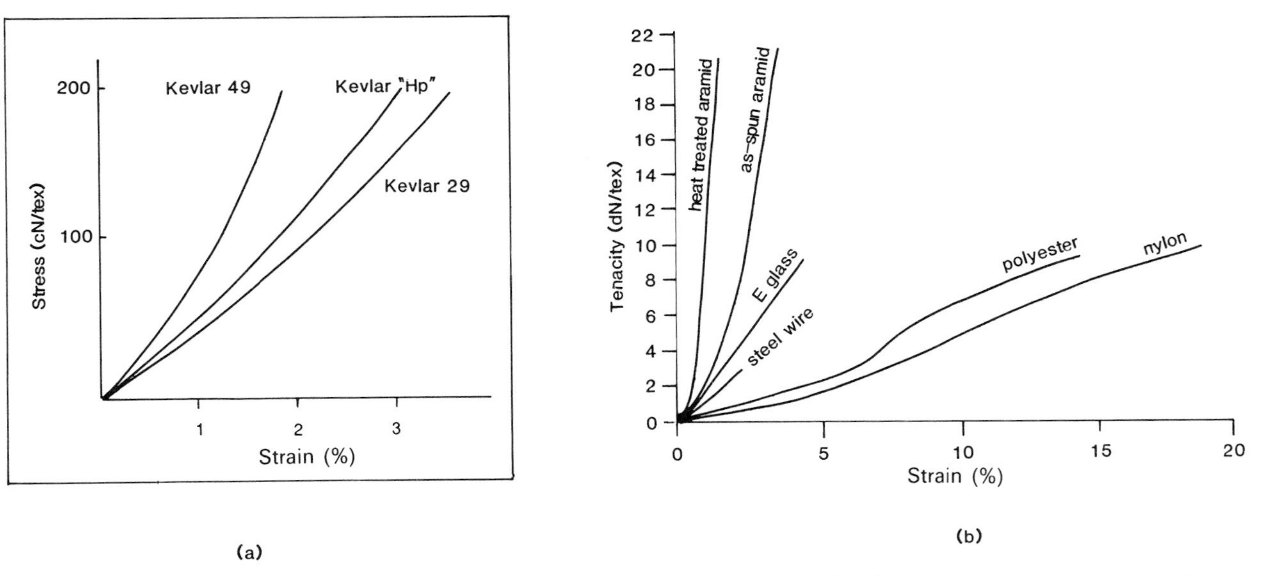

FIGURE 15.11. (a) Comparison of the stress-strain behaviour of the three grades of Kevlar. (b) Comparison of the tenacity of Kevlar fibres with glass, steel, polyester and nylon, as a function of percentage strain. (Reproduced from D. Tanner, J. A. Fitzgerald and B. R. Phillips (1989) with permission from *Verlag Chemie*).

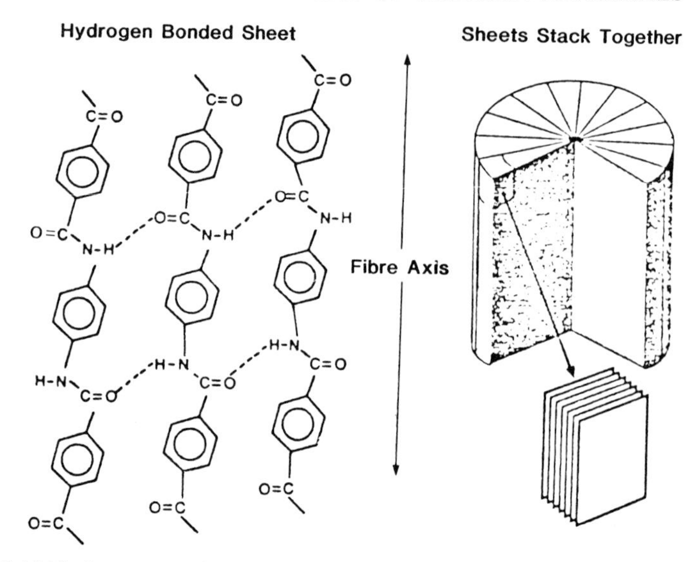

FIGURE 15.12. A representation of the arrangement of sheets of poly(*p*-phenylene terephthalamide) to form the fibre structure. (Adapted from D. Tanner, J. A. Fitzgerald and B. R. Phillips (1989) with permission from *Verlag Chemie*.)

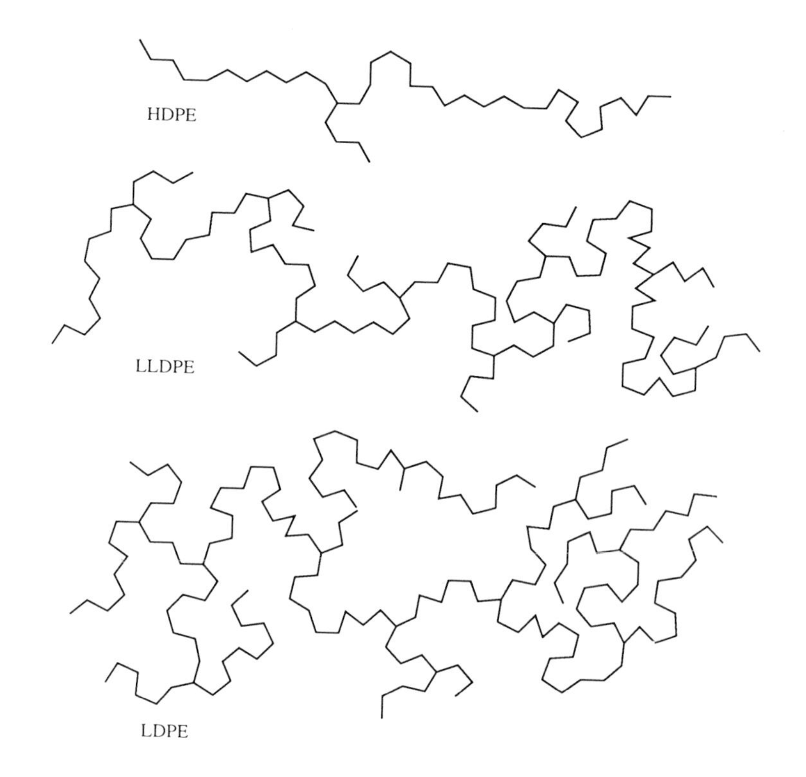

TABLE 15.7. Comparison of various polyethylene grades

Property	*LDPE*	*LLDPE*	*HDPE*
Melting point (K)	383	393–403	> 403
Density (g/cm^3)	0.92	0.92–0.94	0.94–0.97
Film tensile strength (MPa)	24	37	43

crystalline material with a correspondingly high density. The polymer is used to manufacture bottles, crates, and pipes.

LDPE, prepared by a high-pressure, radical-initiated polymerization process, is a highly branched polymer with approximately 60 branch points per 1000 carbon atoms. It has a much lower crystalline content and density, and has good film forming properties so that the largest application is as film for packaging and cable coatings. It has, however, a greater permeability to gases (CO_2, O_2, N_2) than HDPE.

The property gap that exists between HDPE and LDPE has been filled by LLDPE. This polymer can be prepared by solution or gas phase polymerization, and is actually a copolymer of ethylene with 8–10 per cent of an α-olefin such as but-1-ene, pent-1-ene, hex-1-ene, or oct-1-ene. This produces a chain with a controlled number of short chain branches, and densities intermediate between HDPE and LDPE, thereby allowing it to be prepared in various grades by controlling the type of the comonomer. Thus the use of oct-1-ene gives a lower density product than that obtained when but-1-ene is incorporated in the chain, because the longer (hexyl) branch in the former pushes the chains further apart than the ethyl branch of the latter, hence lowering the packing efficiency of the chains.

LLDPE is now beginning to compete with LDPE in film blowing and casting applications because of its superior resistance to puncture by hard particles. It also has better qualities of toughness, and lower brittle temperatures than LDPE, and is now used to replace blends of HDPE and LDPE.

Table 15.7 shows a number of important property comparisons, and indicates how the branching plays a significant role in determining the properties, and hence the end uses of the three polymers.

15.13 Elastomers and crosslinked networks

Rubber-like elasticity and its associated properties have already been discussed in some detail (see section 12.8 and chapter 14) and only a brief résumé of the relevant features will be given.

The fundamental requirements of any potential elastomer are that the polymer is amorphous with a low cohesive energy, and that it is used at temperatures above its glass transition. The polymer in the elastic region is characterized by a low modulus (about $10^5\,N\,m^{-2}$) and, for useful elastomers, by large reversible extensions. This reversibility of the slippage of flow units requires a chain in which there is a high localized mobility of segments, but a low overall movement of chains relative to one another. The first requirement is satisfied by flexible chains, with a low cohesive energy, which are not inclined to crystallize (although the development of some crystalline order on stretching is advantageous). The second requirement, prevention of chain slippage, is overcome by cross-linking the chains to form a three-dimensional network.

Crosslinking. Crosslinking provides anchoring points for the chains and these anchor points restrain excessive movement and maintain the position of the chain in the network. This is not confined to elastomers, however, and the improved material qualities which result are also found in the crosslinked phenol-formaldehyde, melamine, and epoxy resins.

When a sample is crosslinked, (1) the dimensional stability is improved, (2) the creep rate is lowered, (3) the resistance to solvents increases, and (4) it becomes less prone to heat distortion, because T_g is raised. All these effects tend to be intensified as the crosslink density is increased and can be controlled by adjusting the number of crosslinks in a sample.

Creep in crosslinked polymers. The creep response depends mainly on the temperature and the crosslink density. At temperatures below T_g, crosslinking has little effect on the properties of the material, but above T_g, secondary creep, arising from irreversible viscous flow, is reduced or eliminated by crosslinking.

Creep is a function of the elastic modulus, the mechanical damping, and the difference between ambient temperature and T_g. The thermosetting resins usually have a high modulus, low damping characteristics, and T_g well above ambient, consequently the creep rate is low and they have good dimensional stability.

The effect of increasing the crosslink density on these parameters, is illustrated for a phenol-formaldehyde resin in figure 15.13.

Above T_g the modulus is a function of the extent of crosslinking; the damping peaks shift to higher temperatures as T_g increases and eventually become difficult to detect.

This shows the extent to which crosslink density can affect the physical behaviour.

Additives. Many elastomers are subject to oxidative degradation and can be protected to some extent by the addition of antioxidants, such as amines and hydroquinones.

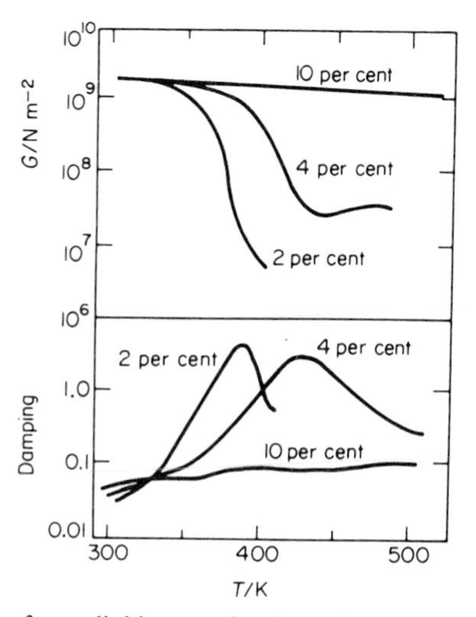

FIGURE 15.13. Influence of crosslinking on the dynamic mechanical response of a phenol-formaldehyde resin. Concentrations of the crosslinking agent hexamethylenetetramine are shown alongside the appropriate curve. (After Nielsen.)

The abrasion resistance can also be improved by adding a filler to reinforce the elastomer, and carbon black is widely used for this purpose. Fillers (glass fibre, mica, sawdust) are also used in the thermosetting resins as reinforcement.

15.14 Plastics

So far attention has been focused predominantly on fibre and elastomer requirements, because it is considerably more difficult to be specific about the qualities desired in a plastic material when the range of applications covered is very much more extensive. The general principles relating to the control of T_m, T_g, modulus, *etc.*, can all be applied to the formation of a specific type of plastic and we shall simply try to illustrate briefly the diversity of problems encountered in the field of plastic utilization.

The conflict between low creep and high impact strength mentioned earlier is not confined to fibres, but is also a problem encountered in plastic selection. It is an important point to consider for the engineering requirements of the material, when the ability to absorb energy is desirable, but is at odds with the equally desirable qualities of high rigidity and low creep. The problem to be faced is then how to make a brittle, glassy polymer tougher, *i.e.* how to limit the modulus or tensile strength. In general, an increase in crystallinity (and consequently the modulus) tends to make a plastic more brittle. Crystallinity can be controlled by copolymerization or branching and the brittleness can be tempered using one or other of these modifications. Alternatively, an elastomeric component can be introduced, which will improve the impact strength by reducing the rigidity and yield stress. This has been used in "high impact" polystyrene (HIPS) or acrylonitrile-butadiene-styrene (ABS) copolymers, where the elastomeric component is above its T_g under prevailing environmental conditions and acts as a second phase. This leads to an increased damping efficiency which is manifest in the appearance of a second low temperature damping maximum in the damping curve. This is seen in figure 15.14 for "high impact" polystyrene-butadiene copolymer (SBR rubber) whose T_g is 213 K. The phenomenon is similar to the toughening effect in semi-crystalline polymers caused by the strengthening of the amorphous regions with crystalline crosslinks, but in the latter case, the two phase aspect arises from the existence of crystalline and amorphous regions.

While orientation is most important in fibre formation it can also improve the response of a brittle polymer and increase its ductility. This is particularly true in film preparation or moulding where viscous flow is inclined to introduce a certain degree of chain alignment at some stage in the process.

Interchain interactions also affect performance and poly(oxymethylene) has a higher modulus in the glassy state than polyethylene presumably because of the polar attractions between the chains.

When faced with the problem of selecting a plastic for a given purpose, a design engineer must then be concerned with the properties of the material, the ease of processing or fabrication, the behaviour under the environmental conditions the product will be subjected to (*i.e.* the thermal range), and, of course, the economic factors. Each problem has to be treated as a specific case and familiarity with structure-property relations aids the selection. The illustrations are limited to two widely differing aspects.

Plastic selection for bottle crate manufacture. The difficulties encountered when

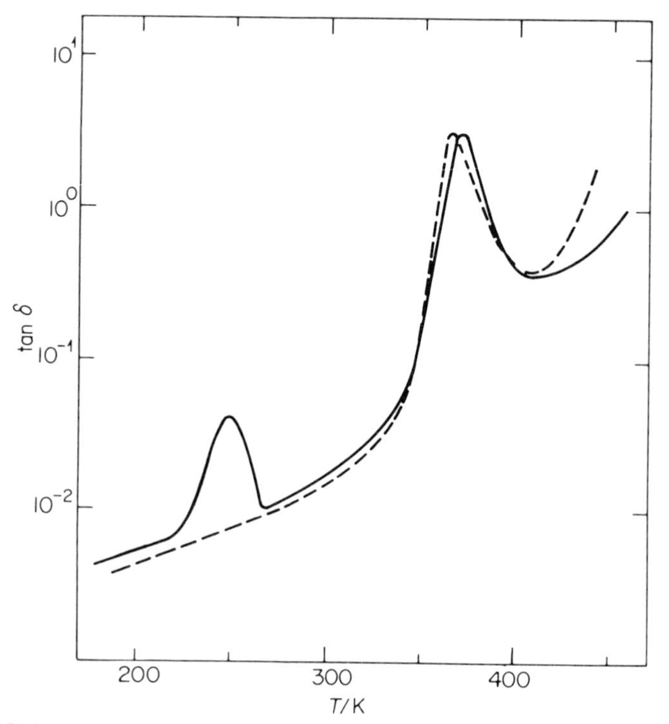

FIGURE 15.14. Damping curves for (- - -) polystyrene, and (———) high impact polystyrene. The latter has an internal friction peak below ambient. The damping tan δ is plotted against temperature T.

choosing a suitable plastic for a particular use arise mainly because each case is associated with a unique combination of properties. A good example, concerning bottle crate manufacture, has been cited by Willbourn.

High density polyethylene was chosen for the manufacture of beer crates in West Germany, because it was the cheapest plastic available which was sufficiently tough and rigid for the purpose. It was also found to have a satisfactory creep response and good impact resistance down to 253 K, which is adequate for continental winter temperatures. This plastic and the crate design were suitable for the use pattern in West Germany, where crates were usually piled 12 high.

When these crates were used in the U.K., where the practice is to stockpile 20 to 36 high for much longer periods, a rapid rate of crate failure was experienced. The change in conditions necessitated a new choice of plastic. This had to have better creep properties and a higher rigidity, but did not have to retain those good qualities at temperatures below 263 K, because of the milder U.K. winters. Poly(vinyl chloride) was considered but is too difficult to mould; polystyrene and polypropylene have good creep characteristics but these deteriorate at lower temperatures. The problem was solved by using poly(propylene-*b*-ethylene) copolymers, which have a good toughness and mechanical response in the required temperature range. This is shown in figure 15.15 where the high density polyethylene failed under a load of 1000 kg in 29 h, but the copolymer survived two months.

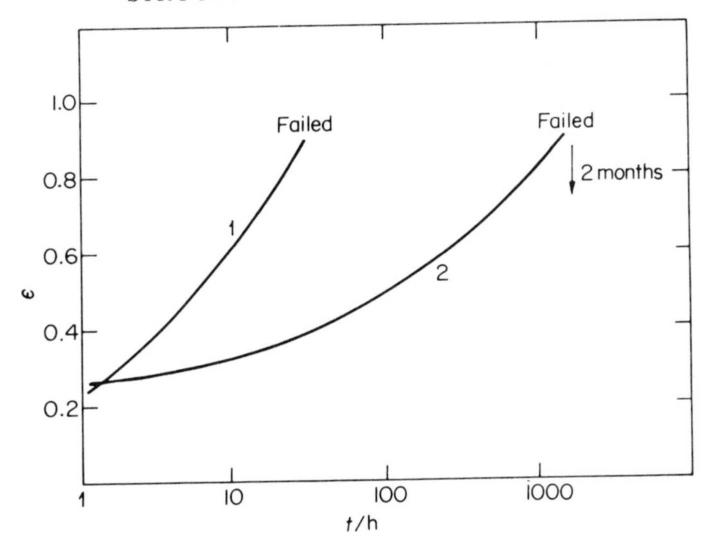

FIGURE 15.15. Comparison of loading tests using 1000 kg load on crates made from (a) high density polyethylene and (2) poly(ethylene *b*-propylene) copolymer. (After Wilbourn, *Plastics and Polymers*, 1969.) The percentage compressive strain ϵ is plotted against time t.

In this instance, both environmental conditions and industrial practice were important factors.

Medical applications. The use of polymeric materials in the medical field is growing and raises problems peculiar to the mode of application. Prosthesis is one of the major medical interests and certain plastic replacement parts are now commonly used. High density polyethylene is a successful replacement part for damaged hip joints and is employed as the socket, which accommodates a steel ball cemented to the femur using poly(methyl methacrylate). Artificial corneas can be prepared from poly(methyl methacrylate), while sections of artery are replaced by woven nylon or terylene tubes. Heart valves have been made from polycarbonates and even artificial hearts, made from silicone rubber, have met with limited success. Plastic replacements for nose and ear cartilage; body absorbing sutures; and the use of polymeric membranes for dialysis in artificial kidney machines are only a few examples in a steadily growing list of uses.

The selection of suitable polymers for medical use focuses attention on the inertness of the polymer, its mechanical properties, and the extent of its biostability. It is useless implanting a polymer in the body which will be rejected or will degrade to produce toxic materials. The sample should also be pure and free of plasticizer, which might leach out and cause harmful side effects. The polymer has to be resistant to mechanical degradation and particularly abrasion, in case the abraded particles act as irritants. These conditions tend to limit the choice.

The use of polymers as adhesives is of particular interest. One reason for using the α-cyano acrylate esters as tissue adhesives has been the observation that there is a progressive, non-toxic, absorption of the substances by the body. Interest in the use of polymers as reagents, which actively take part in the body functions, is now being evaluated.

External to the body, "hydrogels" are used as contact lenses and their use may be extended to implantation. They are composed of crosslinked networks of hydroxyl methacrylate copolymers which swell when in contact with water.

Films and membranes are also used. A patient can be encased in a poly(vinyl chloride) tent while germ-free air is pumped through the canopy. It has also been suggested that films which allow the selective passage of oxygen in one direction could be used as oxygen tents. These would be similar to the silicone membranes which allow predominant passage of oxygen from water to air, and have been used to make a cage capable of supporting underwater, non-aquatic life in an air atmosphere extracted from the water.

This expanding field will no doubt demand new polymers with specific applications.

15.15 High temperature speciality polymers

While many of the more common polymers are remarkably resistant to chemical attack, and stable when subjected to mechanical deformation, few can withstand the destructive effects of intense heat. To overcome this low thermal stability, chains incorporating: (i) thermally unreactive aromatic rings; (ii) resonance stabilized systems; (iii) crosslinked "ladder" structures; and (iv) protective side groups, have been synthesized. Many of these new structures have met with considerable success, and several of the aramids, described in section 15.11, are capable of maintaining 50 per cent of ambient tensile strength at temperatures in excess of 550 K. They are also difficult to ignite and can be used to make excellent fire resistant clothing. Particularly useful in this respect are the poly(benzimidazole)s which are characterized by the presence of the

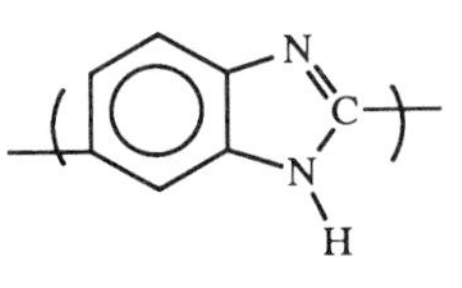

unit in the chain. They have been known since 1961 when they were first synthesized by Marvel and Vogel, but one member in particular has achieved commercial prominence. This is the material prepared by a two step reaction, involving first a melt polycondensation reaction between tetraaminobiphenyl and diphenyl isophthalate to give a prepolymer foam. This is then crushed and heated under N_2 at 530–700 K to close the ring and generate the poly(benzimidazole) structure (PBI).

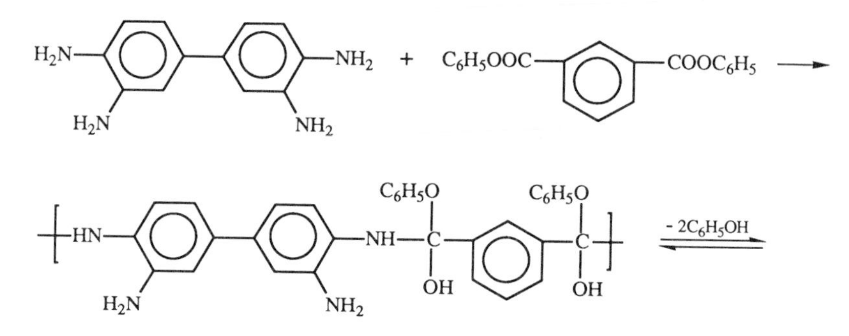

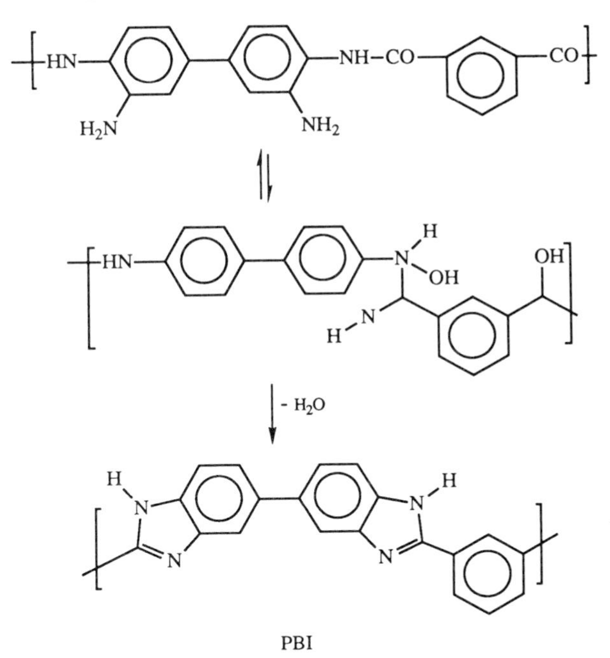

PBI

The polymer is soluble and can be dry spun from dimethyl acetamide/LiCl solutions, as a fibre which has exceptional heat resistant properties. It does not melt or ignite in air at temperatures up to 830 K, and above this temperature, while it degrades, it produces virtually no smoke and undergoes carbonization. Short term heating up to 670 K does not alter the tensile strength and the mechanical properties are retained down to at least 150 K.

The PBI fibre has good moisture absorption characteristics (15 per cent at ambient temperature and 65 per cent relative humidity) making it better than cotton and comfortable to wear. When mixed with aramid fibres, it makes superior heat protective clothing.

PBI can also be moulded to produce castings with high performance qualities over the temperature range 110 to 700 K.

Other examples of semi-ladder-like polymers include the

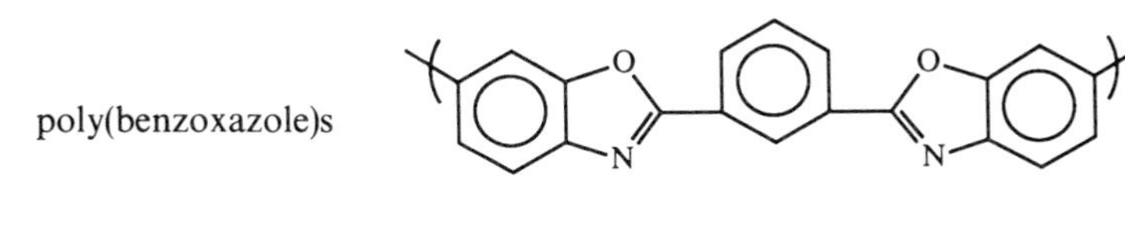

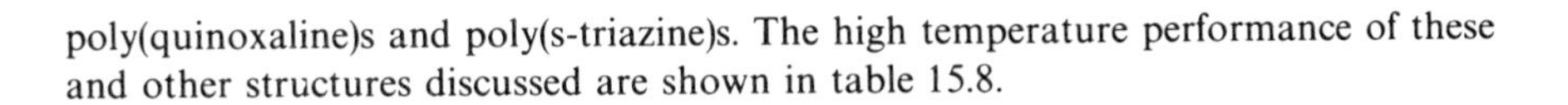

poly(quinoxaline)s and poly(s-triazine)s. The high temperature performance of these and other structures discussed are shown in table 15.8.

TABLE 15.8. Structures of polymers with exceptional high temperature performance

Polymer		*Upper service temperature* [K]
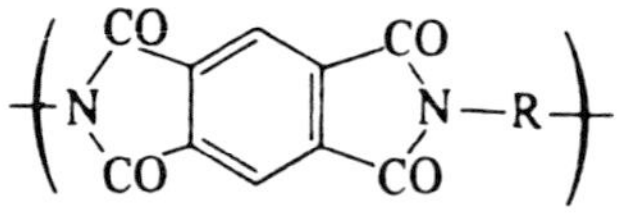	polyimide	570–620
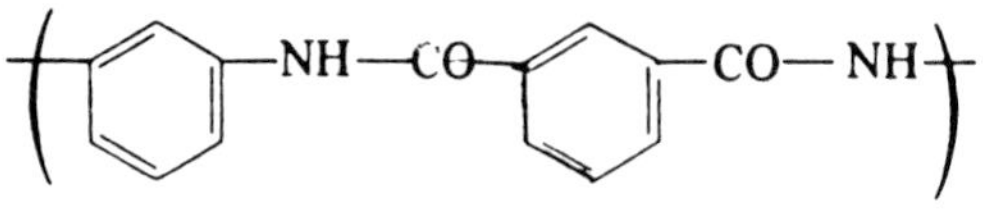	aromatic polyamide (aramid)	470–520
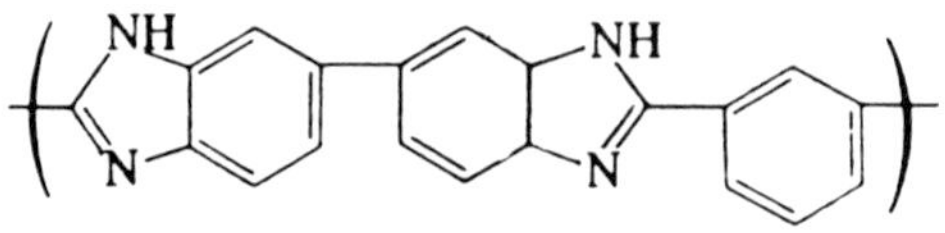	polybenzimidazole	520–570
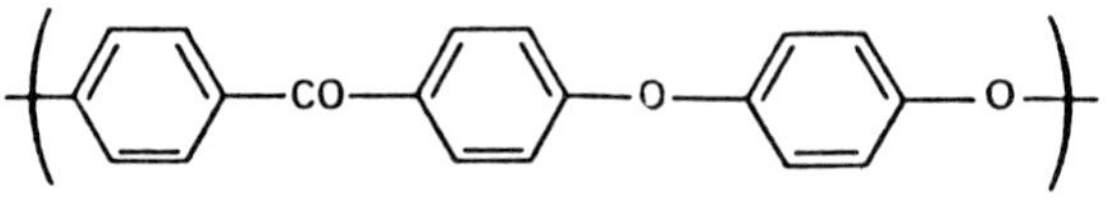	polyetheretherketone	510–530
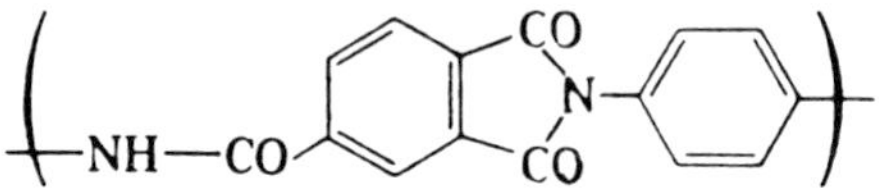	polyamide-imide	490–510
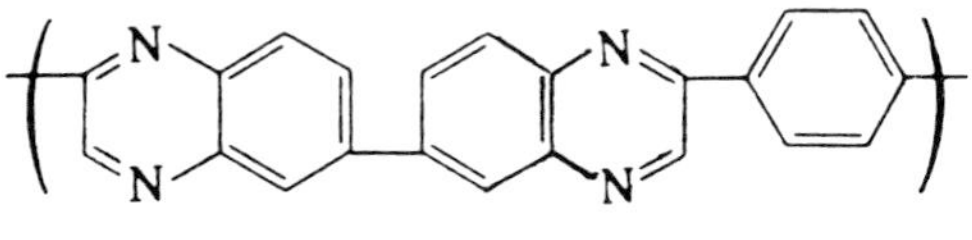	polyquinoxaline	670–720
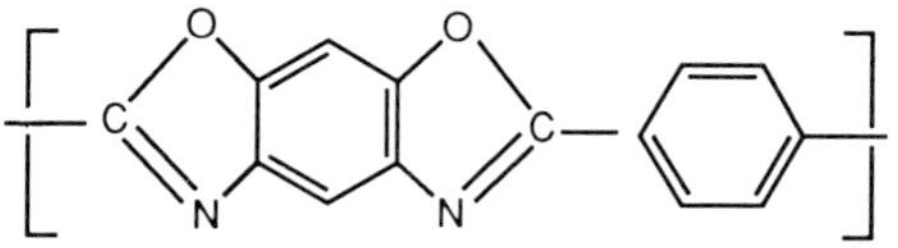	poly (*p*-phenylene-benzobisoxazole)	600–800
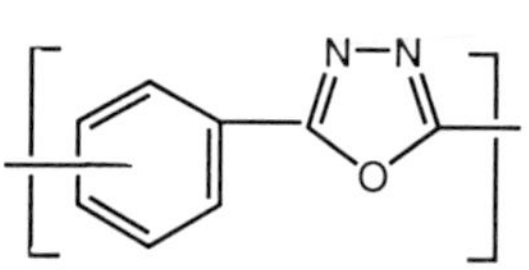	poly(oxadiazole)s	480–600

Greater stability can be achieved if the vulnerable single bonds, which are susceptible to degradation causing chain scission, can be eliminated. This can be done by preparing a ladder structure, a typical example being poly(imidazopyrolone). It is easily seen that single bond scissions at points A along the chain do not lead

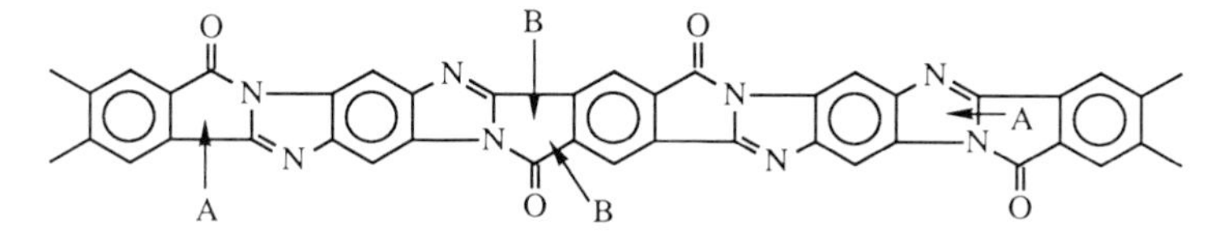

to complete chain scission. This can only be brought about by two single bonds being broken as at points B, *i.e.* two bonds on opposite sides of the chain and between the same two "rungs" of the ladder. As this process has a low probability, many of these polymers can resist temperatures over 850 K, and so can effectively compete with some metals.

During the 1970s a number of non-commodity, speciality polymers have been developed that display superior properties of heat resistance, impact resistance, high tensile strength and stiffness. These are used either as single materials or to form the reinforcing component of a composite, blend or alloy. Several of the more common groups which have been classified as engineering plastics are listed in table 15.9 together with some of their applications.

Acetal is actually polyoxymethylene $\left(\!-O-CH_2-\right)_n$, a highly crystalline ($T_m \approx 450\,K$) polymer prepared from formaldehyde. It can be stabilized by end-capping the polymer using acetic anhydride to acetylate the terminal hydroxyl groups, and prevent polymer degradation by chain "unzipping". The material has good abrasion resistance, a reasonably high heat distortion temperature (383 K) and is not attacked by polar solvents. (In this context the heat distortion or deflection test defines the temperature at which a standard size sample of polymer ($5 \times 1/2 \times 1/8$ inches) distorts under a flexural load of 66 or 264 psi placed at its centre. This is often found to be 10 to 20 K lower than T_g for an amorphous polymer, but may be much closer to T_m in crystalline polymers.) Polyacetal is often used as a composite with a glass filler which raises the heat distortion temperature to 423 K.

TABLE 15.9. Engineering resins, generally used in conditions of high heat, impact, or moisture

Resin	*Typical applications*
Acetal	Sinks, faucets, electrical switches, gears, aerosol bottles, meat hooks, lawn sprinklers, ballcocks, shaver catridges, zippers, telephone push buttons
Polycarbonate	Helmets, power tool housings, battery cases, safety glass, automobile lenses, 5-gal bottles
Polyphenylene-sulphide	Electrical connectors, coil forms, lamp housings
Polysulphone	Electrical connectors, meter housings, coffee makers, camera bodies, automobile switch and relay bases, light-fixture sockets, fuel cell components, battery cases, medical supplies
Modified poly-phenylene oxide	Automobile dashboards, pumps, shower heads, plated automobile grilles and trim, appliance housings, wiring splice devices, protective shields
Polyimide	Radomes, printed circuit boards, turbine blades
Polyamide-imide	Valves, gears, pumps, high-temperature magnet wire enamels

Source: Society of the Plastics Industry

In the polycarbonates, the most widely used member of the group is poly(bisphenol A carbonate)

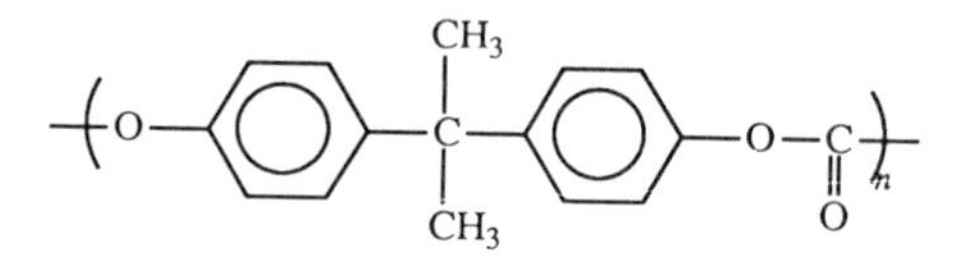

This polymer can be prepared by the interfacial polycondensation of bisphenol A alkali salt dissolved in the water phase, and phosgene ($COCl_2$) dissolved in methylene chloride.

It can be used either as the pure polymer or in blends, particularly with acrylonitrile-butadiene-styrene (ABS) copolymers. The bisphenol A structure appears in other combinations, *e.g.* in a polysulphone copolymer (see table 15.10), in aromatic polyesters with phthalic acid moieties,

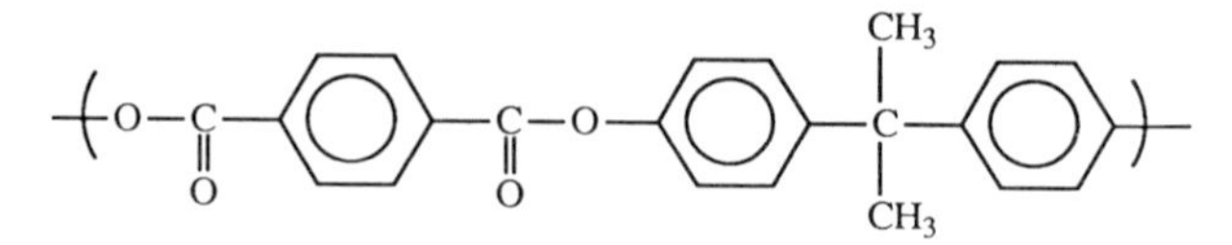

or in poly(ether-imides).

Poly(ester-carbonate)s with the general structure

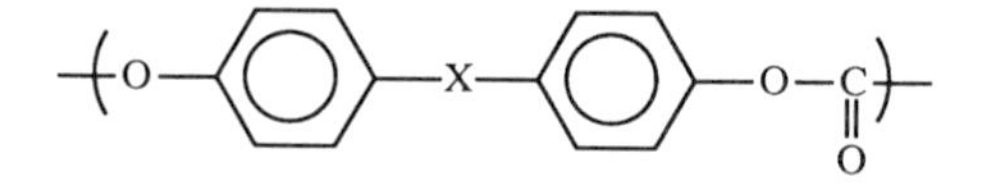

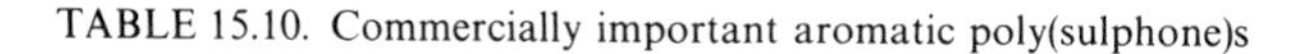

TABLE 15.10. Commercially important aromatic poly(sulphone)s

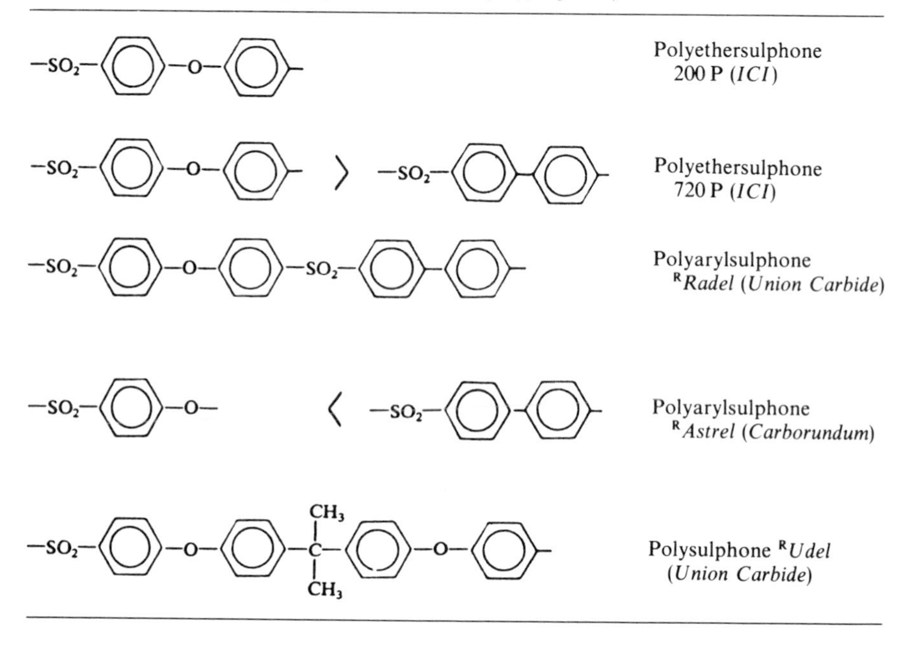

Structure	Name
$-SO_2-C_6H_4-O-C_6H_4-$	Polyethersulphone 200 P (*ICI*)
$-SO_2-C_6H_4-O-C_6H_4-$ > $-SO_2-C_6H_4-C_6H_4-$	Polyethersulphone 720 P (*ICI*)
$-SO_2-C_6H_4-O-C_6H_4-SO_2-C_6H_4-C_6H_4-$	Polyarylsulphone [R]*Radel* (*Union Carbide*)
$-SO_2-C_6H_4-O-$ < $-SO_2-C_6H_4-C_6H_4-$	Polyarylsulphone [R]*Astrel* (*Carborundum*)
$-SO_2-C_6H_4-O-C_6H_4-C(CH_3)_2-C_6H_4-O-C_6H_4-$	Polysulphone [R]*Udel* (*Union Carbide*)

can also be prepared where the group X is an alkylene, ether, sulphide or sulphone group.

The polysulphones form another large group and some of the commercially important structures are shown in table 15.10.

Poly(phenylene sulphone) tends to be too intractable for easy processing and the copolymer structures are more useful. These are usually amorphous materials with high T_g values typically in the range 465 to 560 K. They are thermally stable, show good mechanical properties – particularly the creep resistance – and are resistant to attack by dilute acids and alkalis. They can, however, dissolve in polar solvents, and solvent attack may also cause environmental stress cracking.

The polyarylene sulphones can be synthesized using an electrophilic substitution reaction

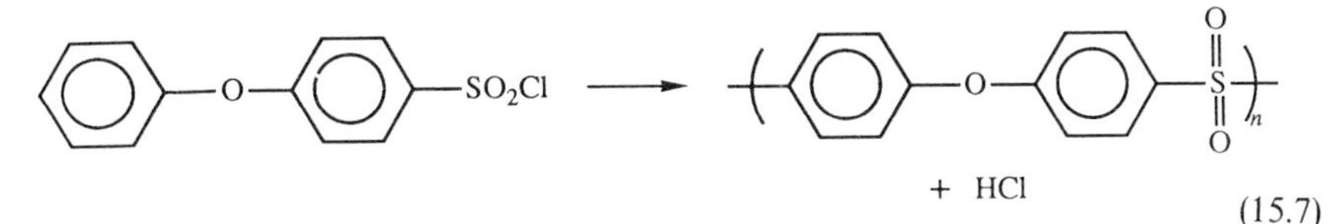

(15.7)

which gives a *p*-substituted product. Although this method uses rather costly starting materials, other routes tend to give a mixture of *o*- and *p*-substitution. Since the brittleness of the product increases with greater degrees of *o*-substitution, more economical variations of the first method have been developed. These are outlined in equations (15.8) and (15.9).

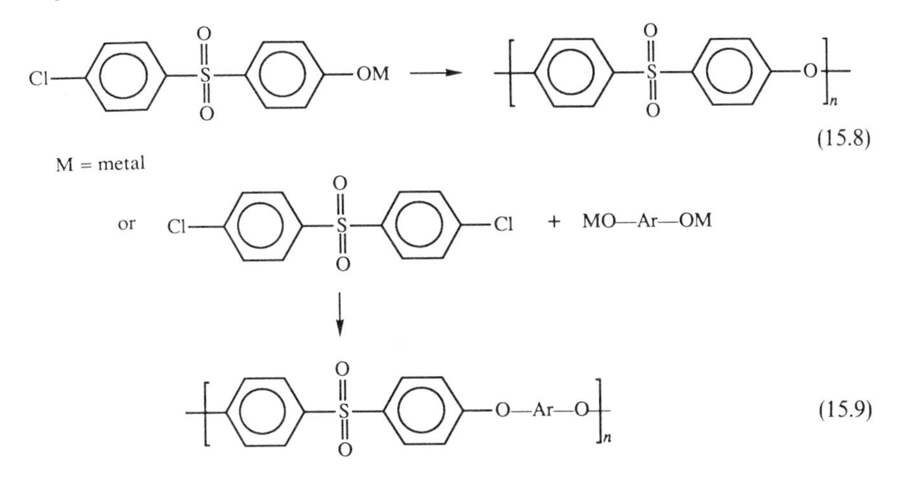

Another sulphur-containing material, poly(phenylene sulphide)

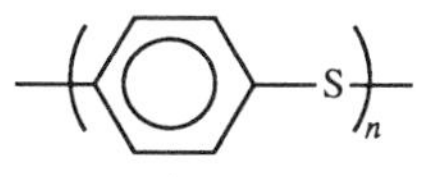

is highly crystalline ($T_m = 563$ K) with a $T_g \approx 470$ K. It has good thermo-oxidative stability, is resistant to solvents, and when prepared as a composite with glassfibre, has a heat distortion temperature of 520 K.

Poly(oxa-2,6-dimethyl-1,4-phenylene) (PPO)

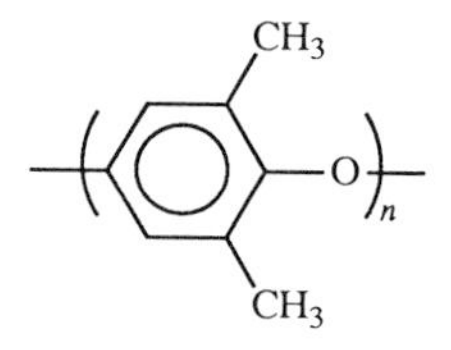

is prepared by the oxidative coupling of 2,6-dimethyl phenol, but it can also be made as a copolymer with styrene grafted on at the synthesis stage. PPO itself does not normally crystallize to any great extent from the melt but it has a good heat deflection temperature (> 370 K), possesses good self-lubricating qualities, and is an excellent, electrical insulating material. It is widely used as a blend with polystyrene or a polyamide to form materials in the NORYL series.

Polyimides form the most important group of thermally stable polymers and are characterized by the presence of the

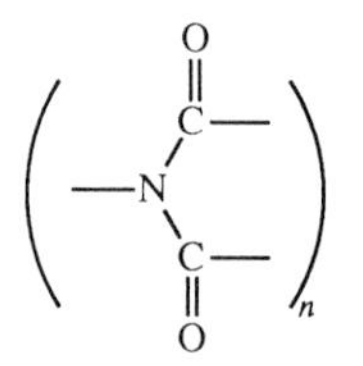

group in the structure. Several different synthetic routes have been developed and a large variety of different materials have been prepared. The standard approach involves the polycondensation of pyromellitic dianhydride with a diamine such as 4,4′-diamino diphenyl ether, and in this case a polyimide marketed under the name KAPTON is produced.

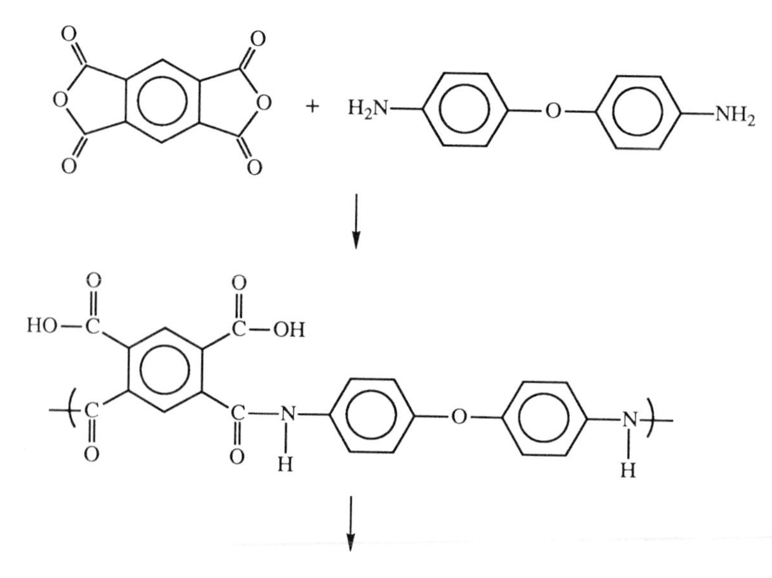

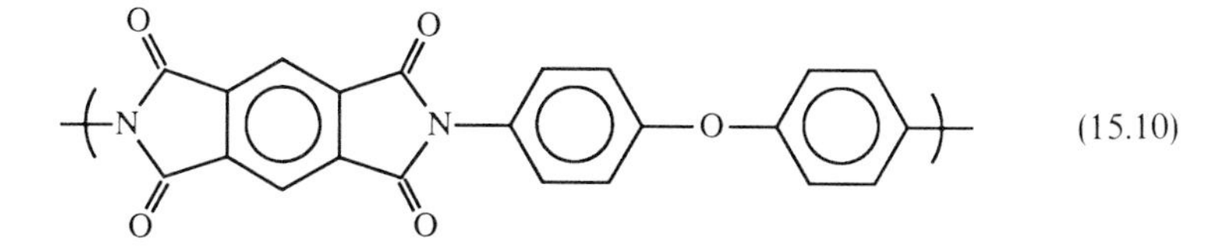

(15.10)

This synthesis involves a two step reaction, the first of which yields an intermediate poly(amic acid), that is soluble in polar solvents. In the second, heating to temperatures of about 570 K effects ring closure to form the insoluble, intractable, polyimide.

Kapton has an extremely high heat distortion temperature of 630 K and shows exceptional thermo-oxidative resistance.

A more direct synthetic method, avoiding the ring closure step, makes use of diisocyanates.

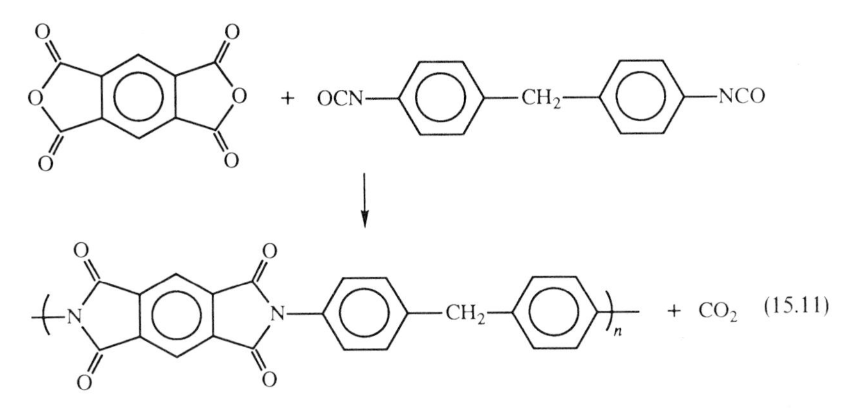

(15.11)

This reaction can take place in aprotic solvents at temperatures below 370 K in the presence of a trace of water and a strong alkaline catalyst such as triethylamine.

NASA have developed a polyimide LARC-TPI (structure V).

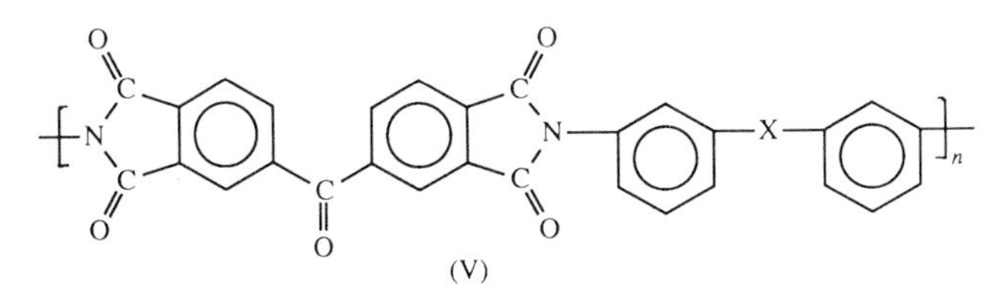

(V)

When X = (C=O) this is a semicrystalline material with $T_m = 623$ K and $T_g = 519$ K. On heating above T_m, the crystallinity is destroyed and a totally amorphous polymer is formed. The film tensile strength and modulus at 298 K are 135.8 MPa and 3.72 GPa respectively. When the group X in structure V is $\text{-}(SO_2)\text{-}$ a totally amorphous material is obtained, with a higher T_g of 546 K, film tensile strength 62.7 MPa and modulus of 4.96 GPa at 298 K.

These high T_g and T_m values can make the polymers difficult to handle, and improved processability can be obtained by introducing flexible groups in the chain. If the group X is an ether oxygen, the T_g is lowered and the crystallinity reduced.

Similarly incorporation of $\left[O \left(CH_2CH_2 \right)_2 O \right]$ in place of X leads to a material which is more easily processed, is amorphous with a $T_g = 428$ K and has a film tensile strength and modulus at 298 K of 86.2 MPa and 2.7 GPa respectively.

Copolymerization can also make the polyimides more tractable. Thus polyamide-imides such as structure VI (KERMEL) can be produced.

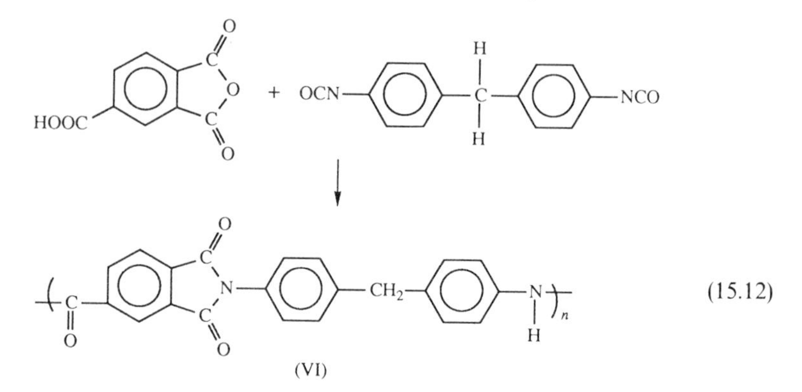

(VI)

Another example of this group, marketed as TORLON, has the structure

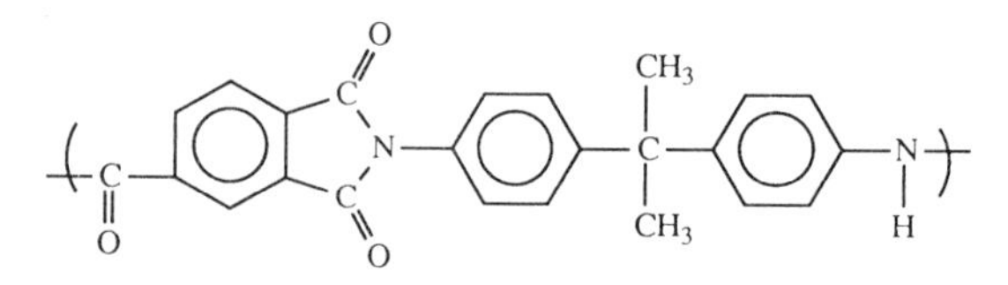

with a film (298 K) tensile strength of 186 MPa, a modulus of 4.6 GPa and a heat distortion temperature of 555 K.

Polyether-imides with good melt processing properties are also available, but tend to have inferior and high temperature mechanical properties when compared with the preceding examples. Typical of this group is structure VII.

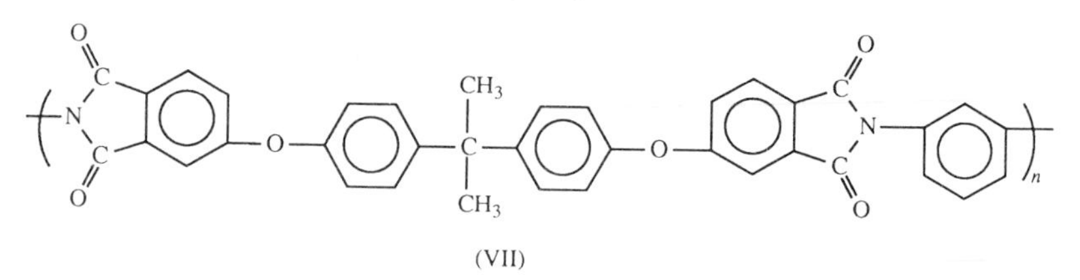

(VII)

This is known as ULTEM but has lower values of the heat distortion temperature (473 K), tensile strength (9 MPa) and modulus (3.2 GPa) at 298 K.

The aromatic polyketones form an important group of high performance plastics as they exhibit extremely good high temperature resistance and are melt processable. A number of examples are shown in table 15.11 and of these the best known is poly(ether ether ketone) or PEEK, structure X. This is a crystalline polymer ($T_m = 607$ K) and $T_g = 416$ K, with a heat distortion temperature of 433 K. The latter can be increased to 588 K for PEEK-glass fibre composites. The material is insensitive to hydrolysis and is particularly stable for long periods in hot water. This contrasts

TABLE 15.11. Typical aromatic poly(ether ketone)s

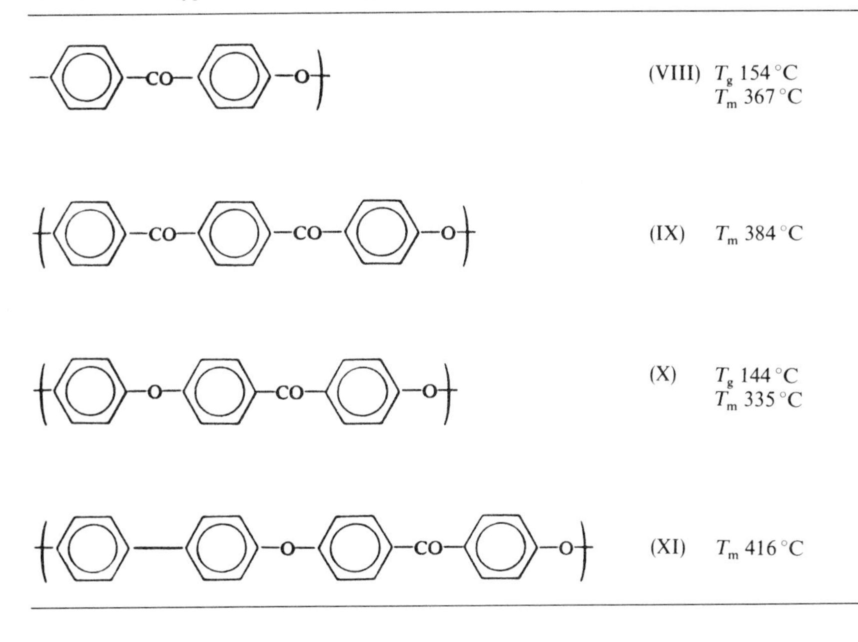

with the polyimides which are sensitive to hydrolytic degradation. PEEK has found very special uses in the nuclear industry as a coating for wires and cables and as blow moulded containers. Other applications include composites with carbon fibres for use in the aerospace industry.

15.16 Carbon fibres

Although originally studied for its high temperature qualities, the carbon fibre is, at present, used to advantage mainly in low temperature situations.

The fibres are prepared by converting oriented acrylic fibres into aligned graphite crystal fibres in a two-stage process. In the first stage the acrylic fibre is oxidized, under tension to prevent disorientation of the chains, by heating in a current of air at 490 K for several hours. This is thought to lead to cyclization and the formation of a ladder polymer.

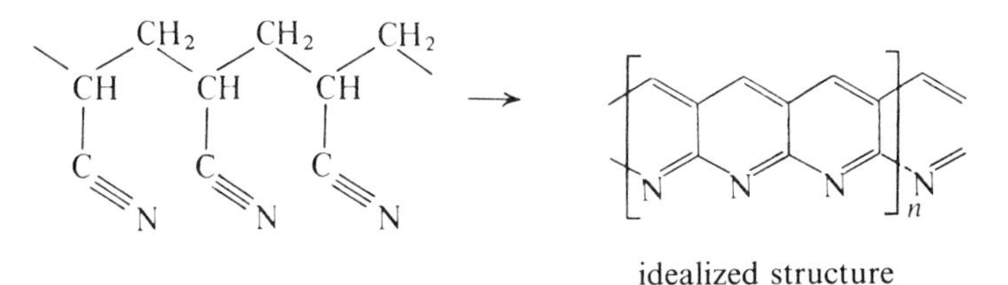

idealized structure

The second stage involves heating the fibres for a further period at 1770 K to eliminate all elements other than carbon. This *carbonization* is believed to involve

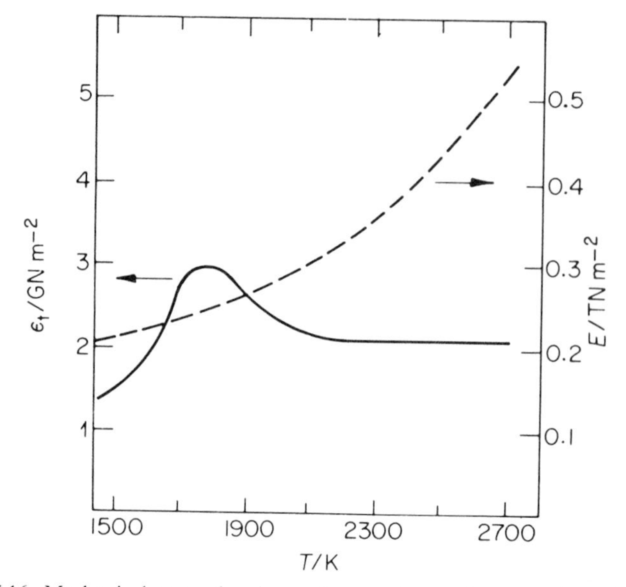

FIGURE 15.16. Mechanical properties, the tensile strength ϵ_t and Young's Modulus E, of carbon fibres as a function of the graphitizing temperature I. (After Bailey and Clarke, *Chem. in Brit.*, 1970.)

crosslinking of the chains to form the hexagonal graphite structure, and this final heat treatment can affect the mechanical properties to a marked extent as shown in figure 15.16. The major application, so far, is in composite structures where they act as extremely effective reinforcing fibres. These reinforced plastic composites find uses in the aircraft industry, in the small boat trade, and as ablative composites.

15.17 Concluding remarks

The systematic study of structure-property relations provides an understanding of many of the fundamentals of the subject and can lead to quick dividends, as shown by the following example of the advanced art of fibre engineering.

A sheep grows a wool fibre which possesses a corkscrew crimp in the dry state, and affords the animal a bulky insulation blanket. When it rains the fibre becomes wet and loses the crimp; the wool strands then bed down to form a close packed, rain-tight, covering capable of reducing body heat loss in the damp conditions. This *evolutionary* wool fibre, whose properties are derived from its bicomponent nature, was simulated in the laboratory by preparing a two component acrylic fibre in which each component had a different hydrophilicity. This feat was achieved by fibre scientists in only a few months by making use of their knowledge of structure and behaviour.

Of course, nature was first, and, at best, only reasonable facsimiles of some natural products can be synthesized in the laboratory. The scientist is not yet able to match the sophistication of many naturally occurring macromolecules, which are no longer simply materials, but working functional units. The complexity of the inter-relation between structure and function in many proteins and nucleo-proteins is the ultimate in molecular design, and while it will be some time before we can hope to reach this level in synthesis, progress in understanding the relations in simpler systems is a step in the right direction.

General Reading

G. Allen and J. C. Bevington, Eds, *Comprehensive Polymer Science*. Vols 2 and 7. Pergamon Press (1989).

C. E. H. Bawn, "Structure and performances", *Plastics and Polymers*, 373 (1969).

B. Bloch and G. W. Hastings, *Plastics in Surgery*. Thomas (1967).

W. Bruce-Black, "Structure-property relationships in high temperature fibres", *Trans. N.Y. Acad. Sci.*, **32**, 765 (1970).

J. P. Critchley, G. J. Knight, and W. W. Wright, *Heat Resistant Polymers*. Plenum Press (1983).

R. W. Dyson, Ed., *Speciality Polymers*. Blackie and Son Ltd (1987).

R. W. Dyson, Ed., *Engineering Polymers*. Blackie and Son Ltd (1989).

H. G. Elias and F. Vohwinkel, *New Commercial Polymers 2*. Gordon and Breach Science Publishers (1986).

M. J. Folkes, Ed., *Processing, Structure and Properties of Block Copolymers*. Elsevier Applied Science Publishers (1985).

I. Goodman, *Synthetic Fibre Forming Polymers*. R.I.C. (1967).

J. W. S. Hearle and R. H. Peters, *Fibre Structure*. Butterworths (1963).

L. Mascia, *The Role of Additives in Plastics*. Edward Arnold (1974).

J. E. McIntyre, *The Chemistry of Fibres*. Edward Arnold (1971).

M. Lewin and J. Preston, *High Technology Fibers*. Parts A (1985) and B (1989). Marcel Dekker Inc.

K. L. Mittal, Ed., *Polyimides: Synthesis, Characterization and Applications*. Vols 1 and 2. Plenum Press (1984).

R. W. Moncrieff, *Man-made Fibres*. John Wiley and Sons (1963).

R. B. Seymour and C. E. Carraher, *Structure-Property Relationships in Polymers*. Plenum Press (1984).

R. B. Seymour and G. S. Kirshenbaum, Eds, *High Performance Polymers: Their Origin and Development*. Elsevier (1986).

A. V. Tobolsky and H. Mark, *Polymer Science and Materials*, Chapters 14 and 15. Wiley-Interscience (1971).

D. Wilson, Ed., *Polyimides*. Blackie and Son Ltd (1989).

References

1. J. E. Bailey and A. J. Clarke, *Chem. in Britain*, **6**, 484 (1970).
2. M. F. Drumm, C. W. H. Dodge, and L. E. Nielsen, *Ind. Eng. Chem.*, **48**, 76 (1956).
3. R. A. Gaudiana, R. A. Minns, R. Sinta, N. Weeks and H. G. Rogers, "Amorphous Rigid Rod Polymers." *Prog. Polym. Sci.* **14**, 47 (1989).
4. P. M. Hegenrother, *Polym. J.* **19**, 73 (1987).
5. L. C. Lopez and G. L. Wilkes, "Poly(phenylene sulphide)" *Rev. Macromol. Chem. Phys.* **C29**, 83 (1989).
6. R. B. Richards, *J. Appl. Chem.*, **1**, 370 (1951).
7. D. Tanner, J. A. Fitzgerald and B. R. Phillips, "The Kevlar Story." *Angew. Chem. Int. Ed. Engl. Adv. Mater.* **28**, 649 (1989).
8. A. H. Willbourn, *Plastics and Polymers*, 417 (1969).

CHAPTER 16

Polymer Liquid Crystals

16.1 Introduction

The liquid crystalline state was first observed by an Austrian botanist, Friedrich Reintzer, in 1888, when he noted that cholesteryl esters formed opaque liquids on melting which, on heating to higher temperatures, subsequently cleared to form isotropic liquids. This behaviour was interpreted by Lehmann as evidence for the existence of a new phase lying between the solid and isotropic liquid states. After further work by Friedel this new state became known as a mesophase, from the Greek *mesos* meaning in-between or intermediate. These mesophases are quite fluid but also show birefringence and as they appear to have properties associated with both crystals and liquids, they were called *liquid crystals* by Lehmann.

Liquid crystals can be divided into two main classes; those, like the cholesteryl derivatives, whose liquid crystalline phases are formed when the pure compound is heated are called *thermotropic*, and those where the liquid crystalline phase forms when the molecules are mixed with a solvent are referred to as *lyotropic*. The thermotropic class also includes enantiotropic types where the liquid crystalline phases can be seen on both the heating and the cooling cycles, and monotropic types where the mesophase is stable only on supercooling from the isotropic melt.

Continued investigations led to the identification of three main types of mesophase; a *smectic* state (Greek: *smegma*, meaning soap), a *nematic* state (Greek: *nema* meaning thread) and a *cholesteric* state observed in systems containing molecules with a chiral centre.

While the early work and much of the recent studies have identified and investigated the liquid crystalline properties of many small molecules, it was suggested that polymeric forms could also exist. In 1956, Flory postulated that concentrated solutions of rigid rod-like polymers should form ordered structures in solution at some critical concentration. This phenomenon was observed initially in 1937 for solutions of tobacco mosaic virus, but the first systematic experimental verification of this prediction came from work on concentrated solutions of poly(γ-methyl glutamate) and poly(γ-benzyl glutamate), where these polymers exist in extended helical forms that can pack readily into ordered bundles with the long axes generally aligned in one direction. This produces a quasi-parallel distribution of chains in their solutions and anisotropic liquid crystalline properties. Later it was shown that anisotropic solutions are formed by some aromatic polyamides and cellulose derivatives, where again the molecules are relatively rigid. These are lyotropic systems, but in the 1970s thermotropic liquid crystalline polymers were also synthesized and this latter group has been developed rapidly since then.

16.2 Liquid crystalline phases

Molecules which have a tendency to form liquid crystalline phases usually have either rigid, long lathe-like shapes with a high length to breadth (aspect) ratio, or disc shaped molecular structures. Chemically these may be composed of a central core comprising aromatic or cycloaliphatic units joined by rigid links, and having either polar, or flexible alkyl and alkoxy terminal groups. Some typical examples of possible small molecule structures that from liquid crystalline phases are shown in table 16.1 and these units are called mesogens. When polymers exhibiting liquid crystalline properties are formed they can be constructed from these mesogens in three different ways: (i) incorporation into chain-like structures by linking them together through both terminal units to form main-chain liquid crystalline polymers; (ii) attachment through one terminal unit to a polymer backbone to produce a side chain comb-branch structure; (iii) a combination of both main and side chain structures. The various possible geometric arrangements are shown schematically in figure 16.1

The mesogenic units can then form the ordered structures that are observed in the small molecule systems (though not necessarily the same type of liquid crystalline phase) and are characterized by long-range orientational order, with the long axes of the mesogenic groups arranged in one preferred direction of alignment, called the *director*. When this spatial ordering is such that the mesogens are arranged in regular layers with respect to their centres of gravity, they are in one of several possible smectic phases. The lateral forces between the molecules in the smectic phases are stronger than the forces between the layers, and so slippage of one layer over another provides the characteristic fluidity of the system without losing the order within each layer. A number of different smectic phases can be identified in which the ordered packing of the mesogens in the layers differs and the mesogens are either orthogonal to, or tilted with respect to, the layer structure. These are identified alphabetically. The most ordered is the smectic B (S_B) with a hexagonally, close-packed structure for the mesogens in the layers. The S_B and smectic E (S_E) phases exhibit three-dimensional order and have tilted modifications smectic H (S_H) and smectic G (S_G). A much less ordered structure within the layers produces a phase called the smectic A (S_A) phase, where there is a random lateral distribution of the mesogens in the layers. The tilted modification of this is called the smectic C (S_C) phase, and both S_A and S_C behave like true two-dimensional liquids. Intermediate in order are the smectic F (S_F) and smectic I (S_I), but the most commonly observed phases are S_A, S_B and S_C.

The nematic phase is much less ordered than the smectic phases. While the directional ordering of the mesogen long axes is maintained, the centres of gravity are no longer confined in layers, but are distributed randomly in the phase. The nematic state is much more fluid than the smectic phases but still exhibits birefringence.

The third important category is a variation of the nematic phase and is called the *chiral nematic* state. It is observed when mesogens that enter a nematic phase also have a chiral centre. This imparts a twist to each successive layer in the phase where the orientation of the director changes regularly from layer to layer, forming a helical arrangement of the directors in three dimensional space. The chiral nematic state was first observed when cholesteryl derivatives were studied, but has now been detected in

TABLE 16.1. A selection of small molecule mesogens and the associated liquid crystalline behaviour

Mesogen		*Transition temperature* (°C)
R—(C_6H_4)—N=CH—(C_6H_4)—O—CH_3	R = —CN	k 106 n 117 i
	—n-C_4H_9	k 20 n 48 i
	—O—C_2H_5	k 83 n 107 i
	—C(=O)—O—CH_3	k 79 n 102 i
H_5C_2—O—(C_6H_4)—N=CH—(C_6H_4)—CH=N—(C_6H_4)—O—C_2H_5		k 200 n 320 i
H_3CO—(C_6H_4)—CH=N—(C_6H_4)—CH=CH—(C_6H_4)—N=CH—(C_6H_4)—OCH_3		k 274 n 340 i
H_3C—O—(C_6H_4)—CH=N—$(C_{10}H_6)$—N=CH—(C_6H_4)—O—CH_3		k 189 n 356 i
H_3C—O—(C_6H_4)—CH=N—(C_6H_3Cl)—(C_6H_3Cl)—N=CH—(C_6H_4)—O—CH_3		k 154 n 344 (dec) i

TABLE 16.1. A selection of small molecule mesogens and the associated liquid crystalline behaviour (*continued*)

Mesogen		*Transition temperature* (°C)
R—C6H4—CH=N—C6H4—C6H4—N=CH—C6H4—R	R = —H	k 239 n 265 i
	—O—CH_3	k 266 n 390 i
R—C6H4—CH=N—C6H4—$(CH_2)_2$—C6H4—N=CH—C6H4—R	R = —CH_3	k 197 n 287 i
	—Cl	k 232 n 318 i
	—CN	k 227 n 367 i
	—O—CH_3	k 181 n 337 i
	—O—n-C_4H_9	k 159 s 186 n 303 i
	—O—n-C_6H_{13}	k 127 s 229 n 276 i
	—C6H5	k 227 n 403 i
	—C6H4—O—CH_3	k 253 n 270 (dec) i
	—C6H4—O—n-C_4H_9	k 232 n 331 i
	—C6H4—O—CH_2—C6H5	k 270 n 346 i
H_5C_2—O—C6H4—N=N—C6H4—O—C(=O)—CH=CH—CH_3		k 110 n 197 i

(continued)

TABLE 16.1. A selection of small molecule mesogens and the associated liquid crystalline behaviour (*continued*)

Mesogen		*Transition temperature* (°C)
		k 137 n* 155 i
	R = —Cl	k 118 n* 125 i
	—CH_3	k 95 n* 117 i
	—C_2H_5	k 97 n* 114 i
	—n-C_6H_{13}	k 96 n* 112 i
	—n-C_7H_{15}	k 97 n* 110 i
	—n-C_8H_{17}	k 78 s* 81 n* 92 i
	—$CH(C_2H_5)(CH_2)_3CH_3$	k 30 n* 50 i
	—n-$C_{13}H_{27}$	k 71 s* 79 n* 83 i
	—n-$C_{14}H_{29}$	k 77 s* 79 n* 83 i
	—n-$C_{16}H_{33}$	k 76 s* 80 n* 83 i
	—$(CH_2)_7(CH{=}CH{-}CH_2)(CH_2)_6CH_3$	k 39 s* 44 n* 49 i
	—$(CH_2)_7(CH{=}CH{-}CH_2)_2(CH_2)_3CH_3$	k 20 s* 44 n* 49 i
	—$(CH_2)_7(CH{=}CH{-}CH_2)_3CH_3$	k 35 s* 45 n* 48 i
	—O—$(CH_2)_8(CH{=}CH{-}CH_2)(CH_2)_6CH_3$	k-10 s* 18 n* 31 i
	—O—$(CH_2)_2$—O—$(CH_2)_2$—O—C_2H_5	k-2 n* 15 i
		k 150 n* 178 i
		k 178 n* 290 i

TABLE 16.1. A selection of small molecule mesogens and the associated liquid crystalline behaviour (*continued*)

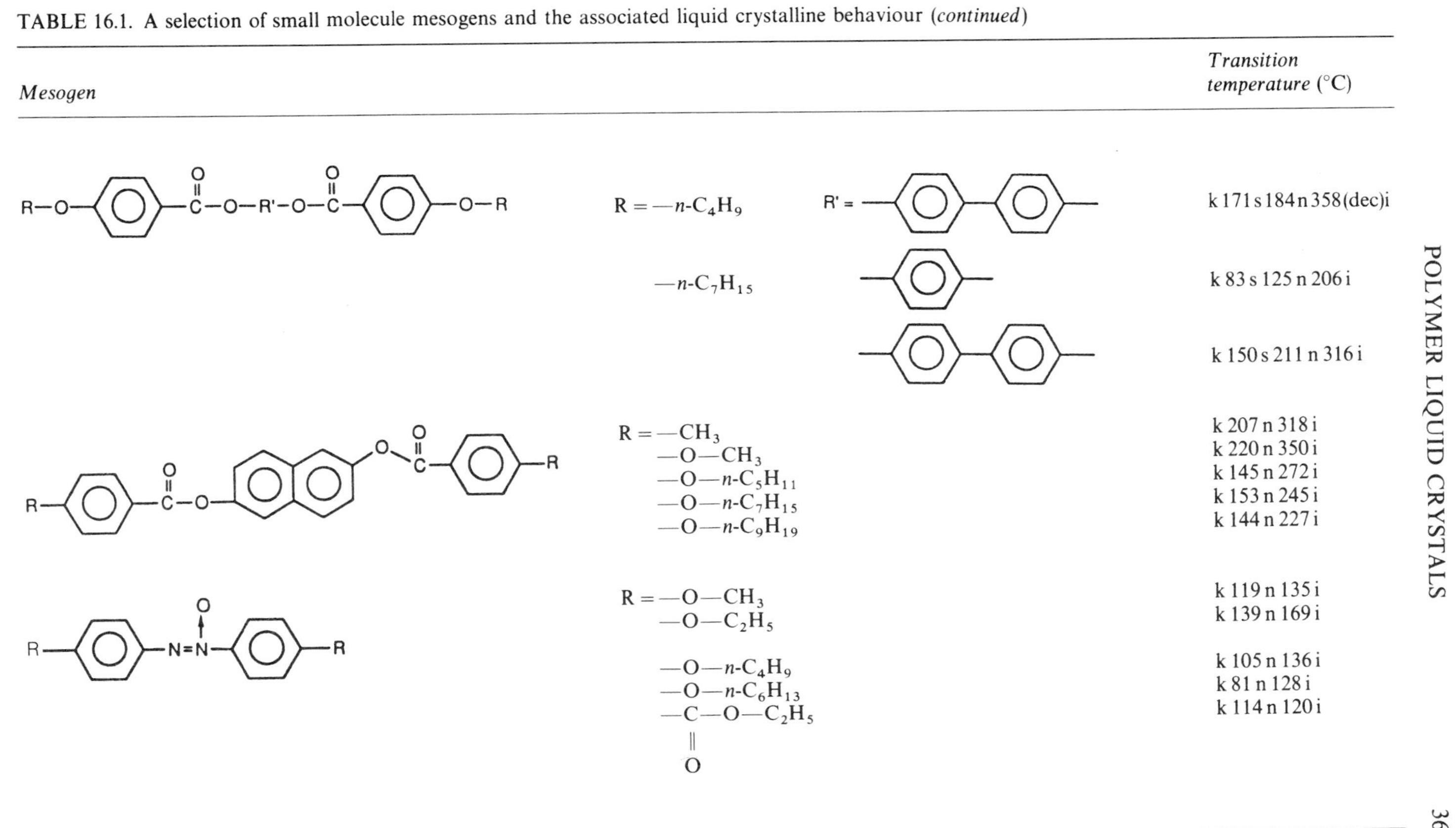

Mesogen			*Transition temperature* (°C)
	R = —n-C_4H_9	R' =	k 171 s 184 n 358 (dec) i
	—n-C_7H_{15}		k 83 s 125 n 206 i
			k 150 s 211 n 316 i
	R = —CH_3		k 207 n 318 i
	—O—CH_3		k 220 n 350 i
	—O—n-C_5H_{11}		k 145 n 272 i
	—O—n-C_7H_{15}		k 153 n 245 i
	—O—n-C_9H_{19}		k 144 n 227 i
	R = —O—CH_3		k 119 n 135 i
	—O—C_2H_5		k 139 n 169 i
	—O—n-C_4H_9		k 105 n 136 i
	—O—n-C_6H_{13}		k 81 n 128 i
	—C(=O)—O—C_2H_5		k 114 n 120 i

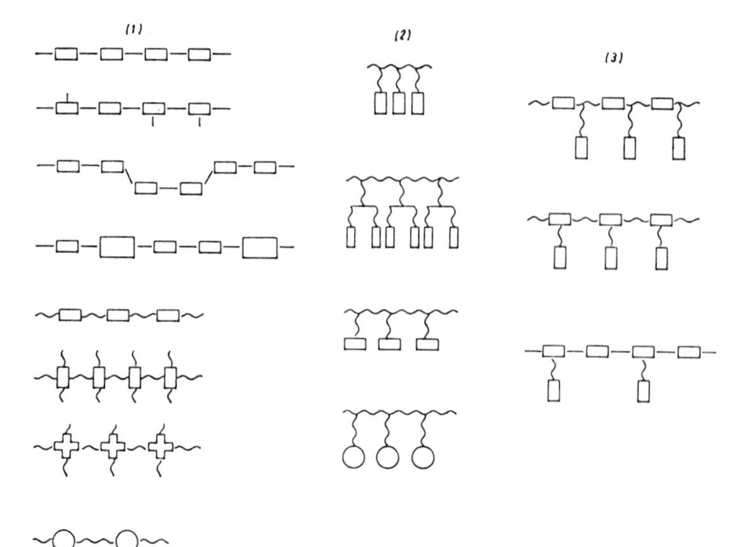

FIGURE 16.1. Schematic representation of various possible arrangements of mesogens in polymer chain structures (1) main chain, (2) side chain, and (3) combinations of main and side chain. (Adapted from D. Sek (1988) with permission from Akademie-Verlag).

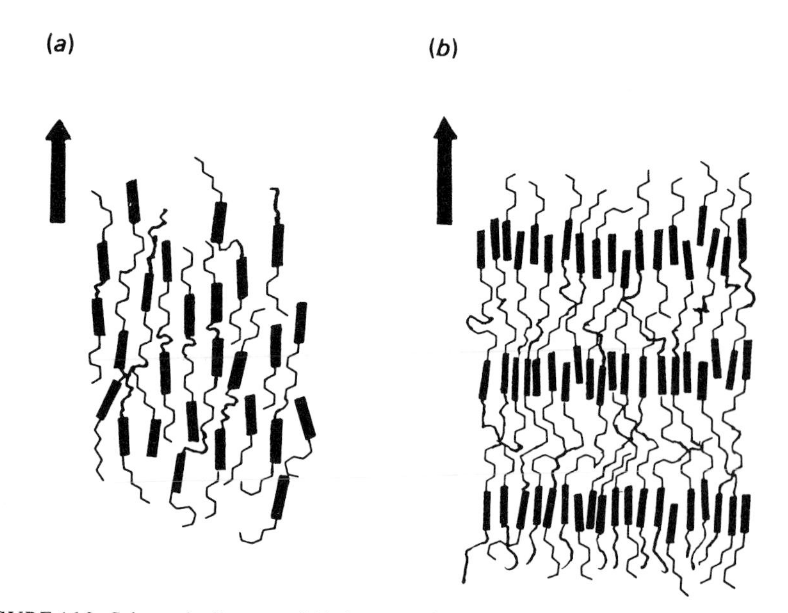

FIGURE 16.2. Schematic diagram of (a) the nematic phase and (b) the smectic phase for main chain liquid crystalline polymers, showing the director as the arrow. The relative ordering is the same for side chain polymer liquid crystals.

other chiral mesogens, and can also be induced by adding small chiral molecules to a host nematic liquid crystalline polymer. The main phase types are shown schematically in figure 16.2

In some polymer liquid crystals, several mesophases can be identified. In main chain liquid crystal polymers there is usually a transition from the crystal to a mesophase, whereas in more amorphous systems when a glass transition is present, the mesophase may appear after this transition has occurred. In multiple transition thermotropic systems, the increase in temperature leads to changes from the most ordered to the least ordered states, *i.e.* crystal(k) → smectic(S) → nematic(N) → isotropic (i), *e.g.*

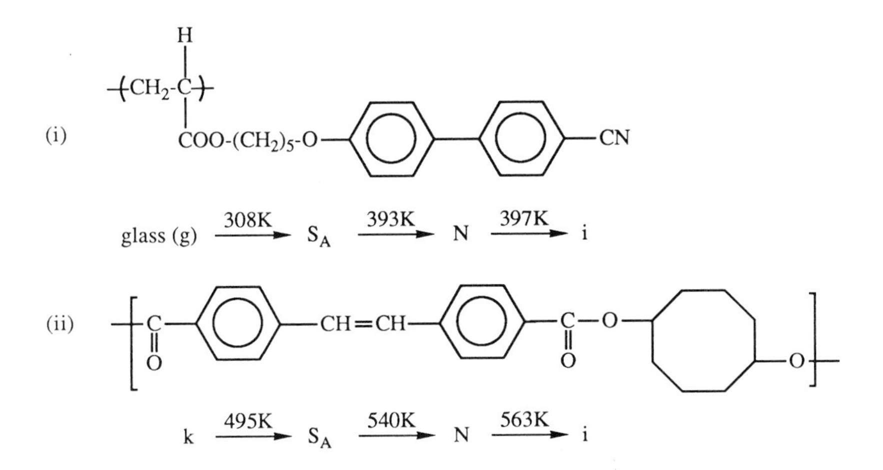

16.3 Identification of the mesophases

The liquid crystalline phases in polymeric materials are sometimes difficult to identify unequivocally, but several techniques can be used that provide information on the nature of the molecular organization within the phase. If used in a complementary fashion these can provide reliable information on the state of order of the mesogenic groups.

POLARIZING MICROSCOPE

The phases can often be identified by observing the characteristic textures developed in thin layers of the polymer when viewed through a microscope using a linearly polarized light source.

The preparation of the glass slides for sample observation can be important and a homogeneous (or planar) texture, where the mesogens all lie parallel to the surface, can only be obtained if the slide is rubbed in one direction with cotton or similar material. This gives a uniform birefringence, unlike the effects obtained from untreated slides, where the long axes of the mesogens are all oriented at right angles to the glass slide, and homeotropic alignment results, giving a uniformly dark field; touching the cover slip can then cause scintillation by tilting the mesogens under the applied pressure. These homeotropic textures can also be prepared by treating the slide with concentrated nitric acid followed by a water and acetone rinse.

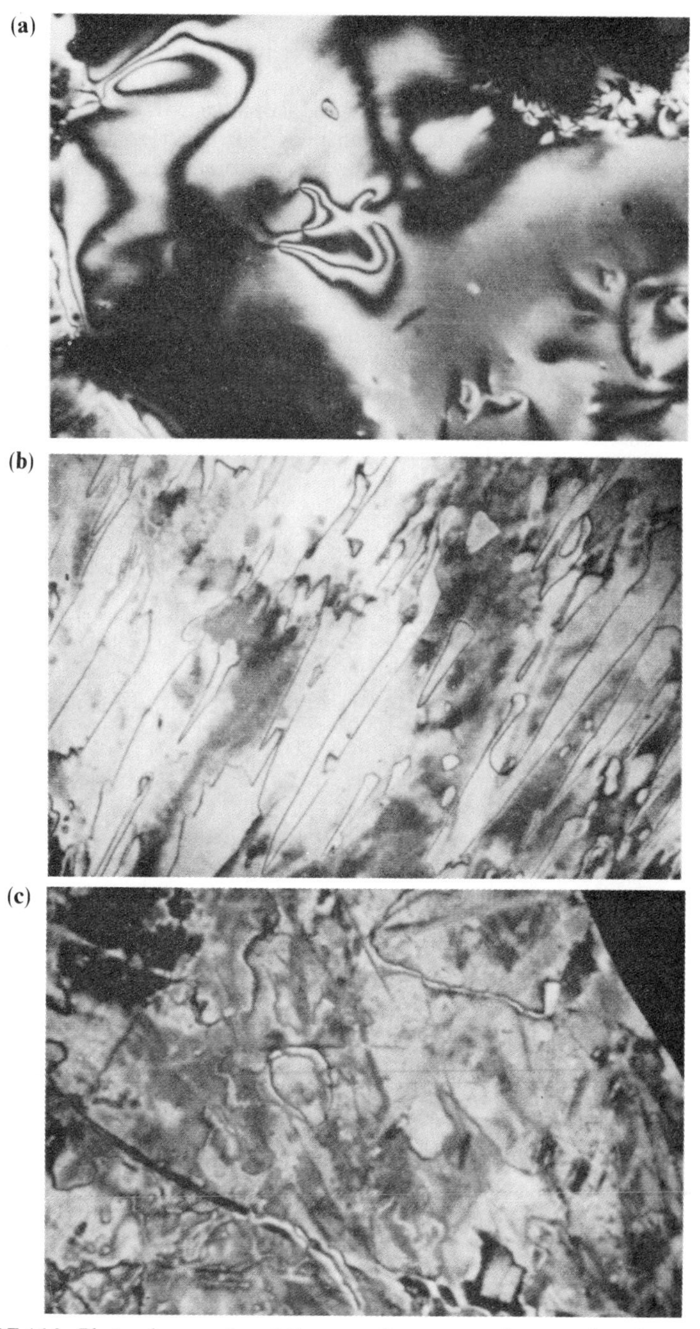

FIGURE 16.3. Photomicrographs of the nematic textures that can be observed using a polarizing microscope (a) Schlieren, (b) threaded, (c) marbled textures. ((c) Reproduced with kind permission from C. Nöel.)

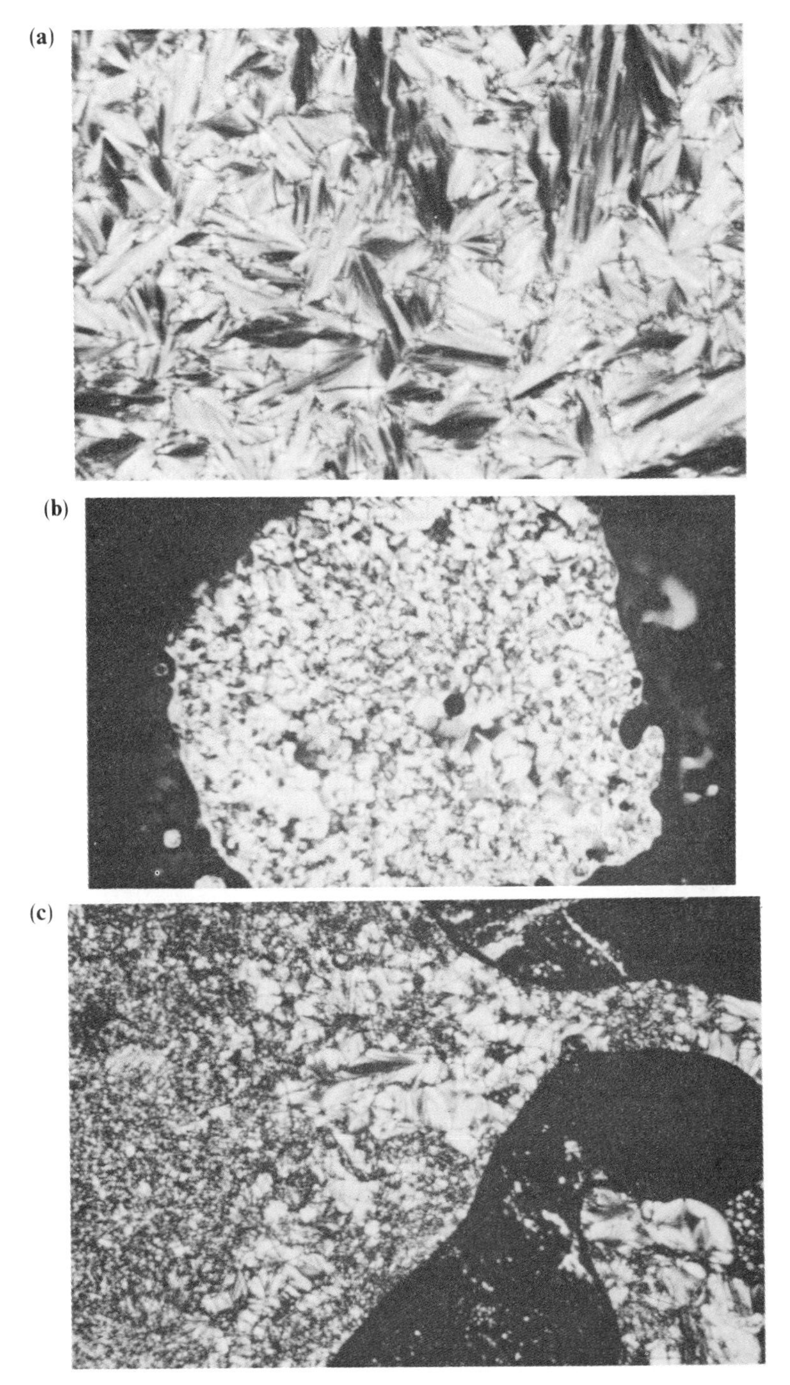

FIGURE 16.4. Photomicrographs of textures characteristic of smectic phases (a) focal conic and fan (smectic A), (b) mosaic (smectic B) and (c) broken focal conic (smectic C). Observed using a polarizing microscope. ((c) Reproduced with kind permission from C. Nöel.)

When a nematic texture appears on cooling from an isotropic melt it is formed from the coalescence of droplets that separate from the liquid. This mechanism is indicative of a nematic phase and one of three possible characteristic textures may result, depending on the way the droplets form larger domains. These are the Schlieren, the threaded, or the marbled nematic textures. Examples of each texture are shown in figure 16.3 (a), (b) and (c).

The Schlieren textures show large dark "brush" patterns, corresponding to the extinction zones where the mesogens are aligned perpendicular to the glass slide. Also noticeable are the points where two or four of these brushes meet and if the texture shows points where only two brushes meet this is an unambiguous indication of a nematic phase. It should be noted that if a Schlieren pattern shows only points where four branches meet then this represents a smectic C texture or its chiral modification. Textures with long thread-like structures are also typical of nematic phases but these tend to be unstable and as the temperature is raised the threads form closed loops which may eventually disappear. The marbled texture is seen when adjoining domains in which the molecular orientation differs are formed, thereby producing different interference colours.

Smectic phases show a number of characteristic textures including (i) the focal conic and fan texture, characteristic of smectic A and often formed from the coalescence of bâtonnets; (ii) the mosaic texture observed when a smectic B phase is formed and (iii) the broken focal conic structure resulting from smectic C phases that can also show the Schlieren pattern described above. Examples of textures (i) to (iii) are shown in figure 16.4.

The chiral nematic phases can show a planar Grandjean texture, with oily streaks caused by defects, but they can also show strong reflection colours depending on the pitch of the helical structure within the phase.

DIFFERENTIAL SCANNING CALORIMETER

This technique is widely used as a means of detecting the temperatures of thermotropic mesophase transitions. These are identified in most cases as first-order endothermic transitions. A schematic diagram of a thermogram, showing the possible transitions for a heating cycle is presented in figure 16.5 but it is advisable to carry out both heating and cooling cycles to confirm the transitions.

X-RAY DIFFRACTION

While it is possible to use the Debye-Scherrer powder technique to characterize the mesophases it is not always easy to distinguish between the various types using this approach, and measurements on aligned samples are preferable and more reliable. The powder diagrams can, however, provide reliable information on the number of phases present.

For the least ordered phases, *e.g.* nematic, S_A and S_C one diffuse halo is seen at large diffraction angles, indicative of a disordered lateral arrangement of the mesogens. In contrast, the more ordered smectic phases exhibit one or more Bragg reflections. At smaller diffraction angles ($\sim 3°$) a diffuse inner ring is formed in nematic samples. This is somewhat sharper than that from the isotropic melt. When the smectic phases are

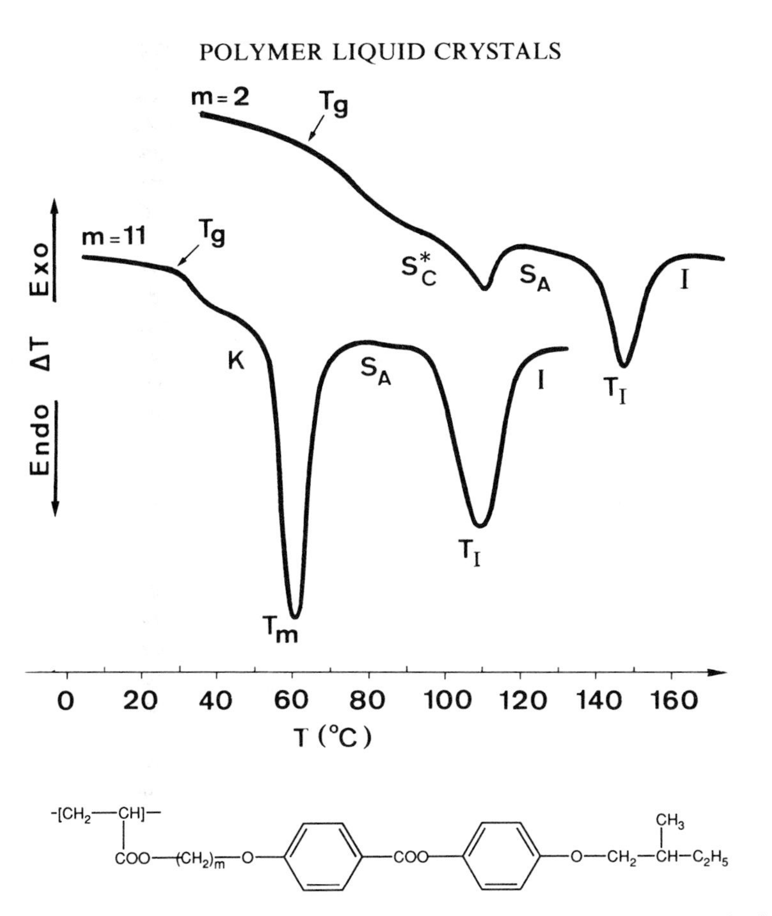

FIGURE 16.5. Representative d.s.c. heating traces for two samples of a side chain liquid crystalline polymer with different lengths of spacer unit. (Reproduced from Decobert *et al.* (1986)).

present, one or more sharp inner rings, arising from the more ordered lamellar structure, are seen.

More discriminating information can be obtained from samples which have been oriented by cooling from the isotropic liquid phase under the influence of a strong magnetic field or by drawing a fibre from the mesophase and quenching it in the oriented structure. Typical X-ray diffraction patterns for the nematic and smectic A and C phases are shown in figure 16.6.

MISCIBILITY STUDIES

The type of phase formed in a polymer liquid crystal can often be identified by examining the manner in which it mixes with a small molecule mesogen of known mesophase type. If these textures are the same then a mixed liquid crystal phase is formed with no observable transition between the two types of molecule. Temperature-composition diagrams can be established for some mixed systems and these may show the presence of a eutectic point. This is illustrated in figure 16.7 for a side chain liquid crystal polymer, with a siloxane backbone, mixed with a low molar mass analogue to

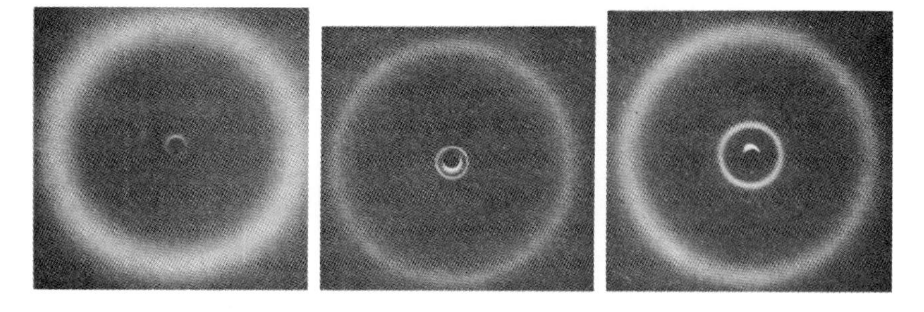

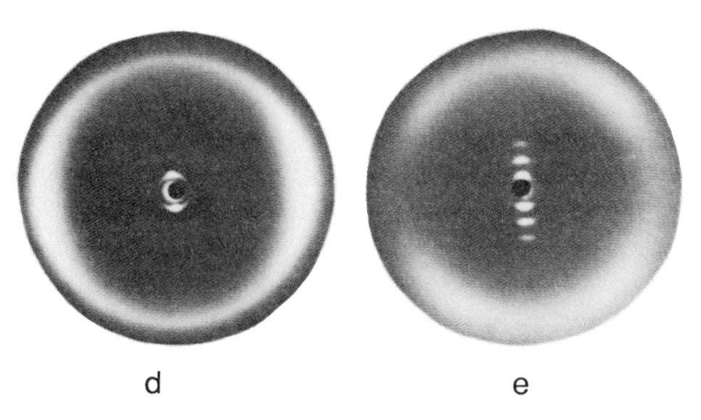

FIGURE 16.6. Typical X-ray diffraction patterns for unoriented (a) nematic, (b) smectic A, (c) smectic C, oriented (d) smectic A and (e) smectic C phases.

give a uniform nematic phase. Similar diagrams have been reported for main chain liquid crystalline polymer–low molar mass mesogen mixtures. However, it should be noted that whereas isomorphism is identified when a mixed system is miscible, the reverse is not necessarily true. If the two liquid crystal phases in the mixture are different then they will be separated by an observable transition.

16.4 Lyotropic main chain liquid crystalline polymers

Certain rod-like polymers, when mixed with small amounts of solvent, form a birefringent fluid that is eventually converted into a true isotropic solution when an excess of the solvent is added. These are called lyotropic liquid crystalline systems and are formed by the dissolution of amphiphilic molecules, in appropriate solvents. The development of liquid crystalline solutions depends on the molar mass of the molecules, the solvent, and the temperature, but most importantly on the structure of the polymer, which should be quite rigid. Polymeric materials that tend to form helical structures, such as the stable α-helical form in polypeptides, are found to be suitable, and esters of poly(L-glutamic acid) are good examples of lyotropic systems, but one of the most important groups of synthetic polymers is the aromatic polyamides. Those that are lyotropic have a variety of structures, some of which are shown in figure 16.8, and

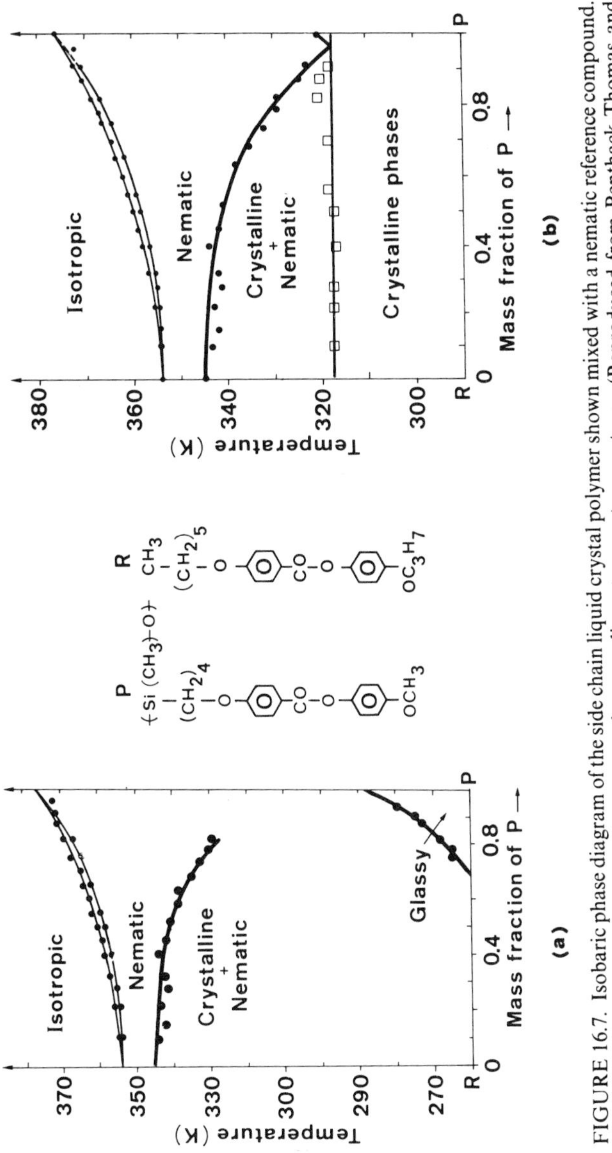

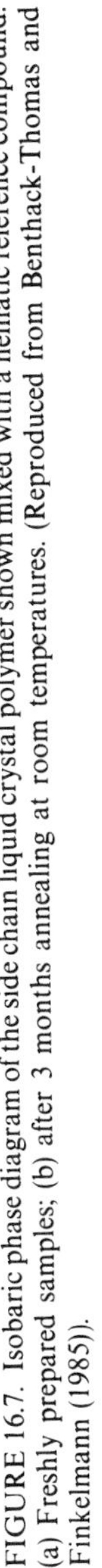
FIGURE 16.7. Isobaric phase diagram of the side chain liquid crystal polymer shown mixed with a nematic reference compound. (a) Freshly prepared samples; (b) after 3 months annealing at room temperatures. (Reproduced from Benthack-Thomas and Finkelmann (1985)).

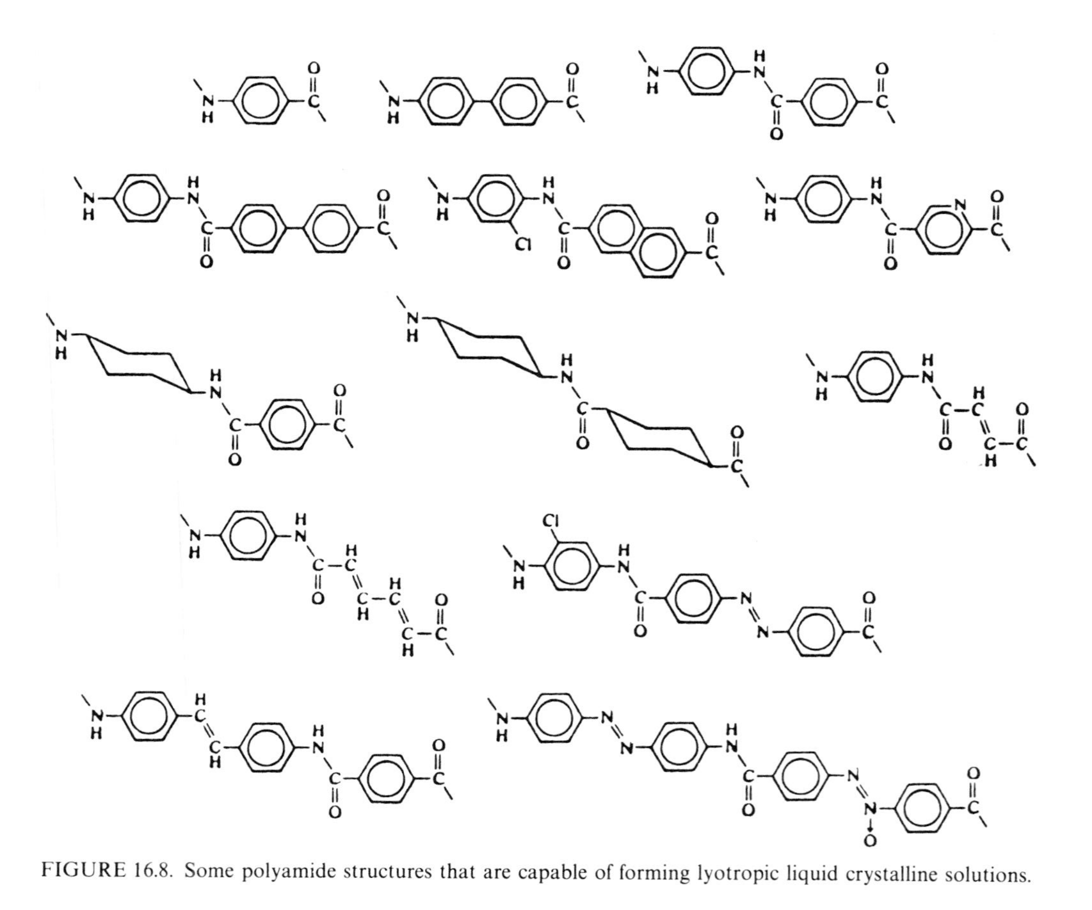

FIGURE 16.8. Some polyamide structures that are capable of forming lyotropic liquid crystalline solutions.

appear to gain their rigidity of structure from the ring systems that are coupled by the amide link. This coupling unit adopts a *trans* conformation and can conjugate with the phenyl rings, when adjacent, to produce an extended rod-like structure in the polymer chain.

In this respect the necessary rigidity of the polymers makes them inherently less soluble in common solvents and dissolution often requires the use of more strongly interacting liquids. While the poly(L-glutamate)s can form lyotropic solutions in dioxane or methylene chloride, the aromatic polyamides require the more agressive, strong, protonating acids (H_2SO_4, CF_3SO_3H, CH_3SO_3H) or aprotic solvents such as dimethylacetamide or N-methyl pyrrolidone in conjunction with LiCl or $CaCl_2$ in small (2 to 5) percentages, to effect solution. Hexamethylene phosphoramide has also been used but is carcinogenic and should be avoided.

In solution, flexible polymers assume a random coil conformation, but the rigid polymers are more rod-like and as their concentration in solution is increased they tend to cluster together in bundles of quasi-parallel rods. These form domains that are anisotropic and within which there is nematic order of the chains. There is, however, little or no directional correlation between the directors within these domains unless the solutions are sheared. When shearing takes place the domains tend to become aligned parallel to the direction of flow thereby reducing the viscosity of the system below what would be expected from a solution of random coils. Lyotropic liquid crystalline polymers exhibit quite characteristic viscosity behaviour as the con-

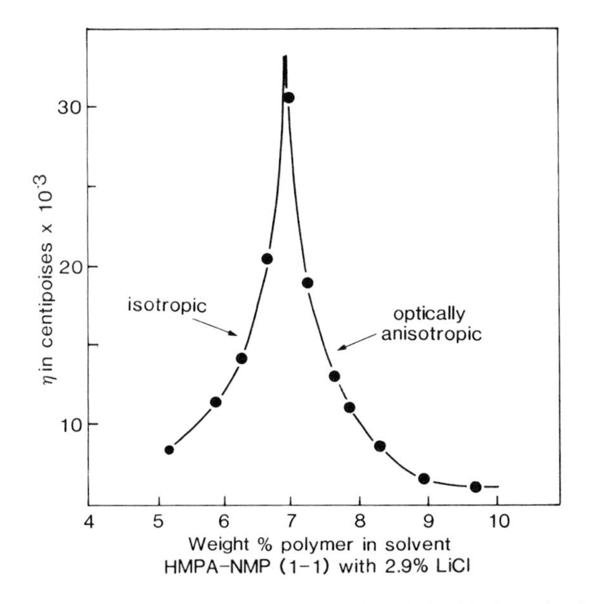

FIGURE 16.9. Variation of the viscosity of solutions of partially chlorinated poly(1,4-phenylene-2, 6-naphthalamide) dissolved in a solvent mixture of hexamethylene phosphoramide and N-methylpyrrolidone containing 2.9 per cent LiCl, as a function of the solution concentration showing the transition from isotropic to anisotropic solutions. (Reproduced from P. W. Morgan (1979) with permission. American Chemical Society, Washington, D.C.)

centration of the solutions is changed. Typically the viscosity follows the trend shown in figure 16.9 for partially chlorinated poly(1,4-phenylene-2,6-naphthalamide) dissolved in a (1:1) mixture of hexamethylene phosphoramide and N-methyl pyrollidone containing 2.9 per cent LiCl. As polymer is added to the solvent the viscosity increases but the solution remains isotropic and clear. At a critical concentration (which depends on the system) the solution becomes opaque and anisotropic, and there is a sharp decrease in the viscosity with further increase in the polymer concentration. This reflects the formation of the oriented nematic domains in which the chains are now aligned parallel to the direction of flow, thereby reducing the frictional drag on the molecules. The critical concentration that must be reached before the nematic phase is achieved is a function of the solvent, and tends to decrease with increase in molar mass of the polymer.

A number of the polyamides have achieved commercial importance because of the very high tensile strengths of the fibres that can be spun from the nematic solutions. The additional chain orientation in the direction of the fibre long axis, obtained from the nematic self-ordering in the system, leads to a dramatic enhancement of the properties and makes them attractive alternatives to metal or carbon fibres for use in composites as reinforcing material.

The most significant of these aramid fibres are:

(i) poly(m-phenylene isophthalamide), trade name Nomex,

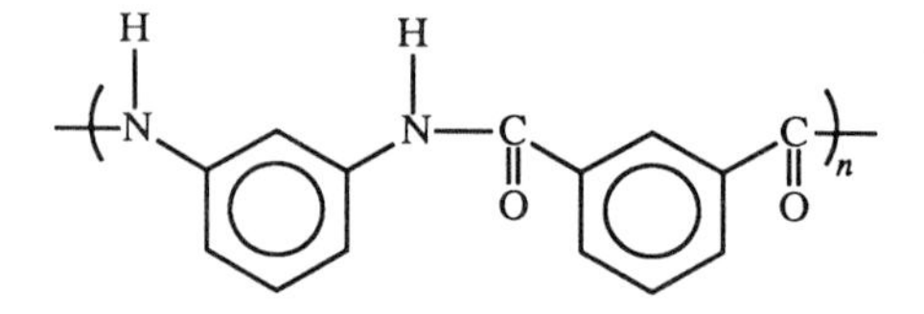

(ii) poly(*p*-benzamide) or Fibre B

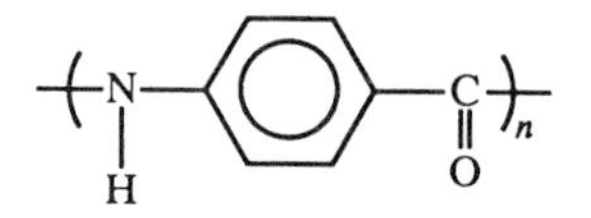

(iii) poly(*p*-phenylene terephthalamide), trade name Kevlar.

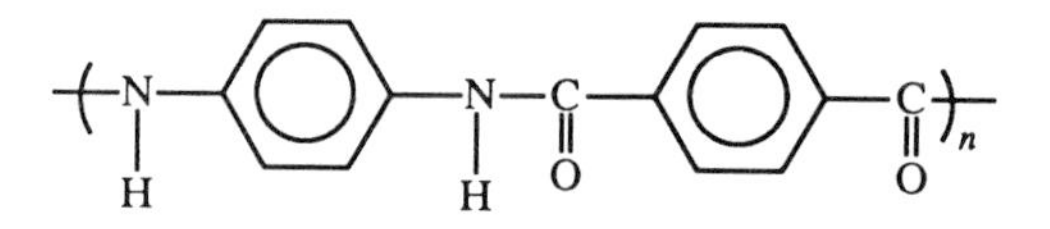

A more extensive description of aramid fibres is given in section 15.11.

Lyotropic polymers incorporating heterocyclic structures are also known, *e.g.*

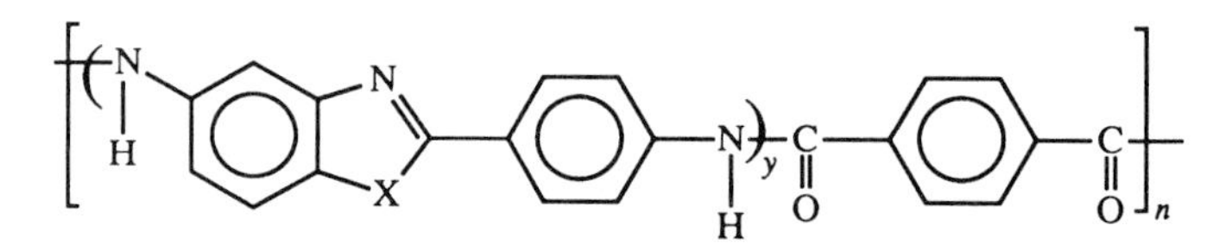

where X = O, S, or –NH. These can form nematic phases in strongly protonating acids and will also spin into high tensile strength fibres.

Other structures capable of forming lyotropic solutions are:

(a) polyisocyanates

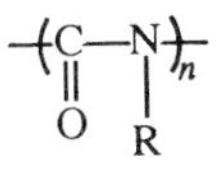

(b) poly(alkyl isonitrile)s

$$\left[\underset{\substack{\| \\ N-R}}{C} \right]_n$$

(c) poly(organophosphazine)s

$$\left[\underset{OR}{\overset{OR}{P}} = N \right]_n$$

where R = alkyl or aryl.

16.5 Thermotropic main chain liquid crystal polymers

The first examples of thermotropic main chain liquid crystal polymers were produced by Roviello and Sigiru in 1975 when they reacted alkyl acylchlorides with p,p'-dihydroxy-α,α'-dimethylbenzalazine giving the structure (I) which exhibited anisotropic fluid phases after melting.

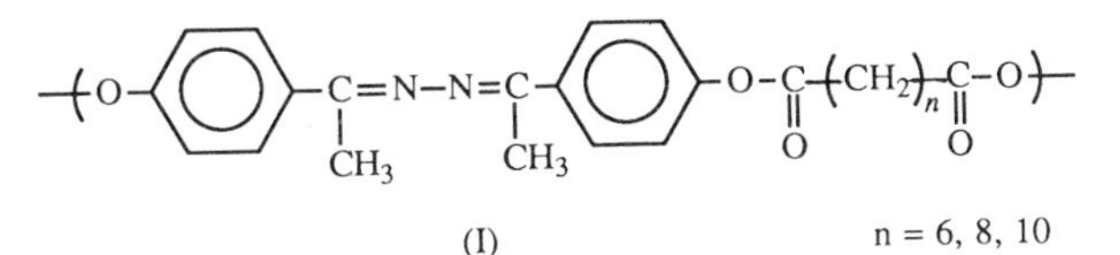

(I) n = 6, 8, 10

Subsequently it was found that by linking mesogens together to form polymeric chains, these chains often showed the presence of a liquid crystal phase appearing at temperatures just above the melting point of the polymer. Many of these materials are polyesters that are synthesized by condensation reactions including interfacial polymerizations, or by high temperature solution polymerizations using diols and diacid chlorides. However, the preferred method is often an ester interchange reaction in the melt. Various combinations of rigid units have been used and some of the structures investigated are shown in table 16.2.

Amongst the commonly used monomer units are hydroxybenzoic acid, p-terephthalic acid, 2,6-naphthalene dicarboxylic acid, 2-hydroxy-6-naphthoic acid and 4,4′-biphenol. Thus polymers can be prepared from simply one unit as with poly(p-hydroxybenzoic acid)

$$(n+1)\ HO-C_6H_4-\underset{\substack{\| \\ O}}{C}-OH \longrightarrow HO-C_6H_4\left[\underset{\substack{\| \\ O}}{C}-O-C_6H_4\right]_n\underset{\substack{\| \\ O}}{C}-OH$$

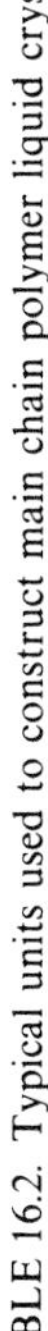

TABLE 16.2. Typical units used to construct main chain polymer liquid crystals

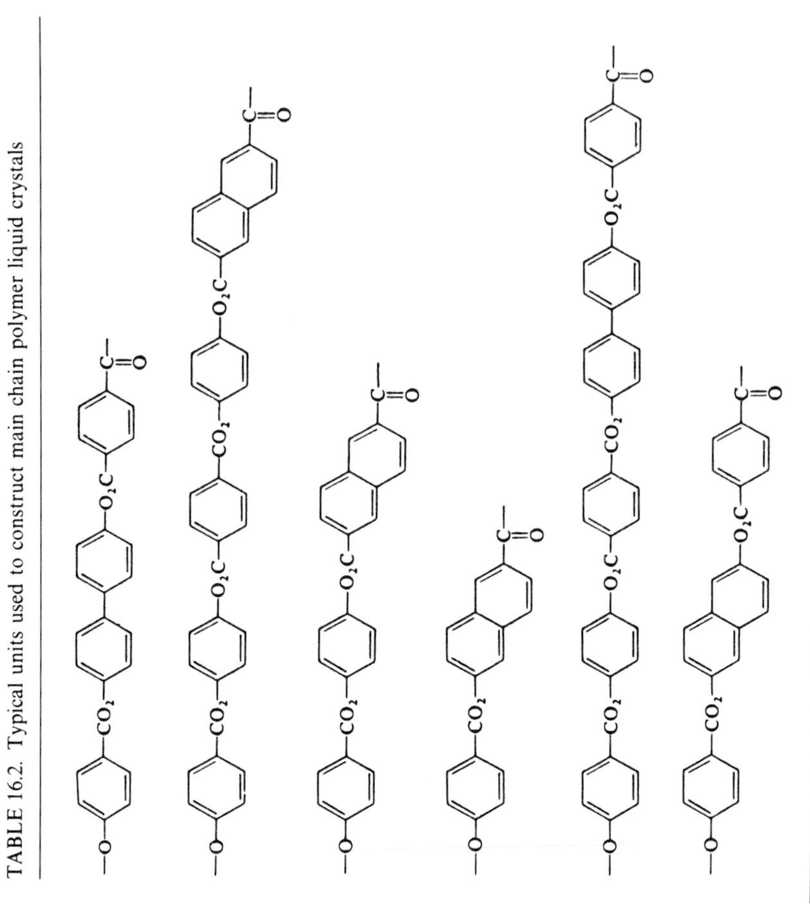

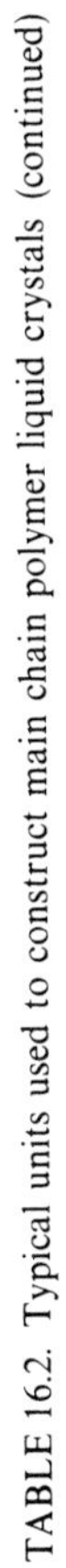

TABLE 16.2. Typical units used to construct main chain polymer liquid crystals (continued)

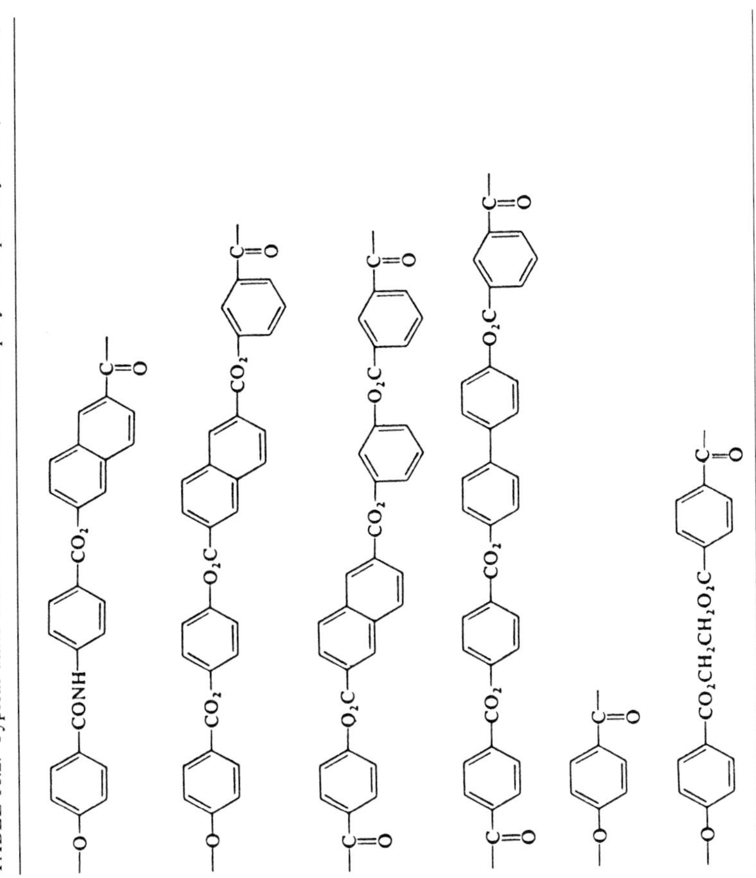

or from more than one, *i.e.* structure (II)

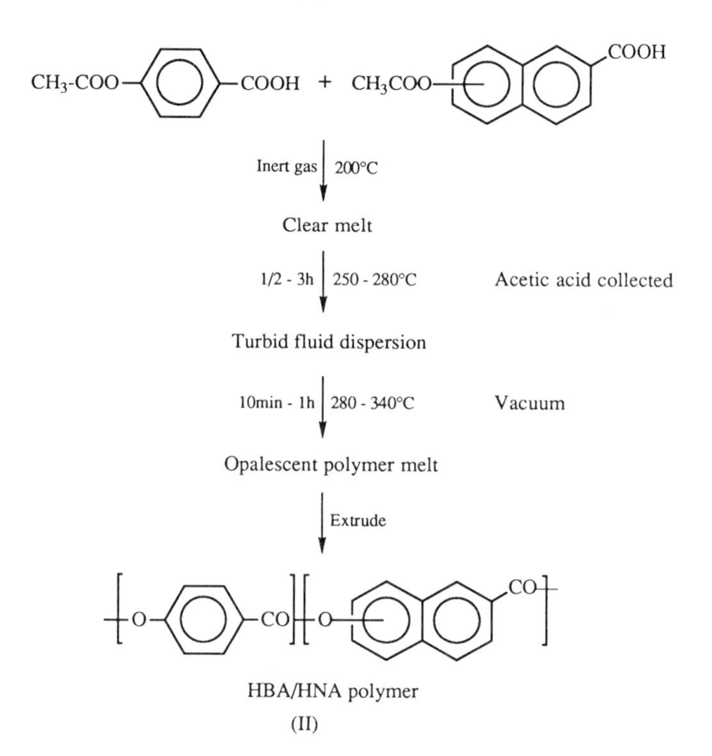

HBA/HNA polymer
(II)

where the ratios of the components may be varied to alter the properties of the product.

The materials prepared in this way tend to be very insoluble polymers with high melting points and mesophase ranges, *e.g.* poly(*p*-hydroxybenzoic acid) melts at ~ 883 K. This makes them difficult to process and alternative structures with much lower melting points are more useful. The melting points of the main chain liquid crystal polymers can be reduced in a number of different ways, *viz.*,

(i) incorporation of flexible spacer units;
(ii) copolymerization of several mesogenic monomers of different sizes to give a random and more irregular structure;
(iii) introduction of lateral substituents to disrupt the chain symmetry;
(iv) synthesis of chains with kinks, such as unsymmetrically linked aromatic units.

The use of flexible spacers is a popular approach and consists of two cyclic units, normally joined by a short rigid bridging unit to form the mesogenic moiety. These are then linked through functional groups to flexible units of varying length which space the mesogens along the chain and reduce the overall rigidity. The schematic chemical constitution of these chains, together with examples of the various types of groups that have been used are shown in table 16.3. The bridging groups must be rigid to maintain the overall stiffness of the mesogens, and they are usually multiple bond units. Ester groups also serve this purpose, particularly when in conjunction with aryl rings where

TABLE 16.3. Group arrangements typical of thermotropic main chain polymer liquid crystals

Cyclic unit	*Linking group*	*Functional group*	*Spacer*
x = 1-3	$-\overset{\text{O}}{\overset{\Vert}{\text{C}}}-\text{O}-$	$-\overset{\text{O}}{\overset{\Vert}{\text{C}}}-\text{O}-$	$-(\text{CH}_2)_n-$
1,4 1,5 2,6	$-\underset{\text{R}}{\text{C}}=\text{N}-\text{N}=\underset{\text{R}}{\text{C}}-$	$-\text{O}-\underset{\text{O}}{\underset{\Vert}{\text{C}}}-$	$-(\text{CH}_2-\underset{\text{R}}{\text{CHO}})_n-$
H H	$-\text{CH}=\underset{\text{R}}{\text{C}}-$	$-\text{O}-$	$-\text{S}-\text{R}-\text{S}-$
	$-\text{CH}=\text{N}-$ $-\text{N}=\text{N}-$ $-\underset{\text{O}}{\text{N}}=\text{N}-$	$-(\text{CH}_2)_n-$	$-\overset{\text{R}}{\underset{\text{R}}{\text{Si}}}-\text{O}-$

[CYCLIC UNIT — LINKING GROUP — CYCLIC UNIT — FUNCTIONAL GROUP — SPACER]

MESOGENIC GROUP

the conjugation leads to a stiffening of the overall structure (III).

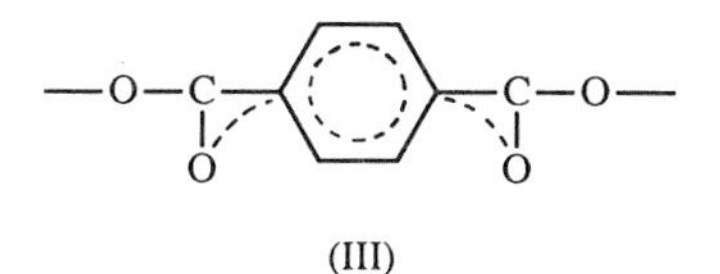

(III)

The majority of these main chain liquid crystal polymers show a nematic phase after melting but in some cases small variations in structure can lead to formation of a smectic mesophase. Thus for polyesters with structure

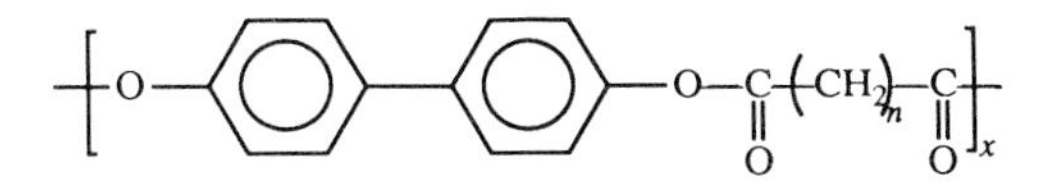

and when the number of methylene units (n) in the spacer is odd, a nematic phase is observed but when n is even a smectic mesophase results. These observations agree with de Genne's predictions. Similarly, for polyesters with multiple rings but different orientation of the ester units, the phases can also be changed, *i.e.*

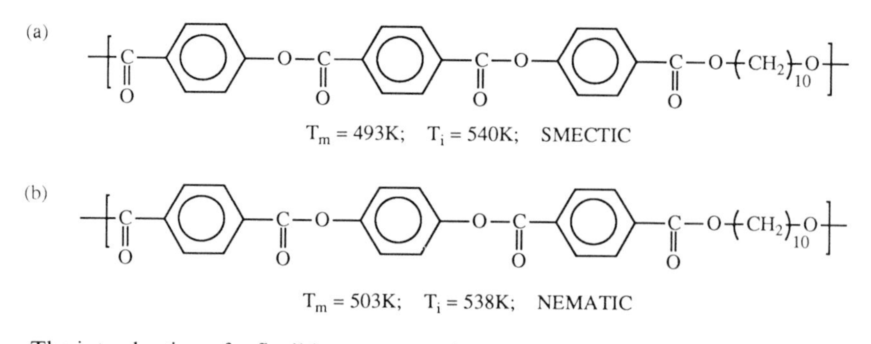

The introduction of a flexible spacer can lower the melting point and increase the temperature range in which the mesophase is stable. The magnitude of the effect will depend on the length of the spacer unit. Thus in systems such as the poly(α-cyanostilbene alkanoate)s, structure (IV),

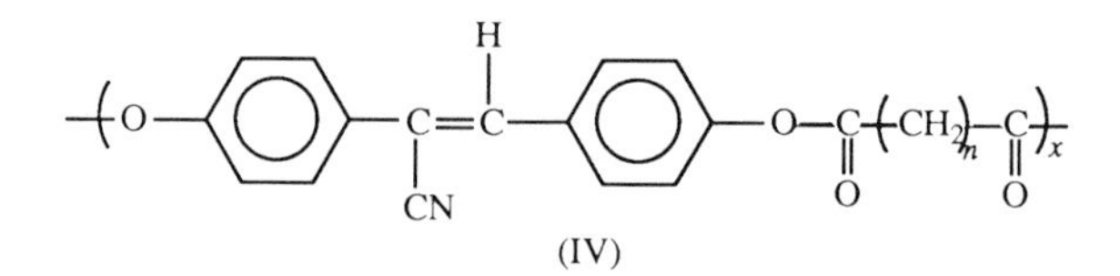

(IV)

in which a nematic phase is detected, both T_m and the nematic to isotropic transition temperature T_i decrease as the length of the methylene sequence (n) increases. This is clearly seen in figure 16.10 where an odd-even alternation is also evident. The polymers with spacers having an even number of (CH_2) units usually have higher melting and clearing temperatures T_i, than those with an odd number and this suggests that the spacer length influences the ordering in the liquid crystal phase. The long range ordering will tend to try and maintain the orientation of the mesogen parallel to the director axis and this may be easier for even numbered methylene unit spacers if they are in the all *trans* zig-zag conformation as shown in figure 16.11.

The spacer units are usually introduced by a copolymerization reaction and the proportion of the units relative to the mesogens can be varied. If a fixed spacer length is selected, as in poly(azophenol hexanoate), structure (V)

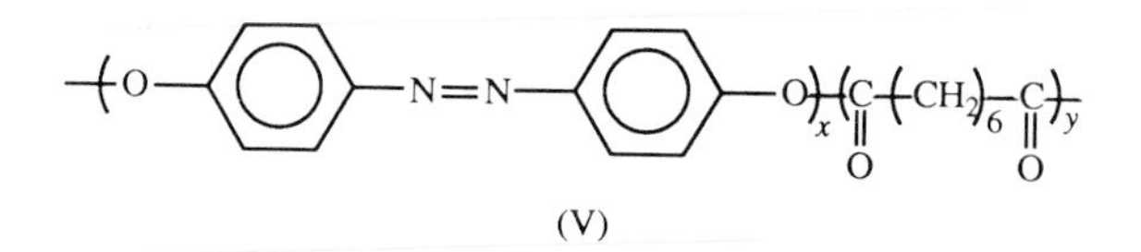

(V)

then T_m is found to decrease up to the (1:1) ratio of the two components and this is accompanied by a corresponding broadening of the mesophase stability range. These effects are seen in figure 16.12.

Copolymerization reactions can use many other combinations but one interesting reaction involves the modification of poly(ethylene terephthalate) by reacting the

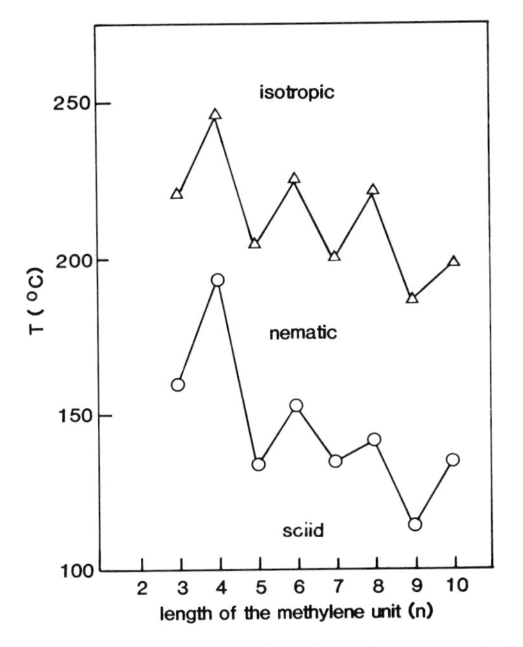

FIGURE 16.10. The transition temperatures $T_i \rightarrow T_N(-\triangle-)$ and $T_N \rightarrow T_K$ $(-\circ-)$ for structure IV as a function of the length of the methylene sequence. (Reproduced from K. Imura, N. Koide and M. Takeda (1987) with permission from Ottenbrite, Utracki and Inoue (Eds), *Current Topics in Polymer Science*, **1**, © Carl Hanser Verlag, Munich.)

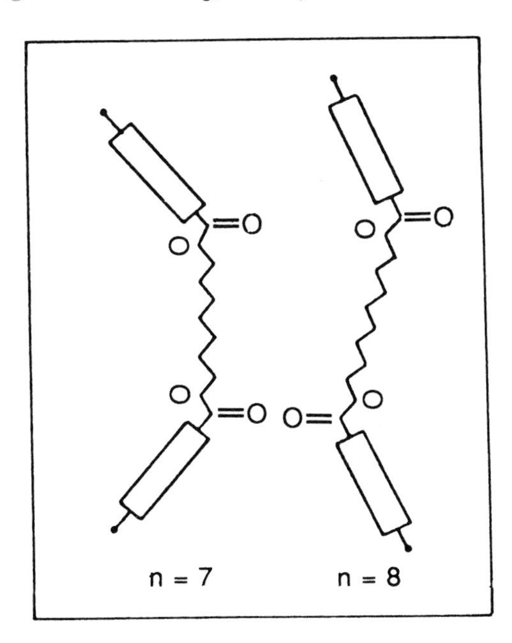

FIGURE 16.11. Schematic diagram showing the effect of odd and even numbers of spacer units on the relative orientation of the mesogenic units in a main chain liquid crystal polymer. (Adapted from W. R. Krigbaum, J. Watanabe and T. Ishikawa, *Macromolecules*, **16**, 1271 (1983) with permission from the American Chemical Society.)

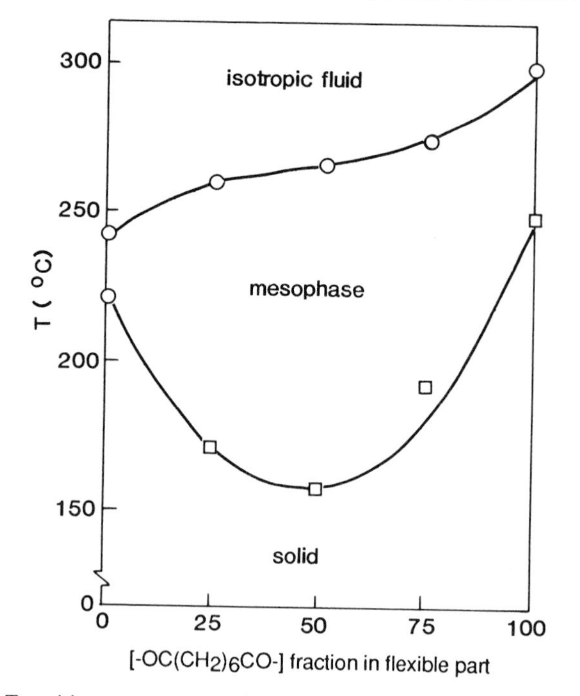

FIGURE 16.12. Transition temperatures plotted as a function of copolymer composition for poly(azophenol hexanoate) structure V. The transition $T_i \rightarrow T_N$ is (–○–) and (–□–) represents T_m. (Reproduced from K. Imura, N. Koide and M. Takeda (1987) with permission from Ottenbrite, Utracki and Inoue (Eds), *Current Topics in Polymer Science*, **1**, © Carl Hanser Verlag, Munich).

preformed polymer with *p*-acetoxybenzoic acid. This has the effect of introducing a mesogenic unit to the structure at the points where the two units combine, producing a thermotropic liquid crystal polymer with a flexible spacer.

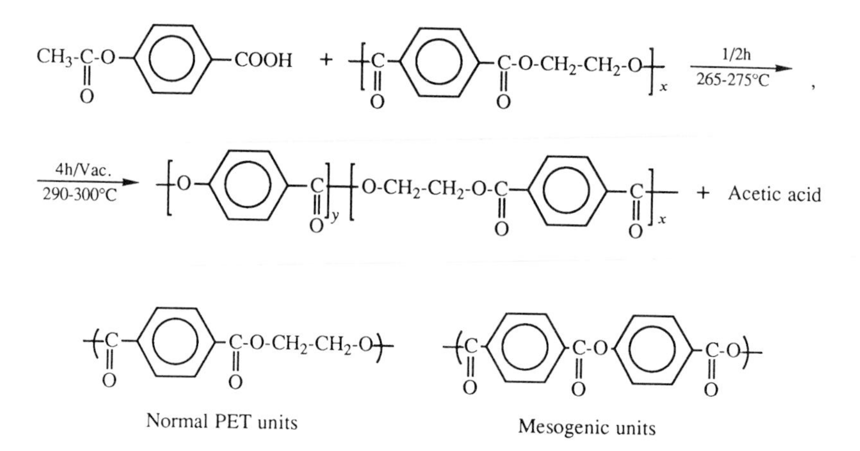

The liquid crystalline phase can then be controlled by introducing specific amounts of the oxybenzoate units, and at about 30 mole % incorporation a nematic phase appears in the melt. The optimum mechanical properties are obtained when 60 to 70 mole % of the oxybenzoate is present in the chain, as reflected by a large increase in the tensile strength of the material which is accompanied by a corresponding decrease in the melt viscosity (see figure 16.13).

Introduction of a lateral substituent into the mesogen will also lower T_m and T_i, as the bulky side groups will tends to force the chains apart thereby reducing the intermolecular forces of attraction. Ring substitution is the easiest way to achieve this and Lenz has shown that the structure (VI),

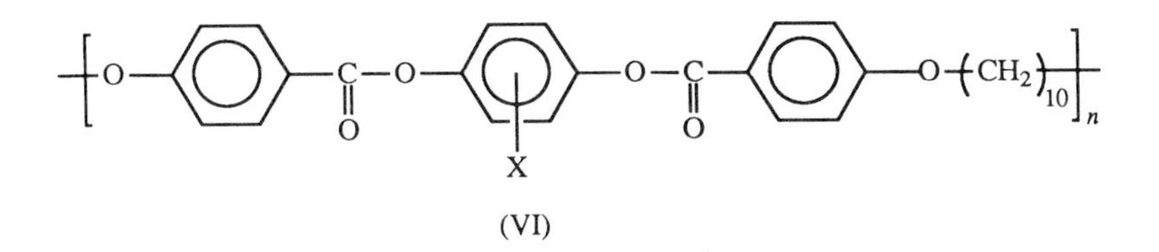

(VI)

which forms a nematic phase, can be altered by varying the group X, to give lower T_m and T_i values as shown in table 16.4. Similarly in structures such as poly(hydroquinone terephthalate) (VII)

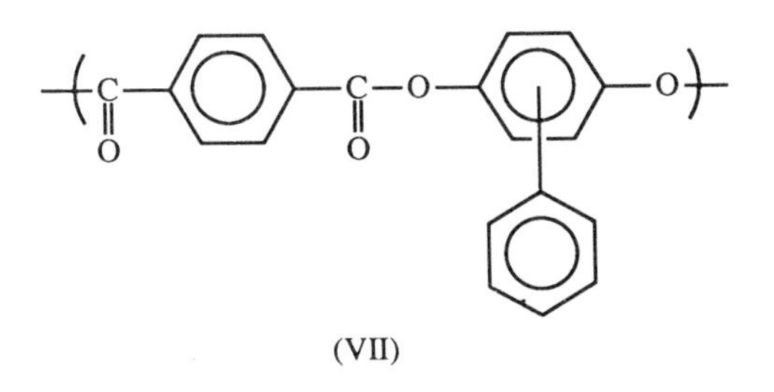

(VII)

substitution of a phenyl ring decreases T_m from > 870 K to approximately 610 K.

The introduction of kinks by using *meta* substituted monomers or a crankshaft monomer such as 6-hydroxy-2-naphthoic acid (HNA) can be equally effective. Hence when HNA is incorporated by copolymerization in the poly(hydroquinone terephthalate) structure this also lowers T_m as shown in figure 16.14.

TABLE 16.4. Effect of a lateral substituent on the transition temperatures of structure (VI)

X	T_g	T_m	T_i	ΔT
H	340	509	540	31
$-CH_3$	317	427	463	36
$-C_2H_5$	308	344	400	56

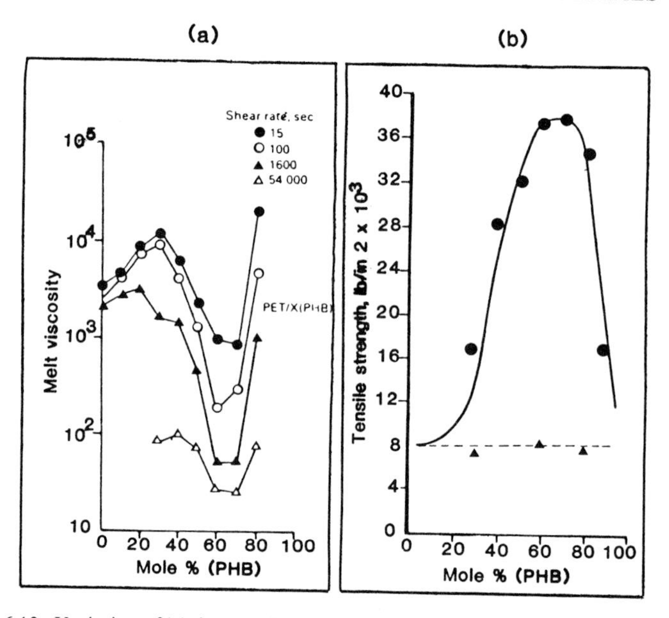

FIGURE 16.13. Variation of (a) the melt flow viscosity and (b) the tensile strength with mole % of oxybenzoate units in the copolymer formed by the reaction of poly(ethylene terephthalate) with *p*-acetoxybenzoic acid. (Reproduced from R. W. Lenz and J. J. Jin (1986) with permission from Gordon and Breach Science Publishers Ltd.)

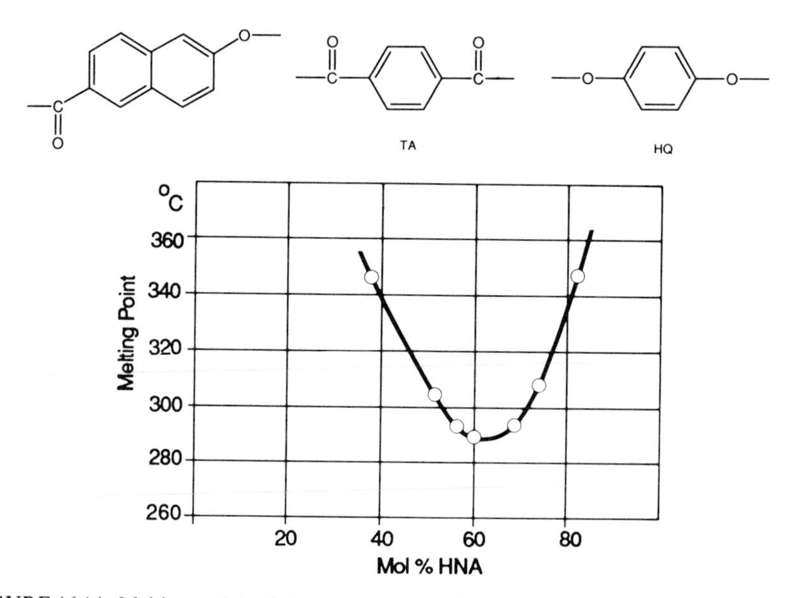

FIGURE 16.14. Melting point of thermotropic copolyesters of 6-hydroxy-2-naphthoic acid (HNA) with terephthalic acid (TA) and hydroquinone, showing the minimum of T_m at about 60 mol % HNA. (Reproduced with permission from Hoechst High Chem. Magazine.)

Other strategies for disrupting chain symmetry include the use of cross-shaped molecules (structure VIII)

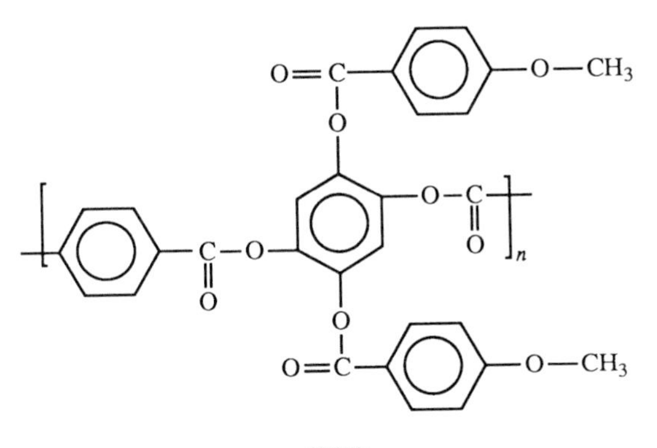

(VIII)

or discotic mesogens (structure IX).

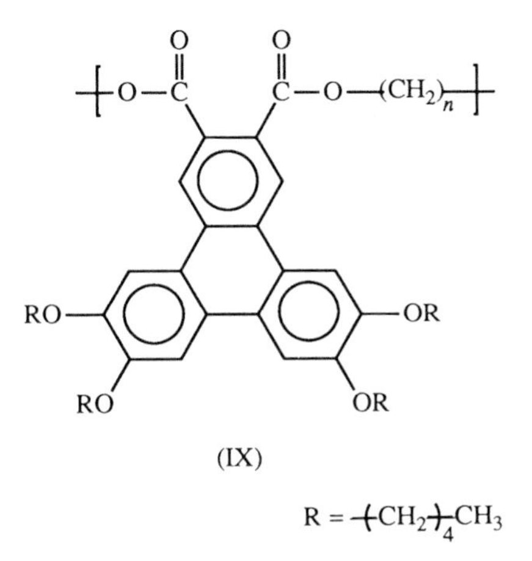

(IX)

$R = \left(CH_2\right)_4 CH_3$

A novel method of lowering the processing temperature has been reported by Porter, who prepared a binary mixture of poly(bisphenol E isophthalate-*co*-naphthalate)

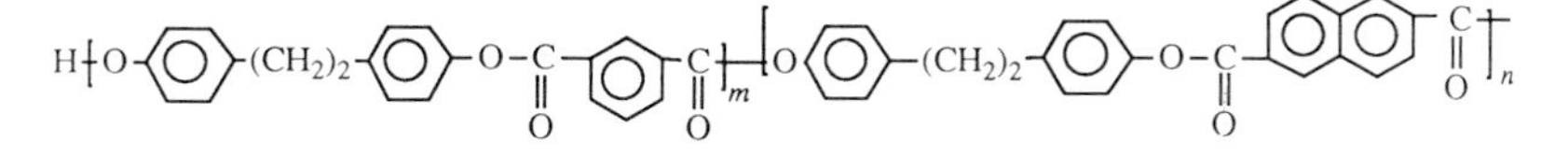

with a small liquid crystalline molecule which plasticized the polyester.

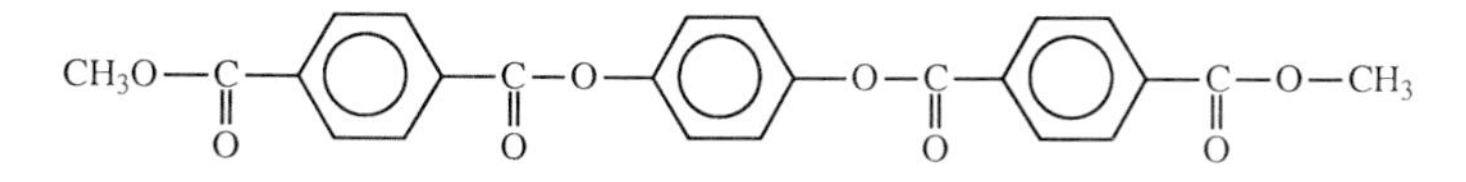

By adjusting the ratio of the two, a decrease of 20 K in T_m could be achieved, thereby allowing easier processing. After orientation of the blend by melt extrusion and cooling to a temperature just below the transition, a post transesterification reaction was carried out which linked the small molecule into the chain while retaining the orientation.

BRIDGING GROUPS

While many of the main chain thermotropic liquid crystalline polymers are polyesters, other bridging groups can be used. Some have been observed to enhance the stability of the mesophase region in the order

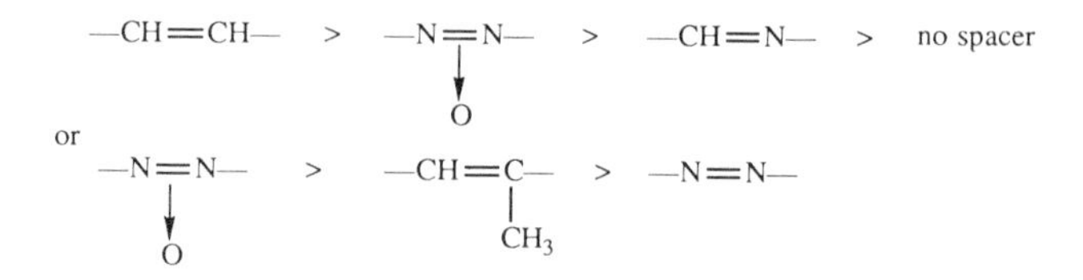

The ester group is difficult to place with respect to these series because of the differing interactions depending on the orientation with respect to the phenyl ring where conjugation increases rigidity, *i.e.*

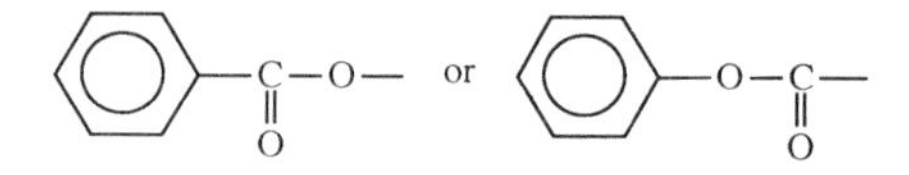

16.6 Side chain liquid crystalline polymers

It has been demonstrated that polymers with mesogens attached as side chains can exhibit liquid crystalline properties and much of the basic knowledge concerning side chain liquid crystal polymers has emanated from the work of Ringsdorf, Finkelmann, Shibaev and Platé. The extent to which the mesophases can develop in these systems is influenced by the flexibility of the backbone chain and whether the mesogen is attached directly to the chain or is pushed further away by the insertion of a flexible spacing unit.

The polymer chain to which the mesogens are bonded can have different degrees of flexibility and this can affect both the T_g and the temperature of the liquid crystal to isotropic phase transition T_i. This is illustrated in table 16.5 for a series of polymers with a constant mesogen unit but a chain flexibility that decreases in the order methacrylate > acrylate > siloxane. The temperatures of the transitions are seen to decrease in the same order.

The conclusions drawn from these trends are that the thermal range of the mesophase (ΔT) is greatest when the chain is most flexible and its conformational changes largely do not interfere with, or disrupt, the anisotropic alignment of the mesogens in the liquid crystalline phase. The influence of the backbone can be minimized by decoupling the motions of the chain from those of the side chain mesogens. This can be accomplished by introducing long flexible spacing units between the backbone and the mesogen, so that a typical side chain liquid crystalline polymer structure would be that shown schematically in table 16.6. Structures of this type can be

TABLE 16.5. Effect of chain flexibility on the transition temperatures of side chain liquid crystal polymers having a common mesogen.

Polymer	*Transitions*/K	ΔT
$-[CH_2-C(CH_3)(COOR)]_n-$	$g \xrightarrow{369} N \xrightarrow{394} i$	25
$-[CH_2-CH(COOR)]_n-$	$g \xrightarrow{320} N \xrightarrow{350} i$	30
$-[O-Si(CH_3)(CH_2R)]_n-$	$g \xrightarrow{288} N \xrightarrow{334} i$	46

$$R = -(CH_2)_2-O-C_6H_4-C(=O)-O-C_6H_4-O-CH_3$$

synthesized in a number of ways and one such scheme is outlined below.

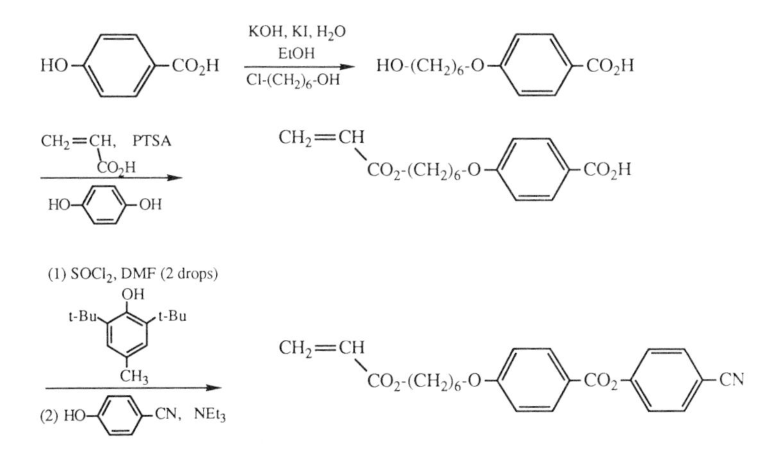

When longer spacer units are introduced the observed effects are that the T_g of the polymer is usually lowered by internal plasticization and the tendency for the more ordered smectic phases to develop is increased.

The examples in table 16.7 show that as the spacer length is increased there is an enhancement of the ordering and the nematic phase gives way to a smectic phase. This ordering effect can also be encouraged by lengthening the alkyl tail unit. Both of these alterations reflect the tendency for long alkyl side chains to order and eventually to

TABLE 16.6. Schematic representation of the organization of a side chain liquid crystal polymer (SCLCP).

SCLCP: [Flexible Tail]—[Cyclic Unit]—[Bridging Group]—[Cyclic Unit]—[Functional Unit]—[Spacer]—[Functional Unit]—[Flexible Backbone]

Flexible Tail	*Cyclic Unit*	*Bridging Group*	*Functional Unit*	*Spacer*	*Flexible Backbone*
none	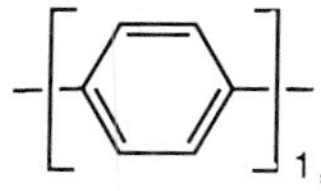 1,3 1,4	none	none	none	—CH—CHR—
R		—CO—O—	—O—	$-(CH_2)_n-$	
OR	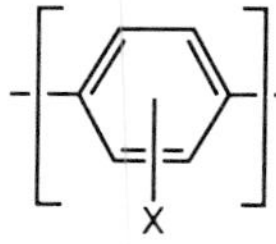 X = Me Ph Cl	—CR=CR—	—CO—O—	—S—R—S—	—SiR—O—
CN		—CR=NO—	—O—CO—	$—SiR_2—O—$	
				$-(CH_2—CHR)_n-$	$—SiR—O—SiR_2—O—$
	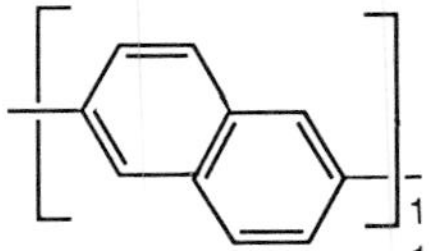 1,4 1,5 2,6	—NO=N—		—NR′—R—NR′	
		—C≡C—			—P=N—
		—CR=N—N=CR—			
	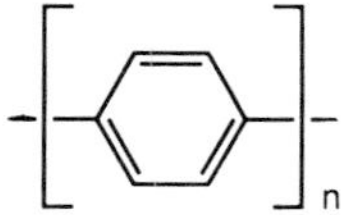 n = 1,2,3				
	 Cholesteryl				

TABLE 16.7. Effect of spacer and tail unit length on the ordering within the mesophase.

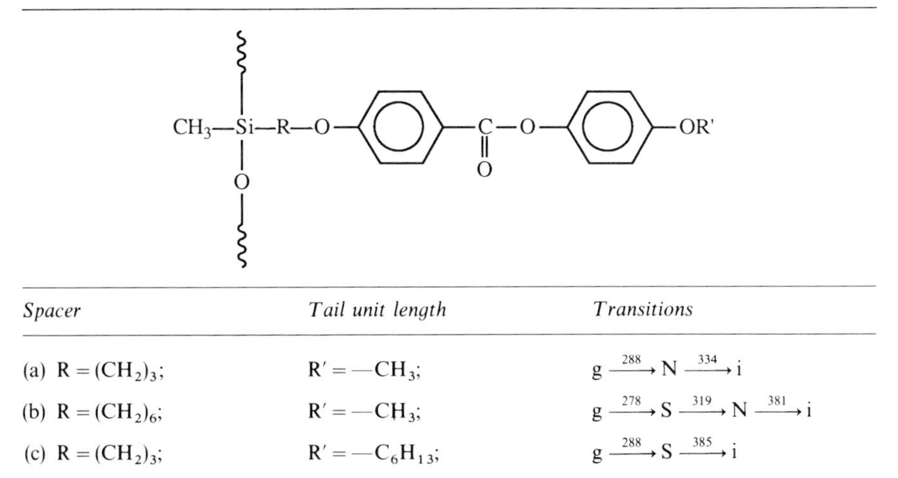

Spacer	*Tail unit length*	*Transitions*
(a) $R = (CH_2)_3$;	$R' = —CH_3$;	$g \xrightarrow{288} N \xrightarrow{334} i$
(b) $R = (CH_2)_6$;	$R' = —CH_3$;	$g \xrightarrow{278} S \xrightarrow{319} N \xrightarrow{381} i$
(c) $R = (CH_2)_3$;	$R' = —C_6H_{13}$;	$g \xrightarrow{288} S \xrightarrow{385} i$

crystallize, when long enough, and show that this is also imposed on the liquid crystal state.

The orientational order in nematic polymers can be described by a parameter S defined by

$$S = 3/2(\overline{\cos^2 \theta} - 1/3) \tag{16.1}$$

where θ is the angle of mean deviation of the molecular axes of the mesogens with respect to the director, *i.e.* $S = 1$ means perfect parallel orientation. The value of S for side chain liquid crystal polymers is approximately 75 per cent of that obtained for the corresponding low molar mass mesogens and decreases with rising temperature as thermal agitation disturbs the orientation of the mesogenic groups. While extension of the flexible spacer length does not appear to change S significantly when the system is in the nematic state, there is a marked change at a smectic-nematic transition, as might be expected.

The ordered state of the mesophase in these side chain liquid crystalline polymers is readily frozen into the glassy state if the temperature is dropped rapidly below the T_g and the value of S remains unchanged. This means that the liquid crystalline phase can be locked into the glassy polymer and remain stable until disturbed by heating above T_g again. This phenomenon offers several interesting applications in opto-electronics and information storage (see sections 17.19 and 17.22). These applications often depend on the ability of mesogenic groups to align under the influence of an external magnetic or electric field.

It was stated earlier (section 16.3) that liquid crystal polymers can be oriented by interaction with the surfaces of the measuring cell which may be as simple as two glass slides with the polymer layer sandwiched between. This can give two extreme cases of either (a) homogeneous or (b) homeotropic alignment where the

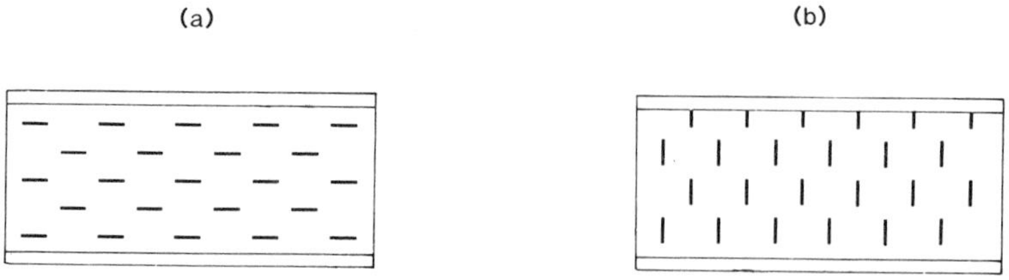

(a)

(b)

FIGURE 16.15. Schematic representation of (a) homogeneous and (b) homeotropic alignment of mesogens in a measuring cell.

long axes of the mesogens are respectively parallel to or perpendicular to the cell surface, shown schematically in figure 16.15, and, when viewed vertically to the glass plates through crossed polarizers, the systems are then opaque or transparent respectively. Because of the high viscosity experienced in polymer systems this alignment may take some time or be incomplete. As the dielectric constant, and diamagnetic susceptibility of many mesogens are anisotropic, side chain liquid crystal polymers in the nematic state can also be oriented quite rapidly by the application of a magnetic or electric field. Now the parameter of interest is the magnitude of the critical field which is required to effect the Fredericks transition. This is the transition from the homogeneous to the homeotropic aligned state. Whereas the relaxation time for this transition to take place in low molar mass mesogens is of the order of seconds, viscosity effects in polymer systems can push this up several orders of magnitude larger. This may make the use of polymeric liquid crystals less attractive in rapid response display devices but the additional stability that can be gained in polymeric systems can be advantageous in other ways, as will be illustrated in chapter 17.

16.7 Chiral nematic liquid crystal polymers

A special form of the nematic phase, first observed in low molar mass esters of cholesterol, and originally called the "cholesteric mesophase" can often be detected in mesogenic systems containing a chiral centre. The structure is a helically-disturbed nematic phase, shown schematically in figure 16.16, where the nematic order is preserved in each alternate layer but where the director of each layer is displaced regularly by an angle θ relative to its immediate neighbours. This imparts a helical twist with a pitch p to the phase. This type of ordering results in a system with a very high optical activity and the ability to selectively reflect circularly polarized light of a specific wave length λ_R, when irradiated by normal light. The wavelength of this reflected light is related to the pitch of the helical structure by

$$\lambda_R = \tilde{n} p \tag{16.2}$$

where $\tilde{n}$ is the average refractive index of the liquid crystalline phase.

The synthesis of polymers capable of entering into a chiral nematic phase initially proved difficult as many of the acrylate and methacrylate comb-branch polymers to which a cholesterol unit was attached as a side chain tended to give a smectic phase. This was overcome by either copolymerizing the cholesterol-containing monomers with another potential mesogenic monomer, or by synthesizing mesogens with a chiral unit in the tail moiety. Examples of both types are shown as structures X and XI.

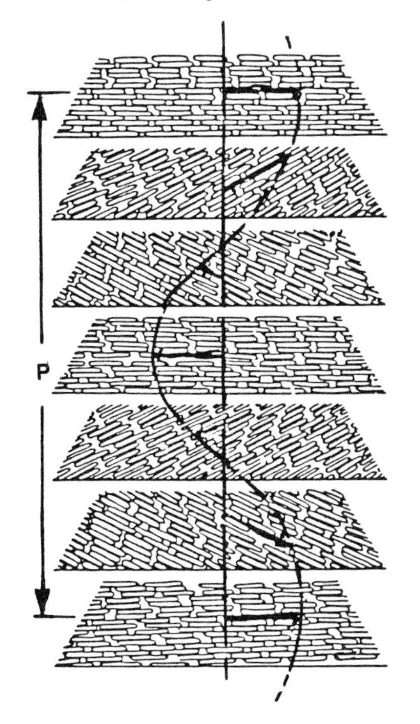

FIGURE 16.16. Schematic representation of the cholesteric or chiral nematic phase, where p is the helix pitch length.

TABLE 16.8. Variation of λ_R with copolymer composition for structure XI.

%*Chol.*	T_g/K	T_i/K	λ_R(nm)
34	50	90	850
40	50	102	660
55	55	105	555
65	55	150	500

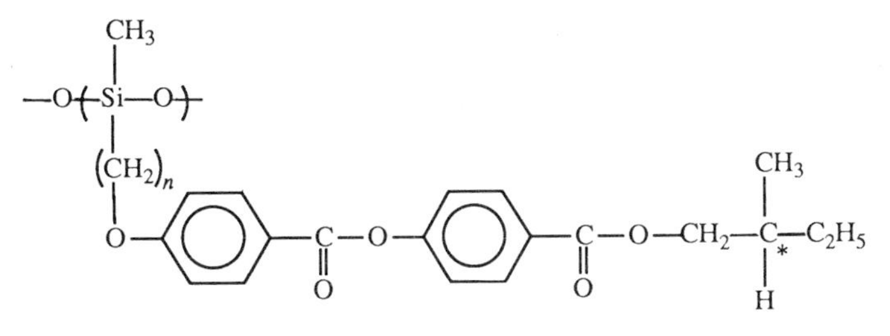

(X)

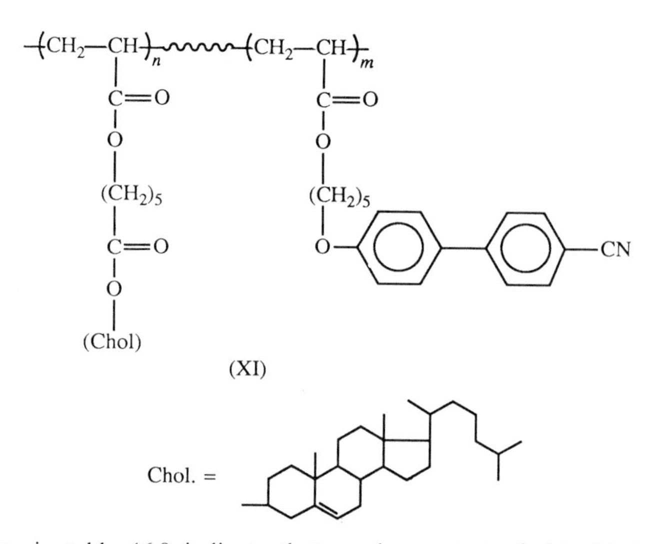

The data in table 16.8 indicate that as the content of the chiral monomer (cholesterol) is varied in the copolymer the wavelength of the reflected light changes, for an average temperature region between T_g and T_i. This suggests that the composition changes alter the pitch of the helical structure in the chiral nematic phase and so the wavelength of reflected light will also change. For XI, the increase in % chol. causes the helix twist to become tighter, *i.e.* p becomes smaller so λ_R also moves to shorter wavelengths. The pitch is also sensitive to temperature and when this is raised the helix tends to unwind with a consequential increase in λ_R.

Like the other liquid crystal polymers described in section 16.6, these materials offer the possibility of locking the chiral nematic phase into the glassy state by rapid supercooling to temperatures below T_g. This leads to a preservation of the structure, and of course the reflected colour, thereby leading to the formation of stable, light-fast, monochromatic films when suitable systems are used.

16.8 Miscellaneous structures

A number of structural variations of side chain liquid crystal polymers can be prepared and include:

(a) Mesogenic groups in both the main and side chain

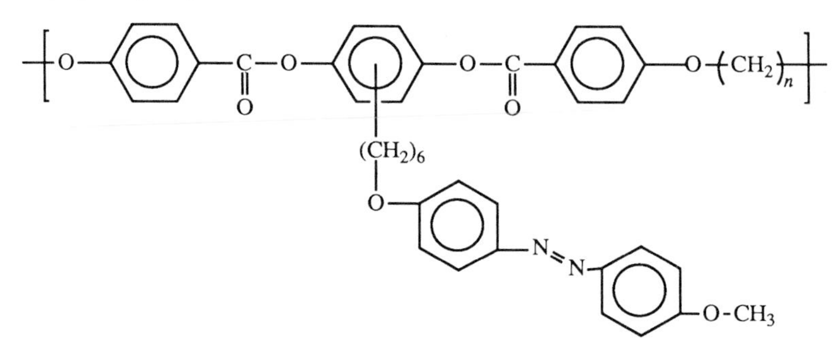

where a nematic phase was detected,

(b) Chains with lateral oriented side chain mesogens

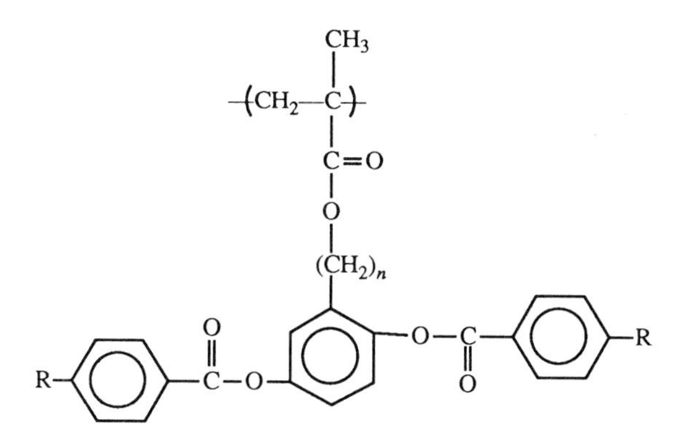

which again produced a nematic phase, and

(c) Materials with discotic side chains

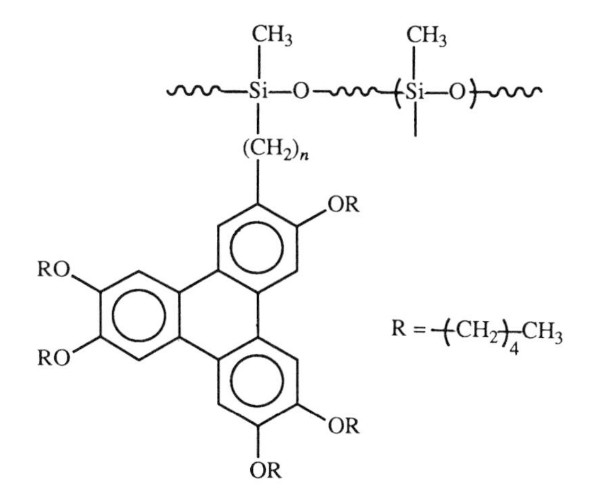

which after slow annealing formed an anisotropic discotic phase.

General reading

A. Blumstein, Ed. *Polymeric Liquid Crystals.* Plenum Press (1985).

A. Ciferri, W. Krigbaum and R. Meyer, Eds, *Polymer Liquid Crystals.* Academic Press (1982).

M. Gordon and N. A. Platé, *Liquid Crystal Polymers. Advances in Polymer Science,* Vols 59/60/61. Springer-Verlag (1984).

G. Gray, Ed. *Thermotropic Liquid Crystals.* CRAC series, Vol. 22. John Wiley and Sons (1987).

G. Gray and J. Goodby, *Smectic Liquid Crystals: Textures and Structures*. Leonard Hill (Blackie) (1984).

C. B. McArdle, Ed. *Side Chain Liquid Crystal Polymers*. Blackie and Sons Ltd (1989).

N. A. Platé and V. P. Shibaev, *Comb-Shaped Polymers and Liquid Crystals*. Plenum Press (1987).

References

1. H. Benthack-Thomas and H. Finkelmann, *Makromol. Chem.*, **186**, 1895 (1985).
2. G. Decobert, J. C. Dubois, S. Esselin and C. Nöel, *Liquid Crystals*, **1**, 307 (1986).
3. W. R. Krigbaum, J. Watanabe and T. Ishikawa, *Macromolecules*, **16**, 1271 (1983).
4. K. Imura, N. Koide and M. Takeda, "Synthesis and characterization of some thermotropic liquid crystalline polymers." In *Current Topics in Polymer Science*, **1**, (1987).
5. R. W. Lenz and J. I. Jin, "Liquid crystal polymers: a new state of matter." *Polymer News*, **11**, 200 (1986).
6. P. W. Morgan, "Aromatic polyamides." *Chem. Tech.*, 316 (1979).
7. C. Nöel, "Synthesis, characterization and recent developments of liquid crystalline polymers." *Makromol. Chem. Macromol Symp.*, **22**, 95 (1988).
8. D. Sek, "Structural variations of liquid crystalline polymer macromolecules." *Acta Polymerica*, **39**, 599 (1988).

CHAPTER 17

Polymers for the Electronics Industry

17.1 Introduction

The application of speciality polymers in electronics and photonics is extensive, both in a "passive" role – when they act as insulators, encapsulating agents, adhesives, and materials for integrated circuit fabrication – and in an "active way" as electronic and photonic conductors, or as active material in non-linear optics.

Some of these uses are rather mundane, in a chemical sense, but are necessary as an integral part of the whole process, *e.g.* in the packaging and protection of fragile integrated circuits (IC) to avoid damage or to prevent the detrimental effects of humidity and corrosion. Epoxy-novolacs, or silicone-epoxy thermosetting resins are ideal for these purposes and are used to encase the IC which may already have been treated with a barrier coating of room temperature vulcanizing (RTV) silicone rubber to prevent moisture absorption. Other applications are more interesting to the chemist as they require thought and ingenuity in the molecular design of the polymer.

The properties of polymers that make them an essential part of microelectronics engineering will be discussed under two main headings, polymer resists and conducting polymers. These will best illustrate why progress in this area could not have been made without exploiting the unique features of polymeric materials. This will be followed by a brief discussion of some photonic applications.

17.2 Polymer resists for integrated circuit fabrication

Integrated circuits are arguably amongst the most important products of the modern electronics industry. They are built up from various arrangements of transistors, diodes, capacitors and resistors, that are individually constructed on a flat silicon or gallium arsenide substrate by selective diffusion of small amounts of materials into particular regions of the semiconductor substrate, and by metallization of the paths linking the active circuit elements. The patterns defining these regions and the linking pathways must first be drawn by a lithographic process on a layer of resist material, and then transferred onto the substrate by an etching process. In this context, the *lithographic process* is the art of making precise designs on thin films of resist material by exposing them to a suitable form of patterned radiation, *e.g.* ultraviolet, electron beam, X-ray, or ion beam, with the formation of a latent image on the resist that can subsequently be developed by treatment with solvents or plasma. The *resist* is a material, usually polymeric, that is sensitive to, and whose properties (either chemical or physical) are changed by, exposure to the electromagnetic radiation used. It must also be resistant (hence the name), after development, to the etching process, and protect the areas it still covers while allowing the exposed regions of the substrate to be attacked. In this way a pattern is transferred onto the substrate and the remaining resist material is removed.

17.3 The lithographic process

Several steps are involved in the lithographic process and these are shown schematically in figure 17.1. If the substrate chip is silicon it is first oxidized to produce a thin surface layer of SiO_2 (step 1). A solution of the polymer resist is then spun evenly onto this surface and baked to remove the solvent and form a thin film of the resist (step 2) approximately 0.5 to 2 μm thick. The next stage (step 3) is the exposure of the resist to electromagnetic radiation either through a patterned mask or by direct "writing" if the radiation source is an electron beam. Depending on the radiation and the nature of the polymer resist, the exposed regions are either rendered soluble if the polymer is degraded – this is called a *positive* acting resist – or insoluble if the polymer is crosslinked – this is called a *negative* acting resist. A positive or negative pattern can then be developed by treatment with solvents that dissolve the exposed regions in the positive resist (step 4a) or the unexposed regions in the case of the negative acting resist (step 4b), so producing the template for the pattern that is to be etched onto the substrate. Etching (step 5) can be achieved by treatment with buffered HF or by using dry plasma etching methods. In either case the polymer must protect the regions of the chip it still covers while allowing the exposed areas of the substrate to be attacked. Once the pattern is transferred in this way the remaining resist is stripped off and discarded (step 6).

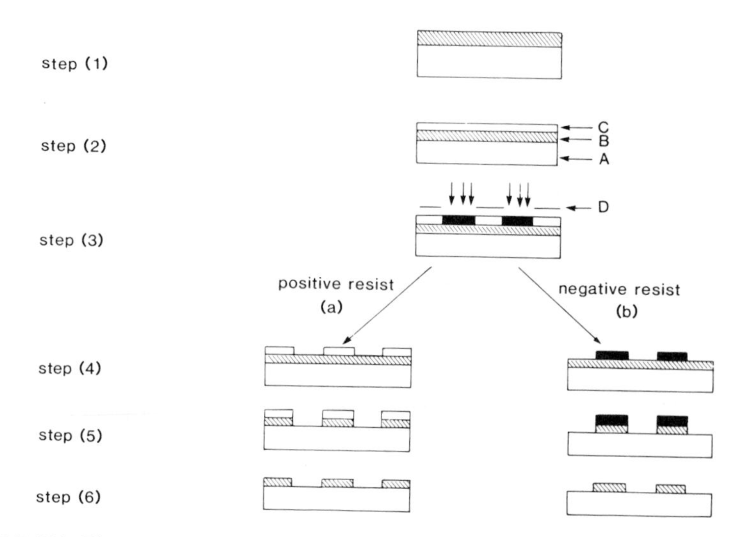

FIGURE 17.1. The steps in the lithographic process. (1) Form dielectric layer, (2) coat with polymer resist, (3) expose to electromagnetic radiation, (4) develop pattern, (5) etch, (6) strip resist.
A: the substrate (silicon, *etc.*); B: thin dielectric layer (*e.g.* SiO_2); C: polymer resist layer, and D: mask.

As designers move towards greater miniaturization and increasing device complexity, there is a need for smaller feature sizes. Thus the minimum feature sizes on MOS random access memory devices may be $\sim 3.5\,\mu m$ if a "Very Large-Scale Integrated Circuit" (VLSI) 256K RAM device is to be made. For such a chip, which is only a 1/2″ square of silicon, to be constructed, 256 000 places where an electrical charge can be located must be defined together with the connecting circuitry. This puts a severe strain on the ability to define these features accurately, and is a function of both the wavelength of the radiation and, more importantly, the response of the resist to the radiation. Now the ingenuity of the polymer chemist is called upon to design suitable resist materials. The manufacture of VLSI devices requires the use of short wavelength radiation but many circuits are still made by photolithography involving irradiation with UV light. An example of a VLSI device is shown in figure 17.2.

17.4 Polymer resists

The response of a polymer to radiation and the resolution obtained are not easy to predict, but there are certain criteria that the prospective resist should meet. The most important of these are:

(a) adequate sensitivity to the radiation used,
(b) the ability to adhere to the substrate and be easily removed after etching,
(c) possession of a high T_g, particularly if it is a positive working resist, to prevent flow and distortion of developed patterns,
(d) resistance to etching reagents.

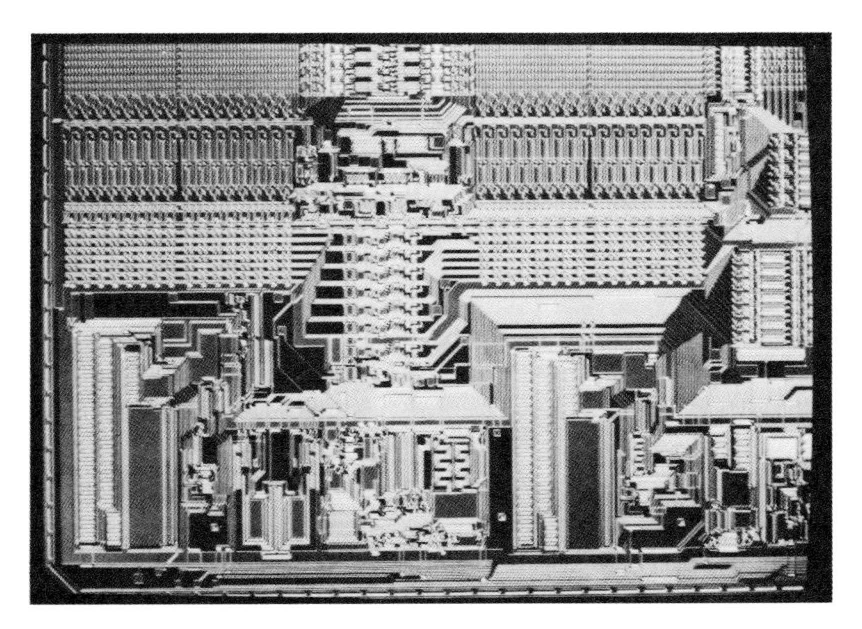

FIGURE 17.2. An example of a Very Large-Scale Integrated circuit chip manufactured by IBM. (Reproduced with permission from Hoechst High Chem. Magazine (1989)).

Resists can be judged using the performance criteria of: (1) Sensitivity and (2) Resolution (contrast).

SENSITIVITY

This can be defined as the amount of incident energy, that is the flux/unit area measured in C/cm^2, required to effect enough chemical change in the resist to ensure that after development the desired relief image is obtained. This can be measured by plotting the log of the radiation dose, D, against the normalized film thickness after development. The response curves for both positive and negative working resists are shown in figure 17.3.

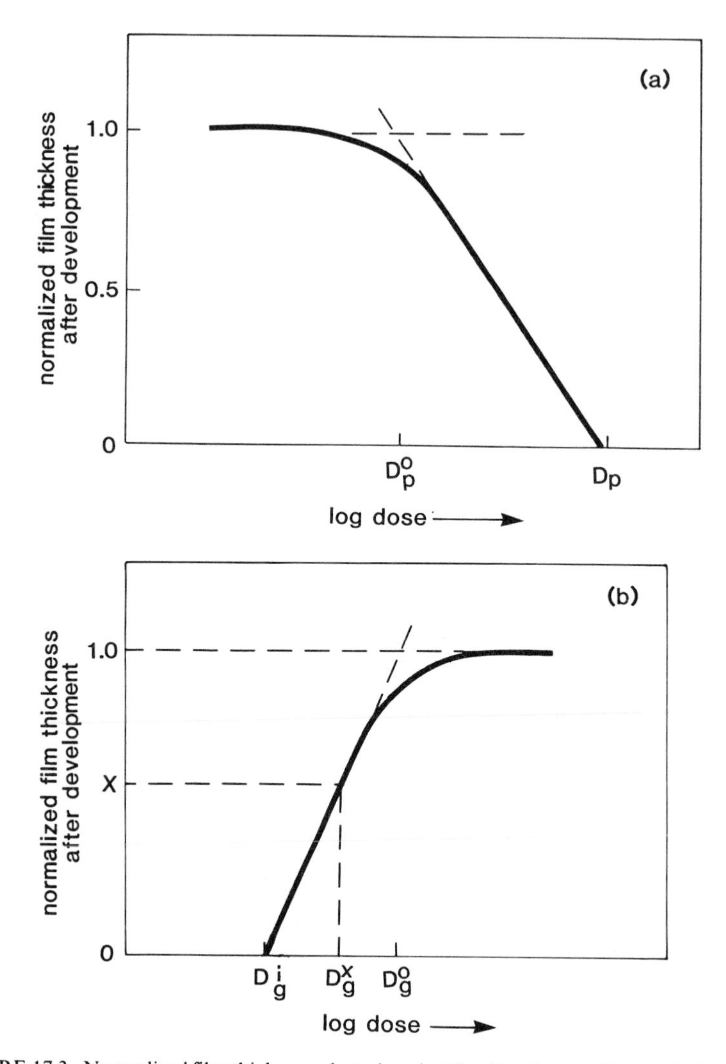

FIGURE 17.3. Normalized film thickness plotted against log (dose) to produce typical sensitivity curves for (a) a positive and (b) a negative acting polymer resist.

Sensitivity increases as the dose required to produce the image decreases, *i.e.* the lower the dose required to produce the desired image, the greater the sensitivity. The sought-after range of sensitivities is approximately (0.01 to 1.0) $\mu C/cm^2$.

RESOLUTION

Resolution (γ), or contrast, has different definitions for positive and negative resists. For positive resists, γ_p is a function of both the rate of degradation and the rate of change of solubility of the resist on exposure, while for a negative resist, γ_n is a function of the rate of gel formation. Numerical values are obtained from the slope of the linear portion of the response curve and are given by:

$$\gamma_p = \log\left[\frac{D_p}{D_p^o}\right]^{-1} \tag{17.1}$$

$$\gamma_n = \log\left[\frac{D_g^o}{D_g^i}\right]^{-1} \tag{17.2}$$

where D_g^i is the onset of gel formation, D_g^o is the dose required to produce 100 per cent initial film thickness, D_p is the dose required to effect complete solubility in the exposed regions while leaving the unexposed region insoluble.

We can now examine some of the systems which have proved to be successful in this application.

17.5 Photolithography

Many polymers are altered on exposure to ultraviolet radiation and this has led to the development of photolithographic techniques using conventional UV radiation from a mercury vapour lamp with an emission spectrum of $\lambda = 430$ nm, 405 nm and 365 nm. When resolution of features of less than $\sim 2\ \mu m$ is required then deep UV sources with $\lambda \approx 150$ to 250 nm may be used, provided resists which absorb in this wavelength region can be selected.

POSITIVE PHOTORESISTS

A positive working resist normally depends for pattern development on an increase in the solubility of the resist in the exposed region relative to the unexposed areas. As the wavelengths used in the near-UV are not sufficiently energetic to induce bond scission, the solubility must be altered by some other means.

A photoresist widely used in the electronics industry is the two component system comprising a short chain Novolac resin, which acts as the film-forming agent, mixed with 20 to 50 wt% of a naphthoquinone diazide photosensitive compound. This sensitizer is insoluble in basic solutions and is a sufficiently large molecule to act as a "dissolution inhibitor" preventing the acidic Novolac film from being dissolved by aqueous alkaline solutions in which the resin is normally soluble. Exposure to UV radiation converts the diazide into indene carboxylic acid that is now soluble in base solution, thereby rendering the whole of the exposed region soluble, while the unexposed regions remain resistant and insoluble. The reaction involves elimination of

N_2 followed by a Wolff rearrangement, after which the presence of small amounts of water in the resist completes the conversion of the ketene to the acid.

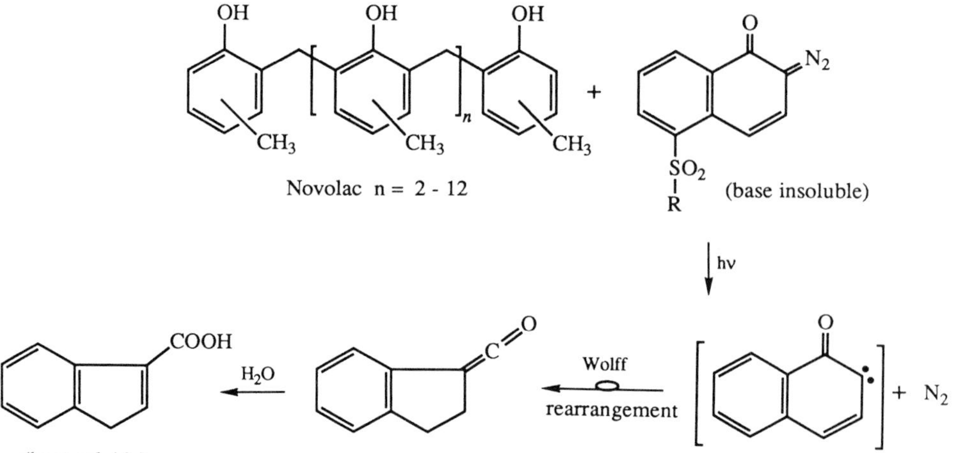

Novolacs tend to absorb too strongly in the deep UV region (~ 250 nm) but other systems have been developed specifically for this purpose. A base-soluble, poly(methyl methacrylate-*stat*-methacrylic acid) copolymer that is transparent in the deep UV, can be mixed with a photosensitive, base-insoluble, dissolution inhibitor *o*-nitrobenzyl-cholate(I). Exposure to deep UV photolyses the ester to produce cholic acid and *o*-nitrosobenzyl alcohol as shown:

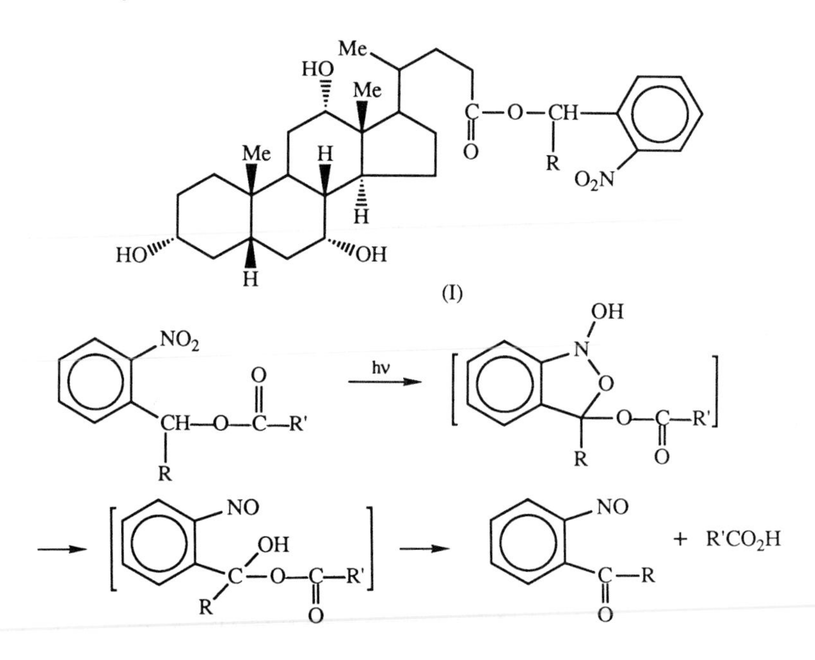

This system functions in the same way as the Novolac resist, and gives a positive tone pattern.

NEGATIVE PHOTORESISTS

This type of resist has been the mainstay of the microelectronics industry when resolution down to 2 μm is adequate, but is not so useful for finer work. The most commonly used systems are prepared from mixtures of cyclized polyisoprene mixed with a suitable photosensitive compound such as an aromatic diazide. The acid-catalyzed cyclization of polyisoprene produces a complex mixture of structures but leads to a higher T_g material, with improved film-forming properties.

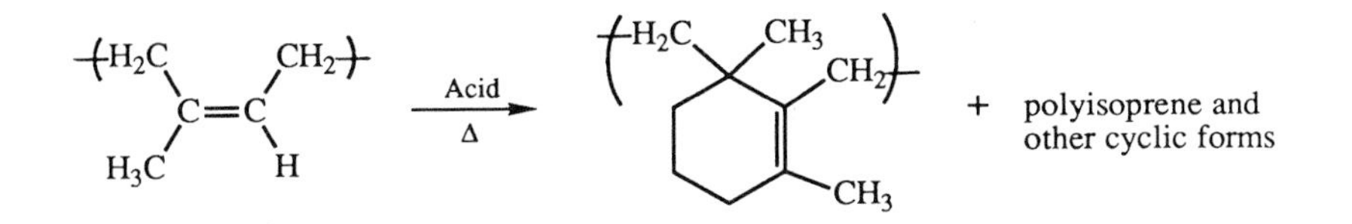

The bisazide forms a bisnitrene on irradiation which then crosslinks the polyisoprene by reacting with the double bonds or the allylic hydrogens in the exposed regions to produce an insoluble matrix, *e.g.*

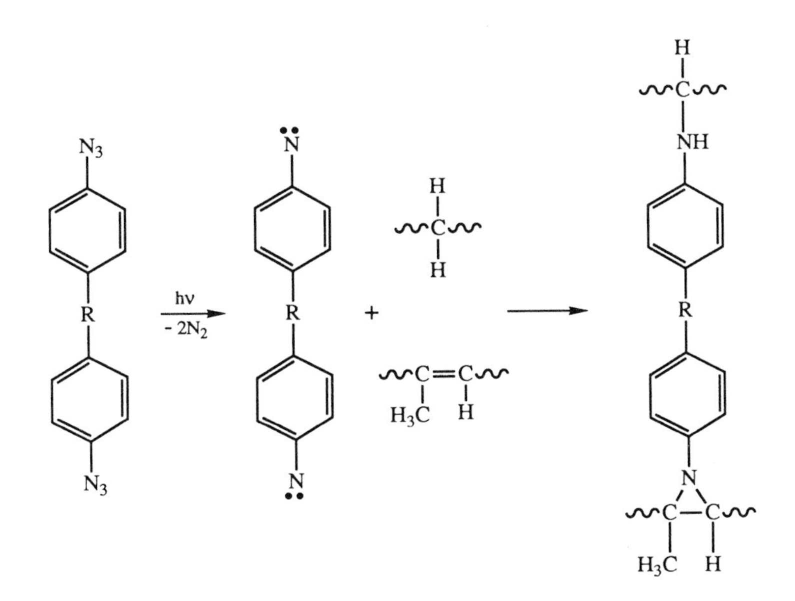

A second system makes use of the potential for poly(vinyl cinnamate) to form crosslinks upon radiation, *e.g.*

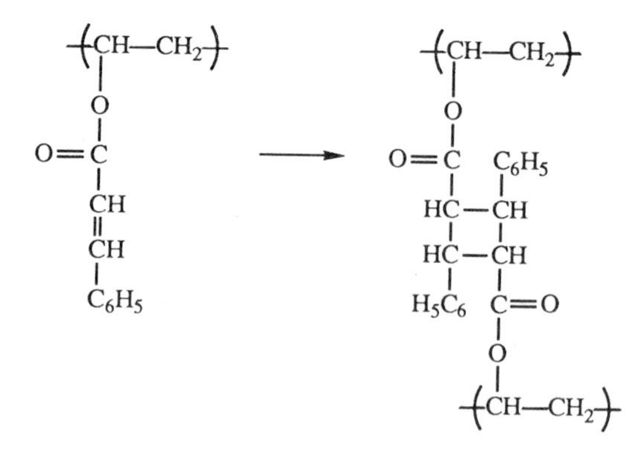

This photodimerization can be sensitized by Michler's ketone, 4,4'-bis(dimethylamino)benzophenone.

Greater resolution can be obtained using deep UV radiation but now other resists must be used as the conventional ones are optically opaque in this region.

Poly(methylmethacrylate) can be used as a positive resist, as can several of its derivatives. In each case the carbonyl groups absorb at 215 nm, and this leads to chain scission and degradation.

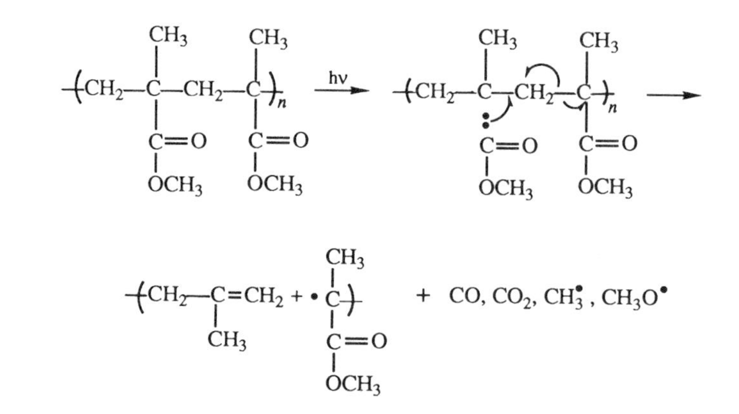

A negative tone resist can be obtained by using an image reversal technique with the Novolac–naphthoquinone diazide system. The procedure makes use of the normal steps for creating a positive resist, but on treating with base and baking at temperatures $> 350\,K$ a base-catalysed decarboxylation of the indene carboxylic acid occurs, forming an indene derivative which is a photo-insensitive dissolution inhibitor. The complete resist is now subjected to a flood exposure of all areas by the UV source and this converts the naphthoquinone diazide in the previously unexposed regions into the acid form. This renders these regions soluble, and development produces the negative tone pattern. These events are summarized in figure 17.4.

Other positive resists, sensitive to these wavelengths include the alkyl and aryl sulphones.

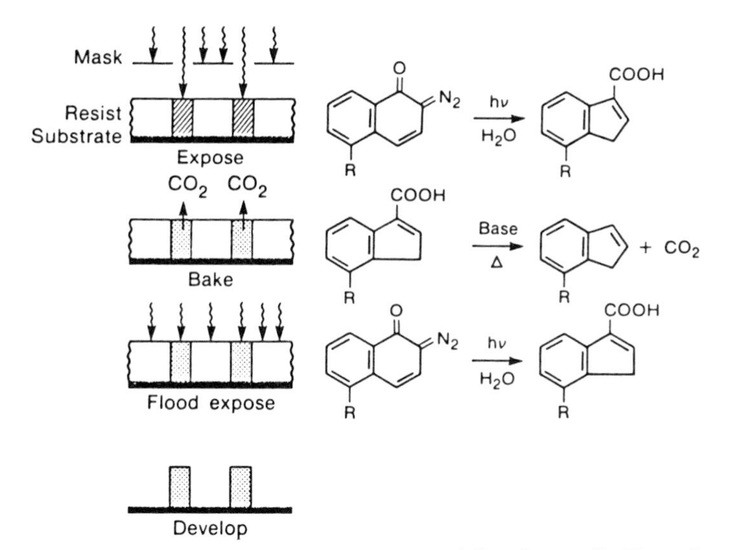

FIGURE 17.4. Reverse imaging using a Novolac-naphthoquinone diazide resist mixture exposed to UV radiation. (Reproduced with permission from M. J. Bowden and S. R. Turner, Eds, *Electronic and Photonic Applications of Polymers*. American Chemical Society, Washington, D.C. (1988)).

17.6 Electron beam sensitive resists

The inherent limitations of photolithography, caused by diffraction problems when resolutions of less than 1 μm are required, is overcome, at least in part, by turning to electron beam and X-ray lithography with much shorter wavelengths of the order 0.5 to 5 nm. As the photon energy of an electron beam is high enough to break virtually all the bonds likely to be found in a polymer resist, the reactions involved are much less selective than those encountered in some of the photoresists. Thus both degradation and crosslinking may take place in the same polymer on exposure to an electron beam, and the behaviour of the resist as a positive or negative working system will depend on which of these processes dominates. This may be a function of the exposure time and intensity of the radiation such that a positive acting resist may begin to crosslink and transform into a negative acting resist on prolonged exposure.

POSITIVE RESISTS.

Most polymers that are positive resists tend to depolymerize via a monomer unzipping action when degraded, and PMMA is typical of this type. Unfortunately the sensitivity of PMMA to electron beam radiation is low, and in an attempt to improve this feature, PMMA derivatives have been prepared by replacing the α-methyl group with more polar electron withdrawing substituents, *e.g.* Cl, CN and CF_3, to assist electron capture (figure 17.5). Modification of the ester group has also been a strategy, but in each case the presence of the quaternary carbon atom is perhaps the single most important feature for the resist to be a positive working system, because of its susceptibility to chain scission. Poly(alkene sulphone)s are also very sensitive positive

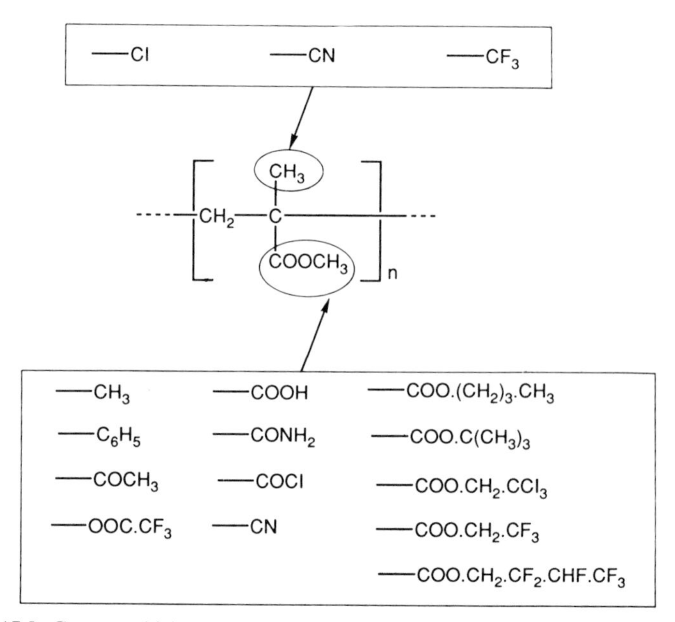

FIGURE 17.5. Groups which have been used to increase the sensitivity of positive working electron resists.

resists, prepared by an alternating copolymerization reaction between sulphur dioxide and an appropriate alkene.

$$\left(CH_2-\underset{\underset{CH_3}{|}}{\underset{(CH_2)_n}{\underset{|}{CH}}}-\overset{O}{\overset{\|}{\underset{\underset{O}{\|}}{S}}}\right) \xrightarrow{\varepsilon\text{-beam}} CH_2=\underset{\underset{CH_3}{|}}{\underset{(CH_2)_n}{\underset{|}{CH}}} + SO_2$$

Exposure to an electron beam source cleaves the polymer chains at the weak C–S bond with liberation of SO_2, and in certain cases such as poly(2-methyl pentene sulphone) there is almost complete vaporization of the exposed regions, when a 20 kV electron beam source is used. The major limitation of this group of resists is their poor resistance to dry etching.

NEGATIVE RESISTS

In general, negative acting resists tend to give poorer resolution, but are faster and tougher than positive working resists. This difference in speed arises from the fact that, although only a few crosslinks will make a polymer insoluble, a positive resist may require extensive fragmentation before it can be developed successfully. Good negative working resists should have crosslinking sites such as double bonds, epoxy groups, and possibly phenyl rings to delocalize and absorb the energy of the electron beam, thereby protecting the chain from scission. Some examples of useful systems

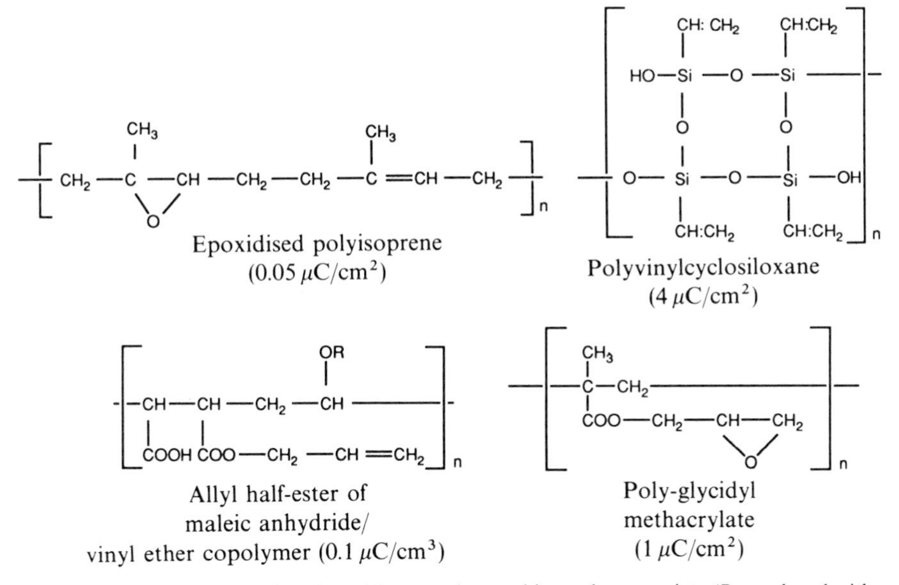

FIGURE 17.6. Some examples of sensitive negative working polymer resists. (Reproduced with permission from Chemistry and Industry (1985)).

TABLE 17.1. Sensitivity of some halogenated aromatic polymers

Polymer	*Sensitivity*
poly(styrene-*stat*-4-chlorostyrene)	6 $\mu C/cm^2$
poly(styrene-*stat*-4-chloromethylstyrene)	1 $\mu C/cm^2$
poly(3-bromo-9-vinylcarbazole)	2 $\mu C/cm^2$

are shown in figure 17.6. Halogenated aromatic polymers form another group of interest (table 17.1), where the introduction of a halogen atom increases the sensitivity of structures based on polystyrene.

17.7 X-ray and ion sensitive resists

One drawback in electron beam lithography stems from the fact that much of the interaction between the polymer and the electron beam occurs as a result of low energy secondary electrons being produced in the film. These are scattered beyond the definition of the exposing beam and can produce unwanted reactions in regions adjacent to the primary exposure. This is known as the "Proximity Effect" and can cause undercutting and overlapping of closely-spaced features. In X-ray and ion beam lithography the secondary electrons produced have lower energies and shorter path lengths, and consequently the proximity effect is less pronounced. These techniques may find greater use in the future. The electron micrograph of an exposed and developed X-ray resist is shown in figure 17.7.

FIGURE 17.7. Pattern cut in a polymer resist using X-ray lithography, showing the steep edges and resolution capable with this technique. (Reproduced with permission from Hoechst High Chem. Magazine (1989)).

17.8 Electroactive polymers

Inherently, organic polymers with all-carbon backbones are insulators and can be used as encapsulating materials when a medium of high resistivity is required, such as in coatings for cables and electrical wiring. It has been found that the resistivity can be decreased if a composite of the polymer with carbon black or finely-divided metal is fabricated, but the conduction in these cases takes place via the filler and not through the polymer which merely acts as a supporting matrix. Incorporation of a filler can also reduce the mechanical strength of the polymer.

In 1977 the first major breakthrough was achieved when it was discovered that polyacetylene, which is a very poor conductor in the pure state, could be turned into a highly conductive polymer by conversion to the salt on reacting it with I_2. The result was a dramatic increase of over 10^{10} in conductivity. As conduction appears to be due to movement of electrons through the polymer, this discovery has added an exciting new dimension to the rapidly expanding area of synthetic metals. Other polymers that display similar characteristics are usually polyconjugated structures which are insulators in the pure state but when treated with an oxidizing or a reducing agent can be converted into polymer salts with electrical conductivities comparable to metals. Some idea of the possible range of conductivities (σ) is given in figure 17.8, where σ varies from $10^{-18}\,\mathrm{S\,cm^{-1}}$ for a good polymeric insulator (*e.g.* poly tetrafluoroethylene) up to $\sigma \sim 10^{6}\,\mathrm{S\,cm^{-1}}$ for a metallic conductor, copper.

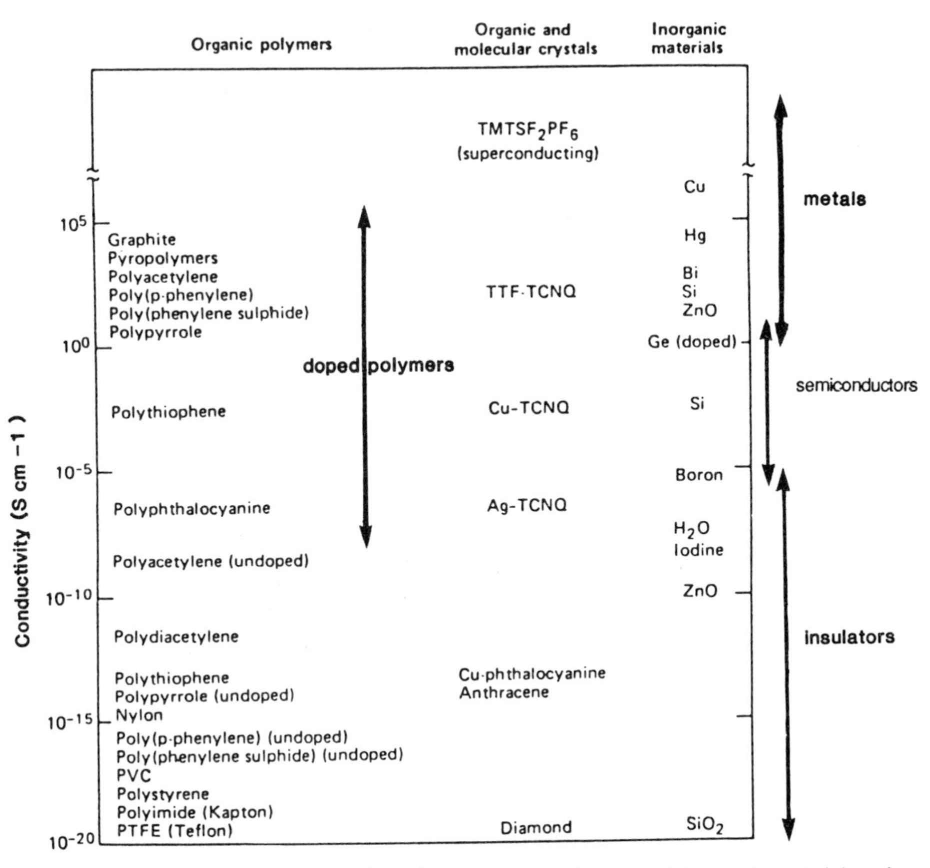

FIGURE 17.8. Conductivity ranges for polymers (doped and undoped), inorganic materials and molecular crystals.

17.9 Conduction mechanisms

Electrical conductivity is a function of the number of charge carriers of species 'i' (n_i), the charge on each carrier (ϵ_i), and carrier mobilities (μ_i) described by the relation $\sigma = \Sigma \mu_i . n_i . \epsilon_i$ where the units of conductivity are S cm^{-1}. Conduction in solids is usually explained in terms of the band theory which postulates that when atoms or molecules are aggregated in the solid state, the outer atomic orbitals containing the valence electrons are split into bonding and antibonding orbitals, and mix to form two series of closely-spaced energy levels. These are usually called the valence band and the conduction band respectively. If the valence band is only partly filled by the available electrons, or if the two bands overlap so that no energy gap exists between them, then application of a potential will raise some of the electrons into empty levels where they will be free to move throughout the solid thereby producing a current. This is the description of a conductor. If, on the other hand, the valence band is full and is separated from the empty conduction band by an energy gap, then there can be no net flow of electrons under the influence of an external field unless electrons are elevated into the empty band and this will require a considerable expenditure of energy. Such materials are either semiconductors or insulators, depending on how large the energy gap may be, and the majority of polymers are insulators. The band model then assumes that the electrons are delocalized and can extend over the lattice (see figure 17.9).

When we come to consider electronic conduction in polymers, band theory is not totally suitable because the atoms are covalently bonded to one another, forming polymeric chains that experience weak intermolecular interactions. Thus macroscopic conduction will require electron movement, not only along chains but also from one chain to another.

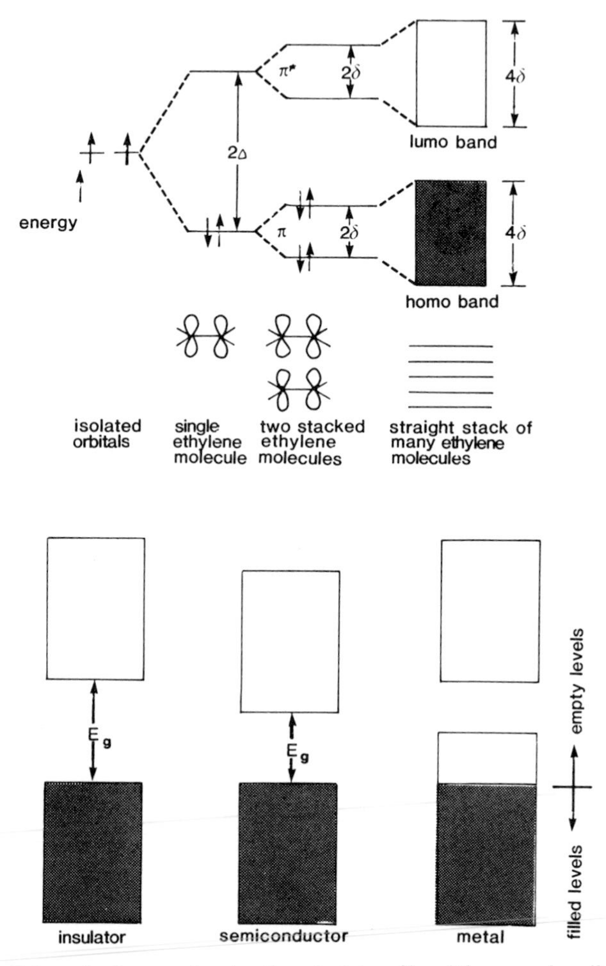

FIGURE 17.9. Schematic diagram showing the principles of band theory as described in the text. The dark regions represent the bands filled with electrons, and the light regions the bands that are available for conduction. The energy gap between filled and empty states is E_g. (Adapted from D. O. Cowan and F. M. Wiygul (1986) with permission. American Chemical Society, Washington, D.C.)

TABLE 17.2. Structures and conductivity of doped conjugated polymers. (Adapted from D. O. Cowan and F. M. Wiygul (1986) with permission. American Chemical Society, Washington, D.C.)

Polymer	*Structure*	*Typical methods of doping*	*Typical conductivity* $(S\,cm)^{-1}$
Polyacetylene	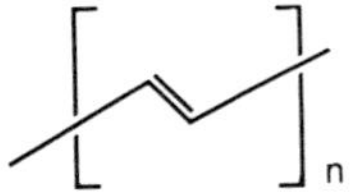	Electrochemical, chemical (AsF_5, I_2, Li, K)	$500–1.5 \times 10^5$
Polyphenylene	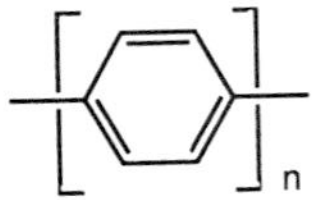	Chemical (AsF_5, Li, K)	500
Poly(phenylene sulphide)	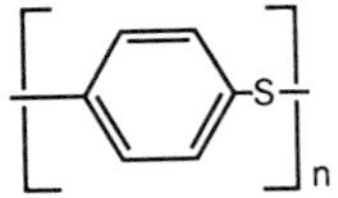	Chemical (AsF_5)	1
Polypyrrole	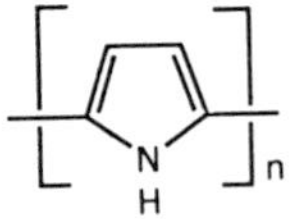	Electrochemical	600
Polythiophene	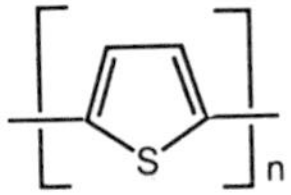	Electrochemical	100
Poly(phenyl-quinoline)	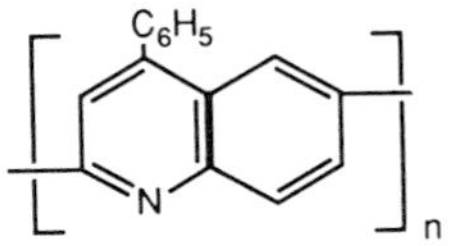	Electrochemical, chemical (sodium naphthalide)	50

17.10 Preparation of conductive polymers

Polymers have the electronic profiles of either insulators or semiconductors; thus the band gap in a fully saturated chain such as polyethylene is 5 eV and decreases to about 1.5 eV in the conjugated system polyacetylene. The respective intrinsic conductivities are $\sim 10^{-17}\,\mathrm{S\,cm^{-1}}$ and $\sim 10^{-8}\,\mathrm{S\,cm^{-1}}$, both very low. Conducting polymers can be prepared either by oxidizing or reducing the polymer using a suitable reagent (see table 17.2). The band theory model would explain the increased conductivity as either removal of electrons from the valence band by the oxidizing agent, leaving it with a positive charge, or donation of an electron to the empty conduction band by a reducing agent. These processes are called p-type doping and n-type doping respectively. This explanation is an over-simplification, as conductivity in polymers is associated with charge carriers that do not have free spins, rather than the expected unpaired electrons detected in metals, so a modified model must be developed. This will be explained when the individual conducting polymers are described.

While the addition of a donor or acceptor molecule to the polymer is called "doping", the reaction which takes place is actually a redox reaction and is unlike the doping of Si or Ge in semiconductor technology where there is substitution of an atom in the lattice. The terminology in common use will be retained here but it should be remembered that the doping of conductive polymers involves the formation of a polymer salt, and that this can be effected either by immersing the polymer in a solution of the reagent, or by electrochemical methods.

The reactions can be represented in the generalized case for oxidation by

$$P_n \underset{\text{Red}}{\overset{\text{Ox/A}^-}{\rightleftharpoons}} [P_n^{\cdot +}\,A^-] \underset{\text{Red}}{\overset{\text{Ox/A}^-}{\rightleftharpoons}} [P^{2+}\,2A^-]$$

where P_n represents a section of polymer chain. The first step is the formation of a cation (or anion) radical, which is called a soliton or a polaron – the distinction will be explained later. This step may then be followed by a second electron transfer with the formation of a dication (or dianion) known as a bipolaron. Alternatively, after the first redox reaction, charge transfer complexes may form between charged and neutral segments of the polymer when possible:

$$[P_n^{\cdot +}A^-] + P_m \longrightarrow [(P_nP_m)^{\cdot +}A^-]$$

These general principles can best be illustrated by examining specific examples, in particular polyacetylene which has been studied intensively.

17.11 Polyacetylene

Polyacetylene can be made by a number of synthetic routes but Ziegler catalysts figured prominently in the early work. Indeed, the synthesis of the material that subsequently opened the way to the discovery of conducting polyacetylene was a fortuitous accident in which acetylene gas was passed through a heptane solution of the Ziegler catalyst $Ti(OC_4H_9)_4/Al(C_2H_5)_3$ that was vastly in excess of the amounts normally used. The polyacetylene which formed at the gas-liquid interface was a lustrous flexible polycrystalline film, rather than the powder usually obtained. This has become known as "Shirakawa" polyacetylene and has a predominantly *cis* conformation when formed at temperatures of 195 K. On raising the temperature of the film, isomerization to the more stable *trans* form takes place. The polymer is infusible, insoluble, usually contaminated by catalyst residues, and tends to become brittle and dull when exposed to air due to slow oxidation. These features make it difficult to process or handle and attempts have been made to either improve the polymer or make derivatives or precursors that are soluble in organic solvents.

Many of these problems have been solved by Feast who developed a very elegant synthetic method, now commonly known as the Durham route. This is a two-stage process in which soluble precursor polymers are prepared by a metathesis ring-opening polymerization reaction and these are subsequently heated to produce polyacetylene by a thermal elimination reaction. An example of one method is given below.

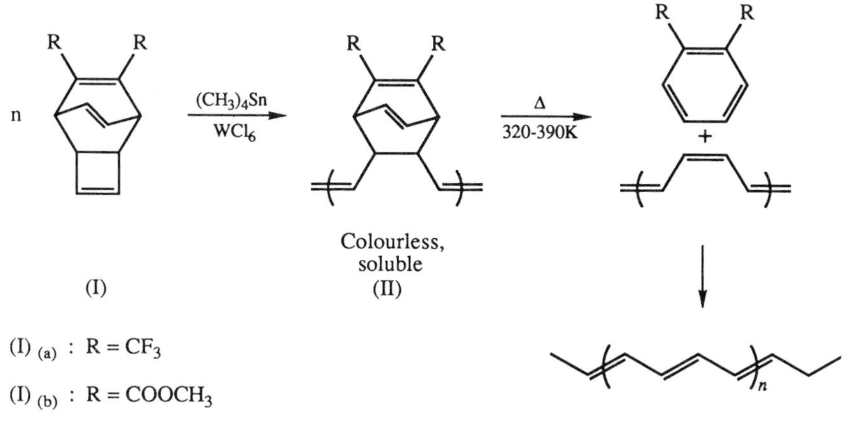

A refinement of the process involves photochemical conversion of I(a) into III

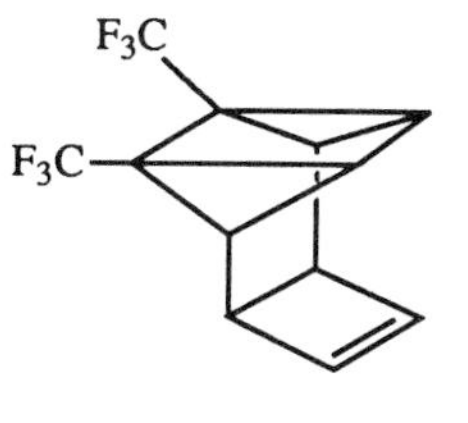

(III)

which on polymerization produces a precursor that is stable at room temperature and can be converted to *trans* polyacetylene on heating at 330 to 340 K. The advantages of the Durham route are (i) contaminating catalyst residues can be removed because the precursor polymers are soluble and can be purified by dissolution and precipitation, (ii) the precursors can be drawn and oriented or cast as films prior to conversion to the all-*trans* form of polyacetylene. This allows some degree of control over the morphology of the final product which in the pristine state appears to be fibrous and disordered. As conductivity can be maximized by alignment of the polymer chains, stretching the film will assist this process and this can best be accomplished using the prepolymer.

We can now examine the mechanism of conduction in polyacetylenes in relation to the structure and doping procedures. In a polyconjugated system the π orbitals are

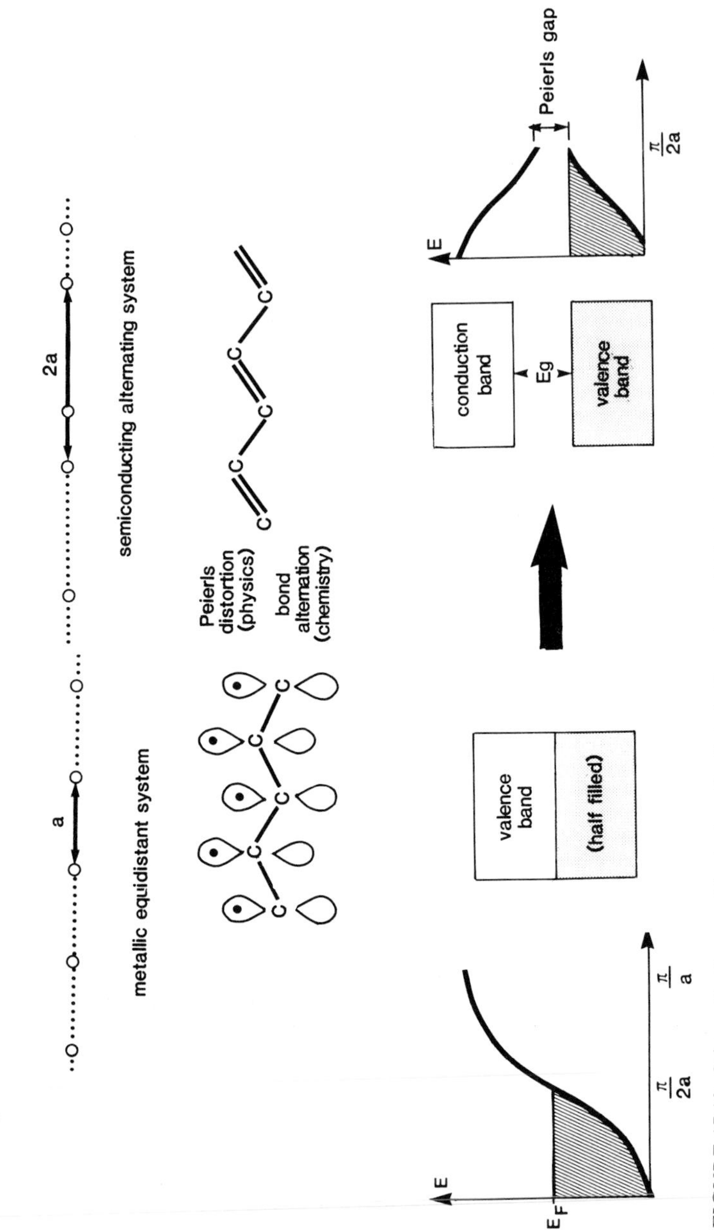

FIGURE 17.10. Schematic diagram illustrating the Peierls distortion which leads to formation of an energy gap and production of a semiconductor rather than a conductor. (Adapted with permission from M. J. Bowden and S. R. Turner, Eds, *Electronic and Photonic Applications of Polymers.* American Chemical Society, Washington, D.C. (1988).)

assumed to overlap, and form a valence and a conduction band as predicted by band theory. If all the bond lengths were equal, *i.e.* delocalization led to each bond having equal partial double bond character, then the bands would overlap and the polymer would behave like a quasi- one-dimensional metal having good conductive properties. Experimental evidence does not substantiate this and reference to the physics of a monatomic one-dimensional metal, with a half-filled conduction band, has shown that this is an unstable system and will undergo lattice distortion by alternative compression and extension of the chain. This leads to alternating atom pairs with long and short interatomic distances found along the chain. The effect is embodied in the Peierls theorem which states that a one-dimensional metal will be unstable and that an energy gap will form at the Fermi level because of this lattice distortion so that the material becomes an insulator or a semiconductor. This break in the continuity of the energy bands is caused by the use of elastic energy during lattice distortion which is compensated by a lowering of the electronic energy and formation of a band gap (see figure 17.10). The analogy with polyacetylene then becomes obvious and it is found that single and double bond alternation persists in the chain leading to an energy gap between the valence and conduction bands. The *trans* structure of polyacetylene is also unique as it has a two-fold degenerate ground state in which sections A and B are mirror images

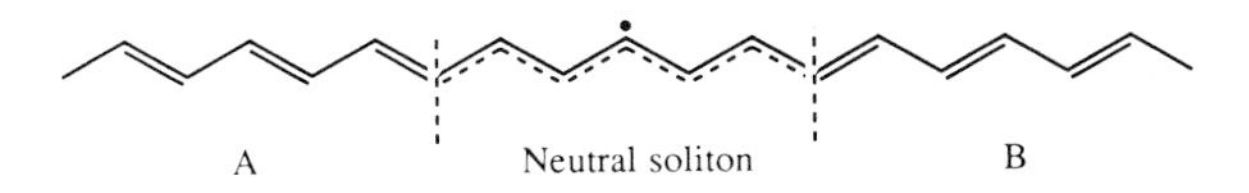

and the single and double bonds can be interchanged without changing the energy. Thus if the *cis* structure begins to isomerize to the *trans* geometry from different locations in a single chain, an A sequence may form and eventually meet a B sequence, as shown, but in doing so, a free radical is produced. This is a relatively stable entity and the resulting defect in the chain is called a neutral *soliton*, which corresponds, in simple terms, to a break in the pattern of bond alternation, *i.e.* it separates the degenerate ground state structures. The electron has an unpaired spin and is located in a non-bonding state in the energy gap, midway between the two bands. It is the presence of these neutral solitons which gives *trans* polyacetylene the characteristics of a semiconductor with an intrinsic conductivity of about 10^{-7} to 10^{-8} S cm^{-1}.

The conductivity can be magnified by doping. Exposure of the film to dry ammonia gas leads to a dramatic increase to $\sigma \sim 10^3$ S cm^{-1}. Controlled addition of an acceptor or p-doping agent such as AsF_5, Br_2, I_2, or $HClO_4$, removes an electron and creates a positive soliton (or a neutral one if the electron removed is not the free electron). In

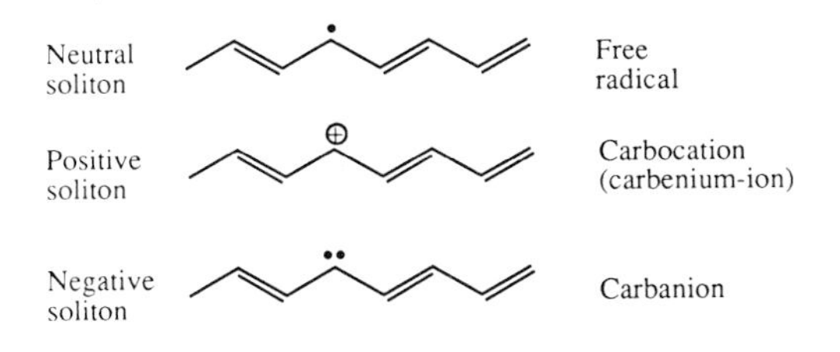

chemical terms this is the same as forming a carbenium ion that is stabilized by having the charge spread over several monomer units. Similarly a negative soliton can be formed by treating the polymer with a donor or n-doping agent that adds an electron to the mid-gap energy level. This can be done by dipping the film in the THF solution of an alkali metal naphthalide or by an electrochemical method.

At high doping levels the soliton regions tend to overlap and create new mid-gap energy bands that may merge with the valence and conduction bands allowing freedom for extensive electron flow. Thus, in polyacetylene the charged solitons are responsible for making the polymer a conductor.

17.12 Poly(*p*-phenylene)

The poly(*p*-phenylene) structure has all the characteristics required of a potential polymer conductor, but it has proved difficult to synthesize high molecular weight material. One method is the polycondensation

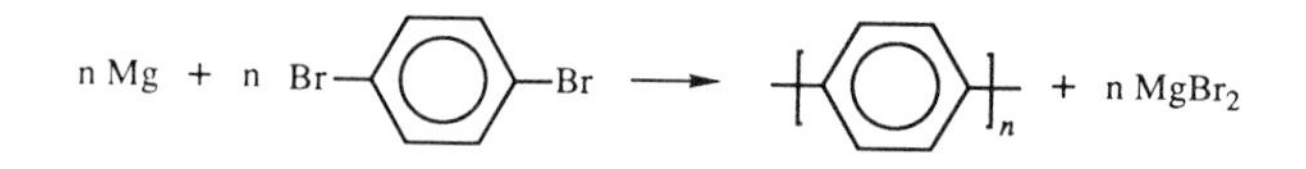

but this only yields oligomeric material and even this is insoluble. A novel route developed by workers at ICI has solved these problems by again making use of a tractable intermediate polymer. Radical polymerization of 5,6-dihydroxycyclohexa-1,3-diene IV leads to a soluble precursor polymer that can be processed prior to the final thermal conversion into poly(*p*-phenylene).

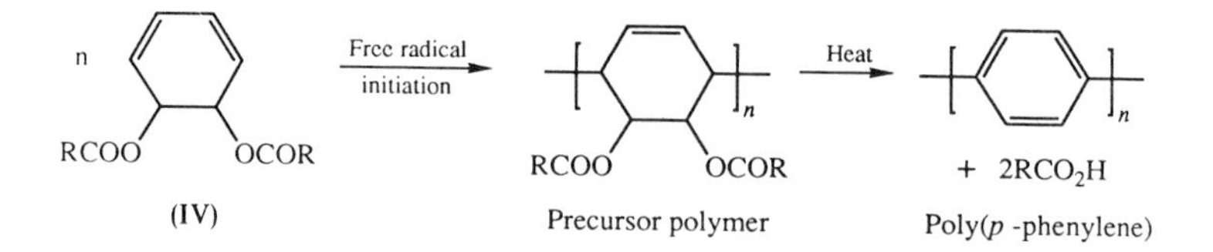

The material is an insulator in the pure state but can be both n- and p-doped using methods similar to those used for polyacetylene. However, as poly(*p*-phenylene) has a higher ionization potential it is more stable to oxidation and requires strong p-dopants. It responds well to AsF_5, with which it can achieve conductivity levels of about $10^2\ S\ cm^{-1}$. In contrast, Br_2 and I_2 are ineffective. An interesting variation in the preparation of conducting poly(*p*-phenylene) involves a one-step reaction where crystalline *p*-phenylene oligomers (*e.g.* *p*-terphenylene) are exposed to AsF_5 vapour. These assume a metallic blue lustre and can then be polymerized to yield a highly conducting polymer. Poly(*p*-phenylene) is very stable and can withstand temperatures up to 720 K in air without degrading.

Examination of the structure shows that the soliton defect cannot be supported in poly(*p*-phenylene) as there is no degenerate ground state. Instead the two nearly-equivalent structures are the benzenoid and quininoid forms which have different energies.

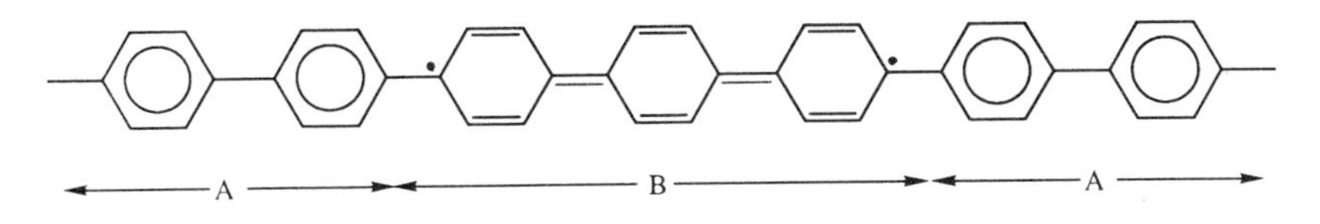

The benzenoid sections A have a lower energy than the quininoid section B which must be limited by the benzenoid structures and so the band gap of 3.5 eV is higher than that in polyacetylene.

In the band theory model it is assumed that conduction occurs because the mean free path of a charge carrier extends over a large number of lattice sites and that the

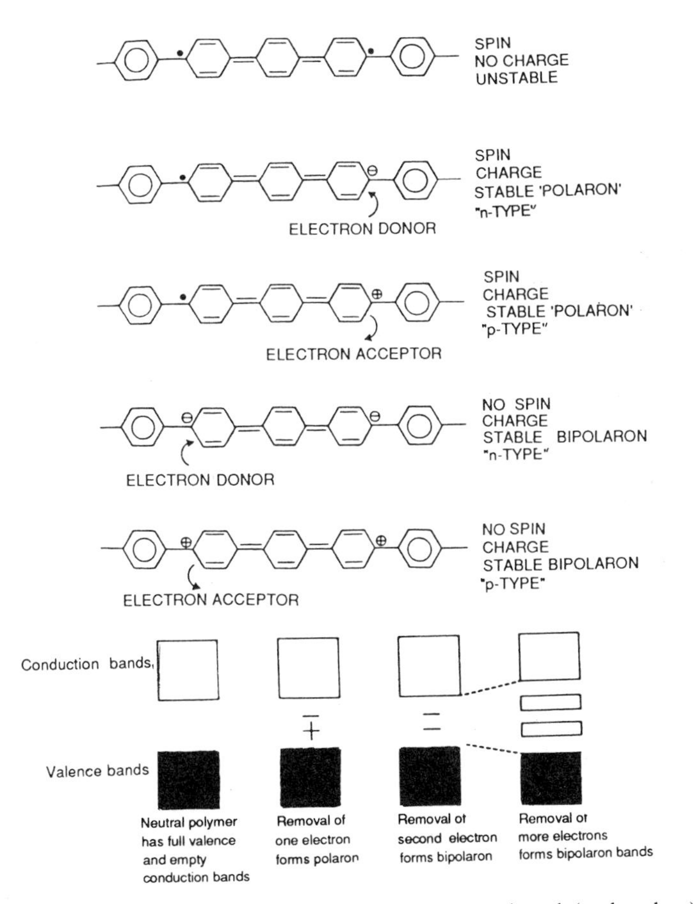

FIGURE 17.11. Illustration of polaron and bipolaron structures in poly(*p*-phenylene) and the proposed band structure for the oxidized (p-type) polymer. (Partially adapted from D. O. Cowan and F. M. Wiygul (1986) with permission. American Chemical Society, Washington, D.C.)

residence time on any one site is small compared with the time it would take for a carrier to become localized. If, however, a carrier is trapped it tends to polarize the local environment which relaxes into a new equilibrium position. This deformed section of the lattice and the charge carrier then form a species called a *polaron*. Unlike the soliton, the polaron cannot move without first overcoming an energy barrier so movement is by a hopping motion. In poly(*p*-phenylene) the solitons are trapped by the changes in polymer structure because of the differences in energy and so a polaron is created which is an isolated charge carrier. A pair of these charges is called a *bipolaron* and, on doping, the chemical equivalents are the radical ion and di-ion respectively. In poly(*p*-phenylene) and most other polyconjugated conducting polymers, the conduction occurs via the polaron or bipolaron (see figure 17.11).

17.13 Polyheterocyclic systems

Several useful polymeric structures based on the repeat unit

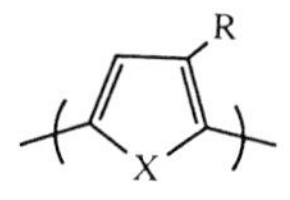

have been studied where R = H, alkyl, *etc*. and X = NH, S.

POLYPYRROLE

The polymerization of pyrrole can be carried out electrochemically by anodic oxidation, during which process simultaneous doping occurs. Typically, electrolysis of a solution of pyrrole (0.06 M) and $(Et)_4N^+BF_4^-$ (0.1 M) in acetonitrile containing 1 per cent water leads to deposition of an insoluble blue-black film of the polymer at the anode. The film contains BF_4^-, has a conductivity of about $10^2\,S\,cm^{-1}$ and the composition shown below.

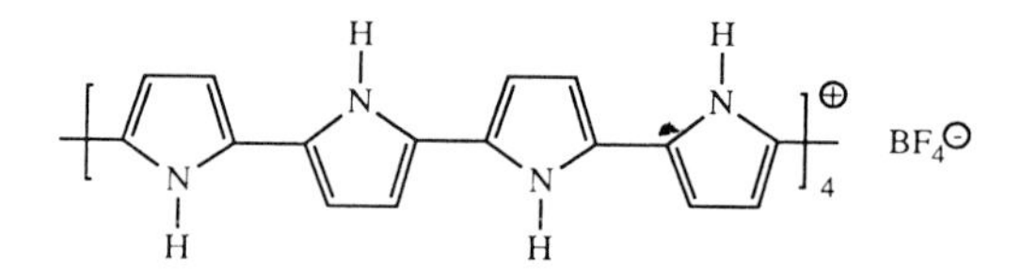

The material is amorphous but is insoluble in organic solvents. It has good stability in air with the conductive properties retained up to about 570 K, with no significant loss in conductivity over prolonged periods. Copper-bronze films of poly(pyrrole perchlorate), with conductivities of $\sim 40\,S\,cm^{-1}$, have been prepared electrochemically under dry box and oxygen-free conditions using $AgClO_4$ in acetonitrile. The stoichiometry of the films was

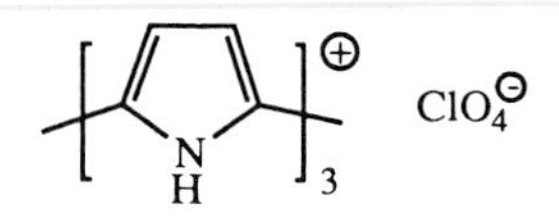

Yellow-green films of neutral polypyrrole can be prepared by the electrochemical reduction of the perchlorate films. This is an insulator with $\sigma \sim 10^{-10}\,S\,cm^{-1}$.

The neutral polymer can be reoxidized by exposure either to air (when the films turn black in about 15 minutes) or to Br_2, I_2, and $FeCl_3$ vapour. Immersion of the polymer in metal salt solutions of Ag^+, Cu^{2+} or Fe^{3+} also makes the polymer conductive.

The N-substituted derivatives of pyrrole produce much poorer conductors and bulky substituents tend to produce powders rather than films.

SULPHUR COMPOUNDS

Polyheterocyclics containing sulphur are also of interest. Poly(2,5-thienylene) can be prepared as a light green powder by electrochemical oxidative coupling. In the neutral state this compound has $\sigma = 10^{-11}\,S\,cm^{-1}$ but on exposure to I_2 the conductivity is raised to $\sigma \sim 10^{-1}\,S\,cm^{-1}$. Of particular interest in this group is poly(isothianaphthalene)

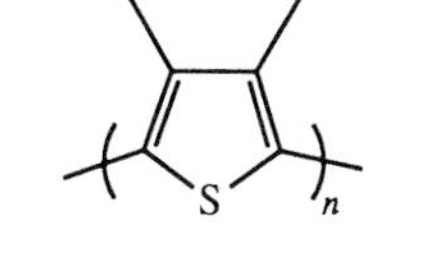

which has an unusual combination of properties. It is a good conductor when doped with a very narrow band gap (1 eV) but it also forms a transparent film.

17.14 Polyaniline

Reaction of aniline with ammonium persulphate in aqueous HCl produces polyaniline as a dark blue powder with a conductivity of $5\,S\,cm^{-1}$. The structure of the conducting form of the polymer is believed to be the di-iminium salt.

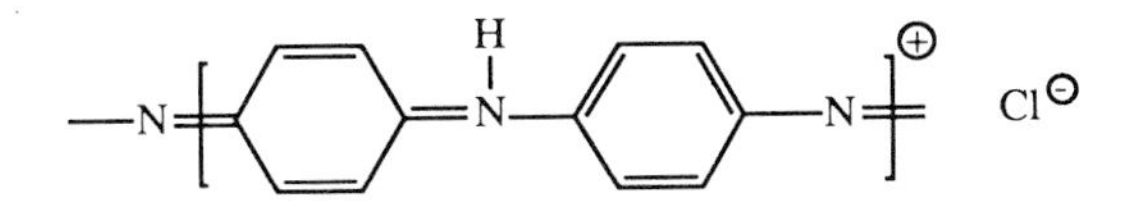

Electrochemical oxidation of aniline in aqueous HBF_4 produces a clear, dark green cohesive film on a platinum foil anode. Reduction with methanolic alkali solution produces the neutral polymer which is an insulator with $\sigma \sim 10^{-11}\,S\,cm^{-1}$.

17.15 Poly(phenylene sulphide)

While some conducting polymers can be processed by synthesis via a soluble precursor, poly(phenylene sulphide) is soluble or can be melt-processed. Doping with AsF_5 (which can be accelerated by the presence of AsF_3) forms a conductive polymer salt with $\sigma \sim 1\,S\,cm^{-1}$, but also tends to make the film brittle.

17.16 Poly(1,6-heptadiyne)

Green-gold lustrous films of poly(1,6-heptadiyne)

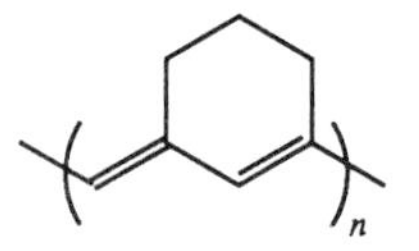

can be prepared by the cyclopolymerization of 1,6-heptadiyne using a Ziegler catalyst. The polymer is amorphous and has the structural repeat unit shown, but doping only achieves $\sigma \sim 10^{-1}$ S cm^{-1} and the films are quite unstable.

17.17 Applications

Doped polyacetylene can act as an electrode and can be used in a rechargeable battery. Metal electrodes are continually subjected to dissolution and redeposition during the charge-discharge cycles, and this results in mechanical wear. This makes the use of polymer electrodes attractive since the ions can enter or leave without significant disturbance of the polymer structure. Although polyaniline electrodes have also been developed, polypyrrole-salt, films are the most promising for practical application. This is due both to their stability, and to their self-supporting characteristics, which are particularly useful in the design of flat, space-saving cells. The use of these films also extends to flexible conductor tracks for contact bridges in switches, and to electrochromic displays for optical memories.

The feasibility of all-plastic batteries has been demonstrated by construction of such a battery from two layers of polyacetylene salt, sandwiching a film of polycarbonate impregnated with $LiClO_4$. All-plastic batteries represent a combination of electrical conductivity with the lightweight corrosion resistant properties of many plastics – an attractive prospect.

As well as battery application, conducting polymers may find uses as electromagnetic shielding since they tend to absorb low frequency radiation, or as parts of solar cells and semiconductors. Their use as heating elements in thin-wall coverings, and in wire and cable applications are also being investigated.

17.18 Photonic applications

Devices designed to transmit information by means of photons now incorporate polymeric materials with the appropriate structures. Passive applications include the coating of optical fibres by UV curable epoxy acrylates, thermal curing silicones or heat shrinkable polyethylene, to protect them from mechanical wear. Polymers have also been used in the manufacture of wave guides. Suitable polymers can now be synthesized having active non-linear optical properties which depend on the electronic excitation of the π-electron system.

17.19 Non-linear optics

The application of polymers as active components can be achieved when they exhibit non-linear optical properties and can participate as an integral part of a device. Non-linear optical properties in a polymer depend on the electronic excitation of a π-electron system that results in the alteration of the phase, frequency or amplitude of the incident radiation to give a new electromagnetic radiation field. Thus if a local electric field E is applied to a molecule, the induced polarization P (a scalar quantity) is expressed as

$$P = \alpha E + \beta E^2 + \gamma E^3 + \dots \tag{17.3}$$

where the tensor quantities are, α the linear polarizability, and β and γ, the second and third order non-linear electronic susceptibilities. The latter non-linear quantities are small but can be detected when intense laser sources are used. A comparable expression for the macroscopic non-linear effects for an assembly of molecules can be written as

$$P = \chi_{IJ}^{(1)} E_J + \chi_{IJK}^{(2)} E_J E_K + \chi_{IJKL}^{(3)} E_J E_K E_L + \dots \tag{17.4}$$

where χ now refers to the properties of the ensemble and the terms E_J *etc.* are components of the electric field strength. Here $\chi_{IJ}^{(1)}$ is related to the refractive index of the medium in linear optics, and the terms $\chi_{IJK}^{(2)} E_K$, and $\chi_{IJKL}^{(3)} E_K E_L$ have the same dimensions. Thus materials with non-zero values of $\chi^{(2)}$ or $\chi^{(3)}$ undergo a change in refractive index when placed in an electric or an optical field.

Application of an electric field polarizes the molecules in the medium and these act as scattering centres for the radiation. If the medium has an asymmetric response to the applied field, then effects such as second harmonic generation, where the frequency of the incident radiation is doubled, or a linear electro-optic (Pockels) effect may be observed. A paramagnetic amplification may also occur where irradiation by two incident fields of frequency ν_a and ν_b results in mixing, with amplification of the weaker wave. These effects arise when the material is $\chi^{(2)}$ active, and for a molecule to exhibit these properties it should (a) have an extended conjugated π-electron system, (b) possess an electron-donating and an electron-accepting group to promote intramolecular charge transfer and (c) crystallize in a non-centrosymmetric fashion to ensure polarization of the molecule in the crystal. A schematic representation would be

where there is a change in the dipole moment between the ground and the excited states. The acceptor A and the donor D sites then provide a push-pull action to distort the electron density and polarize the molecule. Molecules which can form polar crystals with high values of $\chi^{(2)}$ have the following structures:

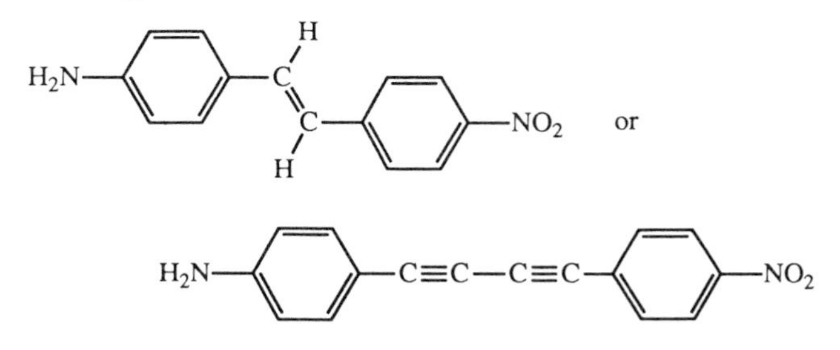

Thus for a polymer to exhibit second order, non-linear optical effects it must satisfy these conditions or alternatively have the asymmetry induced by electric field poling.

In the former case, polydiacetylenes offer an attractive possibility; they can be prepared by solid state polymerization of the crystalline monomers to form polymer crystals, and some derivatives have been found to have quite large $\chi^{(2)}$ values. Provided the packing distances of the monomers in the crystal are suitable, polymerization can be initiated thermally, or by UV or γ radiation, to form polydiacetylenes by 1,4 addition.

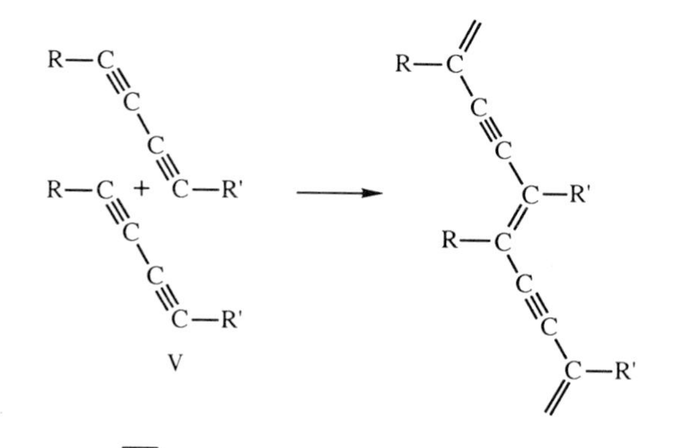

R = CH_2OSO_2–C$_6$H$_4$–CH_3 ; -$(CH_2)_n$-$OCONHCH_2COOC_4H_9$; -$Si(CH_3)_3$; H

If a monomer such as V is used where $R \neq R'$ then the polymer crystals obtained may have second order, non-linear optical properties.

The other route which can be used to prepare polymer films with large $\chi^{(2)}$ is by doping them with molecules possessing large β values. The molecules are aligned in the film by applying an external d.c. field at temperatures above the T_g of the polymer, then quenching this into the glassy state while maintaining the field. These molecularly-doped, poled polymers can produce $\chi^{(2)}$ values of between 10^{-6} and 10^{-8} esu, which is comparable to $LiNbO_3$ or GaAs crystals. Examples of this approach which have been reported are the doping of PMMA with the azo dye Disperse Red I, (VI)

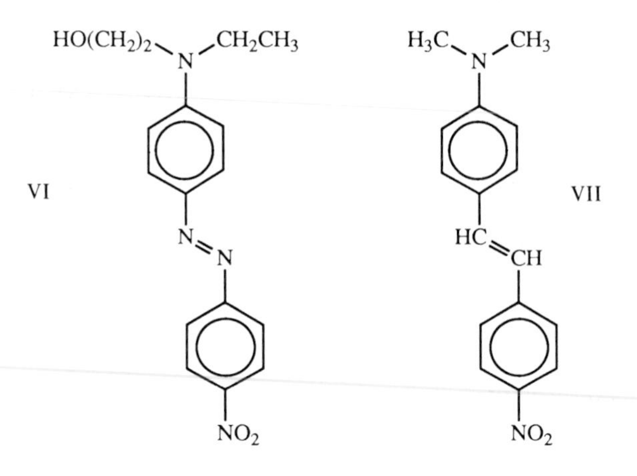

and the dispersion of 4,4′-N,N-dimethyl amino-nitrostilbene (VII) in a liquid crystalline copolymer VIII followed by poling in the nematic liquid crystalline phase prior to quenching.

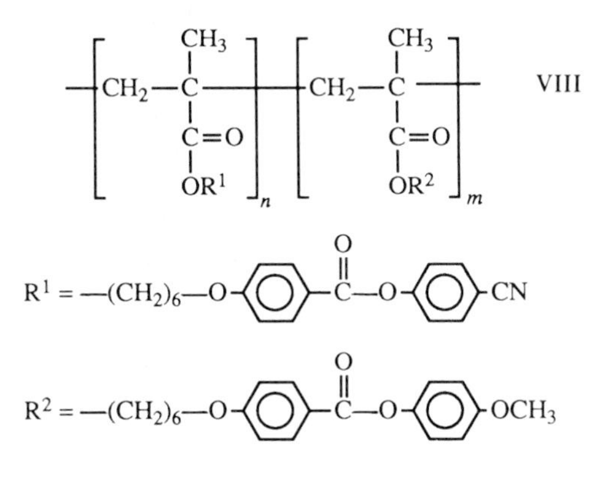

In the latter system there appears to be competition between alignment and thermal motion, so best results were obtained when the poling was carried out close to the T_g $\sim 25°$ C rather than at higher temperatures; the nematic to isotropic transition was $T_i = 100°$ C. More recent attempts have been made to improve the poled systems by incorporating the non-linear, optically-active molecules into the polymer chain structure and comb-branch-liquid crystalline polymers with R′ and the new group

$$R^3 = —(CH_2)_6—O—C_6H_4—X—C_6H_4—NO_2$$

$$X = —\overset{O}{\overset{\|}{C}}—O— \quad \text{or} \quad —CH{=}CH—$$

This provides an all polymeric material for subsequent poling.

While some of the polydiacetylenes show second order, non-linear optical properties, those with R = R′, such as poly[bis(*p*-toluene sulphonate)diacetylene], have a centrosymmetric crystal habit and $\chi^{(2)}$ is zero. They do have finite $\chi^{(1)}$ and $\chi^{(3)}$ values, in common with materials such as polyacetylene, polypyrrole and other conjugated polymers. Important non-linear optical properties occur when a large $\chi^{(3)}$ term is observed, which includes the quadratic electro-optic (Kerr) effect, frequency tripling, optical phase conjugation and optical bistability resulting from changes in the refractive index of the medium. These may prove useful in the development of photonic switches. The values of $\chi^{(3)}$ are found to improve when there is good alignment of the polymer chains in the crystal lattice and this is enhanced in the direction of chain orientation. It should be noted that there are no symmetry constraints on a molecule which can exhibit $\chi^{(1)}$ or $\chi^{(3)}$ properties.

17.20 Langmuir-Blodgett films

The preparation of ordered thin films can be of considerable interest in the construction of electronic devices and in the study of model membrane systems. One method that has gained in popularity is the Langmuir-Blodgett (LB) technique in which molecules with a hydrophilic head and a hydrophobic tail can form a monolayer at an air-water interface, then be transferred onto a solid surface. This latter process can be achieved either by dipping a glass slide (or some other substrate) vertically into the trough containing the monolayer on the surface, as shown schematically in figure 17.12, or alternatively by using a rotating substrate to give a horizontal transfer method. Either way can be used to build up monolayers after one dipping, or multilayers by repeated passage of the substrate into the film. Polymeric films can be prepared by selecting a molecule with a polymerizable unit (double or triple bonds) that can be converted, after film formation, into a polymeric structure by thermal treatment or exposure of the film to UV or γ radiation. Several possible monomer structures are shown schematically in figure 17.13.

Polydiacetylenes with appropriate amphiphilic structures can be used to fabricate thin films in this way, *e.g.* heptadeca-4,6-diyne-1-ol and the corresponding acid. It is often found that the stability of the monolayers is a function of the structure of the molecules. It has also been found advantageous to use the neutralized form of acid derivatives, and diacetylene monocarboxylic acids generally form much more stable films if the cadmium salt is used initially where the Cd^{2+} ion is then present in the aqueous phase as a counter-ion.

Films prepared by the LB method have found application in non-linear optics but they have also found applications in nanolithography. The miniaturization of integrated circuits requires high resolution and electron beams have been used, as noted

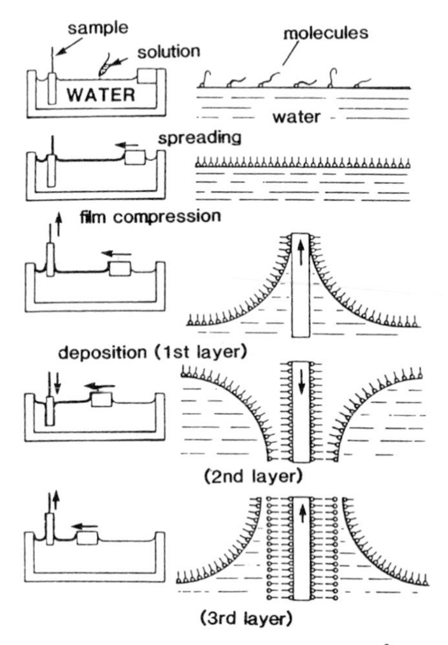

FIGURE 17.12. Schematic representation of the formation of mono-, bi-, and tri-layers of molecules from a Langmuir-Blodgett trough. (Reproduced from Barraud (1987) with permission from Academic Press).

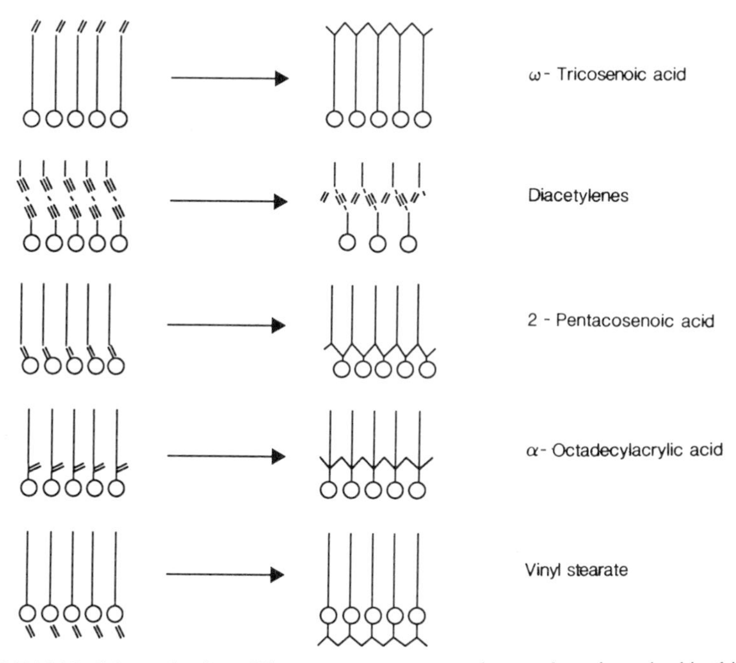

FIGURE 17.13. Schematic of possible monomer structures that can be polymerized in thin films formed using the Langmuir-Blodgett method.

in section 17.6, for this purpose. Some limitations have been experienced arising from the unwanted exposure obtained from the scattering of secondary electrons, which reduces the pattern definition on the resist. This can be improved by using much thinner resist films, and hence shorter exposure times are required. Conventional spin coating techniques do not always guarantee that the resist film will be free from unacceptable defects such as pinholes that can spoil the subsequent pattern, and thin LB films can be prepared that are superior in this respect. Improved resolution has been obtained from resists prepared using polymerized ultrathin (45 nm) LB films of ω-tricosenoic acid, $CH_2{=}CH(CH_2)_{20}COOH$, and α-octadecyl acrylic acid.

The technique in general shows great promise in the area of molecular electronics where precise control of the molecular structure is paramount.

17.21 Optical information storage

Polymeric materials can be used for optical information storage and some are ideally suited for the manufacture of optical video or digital audio discs. The information is normally transferred to the polymer using a monochromatic laser by one of four possible methods:

(a) ablative – "hole burning"
(b) bubble formation
(c) texture change
(d) bilayer alloying.

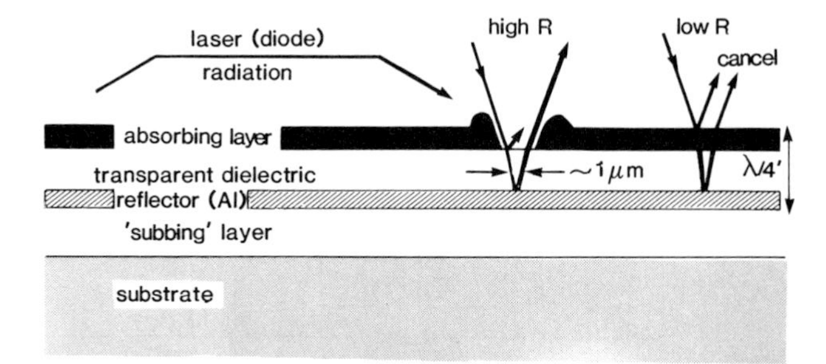

FIGURE 17.14. Tri-layer structure of a typical ablative-mode optical disc. (Reproduced with permission from Chemistry and Industry (1985)).

Technique (a) is commonly employed and involves the creation, on the polymer surface, of a series of small pits which have different lengths and frequency of spacing. The information can be retrieved by measuring the intensity and modulation of light reflected from the pattern of pits on the disc surface.

The discs themselves must be fabricated from materials which have the following characteristics:

(a) dimensional stability
(b) isotropic expansion
(c) optical clarity
(d) low birefringence.

In addition the surface should be free of contaminating particles and occlusions that would interfere with the information retrieval process. A typical ablative-mode optical disc has the structure shown in figure 17.14. The substrate is an optically-transparent material such as polycarbonate, poly(methyl methacrylate) (PMMA), poly(ethylene terephthalate) or poly(vinyl chloride), topped by a "subbing" layer to provide an optically-smooth surface for the recording layer. A metal reflector (typically aluminium) is then incorporated next to a transparent dielectric medium such as poly(α-methyl styrene) and finally the absorbing layer where the information pits are created is added. The latter can be a metal/polymer composite (silver particles in a gel) or a dye molecule dispersed in a polymer matrix, such as squaryllium dyes which act as infrared absorbers for GaAs lasers, typically

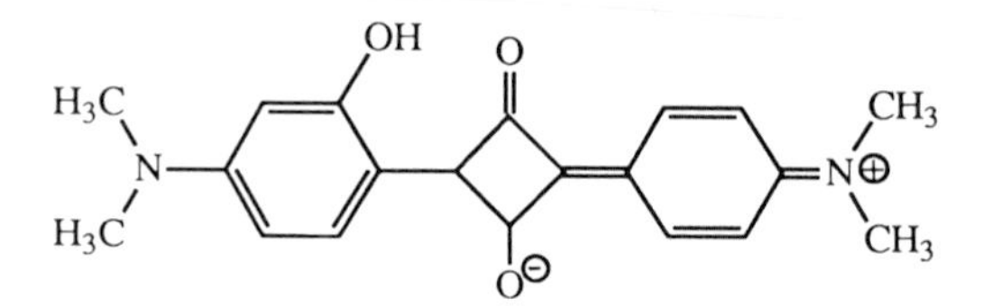

The absorbing layer is then protected by a transparent overcoating of crosslinked poly(dimethyl siloxane). This will produce a "direct read after write" or DRAW disc that is non-erasable. A compact disc profile made in a similar way, and where the information can be read out in digital form, is shown in figure 17.15.

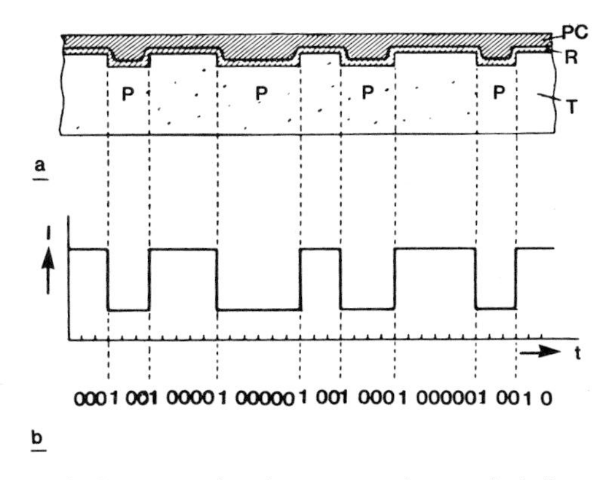

FIGURE 17.15. (a) Typical cross-sectional structure of an optical disc, and (b) the intensity profile of the read-out system in binary code as a function of time (t). Here T is the transparent polymer support; R is the reflective metal layer; PC is the protective coating and P represents a depression created for information storage. (Reproduced with permission from Hüthig and Wepf Verlag).

Polymers can also be used to manufacture lenses and screens for projection television systems. These are most conveniently made from PMMA, or combinations of glass and PMMA to counteract the high thermal expansion of the polymer. The use of ultraviolet curable coatings for lens replication and protective layers is widespread, and these systems are based on diacrylate or dimethacrylate monomers mixed with photoinitiators such as

$$CH_2{=}\overset{\overset{\displaystyle R_1}{|}}{C}-\underset{\underset{\displaystyle O}{\|}}{C}-O-R_2-O-\underset{\underset{\displaystyle O}{\|}}{C}-\overset{\overset{\displaystyle R_1}{|}}{C}{=}CH_2$$

acetophenone or benzilketals, where $R_1 = CH_3$, H and R_2 can often be

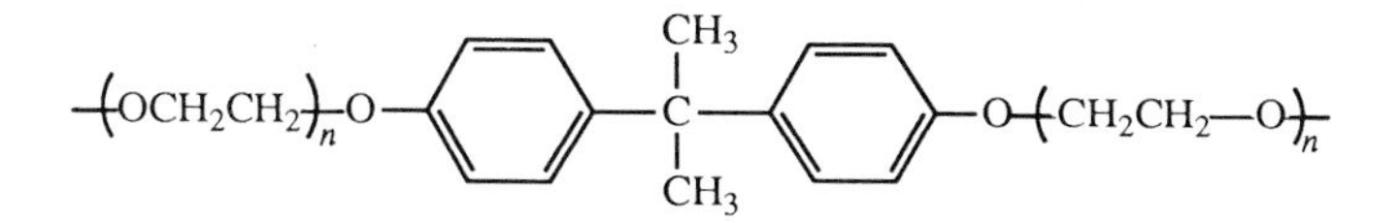

which tends to have low shrinkage after cure while retaining good optical and thermal properties.

17.22 Thermorecording on liquid crystalline polymers

While no systems are yet commercially available, the principles of using side chain liquid crystalline polymers as optical storage systems has been established. This has been demonstrated using a polymer film prepared from a side chain polymer, showing nematic liquid crystalline characteristics, with the structure

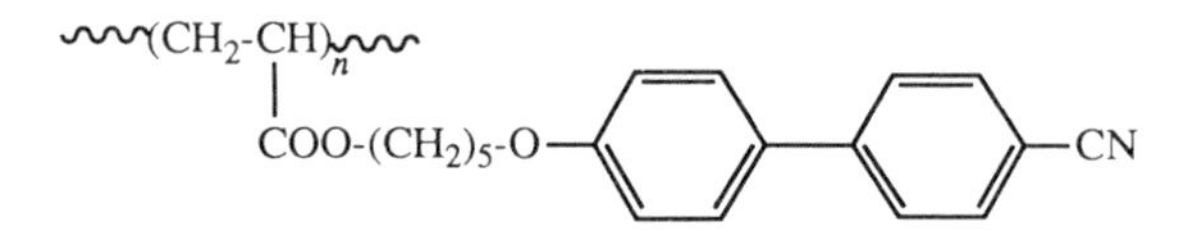

The mesogenic side groups are first oriented by application of an electric field to the polymer above the glass transition temperature, such that homeotropic alignment is obtained. On cooling below the T_g, the alignment is locked into the glassy phase, and a transparent film which will remain stable on removal of the electric field is produced. If this film is now exposed to a laser beam, localized heating occurs at the point where the beam impinges on the film and the material passes into the isotropic melt state. This results in the local loss of the homeotropic orientation and, on cooling, an unoriented region with a polydomain texture, which scatters light and produces a non-transparent spot, forms in the film. Information can then be "written" onto the film, and can subsequently be erased by raising the temperature of the whole film, to regain the isotropic, disordered, melt state. The system is illustrated in figure 17.16.

Polymers are superior to low, molar mass liquid crystalline molecules in this respect as they can retain the orientation longer when cooled into the glassy state before the electric field is switched off, whereas the low molar mass materials lose orientation rapidly.

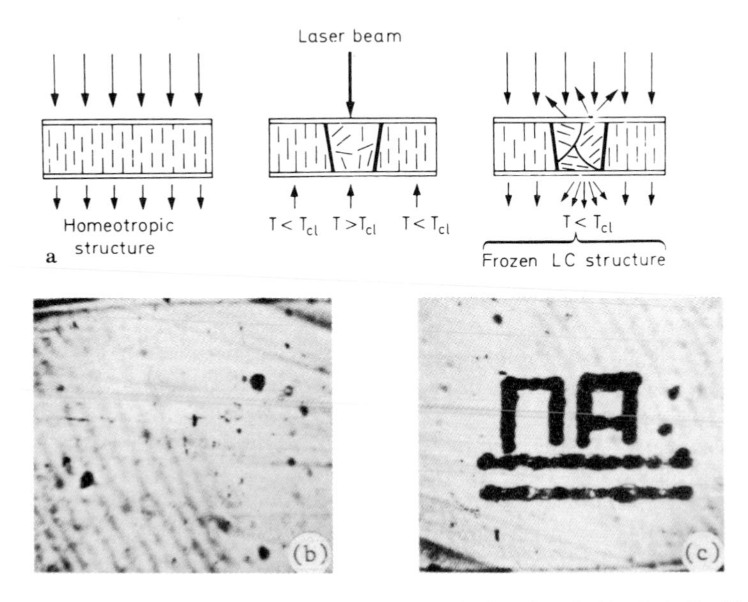

FIGURE 17.16. Thermal recording using a homeotropically-aligned side chain liquid crystal formed as a glassy film with the liquid crystalline state frozen into the glass. A laser beam is used to address the film (b) by producing local heating and disorder which is subsequently frozen in by cooling below T_g (c). (Reproduced from N. A. Platé and V. P. Shibaev (1987) with permission of Plenum Publishers and the authors.)

General Reading

M. J. Bowden and S. R. Turner, Eds, *Electronic and Photonic Applications of Polymers.* American Chemical Society, Washington, D.C. (1988).

T. Davidson, Ed., *Polymers in Electronics.* ACS Symposium, Series 242. American Chemical Society (1983).

M. T. Goosey, Ed., *Plastics for Electronics.* Elsevier Applied Science Publishers (1985).

H. Kuzmany, M. Mehring and S. Roth, *Electronic Properties of Polymers and Related Compounds.* Springer-Verlag (1985).

N. A. Platé and V. P. Shibaev, *Comb-Shaped Polymers and Liquid Crystals.* Plenum Press (1987).

D. A. Seanor, Ed., *Electrical Properties of Polymers.* Academic Press (1982).

T. A. Skotheim, Ed., *Handbook of Conducting Polymers.* Vols I and II, Marcel Dekker Inc. (1986).

L. F. Thompson, C. G. Willson and M. J. Bowden, Eds, *Introduction to Microlithography.* ACS Sympsoium, Series 219, American Chemical Society (1983).

Barraud, A. (1987). In *Non Linear Optical Properties of Organic Molecules and Crystals,* Eds Chemla, D. S. and Zyss, J., Vol. 1, p. 359. Academic Press, No. 4.

L. F. Thompson, C. G. Willson and J. M. T. Frechet, Eds, *Materials for Microlithography.* ACS Symposium, Series 266. American Chemical Society (1984).

References

1. D. G. H. Ballard, A. Courtis, I. M. Shirley and S. C. Taylor, "A biotech route to poly(phenylene)". *J. Chem. Soc. Chem. Commun.,* 954 (1983).
2. M. G. Clark, "Materials for optical storage." *Chem. Ind.,* 258 (1985).
3. D. O. Cowan and F. M. Wiygul, "The organic solid state." *Chem. Eng. News,* 28 (1986).
4. S. Etemad, A. J. Heegor and A. G. MacDiarmid, "Polyacetylene." *Ann. Rev. Phys. Chem.,* **33**, 443 (1982).
5. W. J. Feast, "Synthesis and properties of some conjugated potentially conductive polymers." *Chem. Ind.,* 263 (1985).
6. A. F. Garito and K. Y. Wong, "Non linear optical processes in organic and polymer structures," *Poly. J.,* **19**, 51 (1987).
7. R. G. Gossink, "Polymers for audio and video equipment." *Angew. Makromol. Chem.* **145/146**, 365 (1986).
8. R. S. Potember, R. C. Hoffman, H. S. Hu, J. E. Cocchiaro, C. A. Viands and T. O. Poehler, "Electronic devices from conducting organics and polymers." *Poly. J.,* **19**, 147 (1987).
9. E. D. Roberts, "Resists used in lithography." *Chem. Ind.,* 251 (1985).
10. G. G. Roberts, *Adv. Phys.* **34**, 475 (1985).
11. D. J. Williams, "Organic polymers and non polymeric materials with large optical non linearities." *Angew. Chem. Int. Ed. Enge.* **23**, 690 (1984).

Index